AF784092

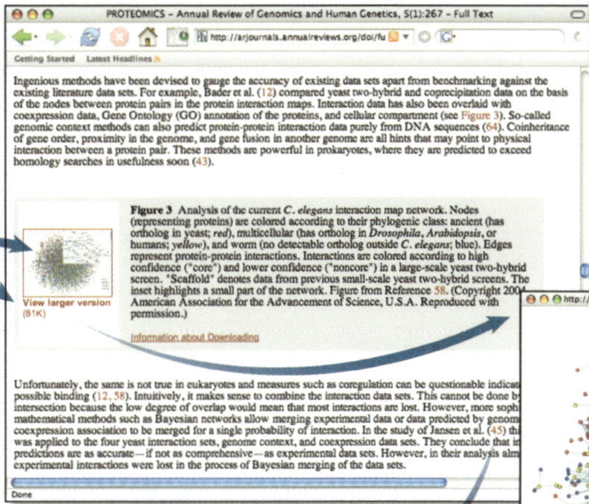

Figures available in three downloadable sizes (resolutions)

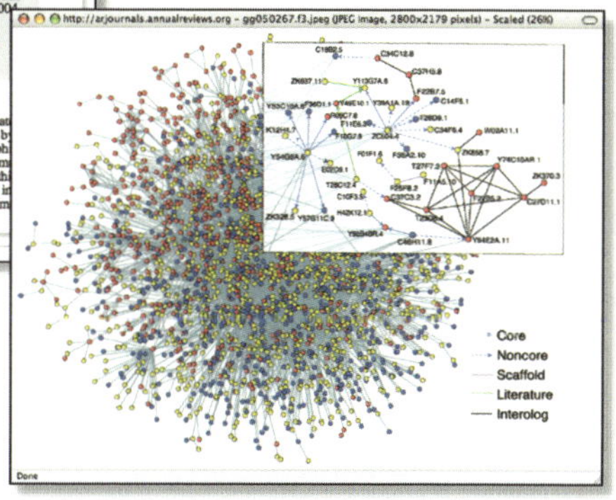

Citations in text link to references in bibliography

References in Annual Reviews article bibliography link out to sources of cited articles online

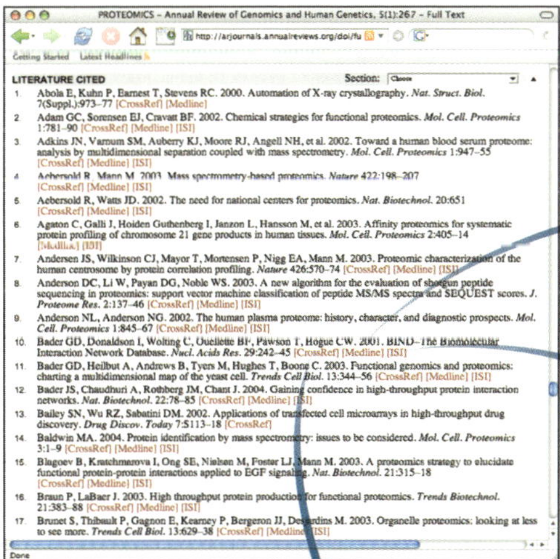

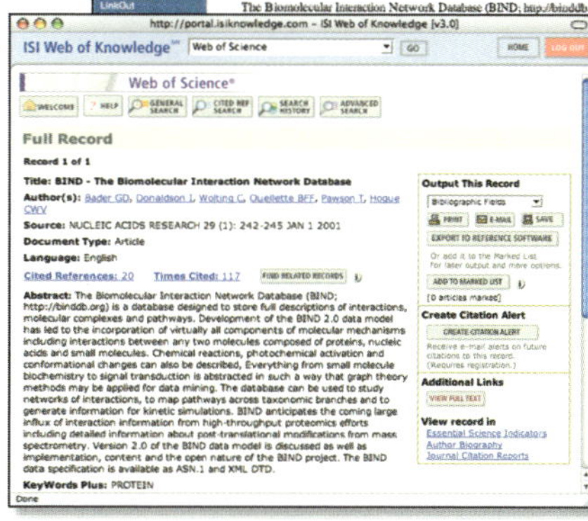

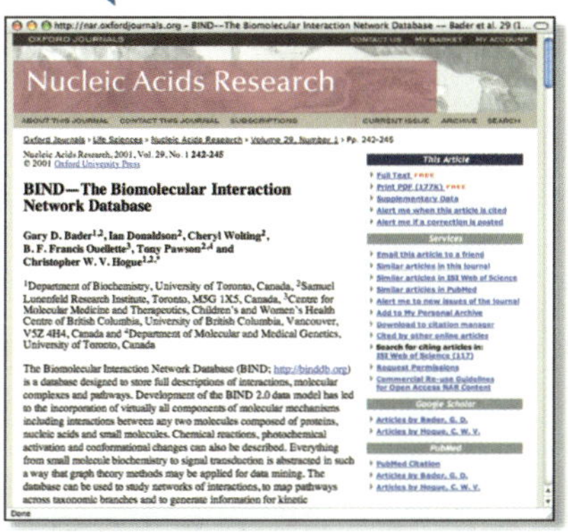

Annual Review of Genomics and Human Genetics

Volume 8, 2007

Aravinda Chakravarti, *Co-Editor*
Johns Hopkins University School of Medicine

Eric Green, *Co-Editor*
National Human Genome Research Institute

www.annualreviews.org • science@annualreviews.org • 650-493-4400

Annual Reviews
4139 El Camino Way • P.O. Box 10139 • Palo Alto, California 94303-0139

Annual Reviews
Palo Alto, California, USA

International Standard Serial Number: 1527-8204
International Standard Book Number: 978-0-8243-3708-7

♾ The paper used in this publication meets the minimum requirements of American National Standards for Information Sciences—Permanence of Paper for Printed Library Materials, ANSI Z39.48-1992.

TYPESET BY APTARA, INC.
PRINTED AND BOUND BY SHERIDAN BOOKS, INC., CHELSEA, MICHIGAN

Annual Review of Genomics and Human Genetics

Volume 8, 2007

Contents

Human Evolution and Its Relevance for Genetic Epidemiology
Luigi Luca Cavalli-Sforza 1

Gene Duplication: A Drive for Phenotypic Diversity and Cause of Human Disease
Bernard Conrad and Stylianos E. Antonarakis 17

DNA Strand Break Repair and Human Genetic Disease
Peter J. McKinnon and Keith W. Caldecott 37

The Genetic Lexicon of Dyslexia
Silvia Paracchini, Thomas Scerri, and Anthony P. Monaco 57

Applications of RNA Interference in Mammalian Systems
Scott E. Martin and Natasha J. Caplen 81

The Pathophysiology of Fragile X Syndrome
Olga Penagarikano, Jennifer G. Mulle, and Stephen T. Warren 109

Mapping, Fine Mapping, and Molecular Dissection of Quantitative Trait Loci in Domestic Animals
Michel Georges 131

Host Genetics of Mycobacterial Diseases in Mice and Men: Forward Genetic Studies of BCG-osis and Tuberculosis
A. Fortin, L. Abel, J.L. Casanova, and P. Gros 163

Computation and Analysis of Genomic Multi-Sequence Alignments
Mathieu Blanchette 193

microRNAs in Vertebrate Physiology and Human Disease
Tsung-Cheng Chang and Joshua T. Mendell 215

Repetitive Sequences in Complex Genomes: Structure and Evolution
Jerzy Jurka, Vladimir V. Kapitonov, Oleksiy Kohany, and Michael V. Jurka 241

Congenital Disorders of Glycosylation: A Rapidly Expanding Disease Family
Jaak Jaeken and Gert Matthijs 261

Annotating Noncoding RNA Genes
Sam Griffiths-Jones 279

Using Genomics to Study How Chromatin Influences Gene Expression
Douglas R. Higgs, Douglas Vernimmen, Jim Hughes, and Richard Gibbons 299

Multistage Sampling for Genetic Studies
Robert C. Elston, Danyu Lin, and Gang Zheng 327

The Uneasy Ethical and Legal Underpinnings of Large-Scale Genomic Biobanks
Henry T. Greely 343

Indexes

Cumulative Index of Contributing Authors, Volumes 1–8 365

Cumulative Index of Chapter Titles, Volumes 1–8 368

Errata

An online log of corrections to *Annual Review of Genomics and Human Genetics* chapters may be found at http://genom.annualreviews.org/

Related Articles

From the ***Annual Review of Biophysics and Biomolecular Structure***, Volume 36 (2007)

Synthetic Gene Circuits: Design with Directed Evolution
Eric L. Haseltine and Frances H. Arnold

From "Simple" DNA-Protein Interactions to the Macromolecular Machines of Gene Expression
Peter H. von Hippel

Gene Regulation: Gene Control Network in Development
Smadar Ben-Tabou de-Leon and Eric H. Davidson

Predictive Modeling of Genome-Wide mRNA Expression: From Modules to Molecules
Harmen J. Bussemaker, Barrett C. Foat, and Lucas D. Ward

Living with Noisy Genes: How Cells Function Reliably with Inherent Variability in Gene Expression
Narendra Maheshri and Erin K. O'Shea

From the ***Annual Review of Genetics***, Volume 40 (2006)

Discovering DNA Encodes Heredity and Prions are Infectious Proteins
Stanley B. Prusiner and Maclyn McCarty

Chromatin Insulators
Lourdes Valenzuela and Rohinton Kamakaka

Mitochondrial Retrograde Signaling
Zhengchang Liu and Ronald A. Butow

Surviving the Breakup: The DNA Damage Checkpoint
Jacob C. Harrison and James E. Haber

The Role of the Nonhomologous End-Joining DNA Double-Strand Break Repair Pathway in Telomere Biology
Karel Riha, Michelle L. Heacock, and Dorothy E. Shippen

DNA Double-Strand Break Repair: All's Well That Ends Well
Claire Wyman and Roland Kanaar

Aspects of Genetic Susceptibility to Human Infectious Diseases
Adrian V. S. Hill

From the ***Annual Review of Medicine***, Volume 58 (2007)

Schizophrenia: New Pathological Insights and Therapies
L. Fredrik Jarskog, Seiya Miyamoto, and Jeffrey A. Lieberman

New Approaches in the Therapy of Cardiomyopathy in Muscular Dystrophy
Elizabeth M. McNally

Toward a Comprehensive Set of Asthma Susceptibility Genes
Yohan Bossé and Thomas J. Hudson

The Multiple Endocrine Neoplasia Syndromes
Vipul T. Lakhani, Y. Nancy You, and Samuel A. Wells

Inflammatory Bowel Disease Genetics: Nod2
Judy H. Cho and Clara Abraham

Macular Degeneration
Edwin M. Stone

From the ***Annual Review of Neuroscience***, Volume 30 (2007)

Trinucleotide Repeat Disorders
Harry T. Orr and Huda Y. Zoghbi

Annual Reviews is a nonprofit scientific publisher established to promote the advancement of the sciences. Beginning in 1932 with the *Annual Review of Biochemistry*, the Company has pursued as its principal function the publication of high-quality, reasonably priced *Annual Review* volumes. The volumes are organized by Editors and Editorial Committees who invite qualified authors to contribute critical articles reviewing significant developments within each major discipline. The Editor-in-Chief invites those interested in serving as future Editorial Committee members to communicate directly with him. Annual Reviews is administered by a Board of Directors, whose members serve without compensation.

Annual Reviews of

Anthropology
Astronomy and Astrophysics
Biochemistry
Biomedical Engineering
Biophysics and Biomolecular Structure
Cell and Developmental Biology
Clinical Psychology
Earth and Planetary Sciences
Ecology, Evolution, and Systematics
Entomology
Environment and Resources
Fluid Mechanics
Genetics
Genomics and Human Genetics
Immunology
Law and Social Science
Materials Research
Medicine
Microbiology
Neuroscience
Nuclear and Particle Science
Nutrition
Pathology: Mechanisms of Disease
Pharmacology and Toxicology
Physical Chemistry
Physiology
Phytopathology
Plant Biology
Political Science
Psychology
Public Health
Sociology

SPECIAL PUBLICATIONS
Excitement and Fascination of Science, Vols. 1, 2, 3, and 4

Human Evolution and Its Relevance for Genetic Epidemiology

Luigi Luca Cavalli-Sforza

Genetics Department, Stanford University Medical School, Stanford, California 94305-5120; email: cavalli@stanford.edu

Annu. Rev. Genomics Hum. Genet. 2007. 8:1–15

First published online as a Review in Advance on April 4, 2007.

The *Annual Review of Genomics and Human Genetics* is online at genom.annualreviews.org

This article's doi:
10.1146/annurev.genom.8.080706.092403

1527-8204/07/0922-0001$20.00

Key Words

population genetics, races, medical genetics

Abstract

The invitation to write the prefatory article to this volume of the *Annual Review of Genomics and Human Genetics* inspired me to collect some thoughts, a few involving ideas that are not new, but perhaps worth resurrecting in light of recent observations made with the data emerging from the Human Genome Diversity Project (HGDP). Data from the many relevant studies based on the HGDP have been made public, as was originally the hope and plan of the project. Here I try to give a short summary of the evolution of modern humans, a unique species in many respects but of special interest to readers of this volume, and a few thoughts on the general rates of evolution that might be relevant to medical genetics and genetic epidemiology. I have made no attempt to give a general bibliography, not even of results from the HGDP, since most authors' conclusions are still unpublished. Citations are limited to very few general concepts and articles discussed in this preface.

GENOMES AND EVOLUTION

Until recently, the study of entire genomes appeared a great but remote dream. It has now begun in earnest and the number of species whose full genome has become publicly available is increasing rapidly. Comparison among species as distant as bacteria and eukaryotes including humans is making it possible to reconstruct with increasing accuracy the steps that led from lost ancestral genomes to those that survive. Genome analysis is clearly giving very strong support to Lamarck's original supposition, widely made known at the beginning of the nineteenth century, that all living beings had a single origin, even if the exact nature of this original organism will have to wait for more research. Science has had to fight skepticism and fear during all of its existence, and thus it is not too surprising that some do not accept evolution, not even as a hypothesis, and try to replace it with religiously based alternatives, at a time when genomics is turning the theory of evolution into one of humanity's most powerful theories.

Doubt and discussion of observations and their interpretations are, of course, the basic stimulus of scientific progress, but one must admit that the very high proportion of Americans who do not accept the evidence in favor of evolution is humiliating and scary, especially considering they are the inhabitants of the country that has been the cradle of much of the latest technology and science. Of course, we are aware that this error is anchored in the belief based on the famous statement implicit in Genesis, that the Earth and life has existed no more than six thousand years or so. Like all innovations, religions have costs and benefits. It is probably no coincidence that Lamarck's publication of his book *Philosophie zoologique* took place after the French revolution had freed the intellectual world of some of the brakes that retarded the development of science. Obviously, credulity and taste for legends about human origins are widespread among all human populations, and are not confined to early ages and cultures. Especially when powered by political interests, belief in such fairy tales can continue well into adulthood.

One of the problems of communication by language is that it is highly ambiguous, and subject to change in time and space. Words are often used metaphorically, and this may add to the confusion. The major bar to accepting evolution arises from the initial statement in Genesis that the Earth was created in six days, which is taken literally by a few fundamentalist churches. There are basic units of measurements of all important quantities like time, length, area, volume, and weight, but their meaning is subject to considerable variation in time and place, as are the words used for them. For obvious reasons, time is measured on the basis of cyclical intervals, of which there are several, the day being the sharpest, while month and year are less clear-cut, depending on geographic location. The "day" of the week of creation in Genesis is probably a poetical metaphor for a very long time period, like scientific eras. After all, what is a day for a timeless God? In another part of the Bible the word used for the years for patriarch's ages seems to have meant months. The use of words meaning time in the Bible is not the only case in which linguistic ambiguity has created major conflict.

THE BASIC IMPORTANCE OF DEMOGRAPHY FOR UNDERSTANDING AND MEASURING EVOLUTION

My own research interests have been largely centered on evolution, and the organisms about which I can claim some knowledge are humans and bacteria. It is well known that Darwin's attention was drawn to demographic thinking by Malthus' emphasis on the simple consequences of exponential growth of living organisms. All self-reproducing entities like living beings go through phases of exponential growth that soon reach limiting conditions, imposing serious constraints. One of the great advantages of the study of human organisms is that demographic analysis is easier in our species than any other. Indeed, three of the four major factors of evolution are described quantitatively in demographic terms. Natural selection is a matter of differences among inherited types in probability of survival to reproduction age, and fertility (the number of children, extended to a full generation cycle). Fisher's fundamental theorem (8) uses these quantities to calculate what he calls the Darwinian fitness of different hereditary types. Special cases, like frequency-dependent selection and inclusive fitness, demand modifications of the theorem. Genetic drift is also a matter of the number of reproducing individuals in a population and of variation in the number of children. It is independent of natural selection. Migration is another straightforward demographic quantity, although for genetics it is important to distinguish two types of migration (2). Type 1 is individual migration, that is (or at least was historically) usually limited to short ranges, and due to the displacement of at least one of the two spouses made necessary by marriage. It typically decreases the variation among populations generated by drift, or by natural selection in different environments. Type 2 is group migration leading to colonization of new areas, which may have opposite effects, being likely to generate new drift through what is called "founder effect" (the usually small numbers of the founders of colonies). Founding generates further opportunities for drift because of the usually small numbers of individuals in generations following that of the founders. It may also give rise to new patterns of genetic variation because of natural selection in the new environments reached by the migrants.

The fourth evolutionary factor, the source of all the evolutionary novelties, mutation in its umpteen forms, can also be described in demographic terms, although it is the one for which it is most difficult to get good estimates because it occurs rarely, and its precise measurement would demand greater numbers than we are usually prepared to handle. Genomic analysis, if extended to a large number of individuals and families, may provide precise measurements as the cost of testing decreases.

The kinetics of population changes under these four evolutionary factors is the main object of the mathematical theory of evolution developed especially by Fisher, Haldane, and Wright in the 1920s and 1930s. It rapidly became a stimulus for mathematical discoveries in stochastic processes, to which Fisher, Wright, and Kimura gave important contributions, and has continued to develop into a more and more complex mathematical theory. We are approaching a stage at which data will be available that allow these theories to produce estimates of the relative importance of these factors in a variety of real situations, and to further develop the general theory of evolution as well as of genetic epidemiology.

THE FREQUENCY OF POLYMORPHISMS AND HUMAN EVOLUTIONARY HISTORY

The analysis of individual genomic variation is clearly going to be one of the major next research steps, and it has already revealed a high frequency of polymorphisms. The main limitation is the cost of analysis of individual genomes. The recent availability of relatively cheap analysis has shown that many nucleotide sites of the order of millions are polymorphic in humans. Until the high cost of the full study of individual genomes comes down, we will have to be content with the polymorphisms now known, all of which are frequent enough that they could be detected on the basis of samples of a few hundred individuals. It will probably still remain true, however, that the most common polymorphisms involve single nucleotides, and those already known indicate that millions of nucleotide sites of our genome are polymorphic, roughly one per thousand of the nucleotide sites (3.14 billion) that form a human genome. The validity of current estimates of distributions of polymorphisms is still rather approximate because of biases of ascertainment. Sites testable today have been found to be polymorphic on the basis of rather small samples of individuals, on which they were selected. The DNA chips now in existence test a somewhat arbitrarily preselected set of polymorphisms, and the criteria for choice are not always clear. With these limitations in mind, it remains exhilarating that we can now examine the polymorphisms at over a million nucleotide sites in the human genome. Naturally, when we are able to examine much larger numbers of individual genomes, the picture of polymorphisms may become much more complicated, but also much more instructive.

A remarkable finding is that the great majority of nucleotide polymorphisms are bi-allelic. A further constraint is that in mitochondrial DNA and Y chromosomes 90% or more of the sites are transitions, due to mutations A←→G, or T←→C, whereas in autosomes there may be a greater relative frequency of transversions. Bi-allelism is still rather frequent, whereas triallelism is relatively rare. This agrees with the simple hypothesis that transitions are more likely to occur than transversions as they involve less chemical difference, although the exact reason must involve a more precise biochemical explanation. The difference in the ratio of transitions to transversions between mtDNA and Y on one hand, and the autosomes on the other, is likely to reflect the differences between uni-parental and bi-parental transmission. DNA under uni-parental transmission has a fourfold higher evolutionary rate under drift, as the basic unit of drift: effective population size is four times smaller than for autosomes.

Another remarkable behavior of mutations that was first exemplified by the "African Eve" story is that all mitochondria of living humans descend from those of a single woman, not because there was a single woman in existence at a given time, but because mitochondria of all other women living at the same time were different, and distinguishable, from that of Eve, by one or more new mutations that occurred in Eve's mother (or close maternal ancestor that left no other descendants but Eve and Eve's descendants), and were not present in the other mtDNAs that existed at the time of Eve but are now extinct. In the famous paper of Allan Wilson's team (1), the "mitochondrial Eve"'s birth date was estimated roughly between 150,000 and 300,000 years ago. Neither the statement of Eve's time of origin nor her African origin had been rigorously proved at the time of their discovery, but much later research has given strong support to both, and the African origin of modern humankind has also received confirmation by archeology. The current estimate of the time of occurrence of Eve's distinguishing mutations is not far from the more recent extreme of the original estimate, 175,000, with a standard error around 10%.

The other uni-parentally transmitted chromosome, Y, has produced a (slightly more approximate) 50,000 years younger estimate. The difference between the birthdates of Adam and Eve is likely the consequence of polygyny, and in fact the ratio of the two dates is not far, as expected, from the ratio of the number of wives per man still observed in many parts of the world.

That all clones of these uni-parentally transmitted chromosomes existing today descend from one ancestral chromosome living some time ago, and all other clones that existed at the time of origin of the only ancestral surviving type are extinct, is a simple consequence of drift due to the variation of the number of children per individual. This expectation cannot be easily extrapolated to all other chromosomes, autosomes, and the X chromosome, or parts of them, because crossing over mixes up the DNAs of homologs and destroys the original genealogical sequence of nucleotides of the DNA strand. However, short DNA segments that had no crossing over still obey the law, except that it is not easy to distinguish the exact extremes of such segments. Most regular genes are too large to satisfy this requirement. When we are able to discover the extremes of segments that have not undergone any crossing over since the single nucleotide mutations known in them originated, we may expect the time of origin of such short autosomal DNA segments to be four times as large (and three times for the X chromosome), on average, as that of uni-parentally transmitted, haploid chromosomes: i.e., 500,000 years (three times as large for the X chromosome).

The average time of these nonrecombined short DNA segments in bi-parentally transmitted chromosomes is likely older than 500,000 years on average, because an unknown, but, as we discuss later, potentially large proportion of them must be under selective heterozygous advantage. If this does not belong to a specific DNA segment, some heterozygous advantage will be borrowed from that of neighboring DNA: "associative overdominance through linkage disequilibrium" (13). In fact, in HLA there are polymorphisms that might be 15 million years old, and examples that have a longer evolutionary age have begun to appear in the literature. There is, of course, a large variation in individual evolutionary times under drift alone (10), but a two-million-years age would be a relatively rare occurrence with an expectation of 500,000 years, and there are other reasons for suspecting heterozygous advantage in many polymorphisms.

A few anthropologists still support an old anthropological hypothesis, called multiregional theory, that the current subdivision into four or five different races arose at least 1.7 million years ago (mya), at the time of the first expansion of *Homo erectus*, and that subsequent events have not erased the evidence of this ancient subdivision. These anthropologists hope to save the multiregional theory by showing that some of the genes currently polymorphic in modern humans originated long ago, away from Africa (a very difficult job), but this approach will not likely be of great help. It is perfectly possible, of course, that mutations originating in Asia that took place before the more recent expansion may have traveled back to Africa a long time ago, and there are already examples of them, although clear ones are clearly posterior to the latest African expansion.

HUMANS AS A COSMOPOLITAN INVASIVE SPECIES

Other complications arise in the application of population genetics to our evolutionary history because our species has gone through major bottlenecks in the total number of individuals, followed by later major expansions. They are mostly effects of major cultural innovations, permitted by the continuous increase of our capacities to communicate and to invent, that have enormously increased the power of cultural evolution compared with other animals, even our nearest cousins. This is reflected in the archeologically observed increase of the volume of our brain, which is now about four times the size of the ancestor common to our nearest Primate cousin, chimpanzees. By contrast, the brain of chimpanzees did not change much. The separation from chimps occurred at least 5–6 mya, somewhere in Africa, and a large number of species originated from that separation especially, if not only, in the human line, of which only one, ours, survives, and it is genetically very homogenous (*H. sapiens sapiens*). Therefore, it must have had a very recent evolutionary origin, as archeologists also show (approximately around 150,000 years ago, in East Africa).

The new species that originated in the first half of the interval of time separating us from chimps are classified into a genus different from ours, *Australopithecus* (so named because it was found predominantly in South Africa). It has been decreed that our genus, *Homo*, descended from it between 2.5 and 3 mya with the species called *habilis*, because it showed the first evidence of making stone tools, however rough. Around 1.7 mya tools had increased in number and complexity, allowing the species now called *H. erectus* to spread from Africa to the whole Old World (Eurasia). Very probably fire (first archeological evidence 1.6 mya) also helped to generate this early expansion, being used for a variety of jobs, including defense against cold, and against wild animals, as well as for cooking food and making tools. Many species were formed at that time by allopatric speciation, but all Australopithecines and species of the genus Homo other than ours have been extinct for quite some time. There have been recently changes of these species names that are not entirely stabilized.

The next great expansion, this time to the whole world, was much later, and was started by a small population—genetic data indicate it may have been made of only 1000 individuals—living in East Africa. Genetic evolutionary time estimates indicate there may have been a slow beginning of growth and expansion perhaps 100,000 years ago, but there was a much more impressive, archeologically proved, successful beginning 50–60,000 years ago of an expansion to Asia and beyond, that continued until the whole world was settled. By that time, language was fully mature and was one of the key ingredients of a sophisticated communication, very useful for spreading and accumulating cultural knowledge. Language is certainly the chief difference between humans and animals. Tribes are basically (or were originally) social groups speaking the same language and sharing cultural identity. The language spoken by the East African "tribe" that started growing demographically, and continued to do so after it began to expand to the whole world, is most probably the one from which all 6000 languages spoken in the world originate. Linguists find difficulties in accepting statements that there was a single original language, from which all those existing today originated, because linguistic evolution is very fast and has produced profound differences between present languages. But important evidence in its favor is that any modern living human, apart from some rare genetically disabled individuals, can learn equally well any of the languages now in existence. Other major innovations that favored this expansion were inventions like navigation, that allowed far islands like Oceania and New Guinea to be reached over 40,000 years ago, and a new set of tools of increasing complexity and usefulness.

At about or shortly after 13,000 years ago the last great glaciation ended, and ice started retiring from Europe. It was the last of a series of glaciations that had started, with interruptions, some 80,000 years ago. It had transformed the world, and the climate change now favored different types of grass that affected the fauna, and also humans. The number of individuals on Earth is estimated to have been, at this time, between 1 and 15 million. This is, in terms of orders of magnitude, a thousand times less than now, but a thousand times larger than before the expansion. The demographic expansion that had started with modern humans and caused a great geographic expansion gave rise to a number of new innovations, such as the bow and arrow, the use of bamboo where it grew (in East Asia), the use of clothes in cold climate. Especially at the end of the last glaciation there were also improvements in stone tools, which became smaller and more varied (called mesolithic, especially in Europe), but the economy was still one of hunting and gathering (and fishing) or, as it is called, in short, foraging, and population density was growing in several more favorable areas, with temperate climates. In these areas archeologists have noted the apparently independent starts of a major innovation, food production: agriculture, and animal breeding, replacing food collection from nature. Both are obviously applications of the discovery of major secrets of life: its modes of reproduction, and the possibility of genetically modifying plants and animals to suit our needs.

Food production began independently in various areas, at least in the sense that the plants and animals that were domesticated were the local ones already used as food by hunters and gatherers (fish was domesticated only very recently). The major areas were the Middle East and Turkey: the oldest, which is near the border between Turkey and Syria, was an already mixed agro-pastoral site

(Abu Hureyra, 11,500 years ago). Domestications of local plants and animals began somewhat later, in China (independently in north and south), in the Sahara, which was not dry at the time, in New Guinea, in the Mexican plateau, and in West Africa. From the places of origin, food production spread slowly at rates of 1–2 km per year, both carried by farmers that looked for new fields at the periphery of the expansion, and imitated by the local hunter-gatherers, who learned from their neighbors and sometimes intermarried with them.

Agriculture changed the diet considerably. The first crops developed in very different parts of the world (wheat, maize, rice) are still today the most important in the whole world. Agriculture also increased sedentariness and decreased the nomadism associated with foraging, but a later development of relatively pure pastoralism in semiarid areas encouraged some special types of nomadism.

Sedentariness and increase of population density caused wide social stratification, job differentiation, the rise of governments, of chances for big property and power that were essentially impossible with foraging. Still later, new expansions were favored by new inventions: metals. The first was bronze (around 5000 years ago) and some 1500 years later iron developed in a similar area, mostly at the boundary between Europe and Asia above the Caucasus. In the same area the horse was domesticated. Together, metals and the horse generated the beginnings of major wars between neighboring governments, and men, originally hunters, now became warriors. Large numbers of war prisoners were turned into slaves. It became possible for large tribes to resettle in new favorable areas, their families traveling by cattle- or horse-driven carts. Horse-mounted Mongols began major conquests in Asia. The camel was domesticated later. With horse and camel, Arabs rapidly conquered Turkey and North Africa, Sicily, and Spain, beginning in the seventh century AD. Arab merchants also spread to East and South Africa. Some single inventions and innovations proved extremely powerful, and their influence on our life, including our genetics, keeps increasing at a growing rate. Cultural history has thus begun to dominate the genetic picture of our species, and to affect the fate of many other species as a consequence of our actions on the environment [comparisons of archeological and genetic observations of modern human evolution appear in Cavalli-Sforza & Feldman (5) and in Cavalli-Sforza et al. (6, 7)].

THE CONTRIBUTIONS OF MUTATION AND SELECTION TO THE POLYMORPHIC PICTURE

Of the four factors of evolution, mutation is responsible for the origin of polymorphisms, and natural selection for the change of their frequencies in response to the environment. In diploid species selection in the form of heterosis increases the stability of polymorphisms. By contrast, drift tends both in the short term and in the long run to destroy polymorphisms, and to differentiate populations one from the other, whereas individual migration mostly tends to redistribute the existing polymorphisms within the species. Kimura (10) has shown that most mutations are selectively neutral in molecular evolution, so that mutation and drift are the main causes of most differences among species. But what about the millions of polymorphisms that show no major frequency differences in our species, at least between the four populations so far tested by the (National Institutes of Health (NIH) Genome Institute) International HapMap Project? There are, of course, some genes, or more precisely DNA segments, that show strong differences among these populations, signaling differential selection that extends to neighboring short DNA lengths because of linkage disequilibrium, but the great majority of polymorphisms show few differences among the HapMap populations, in agreement with the overall homogeneity across our species compared with most other mammals.

A question of interest is: Why are there so many polymorphisms? The following situations come to mind:

1) The polymorphisms were already in the species before its origin, under differential selection that drove the two alleles to their present relative frequency, but may have ceased to operate, so that they are now neutral and fluctuate under drift alone, in equilibrium with migration.
2) Polymorphisms were driven to their present average frequency by linkage with neighboring genes that are or have been under selection, and the present differences among populations are due to drift, in equilibrium with migration.
3) Polymorphic frequencies may be the result of equilibrium between opposing mutation rates in the absence of other pressures, but gene frequency changes under mutation pressure alone are extremely slow. Moreover, the bi-allelism of almost all polymorphic nucleotide sites shows that we are far from an equilibrium of mutation rates. We would expect that at many sites more than two nucleotides would be present at equilibrium under mutation pressures. In fact, the time it takes for equilibrium to be established under mutation at rate μ, in a very large population, is of the order of $1/\mu$ generations. We have little knowledge of mutation rates, and mostly for very few genes; estimates of average values of mutation rates are mostly derived from evolutionary analysis, but we also know that some individual mutation rates may be very different from the average. The order of magnitude of changes due to mutation rates alone would demand millions of generations. There are also major bars to the observation of equilibria determined by neutral mutation rates alone: They are accidents causing strong drift by founder episodes, like those that generate new species, and differential selection effects in different environments.
4) Selective advantage of heterozygotes will generate stable polymorphic equilibria. Even if one of the alleles is lethal or almost so, heterozygote frequencies near 20% or 30% have been observed at equilibrium, for instance for sickle cell anemia. Heterozygous advantage is also spread to neighboring, closely linked neutral genes. The rate at which such polymorphic equilibria are reached is much higher than under mutation pressure alone. In the first example of heterotic equilibrium studied, sickle cell anemia, the selection coefficient involved was of the order of 10%; that is, the fitness of the normal homozygote was 10% smaller than that of the heterozygote, and was enough to give a 20% frequency of heterozygotes at equilibrium, even though the other homozygote was basically lethal. We have few other heterotic examples as clear as that of sickle cell anemia (or thalassemia, which is very similar). With selection of this strength about 100 generations are necessary to take a population of 10,000 individuals from the first appearance of a mutation to near equilibrium. In the simplest conditions, the time in generations necessary for replacement by a new mutant is proportional to the reciprocal of the selection coefficient. Thus, in the course of the latest modern human expansion (50,000 years is ca. 2000 generations), a new mutation that appeared at the beginning and showed heterozygous advantage might be close to equilibrium today, even with selection coefficients as small as 0.5%. If the equilibrium due to heterozygous advantage existed before the start of the expansion, it may have been due to smaller selection coefficients, even as small as 1/10,000, that would of course take a proportionately longer time to establish [a simple numerical

introduction to this type of analysis is found in Cavalli-Sforza & Bodmer (4)].

The safest way to test for the existence of heterozygous advantage is the demonstration of deviation from Hardy Weinberg equilibrium among adults in favor of heterozygotes, but it took of the order of a thousand individuals to prove this in the case of sickle cell anemia. With smaller selection coefficients the number of individuals required is prohibitive, given the size of samples available today. However, an analysis of 91 genes studied at the genome level, in which selection had major effects, showed that 22, i.e. 24%, of them included heterosis (17). It is likely that heterosis involves an even higher fraction of polymorphisms, considering that the stabilization of a polymorphism also extends to neighboring neutral genes because of linkage disequilibrium. Naturally, this stabilizing effect due to linkage disequilibrium with selected variants is smaller the weaker the linkage (13). It is possible that the similarity of frequencies of neighboring polymorphic sites shown by HapMap populations is due to a near ubiquity of heterosis, including heterosis borrowed from neighboring sites by linkage disequilibrium, but a real test will demand adequate numbers of individuals.

DEMOGRAPHIC BOTTLENECKS FOLLOWED BY DEMOGRAPHIC GROWTH

Particularly important in establishing the genetic picture of a population are demographic bottlenecks, especially if they are followed by major demographic growth. We know that our species had such a history at the time of its origin in the last 100,000 years, with spread limited almost only to Africa in the first 50,000 years of this period, and followed by expansion to the whole world in the second half. Later, there were some major expansion events at the time of the beginning of agriculture, beginning around 10,000 years ago. These generated more local spreadings, which were at least partially responsible for the major four or five clusters that correspond approximately to continents (see below), but there were many further later ones that were much more limited geographically. Transoceanic navigation in post-Columbian times led particularly to the spread of Europeans, but this cultural innovation was also responsible for the spread of Indians and of Chinese to many economically attractive parts of the world.

Social customs have tended to keep most societies fairly endogamous after their migrations away from the mother country, although there have also been examples of partial melting pots, especially in the United States and even more in Brazil. But in Europe and elsewhere, groups of colonizers kept some of their original cultural identity, including their language, for periods of remarkable length. There are also examples of extraordinary conservation of languages. The Basque language may have descended, with considerable transformation, from a family of languages that were spread to Eurasia and to North America more than 20,000 years ago, perhaps with the first arrivals from Africa. However, most language families spread more recently with agricultural developments, i.e., in postglacial times, and largely replaced earlier linguistic family groups.

Still more recently there was a multiplication of minor expansions, beginning with the metal ages and later. The Jewish diasporas brought Jews to various continents 2500 and 1900 years ago, and again in more recent times, and Jews grew in numbers almost everywhere. Less than 2000 founders, mostly Dutch, joined by some German and French protestants settled in Capetown, grew by a factor of almost 1000-fold in 350 years, and now form about half of the white population of South Africa. French Canadians of Quebec originated at about the same time, descendants of approximately 1000 French women who were invited by King Louis XIV to become "Daughters of the King," receiving a dowry if they accepted to marry French trappers who were active in Quebec. Here also a 1000-fold increase took place in 350 years. These populations have all been well studied genetically and examples of genetic rarities or diseases present in one or a few founders, sometimes identified from the genealogies, are now relatively frequent in these and similar populations.

A word of caution: one may wonder if some numerical values related to evolutionary times, including many evolutionary time values estimated by simulation of evolutionary processes, are entirely correct, because they are usually calculated assuming exponential population growth. It is very likely that the population growth of our species during its geographic expansion from East Africa was far from exponential, and rather linear with time, or at most quadratic. The best model for demographic-geographic expansions (we use the short "demic" for their combination) is most likely to follow the extension to population expansions of the model introduced by Fisher (9) for advantageous mutations, called "the wave of advance," which shows that expansion under constant population density results in a constant rate of population growth, i.e., is linear and not exponential in both time and space.

THE SERIAL FOUNDER EFFECT IN THE EXPANSION OF MODERN HUMANS

The examination by Rosenberg et al. (16) of the set of 1064 human DNA samples provided by the Human Genome Diversity Project (HGDP) with ca. 400 microsatellite loci generated a cascade of further research. Prugnolle et al. (14) showed that the average heterozygosity of the 52 populations decreased in linear fashion by about 16% as a function of geographic distance from the place of origin. We confirmed their result with twice as many microsatellites, using 400 additional microsatellites that were made public later, and offered (15) an interpretation by a simulation that predicted well the observed fall of genetic variation, measured by heterozygosity. The simulation used a model of "serial founder effect" that can be summarized as follows: The original East African population grew from an initial small size to saturation density, and then generated a colony at some distance, which also grew to saturation. The first colony then generated a second colony that moved further away; it grew and the process continued until the furthest point inhabited on Earth was reached. The origin of each colony was thus an episode of founder effect that was repeated serially at every new colonization. The simulation showed a linear fall of average heterozygosity at a rate that corresponded closely with that observed for the microsatellite data, using numbers for tribes and growth models that were anthropologically reasonable. According to this model, 100 steps (each step being a successive colonization event) could, in 40,000 years, cover the 25,000 kms separating the origin, put arbitrarily in Addis Ababa, from the terminal colony in the southern tip of South America. We allowed for obligatory waypoints that avoided the direct crossing of oceans. A very similar exercise by Liu et al. (12) suggested 300 steps.

These models can be improved, but at least they show that one can explain the remarkable linearity of the process that was observed, on time and space scales compatible with the real conditions. A simulation using Fisher's wave of advance model (9) also fits the observations and is probably preferable in that it would correspond more closely to anthropological reality, but would demand, for a detailed fit, more specific anthropological knowledge than available for hunter-gatherers' societies.

THE CORRELATION OF GEOGRAPHIC AND GENETIC DISTANCES

The Ramachandran et al. analysis (15) of the Rosenberg et al. data (16) also estimated the correlation between genetic and geographic distances calculated for all the possible pairs of the HGDP populations. It showed an extremely high linear correlation between the two distances (r = 0.89), one of the highest correlations ever observed in biology. It should be emphasized that the 52 HGDP populations are all indigenous, i.e., are today located in the approximate places where they were before the great migrations that took place in post-Columbian times (so that in America the collection includes only Native Americans, etc.). It is therefore obvious that all populations belonging to the same continent are more similar to each other than those from different continents, but this description, which is characteristic of many simple race models, is not sufficient. In fact, populations from different continents that are geographically close are also more genetically similar than expected on the simple hypothesis that they are part of their respective continents, and geographic distance between them is unimportant. Thus, for instance, Arabs and Ethiopians, Spaniards and Moroccans, and Turks and Greeks can be expected to be more similar to each other than two random African or two European or two Asian populations. The few comparisons that deviate from the linearity of the correlation of geographic and genetic distance involve some African populations that are particularly ancient in human evolutionary history, and have had more time to diverge genetically. Also slightly out of place in the linear correlation are populations that are known to be partial admixtures of very different groups, as is also expected.

The simplest interpretation of these results, also taking account of the serial founder effect, is that the human genetic differentiation observed for microsatellites is to a very large extent due to the equilibrium of drift and migration. Major effects of natural selection are not detected, but preliminary observations with other types of mutants or by other approaches also show evidence for natural selection of the directional type, as demonstrated by major differences among different HapMap populations. This summary can cover only the major aspects of these findings, but the original paper shows that drift can account for at least 78% of the variation observed for microsatellites. However, it leaves uncertain what role, if any, heterosis had in this process.

GENETIC VARIATION BETWEEN AND WITHIN POPULATIONS, AND THE RACE PROBLEM

In the early 1980s, Lewontin (11) showed that when genetic variation for protein markers is estimated by comparing two or more random individuals from the same populations, or two or more individuals from the whole world, the former is 85% as large as the latter. This means that the variation between populations is the residual 15%, and hence relatively trivial. Later research carried out on a limited number of populations and mostly, though not only, on protein markers has confirmed this analysis. The Rosenberg et al. data actually bring down Lewontin's estimate to 5%, or even less. Therefore, the variation between populations is even smaller than the original 15%, and we also know that the exact value depends on the choice of populations and markers. But the between-population variation, even if it is very small is certainly enough to reconstruct the genetic history of populations—that is their evolution—but is it enough for distinguishing races in some useful way? The comparison with other mammals shows that humans are almost at the lower extreme of the scale of between-population variation. Even so, subtle statistical methods let us assign individuals to the populations of origin, even distinguishing populations from the same continent, if we use enough genetic markers. But is this enough for distinguishing races? Darwin already had an answer. He gave two reasons for doubting the usefulness of races: (1) most characters show a clear geographic continuity, and (2) taxonomists generated a great variety of race classifications. Darwin lists the numbers of races estimated by his contemporaries, which varied from 2 to 63 races.

Rosenberg et al. (16 and later work) analyzed the relative statistical power of the most efficient subdivisions of the data with a number of clusters varying from 2 to 6, and showed that five clusters have a reasonable statistical power. Note that this result is certainly influenced by the populations chosen for the analysis. The five clusters are not very different from those of a few partitions that had already existed in the literature for some time, and the clusters are: (*a*) a sub-Saharan African cluster, (*b*) North Africa–Europe plus a part of western Asia that is approximately bounded eastward by the central Asian desert and mountains, (*c*) the eastern rest of Asia, (*d*) Oceania, and (*e*) the Americas. But what good is this partition? The Ramachandran et al. (15) analysis of the same data provides a very close prediction of the genetic differences between the same populations by the simplest geographic tool: the geographic distance between the two populations, and two populations from the same continent are on average geographically closer than two from different ones. However, the Rosenberg et al. analysis (16) adds the important conclusion that the standard classification into classical continents must be modified to replace continental boundaries with the real geographic barriers: major oceans, or deserts like the Sahara, or other deserts and major mountains like those of central Asia. These barriers have certainly decreased, but they have not entirely suppressed genetic exchanges across them. Thus, the Rosenberg et al. analysis confirms a pattern of variation based on pseudocontinents that does not eliminate the basic geographic continuity of genetic variation. In fact, the extension by Ramachandran et al. of the original Rosenberg et al. analysis showed that populations that are geographically close have an overwhelming genetic similarity, well beyond that suggested by continental or pseudocontinental partitions. Genetic variation is thus very nearly continuous, as Darwin said, but behind this continuity is another discontinuity that has not yet been analyzed. It is due to barriers among small social groups, based on religious separations and socioeconomic stratifications. These also create differences in the environment of individuals and small groups, through diet and economic or other, self-imposed limitations. This analysis suggests that it is important to recognize and study groups much smaller than the traditional major "races." To clarify what groups we want to distinguish, we first have to very clearly answer the question: What purpose do we have in mind when we try to distinguish races?

The number of groups to be distinguished depends on the differences among populations that are really useful for some valid purpose, and we still have not decided on criteria to choose populations that we want to distinguish. Therefore, the real question remains: What do we want races for? Let us start by agreeing on what could be the most important reason for defining "races." Incidentally, I am inclined to dismiss the word "race" because of its connection with the odious episodes of racism with which we are continuously confronted. The word "populations" is useful in statistics for defining the group from which we draw samples, but is in practice used arbitrarily, and perhaps the most neutral term could be that used by Rosenberg et al. (16): "clusters." We want to define useful genetic clusters. But more than a term to be used, what I am looking for here is a general agreement on a good reason for doing research aiming to define a useful genetic stratification of populations, and it seems to me we can really find it in research that can be of help for medicine, that is for diagnosis and therapy.

The expression "ethnic groups" is also useful, but especially in situations in which it is not clear if the basic difference is of genetic or cultural (including socioeconomic) nature, or both.

SHALL WE TAILOR DIAGNOSIS AND THERAPY TO THE INDIVIDUAL GENOME?

Today it might also be objected to that the long-term answer is actually another one. We do not need any types of races at all, even for medicine. The individual genome is going to be the best information for diagnosis and therapy, especially considering the strong individual variation in response to many drugs. But how long will we need to wait before this is true? The most important diseases (the "complex" diseases) do not behave, except in rare cases, in a simple Mendelian way. They are complex in two ways. One is that many of them are likely polygenic—that is, there may be a great variety of genes that contribute to the disease. Specific Mendelian diseases are often caused by different genes and often different variants within each gene causing different patterns, and in polygenic diseases almost every individual case or pedigree might have a somewhat different polygenic system of its own. The other way is that the environment causes additional variation, and the individual environmental effects are usually poorly known.

It is clear that genomic knowledge will be very useful for rapidly extending our knowledge of Mendelian diseases, and of individual responses to drugs that may be available for a specific disease, but when we come to the most difficult diseases of all, the complex diseases, the environment is also very important. It is clear that we must be prepared to deal with the importance of environmental stratification within the genetic cluster, much of which may be due to socioeconomic, rather than genetic, differences.

We obviously hope that the genome will also be very useful for understanding more of the genetic background of complex diseases (provided environmental contributions are accurately studied at the same time). Here, however, we may have to wait for an entirely new development beyond the genome formed by the DNA of germinal cells, which accounts for inheritance. Information of what happens in somatic cells, i.e., epigenetics, will also be necessary. Thus, for genome knowledge to give a response to all or most individual queries much work remains to be done, and perhaps the only sensible answer to the question "how long will it take?" is that we can hope that there will be important progress in this century.

Even so, for many of these investigations it may still be necessary or useful for a while to study genetic clusters of individuals to solve specific problems, but they will have to be smaller populations than simply one per continent or pseudocontinent. Genetic epidemiology shows that many Mendelian diseases are concentrated in some, usually small, social or ethnic groups, especially for the rarer diseases. The reason is simple for recessive diseases: The recent demographic expansions that have occurred in so many parts of the world in the past few centuries or millennia have created clusters of cases of specific genetic diseases that originated from a single or a few mutations that occurred during the demographic growth of specific populations. These social groups are still meaningful today (for example, the demographic bottlenecks followed by major episodes of demographic growths, mentioned above). It is these genetic clusters that will likely be very useful, perhaps also for studying complex diseases, but the conclusions may remain valid only or mostly for the clusters in which they were obtained.

The number of cases of a well characterized disease may give a rough indication of the time of origin of the relevant mutations and populations. But some diseases can be caused by a great variety of mutations in the same gene, and then there is often considerable clinical variety associated with the various mutations, as is clear, for instance, with cystic fibrosis. The size and importance of the gene contribute to this complexity. The demographic history of a population may help to predict which groups are likely to show more genetic similarity in their disease patterns. The remarkable differences in genetic pathologies found among Ashkenazim and Sephardim (Jews of northern-European and of Mediterranean origin) populations indicate that genetic clusters of individuals of medical interest may have to be small to be really useful; even a relatively small group like Jews has to be split into subclusters for medical genetics research. If the size of these groups should be an example of the size of useful genetic clusters from a medical point of view, then one would need of the order of a thousand genetic clusters for the whole species.

It is probably legitimate to state that, as genetic predisposition to many diseases is important, even for the response to some endemic infections, recognizing and studying useful genetic clusters may help us toward the aims of diagnosis and therapy. In practice, we may be going in the direction of studying smaller and smaller clusters and end up with the individuals, but the use of information from an individual's genome may need many more years to become truly adequate. Until then, it seems to me we need to enlarge and improve on collections of the HGDP type, my next subject, and also try new, different approaches.

THE HGDP PROJECT AND ITS FUTURE

Nature Reviews Genetics published a short history of HGDP-Centre d'Etude du Polymorphisme Humain (CEPH) (3). The present collection is made of 1064 cultures of B-lymphocytes from 52 populations that were already in existence. All continents except Australia are represented, but inevitably not in the way that was originally desired, with populations spread at fairly regular distances. There is some geographic clustering of the populations because some countries (China, Pakistan, Israel) have been especially active and generous in providing cell lines. The project was made possible by the kindness of research workers who donated cell lines they had collected, and by the fact that CEPH already had all the necessary equipment for growing cell lines, producing DNA, and preparing vials for distribution to laboratories, and made it available for the HGDP. The equipment had been assembled earlier for producing the human linkage maps that used the Utah and a few other pedigrees for linkage studies. Thus, HGDP-CEPH required very little money.

The original idea of the HGDP project was to collect 10,000 cell lines of 25 individuals from each of 400 populations. The Genographic Project is planning to collect 100 individuals from each of 1000 populations but just DNA from saliva and, in a small fraction of cases, blood samples. Unfortunately, it will not generate cell lines. With present technology, the analysis of 1000 individuals is still somewhat demanding for a single laboratory unless it is limited to a gene or to a sample of special markers. There is an inevitable trade-off between the number of individuals and that of markers that will be examined, mostly for economic reasons. However, we are now entering a new era, in which we can become more ambitious, but also need greater capitals.

Cultures of B lymphocytes are still the best material to collect and store. DNA will remain for some time the most important product of genome collections, but cells also offer important avenues of research into epigenetics. Their RNA and proteins can give information on the epigenetic processes, and on epigenetic evolution taking place during the history of the individual, at least for all genes relating to general maintenance and reproduction of the cell, and for specific immunological functions that occur in the B lymphocyte. Therefore, they are also informative on the immunologically significant environment of the populations.

It is important to maintain and extend the present HGDP collection. So far, the HGDP has been developed with minimum investment. There are hopes that it will be expanded, even if governmental support for science has unfortunately shrunk almost everywhere. Clearly, countries where genome studies are more likely to be fruitful in the short term are those for which it is reasonable to invest proportionately more at this stage. It is practically inevitable that the two biggest clusters of populations, one including Europe, North Africa, and west Asia including Pakistan and India, and the other East Asia, will be studied preferentially, given that these two areas have the strongest economic development, and that they also intergrade genetically into each other. But it will also be important to dedicate a reasonable part of the effort to the indigenous parts of the rest of the world—Africa, Oceania, and the Americas—otherwise recent immigrants to the more highly developed part of the world will not have access to the biomedical information available to other residents. Not only is this inequality undesirable but our understanding of human evolution will remain incomplete without adequate consideration of the other continents.

LITERATURE CITED

1. Cann RL, Stoneking M, Wilson AC. 1987. Mitochondrial DNA and human evolution. *Nature* 325:31–36
2. Cavalli-Sforza LL. 1973. Some current problems in human population genetics. *Am. J. Hum. Genet.* 25:82–104
3. Cavalli-Sforza LL. 2005. The Human Genome Diversity Project: Past, present and future. *Nat. Rev. Genet.* 6(4):333–40
4. Cavalli-Sforza LL, Bodmer WF. 1999. *The Genetics of Human Populations.* New York: Dover Press
5. Cavalli-Sforza LL, Feldman MW. 2003. Biology as history: population genetic approaches to modern human evolution. *Nat. Genet.* 33:266–75
6. Cavalli-Sforza LL, Menozzi P, Piazza A. 1994. *The History and Geography of Human Genes.* Princeton, NJ: Princeton Univ. Press
7. Cavalli-Sforza LL, Piazza A, Menozzi P, Mountain J. 1988. Reconstruction of human evolution: bringing together genetic, archeological and linguistic data. *Proc. Natl. Acad. Sci. USA* 85:6002–6
8. Fisher RA. 1930. *The Genetical Theory of Natural Selection.* New York: Dover Press
9. Fisher RA. 1937. The wave of advance of advantageous genes. *Ann. Eugen.* 7:355–69
10. Kimura M. 1968. Evolutionary rate at the molecular level. *Nature* 217:624–26
11. Lewontin RC. 1985. Population genetics. *Annu. Rev. Genet.* 19:81–102
12. Liu H, Prugnolle F, Manica A, Balloux F. 2006. A geographically explicit model of worldwide Human-Settlement theory. *Am. J. Hum. Genet.* 79:230–37
13. Ohta T, Kimura M. 1970. Development of associative overdominance through linkage disequilibrium in finite populations. *Genet. Res.* 16:165–77
14. Prugnolle F, Manica A, Balloux F. 2005. Geography predicts neutral genetic diversity of human populations. *Curr. Biol.* 15:1022–27
15. Ramachandran S, Deshpande O, Roseman CC, Rosenberg NA, Feldman MW, Cavalli-Sforza LL. 2005. Support for a serial founder effect originating in Africa, using the relationship between genetic and geographic distance in human populations. *Proc. Natl. Acad. Sci. USA* 102:15492–97
16. Rosenberg NA, Pritchard JK, Weber JL, Cann HM, Kidd HM, et al. 2002. Genetic structure of human populations. *Science* 298:2381–85
17. Sabeti PC, Schaffner SF, Fry B, Lohmueller J, Varilly P, et al. 2006. Positive natural selection in the human lineage. *Science* 312:1614–20

Gene Duplication: A Drive for Phenotypic Diversity and Cause of Human Disease

Bernard Conrad[1,2] and Stylianos E. Antonarakis[1]

[1]Department of Genetic Medicine & Development, University of Geneva Medical School and Geneva University Hospitals, CH-1211 Geneva 4, Switzerland

[2]Division of Human Genetics, Bern University Children's Hospital, CH-3010 Bern, Switzerland; email: bernard.conrad@insel.ch

Annu. Rev. Genomics Hum. Genet. 2007. 8:17–35

First published online as a Review in Advance on March 26, 2007.

The *Annual Review of Genomics and Human Genetics* is online at genom.annualreviews.org

This article's doi: 10.1146/annurev.genom.8.021307.110233

1527-8204/07/0922-0017$20.00

Key Words

gene duplication, copy number variant, haploinsufficiency, gene balance hypothesis, insufficient amount hypothesis

Abstract

Gene duplication is one of the key factors driving genetic innovation, i.e., producing novel genetic variants. Although the contribution of whole-genome and segmental duplications to phenotypic diversity across species is widely appreciated, the phenotypic spectrum and potential pathogenicity of small-scale duplications in individual genomes are less well explored. This review discusses the nature of small-scale duplications and the phenotypes produced by such duplications. Phenotypic variation and disease phenotypes induced by duplications are more diverse and widespread than previously anticipated, and duplications are a major class of disease-related genomic variation. Pathogenic duplications particularly involve dosage-sensitive genes with both similar and dissimilar over- and underexpression phenotypes, and genes encoding proteins with a propensity to aggregate. Phenotypes related to human-specific copy number variation in genes regulating environmental responses and immunity are increasingly recognized. Small genomic duplications containing defense-related genes also contribute to complex common phenotypes.

INTRODUCTION

Ever since Susumu Ohno's insightful suggestion 35 years ago that gene duplication is a key factor shaping evolution, the model and its general predictions continue to attract much attention (70). On an evolutionary scale, gene duplication may result in new functions via different scenarios (**Figure 1**). Although the most likely outcome is loss of function in one of the two gene copies (**Figure 1*a***, nonfunctionalization), in rare instances one gene copy may retain the original function while the other acquires a novel, evolutionarily advantageous (adaptive) function (**Figure 1*b***, neofunctionalization) (30). Alternatively, after duplication, mutations may occur in both genes that specialize to perform complementary functions (**Figure 1*b***, subfunctionalization) (56, 57). The question of how duplicate genes are retained

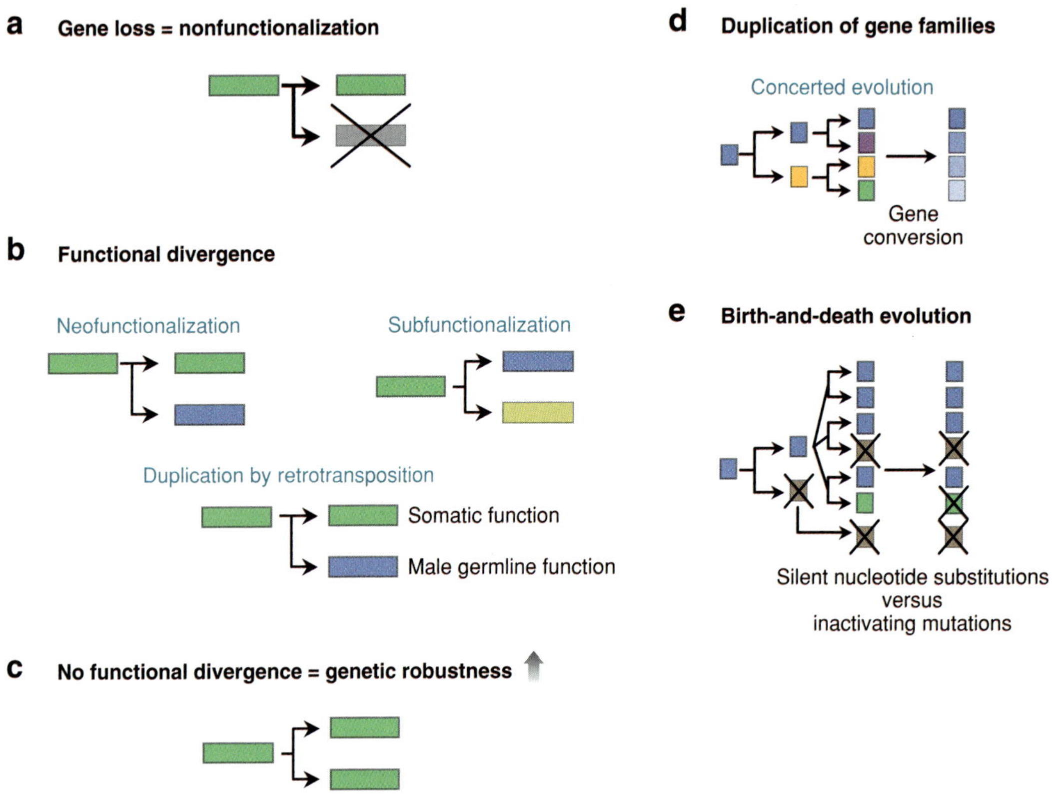

Figure 1

Evolutionary fate of single gene duplications (*a–c*), and duplication of multigene families (*d–e*). Single gene duplication most often results in a nonfunctional duplicate gene copy (*a*, nonfunctionalization). (*b*) In rare instances, the functional duplicate gene copy and the ancestral gene diverge in function; neofunctionalization means that one of the two genes retains the original function, while the other evolves a new, often beneficial function. Subfunctionalization implies that both the original and the duplicate genes mutate and evolve to fulfill complementary functions already present in the original gene. Duplication via retrotransposition represents a particular case of sub- or neofunctionalization. Multigene families evolve in a coordinated fashion, such that the DNA coding sequences and function of the single members of a family remain close to that of the ancestral gene (*d–e*). (*d*) Concerted evolution: After multiple rounds of duplication, gene conversion homogenizes the DNA sequences of the individual members. (*e*) Birth-and-death evolution invokes a process of equilibrium between inactivating mutations and ongoing duplication of functional gene copies.

in a population remains controversial. Classical duplication-degeneration-complementation/subfunctionalization models do not invoke positive selection, but stipulate a higher retention rate of duplicate genes in small rather than larger populations. Considerably more retentions and fewer losses of duplicate genes in rodents as compared with humans indicate that positive selection may play a more important role than originally anticipated (91). If two redundant gene copies were retained in the genome without significant functional divergence, the organism may acquire increased genetic robustness against harmful mutations (**Figure 1*c***). In multigene families descended from a common ancestor, individual genes in the group exert similar functions and have similar DNA sequences (67, 68). One concept, concerted evolution, applies particularly to localized and typically tandem copies of a gene. The concept posits that all genes in a given group evolve coordinately, and that homogenization is the result of gene conversion (**Figure 1*d***). For most multigene families, the currently favored model is birth-and-death evolution, according to which similarity in protein sequence among the members of a family is assured by strong purifying selection, such that individual genes evolve essentially via silent synonymous nucleotide substitutions (67, 68) (**Figure 1*e***). Inactivating mutations in a single member of a multigene family do not necessarily imply an evolutionary dead end, as illustrated by the more is less hypothesis (72). For instance, the human *CASP12* pseudogene located in a cluster of functional *Caspase* genes shows that a protein-truncating mutation can be positively selected, probably because the variant allele confers resistance to severe sepsis (105, 108). A recently recognized primate-specific subgroup of duplications generated by retrotransposition was recruited to enhance male germline function (**Figure 1*b***) (103). This mechanism thus represents a variation of neo- or subfunctionalization. Irrespective of these alternatives, gene duplication is, along with alternative splicing, a major mechanism of gene diversification (48, 99). For instance, along the lineage leading to humans, 689 genes were gained and 86 genes lost since the split from chimpanzees, contributing to a 6% difference in the complement of genes between humans and chimpanzees (1418 of 22,000 genes), largely outnumbering the 1.5% nucleotide difference between orthologous sequences of the two species (23).

Whole-Genome versus Small-Scale Duplications

Recent analysis of eukaryotic genomes shows that gene duplication is widespread (50, 56). A series of arguments suggest that current vertebrate genomes have been shaped by two rounds of whole-genome duplication (22, 100). An alternative hypothesis favors one single round of whole-genome duplication plus continuous small-scale duplications (35). Using *Arabidopsis* as a model system that underwent several rounds of complete genome duplication, the fate of small-scale versus whole-genome duplication was addressed. This analysis not only revealed the dominant evolutionary impact of whole-genome duplication, but also the important role of small-scale duplications for gene categories related to metabolism, stress responses, and cell death (58). The functional gene categories that have been selectively maintained by the two mechanisms differ somewhat (20). For instance, a group of genes also referred to as connected genes, encoding among other transcriptional regulators with limiting downstream partners that participate in protein-protein interactions, and interacting proteins that are part of signal transduction pathways (31), are consistently underrepresented among small-scale duplications, and overrepresented after whole-genome duplication (76). Duplication of such dosage-sensitive genes, required at stoichiometrically precise levels, may be tolerated in whole-genome, but not in small-scale, duplications (20). These findings comply with the gene balance hypothesis

Whole-genome duplication and 2R hypothesis: increased complexity and genome size of vertebrates resulted from two rounds (2R) of whole-genome duplication during early vertebrate evolution

Small-scale duplications: a recent gene family expansion by segmental or tandem duplication 50–150 Mya that may have followed on one single precedent round (1R) of WGD

Gene balance hypothesis: posits that an imbalance in the concentration of individual components of a multiprotein complex is deleterious

(9, 10, 31, 76, 102), which posits that connected genes are particularly dosage sensitive, and are expected to be overretained relative to nondosage-sensitive genes after whole-genome and larger-scale duplications, but not after local duplications.

Recent segmental duplications: sequences that are >1 kb in length and that show >90% sequence identity

Breakpoint in conserved synteny: syntenic segments or blocks ≥100 kb with breakpoints identified as change in orientation or chromosome location based on unique regions of the genome

Copy number variant (CNV) or copy number polymorphism (CNP): a DNA segment ≥1 kb present at variable copy number compared with a reference genome

Copy number variable region (CNVR): CNVs identified in individuals of the HapMap collection, called in more than one individual, and replicated by a second independent platform

Recent Segmental Duplications

Recent segmental duplications comprise about 5% of the euchromatic portion of the human genome (26). Although one current definition includes sequences that are >1 kb in length and that show >90% sequence identity (6, 16), other authors use the term variably, giving rise to different numbers of duplicated segments in the genome. Segmental duplications are particularly enriched at pericentromeric and subtelomeric regions; subtelomeric duplications account for 40% (51), and pericentromeric duplications for ~33% of the total (90). A considerable fraction (25%) of the nonpericentromeric and nonsubtelomeric duplications are associated with syntenic breaks (4, 5), and 98% of primate-specific breakpoints contain segmental duplications (66). This was recently confirmed for the duplication-rich human chromosomes 15 (113) and 17 (114) (HSA15, HSA17). On these chromosomes, the human-specific breakpoints of conserved synteny (breakpoint identified as change in orientation or chromosome location based on unique regions of the genome, and for syntenic segments ≥100 kb) (5) occur mostly in regions containing duplications; 13 out of 15 breakpoints on HSA15 contain duplications, and 74% of duplicated bases on HSA17 reside in breakpoints. At least 50% of duplications are strictly intrachromosomal [50% for HSA15 (110) and 62% for HSA17 (111)]. Between 3.5% and 11% of all duplications contain complete genes (110), and nearly one third of duplicated genes are arrayed in tandem (92). An unexpected feature of areas with tandem duplications is their asynchronous replication, suggesting that duplicate structures alter the epigenetic state of a given locus (33). Ultraconserved elements (UCEs) are rare in segmental duplications and copy number variants (CNVs), except for the UCE-subclass that overlaps exons (24). The mechanistic basis for this exquisite intolerance to copy number changes of UCEs awaits further characterization.

At least 30% of the pericentromeric duplications were duplicatively transposed from euchromatic regions, generating a minimum of 28 new transcripts that are expressed primarily in the testis (90). The overall proportion of duplications is highest close to centromeres, and gradually diminishes within a distance of ~5 Mb from centromeres. Conversely, gene density increases with distance from centromeric α-satellite repeats. The subtelomeres contain 25 small gene families organized in tandemly repeated blocks sharing extensive sequence similarity, and 18 of these have at least one functional member (51). Gene products within these blocks exert highly varied functions, but are predominantly odorant- and cytokine-receptors, tubulins, and transcription factors.

Several recent studies revealed the widespread presence of large-scale CNVs in the normal human population (85, 88). Deletions and insertions were equally represented, often occurring around regions of chromosomal instability (88). In one study, 70 different genes were shown to vary in copy number within CNVs (85). Similarly, extensive CNVs in subtelomeric blocks were documented (51). Approximately 1450 copy number variable regions (CNVRs) encompassing 360 Mbs, or 12% of the genome, were mapped through the study of 270 individuals from the HapMap collection (80), and 5150 CNVs genome-wide are currently recorded in the available databases [**http://paralogy.gs.washington.edu/structuralvariation/** (85), **http://projects.tcag.ca/variation/** (43), **http://www.som.soton.ac.uk/research/geneticsdiv/anomaly%20register/**]. Within human CNVs, a significant overrepresentation of genes associated with environmentally regulated functions and immunity was found, suggesting an adaptive advantage of dosage imbalance in these regions (69). Mechanistically, duplications that share extensive sequence similarity, called low copy repeats (LCRs), occur frequently in regions with large duplicated stretches. One common definition of LCRs is sequences that are ≥10 kb in length, show ≥95% sequence identity, and are separated by 50 kb–10 Mb of intervening sequence (97). Such LCRs serve as substrates for nonallelic homologous recombination (NAHR) that can result in duplications, reciprocal deletions, inversions, and reciprocal translocations (83, 89). The primate-specific burst of Alu-repeats ~35–40 million years ago (Mya) could have been one critical event initiating segmental gene duplications (7). Consistent with this, the generation and structure of LCRs appear to be associated with Alu-elements (89), and 492 human-specific deletions can be attributed to this process (87). A detailed analysis of the proximal short arm of HSA17 indicated that NAHR between LCRs is a major mechanism of recurrent rearrangements, whereas nonhomologous end-joining (NHEJ) can, in many instances, be responsible for nonrecurrent rearrangments (54). The prevailing molecular mechanism also depends on the chromosomal position, because subtelomeric duplications were generated almost exclusively via NHEJ of double-strand breaks (51).

Functional, duplicated genes comprise an important, rapidly evolving euchromatic fraction of our genome that displays extensive polymorphism; it is therefore important to examine their contribution to phenotypic variation and disease.

Evolutionary Forces and Duplicated Genes

The evolutionary forces acting on duplicated genes are diverse. A number of interdependent variables determine whether a gene will be retained after duplication; these include its functional category (46, 61, 76), degree of conservation (18, 19, 45), sensitivity to dosage effects (46), as well as its regulatory and architectural complexity (39). In general, genes encoding proteins that interact with the environment tend to be more frequently retained after duplication than those that interact with intracellular compartments (50, 61). Genes with permanent duplicates in the *Caenorhabditis elegans* and *Saccharomyces cerevisiae* genomes are more constrained before duplication than genes that never duplicated (19). However, human CNVs are consistently enriched in genes with increased rates of synonymous and nonsynonymous codon substitutions (69). Haploinsufficient genes—i.e., genes experiencing a loss of fitness when present in a single copy in diploid species—have more paralogs than haplosufficient genes, supporting the concept that gene dosage may be critical in fixing duplications (gene balance hypothesis) (46). In general, duplicated genes encode for longer proteins containing more domains and more *cis*-regulatory elements than singleton genes (39). These observations indicate that natural selection created a preferential association of duplications with certain gene categories.

Experiments performed on multiple species, from unicellular organisms to mammals, indicate that genes rapidly diverge after duplication (**Figure 2**) (17, 47). Although divergence of duplicated genes was traditionally assessed by comparing the rate of amino acid changes, more recently the focus was on divergence of gene expression, of transcriptional networks, and of protein-protein interactions. It has been shown that the protein-coding sequence and *cis*-regulatory regions of duplicated genes evolve independently (**Figure 2*a***) (104). Shortly after duplication, expression and regulatory divergence exceed changes in the protein sequence (36). In humans and yeast, protein sequence evolution and gene expression diversity significantly correlate and increase with evolutionary time (37, 60). In addition, the mode of duplication and the functions of the genes involved also play important roles in expression divergence

Low copy repeats (LCRs): sequences that are ≥10 kb in length that show ≥95% sequence identity, and are separated by 50 kb–10 Mb of intervening sequence

Haploinsufficient genes: subset of genes experiencing a loss of fitness when present in a single copy in diploid species

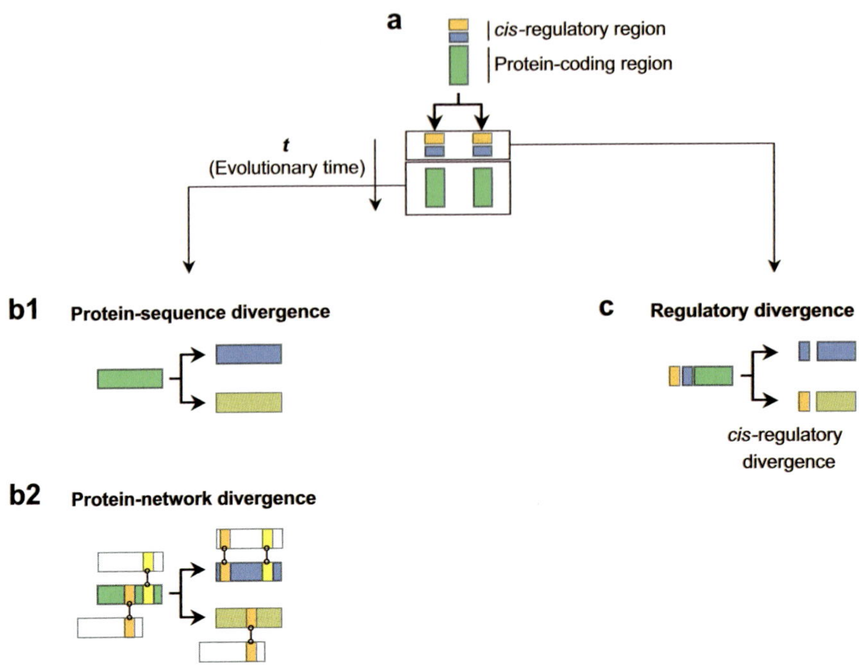

Figure 2

Functional divergence of duplicated genes. (*a*) The *cis*-regulatory and the protein-coding regions evolve independently after duplication. Divergence increases with evolutionary distance. (*b*) Schematic representation of (*b1*) protein sequence divergence after duplication (*green sequence* diverges into *blue* or *orange*); (*b2*) protein network divergence, where the protein interaction domains of the original green sequence evolve by maintenance, gain, or loss of interacting partners. (*c*) Schematic representation of DNA sequence regulatory divergence after duplication (certain regulatory motifs are lost in one copy of the duplicated gene sequence).

(13). Divergence of *cis*-regulatory motifs in the promoter-proximal region (**Figure 2*c***) is probably not the only substrate for expression divergence, suggesting that *trans*-acting factors could also be important (111). In humans, genes diverge asymmetrically after duplication; they rapidly lose their original coexpressed partners and acquire new ones (**Figure 2*c***) (17). Expression levels of duplicate genes diverge significantly during development within and between species, as compared to single-copy genes (37). Paralogous genes in humans and mice tend to become more specialized in their expression patterns (42).

The protein interaction partners (**Figure 2*b***) change at a slower rate than the transcription factors shared by duplicated genes (62). The divergence of protein-protein interactions after duplication depends on the connectivity [i.e., number of binding partners of a protein (8)] of the ancestral gene (112). Proteins with a higher ancient connectivity tend to display an asymmetrical evolution of the duplicates, whereas duplicates with a lower connectivity tend to gain and lose interacting partners at about the same rate. In conclusion, gene expression divergence is a key substrate for functional divergence of duplicate genes, and gene regulatory networks evolve at a higher rate after duplication than protein-protein interactions.

The Phenotypic Spectrum of Duplicated Genes

Gene dosage effects. The phenotypic consequence of duplicating large regions or even whole chromosomes is at least in part determined by the extent of regulatory imbalances. It is a priori expected that duplicate genes (three copies per diploid genome) will exhibit a ~1.5-fold increase in mRNA expression. Consistent with the prediction, ~50% of trisomic genes are overexpressed at the expected 1.5 level or higher; this is true for genes overexpressed as a consequence of an additional whole chromosome copy in human trisomy 21 and in the respective trisomic mouse models (81). Similarly, a recent comparison of chimpanzee and human segmental duplications revealed that among the human-specific duplicates (causing trisomies), 56% show a significant difference in gene expression between the two species, mostly (83%) overexpression in human as compared to chimpanzee (15).

For genes that fail to show such a dose-dependent increase, compensatory mechanisms are usually involved (i.e., dosage compensation). For instance, this can happen when the regulator controlling transcription and the target gene reside together on an aneuploid segment, canceling dosage imbalances (9). Conversely, duplicated regions containing positive or negative regulators can produce negative effects on target genes located outside of the aneuploid area (9, 10). The main conclusion from this is that the most significant alteration in gene expression caused by aneuploidy will be exerted by the target genes that are not in the aneuploid region.

In aneuploidy, it is generally assumed that only a restricted set of dosage-sensitive genes is responsible for the phenotype. In a systematic analysis of overexpression phenotypes in yeast, ~15% of overexpressed genes reduced cell growth, and among those, cell cycle-regulated genes, signaling molecules, and transcription factors were enriched (96). Interestingly, the overexpression phenotypes differed from the deletion mutant phenotypes, indicating that the underlying mechanisms are specific for under- and overexpression, respectively, rather than resulting from a common disruption of protein complex stoichiometry. These results contradict the gene balance hypothesis and are consistent with the insufficient amounts hypothesis (25), in which haploinsufficient genes are needed at abnormally high levels, and therefore are more sensitive to a reduction in dose (see sidebar on Under- and Overexpression Phenotypes: The Gene Balance Hypothesis versus the Insufficient Amount Hypothesis for more information).

UNDER- AND OVEREXPRESSION PHENOTYPES: THE GENE BALANCE HYPOTHESIS VERSUS THE INSUFFICIENT AMOUNT HYPOTHESIS

Another extrapolation of the gene balance hypothesis is that under- and overexpression phenotypes are identical, or at least similar, as they both act via the disruption of an identical multimeric, regulatory protein complex (disrupted stoichiometry of protein subunits encoded by connected genes).

The primary mechanism of haploinsufficiency in yeast is an insufficient protein production, which led to the formulation of the insufficient amount hypothesis, contradicting the predictions of the balance hypothesis. In further support of the insufficient amount hypothesis is the fact that under- and overexpression phenotypes of bona fide connected genes in yeast differ, indicating specific regulatory imbalances, rather than disrupted stoichiometry (see References 25 and 96).

Relationship of gene dosage and fitness (phenotype). The relationship between gene dosage and fitness (phenotype) is complex. Three alternatives have been proposed, each one characteristic of certain gene categories (**Figure 3**) (46). The first describes a linear relationship that is often found for structural and regulatory proteins (**Figure 3*a***); the second fits a "diminishing returns" principle typical of enzymes that function at limiting concentrations (**Figure 3*b***). Consistent with this, disorders caused by genes encoding enzymes are primarily recessive, indicating that these genes are not dosage sensitive (haplosufficient)

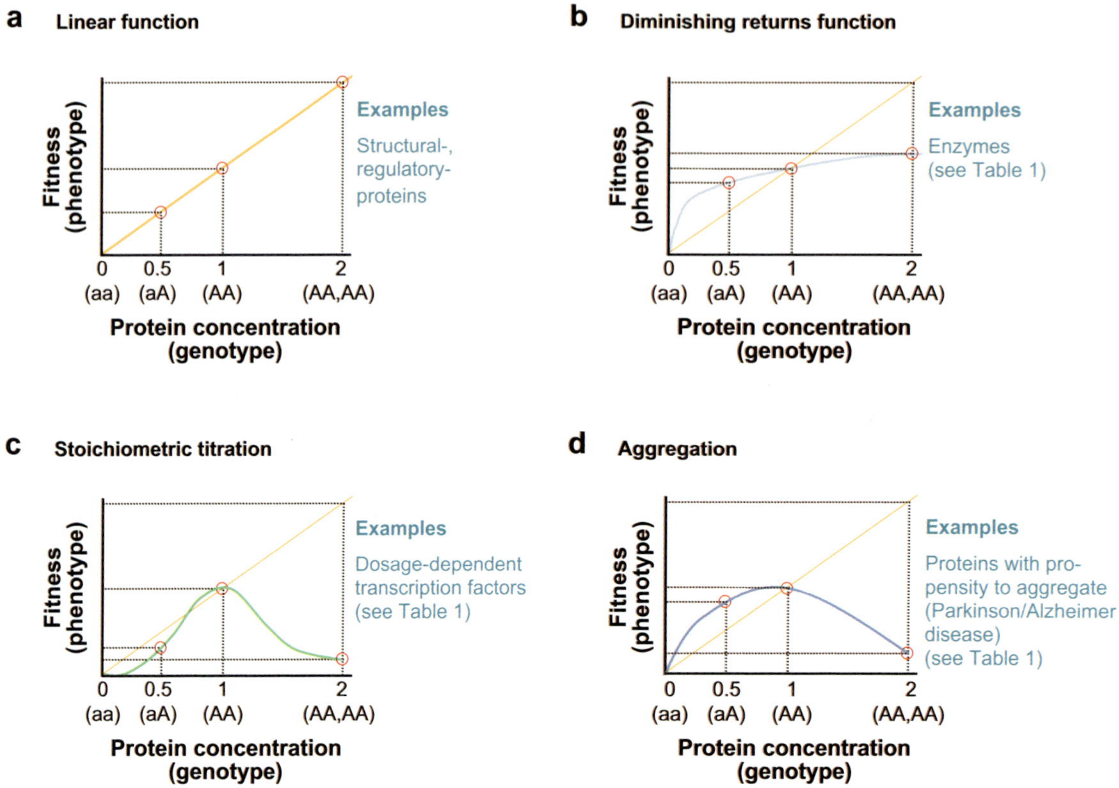

Figure 3

Schematic representations of gene dosage and fitness (phenotype). Four different relationships are shown. (*a*) A linear relationship is typically found for structural and regulatory proteins; (*b*) a diminishing returns function classically involves enzymes that show little variation in function over large dose ranges. (*c*) Certain functional gene classes enriched for transcriptional regulators and signaling molecules cause phenotypes both when under- and overexpressed (haploinsufficiency and pathogenic gene duplication). (*d*) Disease phenotypes caused by protein aggregation. The wild-type protein aggregates at doses beyond the threshold level of two gene copies. Heterozygous mutations can also cause protein aggregation at one gene copy. Uppercase "A" depicts the normal allele; lowercase "a" represents a mutant loss of function allele.

(44). The third alternative corresponds to a diminished fitness for both increased and decreased gene dosage, indicating either multisubunit complexes with a single component that has a tight stoichiometry (46) (gene balance hypothesis), or specific regulatory imbalances as a consequence of under- (insufficient amount hypothesis) and overexpression (25) (**Figure 3*c***). Transcriptional regulators can cause phenotypes both when under- and overexpressed (84), a mechanism responsible for many developmental/malformation syndromes (**Figure 3*c***) (**Table 1**). A fourth alternative concerns proteins with a propensity to aggregate (**Figure 3*d***). These proteins display a dual behavior. The wild-type protein aggregates in a dose-dependent manner once the diploid threshold dose is exceeded (e.g., gene duplication). Alternatively, aggregation occurs in the diploid state in the presence of a dominant mutation (**Figure 3*d***) (49). Genes encoding such proteins include α-synuclein (*SNCA*), responsible for one rare familial

Haploinsufficiency: a dominant phenotype in a diploid organism that is heterozygous for a loss of function allele

form of early-onset Parkinson disease, and the amyloid β precursor protein (*APP*) that causes one form of dominant early-onset Alzheimer disease (38, 93).

Synopsis of diseases and mechanisms associated with duplicated genes. In **Table 1**, we summarized the main features of phenotypes and diseases caused by duplicated genes.

Examples of pathogenic duplications involving dosage-sensitive genes. The general models discussed predict that among the genes overexpressed as a result of duplication, a minor fraction of dosage-sensitive genes enriched for transcription factors, signaling molecules, and cell cycle-regulated genes will be critical for the phenotypic features (96). Competing models stipulate that over- and underexpression phenotypes of such genes are either similar (gene balance hypothesis), or alternatively distinct (insufficient amounts hypothesis). In Charcot-Marie-Tooth (CMT) disease, the most frequent inherited disorder of the peripheral nervous system (55, 65), a 1.4-Mb tandem duplication is found in ~70% of autosomal dominant CMT1A patients. Although the 1.4-MB interval contains ≥30 genes, *PMP22*, encoding the major myelin protein, is responsible for the disorder through a gene dosage effect. In the heterozygous duplication state, *PMP22* trisomy causes a nerve conduction deficit due to demylinization (CMT1A), whereas *PMP22* monosomy (deletion) causes a nerve conduction block [hereditary neuropathy with liability to pressure palsies (HNPP)] in its haploinsufficient state. Because many central and peripheral demyelinating disorders, such as CMT1A, Pelizaeus-Merzbacher disease (PMD) (27), and autosomal dominant leukodystrophy (ADLD) (75), can be caused by gene duplication, it was suggested that myelin formation is particularly susceptible to gene dosage effects (75). The CMT/HNPP-causing region shows how two distinct phenotypes can result from over- and underexpression of a critical dosage-sensitive gene. Similarly, deletion and duplication of the Sotos syndrome region and *NSD1* gene dosage effects cause contrasting phenotypes (14).

Over- and underexpression phenotypes can also be similar, as shown by the methyl-CpG-binding protein 2 (MECP2), a chromatin architectural protein linked to transcriptional repression (52). The progressive neurodegenerative disorder Rett syndrome, which in its classical form affects females almost exclusively, is due to *MECP2* haploinsufficiency. Severe mental retardation and neurological symptoms with features of Rett syndrome in males can also be caused by *MECP2* duplications (101). Likewise, abnormal neurodevelopmental phenotypes linked to both *MECP2* under- and overexpression were described in human and transgenic mouse models (53). Similarly, X-linked hypopituitarism can be caused by inactivating heterozygous *SOX3* mutations, or by duplication of a region containing the developmental transcription factor *SOX3* (**Table 1**) (107).

Pathogenic duplications also occur in regions of genomic microdeletions, such as the velocardiofacial, the Williams-Beuren, the Alagille, and the Smith-Magenis syndrome regions (**Table 1**) (64, 79, 95, 109). These examples partially fit the model that duplication of one or a few dosage-sensitive genes causes overexpression phenotypes that resemble the underexpression phenotype (**Figure 3*c***). On the other hand, the duplication at 1p36 matches an incremental/linear gene-phenotype model for closure of cranial sutures (**Figure 3*a–b***) (32). Haploinsufficiency of genes in this region result in delayed closure of cranial sutures, whereas increased gene dosage via duplication, for instance, results in craniosynostosis. A heterogenous group of duplications, spanning 0.5–1-Mb regions, causes phenotypes (*a*) for which the corresponding monosomy has either not yet been reported (split hand/split foot syndrome, SHFM3) (21); (*b*) for which duplication and haploinsufficiency of as-yet-unidentified genes have been implicated in a similar/identical phenotype

Table 1 Duplication phenotypes

Species	Gene	Gene category	Genomic alteration	Disease	Phenotype	Mechanism	References
Similar under- and overexpression phenotypes							
Hs/Mm	*MECP2*	Methylated DNA binding	Duplication *MECP2*	Progressive neurodevelopmental syndrome in males	Mental retardation, epilepsy	Loss of gene function due to under- and overexpression	(52, 98)
Hs	*SOX3*	Developmental regulator (TF)	Duplication *SOX3*	X-linked hypopituitarism (XLHP)	X-linked hypopituitarism and infundibular hypoplasia	Loss of gene function due to under- and overexpression	(104)
Hs	ND (*TBX1*)	ND	Duplication 22q11.2	Velocardiofacial syndrome (VCFS)	Variable: normal to developmental delay and malformations	Loss of gene function due to under- and overexpression	(106)
Hs	ND (*ELN*)	ND	Duplication 7q11.23	Williams Beuren syndrome (WBS)	Delay expressive language	Loss of gene function due to under- and overexpression	(92)
Hs	ND (*JAGGED1*)	Developmental regulator (TF)	Duplication 20p11	Alagille syndrome (AS)	Cardiovascular-, ocular-, bile duct-, and skeletal anomalies	Loss of gene function due to under- and overexpression	(63)
Hs	*RAI1*	Developmental regulator (TF)	Duplication 17p11.2	Smith-Magenis syndrome (SMS)	Mild mental retardation and dental abnormalities	Loss of gene function due to under- and overexpression	(77)
Hs	*PLP1*	Proteolipid protein	Duplication *PLP1*	Pelizaeus-Merzbacher (PM)	Demyelination disorder CNS	Loss of gene function due to under- and overexpression	(26)
Dissimilar under- and overexpression phenotypes							
Hs	*PMP22*	Myelin protein	Duplication *PMP22*	Charcot Marie Tooth 1A (CMT1A)	Peripheral myelin neuropathy	Loss of gene function due to under- and overexpression	(54, 64)
Hs	*NSD1*	Histone methyltransferase	Duplication *NSD1*	Growth retardation syndrome	Growth retardation	Loss of gene function due to under- and overexpression	(63)
Hs	ND (*MMP23*)	ND	Duplication 1p36	Premature closure cranial sutures	Craniosynostosis	Incremental gene function	(31)
Complex, yet unresolved expression phenotypes							
Hs	*LMB1*	Laminar nuclear envelope protein	Duplication *LMB1*	Autosomal dominant leukodystrophy (ADLD)	Demyelination disorder CNS	ND	(73)
Hs	ND	ND	Duplication 10q24	Split hand/split foot malformation 3 (SHFM3)	Split hand/split foot	ND	(21)
Hs	ND	ND	Duplication 2q13	Orofacial clefting/cleft palate only	Mental retardation and orofacial clefting	ND	(72)

(*Continued*)

Table 1 (*Continued*)

Species	Gene	Gene category	Genomic alteration	Disease	Phenotype	Mechanism	References
Hs	ND	ND	Duplication 16p13	ATR-X-like	X-linked α-thalassemia/mental retardation	ND	(2)
Protein aggregation due to overexpression							
Hs	*SNCA*	Molecular chaperone	Duplication *SNCA*	Parkinson disease	Nigrostriatal neuron degeneration	Protein aggregation and underlying gene mutations	(91)
Hs	*APP*	Amyloid precursor protein	Duplication *APP*	Alzheimer disease	Parenchymal/vascular amyloid deposition	Protein aggregation and underlying gene mutations	(37, 79)
Phenotypes of CNVs related to environment and immunity							
Hs	*CYP2D6*	P-450 isoenzyme	*P-450* CNV	Altered drug metabolism	Adverse drug effects	Incremental/linear gene function model	(12, 40, 75)
Hs	*CCL3L1*	Chemokine receptor	*CCL3L1* CNV	Altered HIV susceptibility	Enhanced HIV/AIDS susceptibility	Incremental/linear gene function model	(33)
Common complex phenotypes of CNVs in defense-related genes							
Hs/Rn	*FCGR3B*/ *Fcgr3-rs*	Fc receptor for IgG	*FCGR3* CNV	Glomerulonephritis/ systemic lupus erymathosus	Susceptibility to glomerulonephritis	Pathogenic underexpression phenotype	(1)
Mm	*TLR7*	Toll-like receptor	*TLR7* CNV	Systemic lupus erymathosus-like disease	Autoantibody-elicted autoimmunity	Pathogenic overexpression phenotype	(76)
Hs	*hBD2*	Antimicrobial peptides	*hBD2* CNV	Crohn's disease of the colon	Inflammatory bowel disease	Pathogenic underexpression phenotype	(28)

(mental retardation/orofacial clefting syndrome) (74); and (*c*) for which haploinsufficiency of a transcription factor (*ATRX* at Xq13.3) and duplication of a second region containing candidate *ATRX* target genes (region at 16p13.11-16p13.3 comprising the α-*globin* gene) result in a similar clinical outcome (X-linked α-thalassemia/mental retardation syndrome) (2). Although these examples are more complex than those previously discussed, the last two cases may be compatible with similar under- and overexpression phentoypes.

Diseases caused by protein aggregation. Protein aggregation is particularly associated with disease genes (106). Two rare forms of neurodegenerative disorders, Parkinson and Alzheimer disease, provide constructive examples of pathogenic gene dosage effects mediated via protein aggregation (**Figure 3*d***) (38, 93). Alpha-synuclein (*SNCA*) duplication and triplication lead to increased expression of α-synuclein, a small protein thought to play a role as a molecular chaperone in vesicular transport and/or turnover of synaptic vesicles. The pathology induced by this protein and the severity of disease both depend on the gene expression level. Although the mechanism is not yet fully understood, α-synuclein aggregation, controlled by mutations in at least three genes including *SNCA* itself, is thought to promote nigrostriatal neurons for degeneration (94).

The amyloid β precursor protein (APP) can be cleaved into either of two smaller peptides, Aβ40 and Aβ42 (38). It is accumulation of the latter causing parenchymal and vascular deposition that is probably responsible for Alzheimer pathogenesis. The disease phenotype can be caused by a variety of mechanisms that lead to an increase in APP protein aggregates, namely duplications of *APP* on chromosome 21 (82), Down syndrome (trisomy 21) with three *APP* copies, or, alternatively, mutations in *APP* itself or in the gene encoding the proteases responsible for APP cleavage (38). Similar mechanisms may account for other neurological diseases that are based on protein deposition (38).

Modified responses to drugs. Genes encoding enzymes are not usually dosage sensitive, yet there are examples of such gene duplications that cause variable phenotypes, particularly those encoding protein complexes that regulate drug metabolism (12). The three cytochrome P-450 isoforms CYP2C9, CYP2C19, and CYP2D6, which all vary in copy number, are responsible for the biotransformation of about 40% of all drugs that are metabolized by P-450. Copy number variation among the 80 distinct alleles of the *CYP2D6* gene partially defines the individual capacity to metabolize drugs. Four major phenotypic categories of drug oxidation have been recognized to date, namely poor, intermediate, extensive, and ultrarapid metabolizers. Because these phenotypic differences have clinical consequences, such as adverse drug reactions or therapeutic failure (77), it will be important to develop accurately predictive genotyping for these and other enzymes that appear to display clinically relevant CNVs (41). CNVs are currently under investigation for their pharmacogenetic relevance (73).

Altered responses to pathogens. Because genes encoding proteins in metabolic pathways in unicellular organisms have a higher duplicability rate than other gene categories (61), this could generally be true for genes responding to and/or regulated by environmental stimuli. Intriguingly, copy number variation of the gene encoding CCL3L1, a potent human immunodeficiency virus-1 (HIV-1)-suppressive chemokine, influences the susceptibility to HIV-1 infection (34). One could speculate that CNVs containing genes related to immune responses may be involved in the control of infections. Of relevance in this regard are CNVs of the novel defensin gene family that function as chemotactic activators of the immune response (3, 86). As discussed below, defensin CNVs may cause inflammatory bowel disease by modulating the host response to pathogens (29). Gene families that constitute the core elements controlling infections, namely the T-cell receptor and immunoglobulin genes, evolved via species-specific sequential rounds of duplications (98). Copy number variation in these gene families may also play a role in complex phenotypes such as autoimmune diseases.

Role of copy number variants in complex diseases. An example of the possible role of CNVs in common complex phenotypes is the demonstration that low copy numbers of the *FCGR3B* gene encoding the activatory Fc receptor for IgG predispose patients with systemic lupus erythematosus to an inflammatory disease of the kidneys (glomerulonephritis) (1). Analogous findings in rats corroborate this observation. Additional examples include duplication of *TLR7*, an innate immune receptor that predisposes mice carrying it to autoreactive B-cell responses (78), and low copy number of the human beta-defensin 2 (*hBD2*) gene that predisposes to Crohn's disease of the colon (29). The importance of CNVs for complex phenotypes and diseases in genes related to immune defense should therefore be generally explored.

Disease-causing gene conversion mediated by duplicated pseudogenes. Among the 1945 nonprocessed pseudogenes located near the ancestral gene (out of a total of 3426 nonprocessed pseudogenes), 11 cases of deleterious gene conversion induced by the pseudogene were recognized (11). Among those is the retinitis pigmentosa 9 pseudogene (*RP9P*) carrying a mutation that produces a nonsynonymous substitution in the wild-type gene associated with the RP9 form of autosomal dominant retinitis pigmentosa (ADRP). Two other processed pseudogenes also contain mutations associated with diseases, an inosine monophosphate dehydrogenase 1 pseudogene (*IMPDH1P1*) that causes the RP10 form of ADRP, and a phosphoglycerate kinase 1 pseudogene (*PGK1P1*) associated with phosphoglycerate kinase deficiency. These observations highlight the pathogenic potential of duplicate gene copies that acquired inactivating mutations (pseudogenes) for the wild-type progenitor gene located nearby.

Partial phenotype rescue by a duplicated gene. The second most frequent autosomal recessive disease in Europeans, spinal muscular atrophy (SMA), adds an important lesson to the phenotypic consequences of gene duplication. The disease, which is usually due to homozygous mutations in the *survival motor neuron 1* gene (*SMN1*), can be partially rescued by increasing copy numbers of its nearly identical and partially functional duplicate *SMN2* (28). This example emphasizes the partial functional redundancy of some duplicate genes (40), and supports the notion that deletion of a duplicate gene results in a less severe phenotype than deletion of a singleton gene.

CONCLUSIONS AND PERSPECTIVES

The frequency and variety of phenotypes induced by gene duplication are diverse, and have not yet been fully appreciated. Several fields deserve more investigation, notably the role of gene duplication in (*a*) monogenic phenotypes, [particularly disorders implicating haploinsufficient dosage-sensitive genes, for instance congenital heart disease (71)]; (*b*) polygenic complex multifactorial phenotypes, as exemplified by the implication of the *FCGR3* and *hBD2* copy number in glomerulonephritis and Crohn's disease, and by the role of *TLR7* duplications in autoimmunity elicited by autoantibodies; (*c*) drug, host, pathogen, and metabolic responses; (*d*) diseases due to protein aggregation; and (*e*) gene regulation, as exemplified by the contribution of microRNA gene duplication to complex patterns of gene regulation (59), and by duplication of conserved noncoding regions that helped diversify expression of developmental regulators (63). Finally, because increased dosage of genes related to olfaction, immunity, and protein secretion may have been positively selected in humans (69), a possibly more general contribution to phenotypic variation of CNVs in genes responding to environmental stimuli deserves a more thorough exploration in the near future. In summary, this review article discusses and emphasizes that gene duplication plays an important role in genome evolution and phenotypic variability, and can cause disease phenotypes.

ACKNOWLEDGMENTS

We thank Drs. Jacques Beckmann, Samuel Deutsch, and Henrik Kaessmann for reading the manuscript. B.C. is supported by the SNSF and Helmut Horten Foundation; S.E.A. is supported by the SNSF, EU, NIH, and Childcare Foundation, and by funds from the University of Geneva.

LITERATURE CITED

1. Aitman TJ, Dong R, Vyse TJ, Norsworthy PJ, Johnson MD, et al. 2006. Copy number polymorphism in Fcgr3 predisposes to glomerulonephritis in rats and humans. *Nature* 439:851–55

2. Akahoshi K, Ohashi H, Hattori Y, Saitoh S, Fukushima Y, Wada T. 2005. A woman with 46,XX,dup(16)(p13.11 p13.3) and the ATR-X phenotype. *Am. J. Med. Genet. A* 132:414–18
3. Aldred PM, Hollox EJ, Armour JA. 2005. Copy number polymorphism and expression level variation of the human alpha-defensin genes DEFA1 and DEFA3. *Hum. Mol. Genet.* 14:2045–52
4. Armengol L, Pujana MA, Cheung J, Scherer SW, Estivill X. 2003. Enrichment of segmental duplications in regions of breaks of synteny between the human and mouse genomes suggest their involvement in evolutionary rearrangements. *Hum. Mol. Genet.* 12:2201–8
5. Bailey JA, Baertsch R, Kent WJ, Haussler D, Eichler EE. 2004. Hotspots of mammalian chromosomal evolution. *Genome Biol.* 5:R23
6. Bailey JA, Gu Z, Clark RA, Reinert K, Samonte RV, et al. 2002. Recent segmental duplications in the human genome. *Science* 297:1003–7
7. Bailey JA, Liu G, Eichler EE. 2003. An Alu transposition model for the origin and expansion of human segmental duplications. *Am. J. Hum. Genet.* 73:823–34
8. Berg J, Lassig M, Wagner A. 2004. Structure and evolution of protein interaction networks: a statistical model for link dynamics and gene duplications. *BMC Evol. Biol.* 4:51
9. Birchler JA, Bhadra U, Bhadra MP, Auger DL. 2001. Dosage-dependent gene regulation in multicellular eukaryotes: implications for dosage compensation, aneuploid syndromes, and quantitative traits. *Dev. Biol.* 234:275–88
10. Birchler JA, Riddle NC, Auger DL, Veitia RA. 2005. Dosage balance in gene regulation: biological implications. *Trends Genet.* 21:219–26
11. Bischof JM, Chiang AP, Scheetz TE, Stone EM, Casavant TL, et al. 2006. Genome-wide identification of pseudogenes capable of disease-causing gene conversion. *Hum. Mutat.* 27:545–52
12. Caraco Y. 2004. Genes and the response to drugs. *N. Engl. J. Med.* 351:2867–69
13. Casneuf T, De Bodt S, Raes J, Maere S, Van de Peer Y. 2006. Nonrandom divergence of gene expression following gene and genome duplications in the flowering plant *Arabidopsis thaliana*. *Genome Biol.* 7:R13
14. Chen CP, Lin SP, Lin CC, Chen YJ, Chern SR, et al. 2006. Molecular cytogenetic analysis of de novo dup(5)(q35.2q35.3) and review of the literature of pure partial trisomy 5q. *Am. J. Med. Genet. A* 140:1594–600
15. Cheng Z, Ventura M, She X, Khaitovich P, Graves T, et al. 2005. A genome-wide comparison of recent chimpanzee and human segmental duplications. *Nature* 437:88–93
16. Cheung VG, Nowak N, Jang W, Kirsch IR, Zhao S, et al. 2001. Integration of cytogenetic landmarks into the draft sequence of the human genome. *Nature* 409:953–58
17. Chung WY, Albert R, Albert I, Nekrutenko A, Makova KD. 2006. Rapid and asymmetric divergence of duplicate genes in the human gene coexpression network. *BMC Bioinformatics* 7:46
18. Conant GC, Wagner A. 2002. GenomeHistory: a software tool and its application to fully sequenced genomes. *Nucleic Acids Res.* 30:3378–86
19. Davis JC, Petrov DA. 2004. Preferential duplication of conserved proteins in eukaryotic genomes. *PLOS Biol.* 2:E55
20. Davis JC, Petrov DA. 2005. Do disparate mechanisms of duplication add similar genes to the genome? *Trends Genet.* 21:548–51
21. de Mollerat X, Gurrieri F, Morgan CT, Sangiorgi E, Everman DB, et al. 2003. A genomic rearrangement resulting in a tandem duplication is associated with split hand-split foot malformation 3 (SHFM3) at 10q24. *Hum. Mol. Genet.* 12:1959–71

22. Dehal P, Boore JL. 2005. Two rounds of whole genome duplication in the ancestral vertebrate. *PLOS Biol.* 3:e314
23. Dermuth JP, De Bie T, Stjich JE, Cristianini N, Hahn MW. 2006. *PLoS* 1:e85
24. Derti A, Roth FP, Church GM, Wu CT. 2006. Mammalian ultraconserved elements are strongly depleted among segmental duplications and copy number variants. *Nat. Genet.* 38:1216–20
25. Deutschbauer AM, Jaramillo DF, Proctor M, Kumm J, Hillenmeyer ME, et al. 2005. Mechanisms of haploinsufficiency revealed by genome-wide profiling in yeast. *Genetics* 169:1915–25
26. Eichler EE. 2001. Recent duplication, domain accretion and the dynamic mutation of the human genome. *Trends Genet.* 17:661–69
27. Ellis D, Malcolm S. 1994. Proteolipid protein gene dosage effect in Pelizaeus-Merzbacher disease. *Nat. Genet.* 6:333–34
28. Feldkotter M, Schwarzer V, Wirth R, Wienker TF, Wirth B. 2002. Quantitative analyses of SMN1 and SMN2 based on real-time lightCycler PCR: fast and highly reliable carrier testing and prediction of severity of spinal muscular atrophy. *Am. J. Hum. Genet.* 70:358–68
29. Fellermann K, Stange DE, Schaeffeler E, Schmalzl H, Wehkamp J, et al. 2006. A chromosome 8 gene-cluster polymorphism with low human Beta-defensin 2 gene copy number predisposes to crohn disease of the colon. *Am. J. Hum. Genet.* 79:439–48
30. Force A, Lynch M, Pickett FB, Amores A, Yan YL, Postlethwait J. 1999. Preservation of duplicate genes by complementary, degenerative mutations. *Genetics* 151:1531–45
31. Freeling M, Thomas BC. 2006. Gene-balanced duplications, like tetraploidy, provide predictable drive to increase morphological complexity. *Genome Res.* 16:805–14
32. Gajecka M, Yu W, Ballif BC, Glotzbach CD, Bailey KA, et al. 2005. Delineation of mechanisms and regions of dosage imbalance in complex rearrangements of 1p36 leads to a putative gene for regulation of cranial suture closure. *Eur. J. Hum. Genet.* 13:139–49
33. Gimelbrant AA, Chess A. 2006. An epigenetic state associated with areas of gene duplication. *Genome Res.* 16:723–29
34. Gonzalez E, Kulkarni H, Bolivar H, Mangano A, Sanchez R, et al. 2005. The influence of CCL3L1 gene-containing segmental duplications on HIV-1/AIDS susceptibility. *Science* 307:1434–40
35. Gu X, Wang Y, Gu J. 2002. Age distribution of human gene families shows significant roles of both large- and small-scale duplications in vertebrate evolution. *Nat. Genet.* 31:205–9
36. Gu X, Zhang Z, Huang W. 2005. Rapid evolution of expression and regulatory divergences after yeast gene duplication. *Proc. Natl. Acad. Sci. USA* 102:707–12
37. Gu Z, Rifkin SA, White KP, Li WH. 2004. Duplicate genes increase gene expression diversity within and between species. *Nat. Genet.* 36:577–79
38. Hardy J. 2006. Amyloid double trouble. *Nat. Genet.* 38:11–12
39. He X, Zhang J. 2005. Gene complexity and gene duplicability. *Curr. Biol.* 15:1016–21
40. He X, Zhang J. 2005. Transcriptional reprogramming and backup between duplicate genes: Is it a genome-wide phenomenon? *Genetics* 172:1363–67
41. Hildebrandt MA, Salavaggione OE, Martin YN, Flynn HC, Jalal S, et al. 2004. Human SULT1A3 pharmacogenetics: gene duplication and functional genomic studies. *Biochem. Biophys. Res. Commun.* 321:870–78
42. Huminiecki L, Wolfe KH. 2004. Divergence of spatial gene expression profiles following species-specific gene duplications in human and mouse. *Genome Res.* 14:1870–79

43. Iafrate AJ, Feuk L, Rivera MN, Listewnik ML, Donahoe PK, et al. 2004. Detection of large-scale variation in the human genome. *Nat. Genet.* 36:949–51
44. Jimenez-Sanchez G, Childs B, Valle D. 2001. Human disease genes. *Nature* 409:853–55
45. Jordan IK, Wolf YI, Koonin EV. 2004. Duplicated genes evolve slower than singletons despite the initial rate increase. *BMC Evol. Biol.* 4:22
46. Kondrashov FA, Koonin EV. 2004. A common framework for understanding the origin of genetic dominance and evolutionary fates of gene duplications. *Trends Genet.* 20:287–90
47. Kondrashov FA, Rogozin IB, Wolf YI, Koonin EV. 2002. Selection in the evolution of gene duplications. *Genome Biol.* 3: RESEARCH0008
48. Kopelman NM, Lancet D, Yanai I. 2005. Alternative splicing and gene duplication are inversely correlated evolutionary mechanisms. *Nat. Genet.* 37:588–89
49. Li J, Uversky VN, Fink AL. 2001. Effect of familial Parkinson's disease point mutations A30P and A53T on the structural properties, aggregation, and fibrillation of human alpha-synuclein. *Biochemistry* 40:11604–13
50. Li WH, Gu Z, Cavalcanti AR, Nekrutenko A. 2003. Detection of gene duplications and block duplications in eukaryotic genomes. *J. Struct. Funct. Genom.* 3:27–34
51. Linardopoulou EV, Williams EM, Fan Y, Friedman C, Young JM, Trask BJ. 2005. Human subtelomeres are hot spots of interchromosomal recombination and segmental duplication. *Nature* 437:94–100
52. Luger K, Hansen JC. 2005. Nucleosome and chromatin fiber dynamics. *Curr. Opin. Struct. Biol.* 15:188–96
53. Luikenhuis S, Giacometti E, Beard CF, Jaenisch R. 2004. Expression of MeCP2 in postmitotic neurons rescues Rett syndrome in mice. *Proc. Natl. Acad. Sci. USA* 101:6033–38
54. Lupski JR, Stankiewicz P. 2005. Genomic disorders: molecular mechanisms for rearrangements and conveyed phenotypes. *PLOS Genet.* 1:e49
55. Lupski JR, Wise CA, Kuwano A, Pentao L, Parke JT, et al. 1992. Gene dosage is a mechanism for Charcot-Marie-Tooth disease type 1A. *Nat. Genet.* 1:29–33
56. Lynch M, Conery JS. 2000. The evolutionary fate and consequences of duplicate genes. *Science* 290:1151–55
57. Lynch M, Force A. 2000. The probability of duplicate gene preservation by subfunctionalization. *Genetics* 154:459–73
58. Maere S, De Bodt S, Raes J, Casneuf T, Van Montagu M, et al. 2005. Modeling gene and genome duplications in eukaryotes. *Proc. Natl. Acad. Sci. USA* 102:5454–59
59. Maher C, Stein L, Ware D. 2006. Evolution of *Arabidopsis* microRNA families through duplication events. *Genome Res.* 16:510–19
60. Makova KD, Li WH. 2003. Divergence in the spatial pattern of gene expression between human duplicate genes. *Genome Res.* 13:1638–45
61. Marland E, Prachumwat A, Maltsev N, Gu Z, Li WH. 2004. Higher gene duplicabilities for metabolic proteins than for nonmetabolic proteins in yeast and *E. coli*. *J. Mol. Evol.* 59:806–14
62. Maslov S, Sneppen K, Eriksen KA, Yan KK. 2004. Upstream plasticity and downstream robustness in evolution of molecular networks. *BMC Evol. Biol.* 4:9
63. McEwen GK, Woolfe A, Goode D, Vavouri T, Callaway H, Elgar G. 2006. Ancient duplicated conserved noncoding elements in vertebrates: a genomic and functional analysis. *Genome Res.* 16:451–65
64. Moog U, Engelen J, Albrechts J, Hoorntje T, Hendrikse F, Schrander-Stumpel C. 1996. Alagille syndrome in a family with duplication 20p11. *Clin. Dysmorphol.* 5:279–88
65. Murakami T, Garcia CA, Reiter LT, Lupski JR. 1996. Charcot-Marie-Tooth disease and related inherited neuropathies. *Medicine (Baltimore)* 75:233–50

66. Murphy WJ, Larkin DM, Everts-van der Wind A, Bourque G, Tesler G, et al. 2005. Dynamics of mammalian chromosome evolution inferred from multispecies comparative maps. *Science* 309:613–17
67. Nei M, Rogozin IB, Piontkivska H. 2000. Purifying selection and birth-and-death evolution in the ubiquitin gene family. *Proc. Natl. Acad. Sci. USA* 97:10866–71
68. Nei M, Rooney AP. 2005. Concerted and birth-and-death evolution of multigene families. *Annu. Rev. Genet.* 39:121–52
69. Nguyen DQ, Webber C, Ponting CP. 2006. Bias of selection on human copy-number variants. *PLOS Genet.* 2:e20
70. Ohno S. 1970. *Evolution by Gene Duplication*. New York: Springer Verlag
71. Olson EN. 2006. Gene regulatory networks in the evolution and development of the heart. *Science* 313:1922–27
72. Olson MV. 1999. When less is more: gene loss as an engine of evolutionary change. *Am. J. Hum. Genet.* 64:18–23
73. Ouahchi K, Lindeman N, Lee C. 2006. Copy number variants and pharmacogenomics. *Pharmacogenomics* 7:25–29
74. Ounap K, Ilus T, Laidre P, Uibo O, Tammur P, Bartsch O. 2005. A new case of 2q duplication supports either a locus for orofacial clefting between markers D2S1897 and D2S2023 or a locus for cleft palate only on chromosome 2q13-q21. *Am. J. Med. Genet. A* 137:323–27
75. Padiath QS, Saigoh K, Schiffmann R, Asahara H, Yamada T, et al. 2006. Lamin B1 duplications cause autosomal dominant leukodystrophy. *Nat. Genet.* 38:1114–23
76. Papp B, Pal C, Hurst LD. 2003. Dosage sensitivity and the evolution of gene families in yeast. *Nature* 424:194–97
77. Phillips KA, Veenstra DL, Oren E, Lee JK, Sadee W. 2001. Potential role of pharmacogenomics in reducing adverse drug reactions: a systematic review. *JAMA* 286:2270–79
78. Pisitkun P, Deane JA, Difilippantonio MJ, Tarasenko T, Satterthwaite AB, Bolland S. 2006. Autoreactive B cell responses to RNA-related antigens due to TLR7 gene duplication. *Science* 312:1669–72
79. Potocki L, Chen KS, Park SS, Osterholm DE, Withers MA, et al. 2000. Molecular mechanism for duplication 17p11.2—the homologous recombination reciprocal of the Smith-Magenis microdeletion. *Nat. Genet.* 24:84–87
80. Redon R, Ishikawa S, Fitch KR, Feuk L, Perry GH, et al. 2006. Gloval variation in copy number in the human genome. *Nature* 444:444–54
81. Roper RJ, Reeves RH. 2006. Understanding the basis for Down syndrome phenotypes. *PLOS Genet.* 2:e50
82. Rovelet-Lecrux A, Hannequin D, Raux G, Meur NL, Laquerriere A, et al. 2006. APP locus duplication causes autosomal dominant early-onset Alzheimer disease with cerebral amyloid angiopathy. *Nat. Genet.* 38:24–26
83. Samonte RV, Eichler EE. 2002. Segmental duplications and the evolution of the primate genome. *Nat. Rev. Genet.* 3:65–72
84. Schedl A, Ross A, Lee M, Engelkamp D, Rashbass P, et al. 1996. Influence of PAX6 gene dosage on development: overexpression causes severe eye abnormalities. *Cell* 86:71–82
85. Sebat J, Lakshmi B, Troge J, Alexander J, Young J, et al. 2004. Large-scale copy number polymorphism in the human genome. *Science* 305:525–28
86. Semple CA, Gautier P, Taylor K, Dorin JR. 2006. The changing of the guard: Molecular diversity and rapid evolution of beta-defensins. *Mol. Divers.* 10:575–84
87. Sen SK, Han K, Wang J, Lee J, Wang H, et al. 2006. Human genomic deletions mediated by recombination between Alu elements. *Am. J. Hum. Genet.* 79:41–53

88. Sharp AJ, Locke DP, McGrath SD, Cheng Z, Bailey JA, et al. 2005. Segmental duplications and copy-number variation in the human genome. *Am. J. Hum. Genet.* 77:78–88
89. Shaw CJ, Lupski JR. 2004. Implications of human genome architecture for rearrangement-based disorders: the genomic basis of disease. *Hum. Mol. Genet.* 13 Spec No 1:R57-64
90. She X, Horvath JE, Jiang Z, Liu G, Furey TS, et al. 2004. The structure and evolution of centromeric transition regions within the human genome. *Nature* 430:857–64
91. Shiu SH, Byrnes JK, Pan R, Zhang P, Li WH. 2006. Role of positive selection in the retention of duplicate genes in mammalian genomes. *Proc. Natl. Acad. Sci. USA* 103:2232–36
92. Shoja V, Zhang L. 2006. A roadmap of tandemly arrayed genes in the genomes of human, mouse, and rat. *Mol. Biol. Evol.* 23:2134–41
93. Singleton A, Gwinn-Hardy K. 2004. Parkinson's disease and dementia with Lewy bodies: a difference in dose? *Lancet* 364:1105–7
94. Singleton AB. 2005. Altered alpha-synuclein homeostasis causing Parkinson's disease: the potential roles of dardarin. *Trends Neurosci.* 28:416–21
95. Somerville MJ, Mervis CB, Young EJ, Seo EJ, del Campo M, et al. 2005. Severe expressive-language delay related to duplication of the Williams-Beuren locus. *N. Engl. J. Med.* 353:1694–701
96. Sopko R, Huang D, Preston N, Chua G, Papp B, et al. 2006. Mapping pathways and phenotypes by systematic gene overexpression. *Mol. Cell* 21:319–30
97. Stankiewicz P, Lupski JR. 2002. Genome architecture, rearrangements and genomic disorders. *Trends Genet.* 18:74–82
98. Su C, Nei M. 2001. Evolutionary dynamics of the T-cell receptor VB gene family as inferred from the human and mouse genomic sequences. *Mol. Biol. Evol.* 18:503–13
99. Su Z, Wang J, Yu J, Huang X, Gu X. 2006. Evolution of alternative splicing after gene duplication. *Genome Res.* 16:182–89
100. Van de Peer Y. 2004. Computational approaches to unveiling ancient genome duplications. *Nat. Rev. Genet.* 5:752–63
101. Van Esch H, Bauters M, Ignatius J, Jansen M, Raynaud M, et al. 2005. Duplication of the MECP2 region is a frequent cause of severe mental retardation and progressive neurological symptoms in males. *Am. J. Hum. Genet.* 77:442–53
102. Veitia RA. 2002. Exploring the etiology of haploinsufficiency. *Bioessays* 24:175–84
103. Vinckenbosch N, Dupanloup I, Kaessmann H. 2006. Evolutionary fate of retroposed gene copies in the human genome. *Proc. Natl. Acad. Sci. USA* 103:3220–25
104. Wagner A. 2000. Decoupled evolution of coding region and mRNA expression patterns after gene duplication: implications for the neutralist-selectionist debate. *Proc. Natl. Acad. Sci. USA* 97:6579–84
105. Wang X, Grus WE, Zhang J. 2006. Gene losses during human rrigins. *PLOS Biol.* 4:e52
106. Wong P, Fritz A, Frishman D. 2005. Designability, aggregation propensity and duplication of disease-associated proteins. *Protein Eng. Des. Sel.* 18:503–8
107. Woods KS, Cundall M, Turton J, Rizotti K, Mehta A, et al. 2005. Over- and underdosage of SOX3 is associated with infundibular hypoplasia and hypopituitarism. *Am. J. Hum. Genet.* 76:833–49
108. Xue Y, Daly A, Yngvadottir B, Liu M, Coop G, et al. 2006. Spread of an inactive form of caspase-12 in humans is due to recent positive selection. *Am. J. Hum. Genet.* 78:659–70
109. Yobb TM, Somerville MJ, Willatt L, Firth HV, Harrison K, et al. 2005. Microduplication and triplication of 22q11.2: a highly variable syndrome. *Am. J. Hum. Genet.* 76:865–76

110. Zhang L, Lu HH, Chung WY, Yang J, Li WH. 2005. Patterns of segmental duplication in the human genome. *Mol. Biol. Evol.* 22:135–41
111. Zhang Z, Gu J, Gu X. 2004. How much expression divergence after yeast gene duplication could be explained by regulatory motif evolution? *Trends Genet.* 20:403–7
112. Zhang Z, Luo ZW, Kishino H, Kearsey MJ. 2005. Divergence pattern of duplicate genes in protein-protein interactions follows the power law. *Mol. Biol. Evol.* 22:501–5
113. Zody MC, Garber M, Sharpe T, Young SK, Rowen L, et al. 2006. Analysis of the DNA sequence and duplication history of human chromosome 15. *Nature* 440:671–75
114. Zody MC, Garber M, Adams DJ, Sharpe T, Harrow J, et al. 2006. DNA sequence of human chromosome 17 and analysis of rearrangement in the human lineage. *Nature* 440:1045–49

DNA Strand Break Repair and Human Genetic Disease

Peter J. McKinnon[1] and Keith W. Caldecott[2]

[1]Department of Genetics and Tumor Cell Biology, St. Jude Children's Research Hospital, Memphis, Tennessee 38105; email: peter.mckinnon@stjude.org

[2]Genome Damage and Stability Center, University of Sussex, Falmer, Brighton BN1 9RQ, United Kingdom; email: k.w.caldecott@sussex.ac.uk

Annu. Rev. Genomics Hum. Genet. 2007. 8:37–55

The *Annual Review of Genomics and Human Genetics* is online at genom.annualreviews.org

This article's doi:
10.1146/annurev.genom.7.080505.115648

1527-8204/07/0922-0037$20.00

Key Words

single-strand break, double-strand break, neurodegeneration, cancer predisposition

Abstract

Each day tens of thousands of DNA single-strand breaks (SSBs) arise in every cell from the attack of deoxyribose and DNA bases by reactive oxygen species and other electrophilic molecules. DNA double-strand breaks (DSBs) also arise, albeit at a much lower frequency, from similar attacks and from the encounter of unrepaired SSBs and possibly other DNA structures by DNA replication forks. DSBs are also created during normal development of the immune system. Defects in the cellular response to DNA strand breaks underpin many human diseases, including disorders associated with cancer predisposition, immune dysfunction, radiosensitivity, and neurodegeneration. Here we provide an overview of the genetic diseases associated with defects in the repair/response to DNA strand breaks.

DNA SINGLE-STRAND BREAK REPAIR

DNA single-strand break repair (SSBR) can be divided into four basic steps and is reviewed in detail elsewhere (21, 157). Briefly, single-strand breaks (SSBs) are rapidly detected by poly (ADP-ribose) polymerase protein (35), which fulfills multiple roles during SSBR. One role is to sequester XRCC1, a scaffold protein that interacts with and accumulates, stimulates, and/or stabilizes multiple enzymatic components of the repair process at SSBs. The termini of most, if not all, SSBs are damaged or modified and must be restored to conventional 3′-hydroxyl and 5′-phosphate moieties if repair is to be completed. End processing is the most diverse enzymatic step of SSBR, due to the variety of termini that can arise. This is reflected by the number of enzymes involved, including polynucleotide kinase, AP endonuclease-1, DNA polymerase β (Pol β), tyrosyl DNA phosphodiesterase-1 (TDP1) (see below), Aprataxin (see below), and flap endonuclease-1 (FEN-1). Following end processing, gap filling can occur. This is usually conducted by Pol β although other DNA polymerases can also be involved. Gap filling typically involves single–nucleotide insertion, but at some SSBs gap filling may continue for 2–10 nucleotides, with FEN-1 endonuclease removing the displaced nucleotides. The final step of SSBR is DNA ligation, which is conducted by DNA ligase I (LigI) or DNA ligase IIIα (Lig3α). In addition to the rapid global SSBR process that operates throughout interphase, S/G2-phase cells may possess an additional, XRCC1-dependent SSBR process that is intimately linked to DNA replication, termed replication-coupled SSBR (RC-SSBR) (21). RC-SSBR may operate in conjunction with homologous recombination because SSBs encountered by a replication fork are most likely converted into double-strand breaks (DSBs) (79).

DNA DOUBLE-STRAND BREAK REPAIR

Two major pathways exist for the repair of DSBs in mammalian cells: homologous recombination (HR) and nonhomologous end joining (NHEJ). HR ensures the accurate repair of DSBs by using a homologous undamaged DNA strand from an intact sister chromatid as a template for DNA polymerase to extend past the break, and is therefore restricted to late S and G2 of the cell cycle (172). HR involves the RAD52 group of proteins (including RAD50, RAD51, RAD52, and RAD54), RPA, XRCC2, XRCC3, and the BRCA proteins (see below) to ensure high-fidelity repair (58, 172). Conversely, NHEJ functions in all phases of the cell cycle and ligates broken, often noncompatible, DNA ends without the need of an undamaged template. Both of these pathways are critical for normal mammalian development and maintaining genome stability (85, 88).

Nonhomologous end joining requires Ku heterodimer (Ku70 and Ku86), DNA dependent protein kinase catalytic subunit (DNA-PK_{cs}), DNA ligase IV (Lig4), and XRCC4. The repair of DSBs by NHEJ requires the coordinated assembly of various proteins at DNA ends, of which one of the first, if not the first, is Ku (130). Ku then recruits DNA-PK_{cs} to DSBs and may serve to bring the two DNA ends in close proximity. DNA-PK is a nuclear serine/threonine kinase and its activity is stimulated by DSBs (32, 136). Ku stabilizes DNA binding by DNA-$PK_{cs,}$ thereby allowing the protein kinase to tether the broken DNA ends. DNA-PK may also recruit and activate proteins involved in DNA end processing, such as Artemis (55, 98, 131). Other proteins most likely involved in end processing during NHEJ are polynucleotide kinase (6, 27, 76), Werner helicase (WRN) (69), and Aprataxin (see below). As is the case for SSBs, it is likely that one or more DNA polymerases are involved in replacing missing nucleotides at DSBs (18, 97). Finally, Ku recruits XRCC4/DNA ligase IV for ligation of the two DNA ends (117).

TDP1 and Spinocerebellar Ataxia with Axonal Neuropathy-1

Takashima et al. (150) recently identified an adult-onset autosomal recessive ataxia, which they denoted spinocerebellar ataxia with axonal neuropathy-1 (SCAN1). Cerebellar ataxia is a neurological abnormality characterized by an inability to properly control or coordinate movement, resulting in an unsteady or shaky gait. The pathology of SCAN1 lacks chromosomal instability and cancer predisposition and appears to be restricted to the nervous system. SCAN1 is associated with a mutation in tyrosyl DNA phosphodiesterase 1 (TDP1), a protein that is primarily involved in the repair of abortive DNA strand breaks created by Topoisomerase 1 (Top1). In yeast, Tdp1 provides one of several mechanisms for removing Top1 peptide from a DSB created by collision of Top1 cleavage complexes with DNA replication forks (39, 129, 166). Once Top1 peptide has been removed the DSB is channeled into HR-mediated DSBR (129, 166). Although it is likely that TDP1 plays a similar role in human cells, it is unlikely that this accounts for the pathology of SCAN1, which is restricted to postmitotic neurons. SCAN1 could result from a defect in the repair of replication-independent DSBs, which might arise from overlapping Top1 SSBs, but such a defect has not yet been detected. However, SCAN1 lymphoblastoid cells exhibit a profound defect in the repair of SSBs following treatment with camptothecin, a drug that promotes the formation of abortive Top1 breaks, and to a lesser extent following oxidative stress. Moreover, TDP1 physically associates with the SSBR machinery through interaction with DNA ligase IIIα (46, 128). Because the occurrence of SSBs is several orders of magnitude more frequent than DSBs, and because SSBs can block transcription (194–196), it is plausible that a defect in the repair of SSBs can account for the neurodegenerative pathology of SCAN1.

One reason why SCAN1 is not associated with genetic instability and cancer may be an ability of proliferating cells to repair SSBs, at physiological levels of damage at least, once they are converted into DSBs by collision with a replication fork using alternative end-processing nucleases such as MRE11 or the ERCC1/XPF heterodimer in conjunction with HR-mediated DSBR (39, 92, 165). This scheme also applies to other types of unrepaired SSBs (see below), because structure-specific nucleases such as MRE11 or ERCC1/XPF may enable removal of a wide variety of damaged termini. It is also unclear why the nervous system might be so sensitive to loss of TDP1 compared with other largely noncycling tissues. One likely factor is that the nervous system consumes ~20% of inhaled oxygen and possesses low levels of antioxidant enzymes (11), and so encounters high levels of oxidative stress. In addition, there is a high transcriptional demand in postmitotic neurons, and unrepaired SSBs can be potent blocks to RNA polymerases, which may subsequently trigger cell death. The limited regenerative capacity of the nervous system, compared with other predominantly noncycling tissues, may also be a factor, and may render this tissue particularly susceptible to cell loss.

Aprataxin and Ataxia Oculomotor Apraxia-1

Two groups have independently associated ataxia oculomotor apraxia-1 (AOA1) with mutations in a novel human gene, denoted *APTX* (37, 107). AOA1 is the most common form of autosomal recessive ataxia in Japan and the second most common in Portugal, after Friedrich ataxia. AOA1 is characterized as ataxia with oculomotor apraxia and, similar to SCAN1, is associated with cerebellar atrophy and axonal motor neuropathy. One difference between AOA1 and SCAN1 is the absence of oculomotor apraxia in the latter, a characteristic of AOA1 that is shared with ataxia telangiectasia (A-T) (see below). Indeed, AOA1 shares a number of neurological features with A-T and was previously classified as ataxia telangiectasia-like syndrome (A-TLS), although the extraneurological features evident in A-T are absent from AOA1 (37, 107). As with SCAN1, AOA1 is a disease of the nervous system.

Aprataxin, the protein encoded by *APTX*, is a member of the histidine triad (HIT) domain superfamily of nucleotide hydrolases/transferases (16). The link between Aprataxin and SSBR was first suggested by a region of homology between the amino terminus of this polypeptide and that of human PNK (107). Close inspection of this region of both proteins identifies the presence of a putative fork-head associated (FHA) domain (20). This domain mediates the interaction of both Aprataxin and polynucleotide kinase with the SSBR scaffold protein XRCC1, and with the NHEJ protein XRCC4 (20, 31, 96, 137). It was recently established that Aprataxin processes abortive intermediates of DNA ligase activity, removing AMP from blocked 5′-termini at SSBs (1), and most likely also DSBs. Although AOA1 could result from defects in the repair of either SSBs or DSBs, a defect in the former seems more likely. This is because mutation of NHEJ factors such as XLF or DNA ligase IV (see below) do not manifest specifically in spinocerebellar ataxia, but rather lead to a broader clinical pathology that includes microcephaly. As discussed above, unrepaired SSBs with blocked termini can inhibit progression of RNA polymerases (194–196), inhibiting transcription and ultimately leading to cell death.

DNA Ligase I and 46BR

DNA ligase I is required for maturation of Okazaki fragments during DNA replication and is mutated in 46BR, a fibroblast cell line established from an immunodeficient female with stunted growth (10, 99). To date, this is the only individual with identified mutations in Lig1. 46BR cells exhibit defects in the rate of rejoining DNA breaks following exposure to UV and alkylating agents, implying a defect in SSBR during DNA nucleotide- and DNA base-excision repair, respectively (86, 119, 144, 153).

Artemis and RS-SCID

Severe combined immunodeficiency (SCID) can arise through mutation of the RAG genes involved in somatic rearrangement of immunoglobulin genes. An alternative source of SCID, denoted severe combined immunodeficiency associated with increased cellular radiosensitivity (RS-SCID), has been associated with mutations in Artemis, a member of the *SNM1/PS02* gene family (110). RS-SCID is associated with cellular hypersensitivity to ionizing radiation, defective V(D)J recombination, immunodeficiency, and lymphoma (110, 111). Initial studies showed that Artemis forms a complex with DNA-PK_{cs} and together this complex possesses endonuclease activity that can resolve the DNA hairpins that form during V(D)J recombination (98). Whereas the defect in V(D)J recombination can account for the immunodeficiency in RS-SCID, it cannot explain the hypersensitivity to IR. It has been reported that Artemis-deficient cells possess a defect in G2/M checkpoint control, but that NHEJ is normal (190). However, it has also been reported that NHEJ is defective in Artemis-deficient cells at a subset of DSBs (132, 135, 168). The precise structures of the termini that are substrates for Artemis remain to be determined.

DNA Ligase IV and Lig4 Syndrome

Missense mutations in DNA ligase IV (Lig4) have also been associated with a disorder characterized by radiosensitivity and immunodeficiency, and additionally with pancytopenia, developmental abnormalities, and microcephaly (13, 120). In this disorder, denoted Lig4 syndrome, hypomorphic *Lig4* mutations lead to an attenuation of DNA ligase activity and a defect in NHEJ (120). The missense mutations found in Lig4 syndrome affect discrete functional motifs in Lig4, including the DNA binding region, the adenylation motif, or the C-terminal BRCT region (54, 121). The severity of Lig4 syndrome is related to the amount of residual Lig4 activity (54). Interestingly, one milder Lig4 mutation (R278H) was exacerbated by a compounding effect of Lig4 polymorphisms that increased the severity of the R278H missense mutation (54). Of six Lig4 syndrome patients identified to date, two have developed leukemia, suggesting that lymphoreticular disease may also be a feature of this syndrome (13).

XLF/Cernunnos and Human Immunodeficiency with Microcephaly

Xrcc4-like factor (XLF)/Cernunnos is structurally related to, and interacts with, XRCC4 (2, 17). As observed with cells from individuals with RS-SCID and Lig4 syndrome, cells lacking or depleted of full-length XLF exhibit hypersensitivity to ionizing radiation, defective V(D)J recombination, and reduced rates of NHEJ. Similarities between these three NHEJ-defective diseases are also evident at the clinical level, with mutations in XLF resulting in immunodeficiency, developmental delay, and microcephaly. The precise role of XLF/Cernunnos is currently unknown, but may involve promoting the DNA ligation step conducted by XRCC4/Lig4 (60). Intriguingly, microcephaly has been reported in another disease associated with defective responses to DNA damage (182). However, in this case, in which microcephaly is the primary phenotype and in which microcephalin is mutated, the abnormality appears to be due to defective ATR/BRCA1/Chk1-dependent cell cycle checkpoints (3, 89).

ATM AND ATAXIA TELANGIECTASIA

A-T is an autosomal recessive neurodegenerative disease that results from inactivation of the ATM protein kinase (30, 53, 103). A-T is apparent early in childhood and is characterized by uncoordinated or ataxic movements that are generally associated with telangiectasia. A-T also features cellular radiosensitivity, immunodeficiency, sterility, and elevated predisposition to cancer. The cancer predisposition is primarily increased lymphoreticular disease, such as leukemia and lymphoma (57), although *ATM* mutations have also been linked to breast cancer (5, 158). As with many other syndromes characterized by DNA damage response defects, neurodegeneration is a hallmark of A-T (20, 134), indicating that the response to DNA damage is crucial for homeostasis of the nervous system. ATM has been directly implicated in the response to endogenous damage in the nervous system as DNA lesions arising from Lig4 deficiency activate ATM to initiate neural apoptosis (84, 139). Therefore, as ATM eliminates neural cells that incur DNA damage, *ATM* mutations might lead to the accumulation of genetic lesions that eventually compromise cellular function and viability, resulting in neurodegeneration.

ATM is a large nuclear protein kinase with sequence similarity to a family of proteins related to the phosphatidylinositol-3-kinases. In response to DNA DSBs, ATM undergoes rapid posttranslational modification at multiple sites coinciding with the dissociation of inactive multimeric ATM to an active monomeric ATM kinase (8, 78, 124, 148). Many ATM substrates are cell cycle regulators that have essential functions in the cellular response to DNA damage and include p53, BRCA1, p53 binding protein 1 (53BP1), and the checkpoint kinase CHK2. For example, p53 is required for the G1 and CHK2 for the G2 DNA damage-induced checkpoints, whereas proteins such as BRCA1 and NBS1 (see below) influence the intra-S-phase checkpoint (70, 112, 141). Although ATM is essential for the DSB response, it functions in concert with other factors. Foremost among these is the MRN damage-sensing complex (see below), named as such because of its three principal component proteins: MRE11, RAD50, and NBS1 (36, 125, 164). Normal ATM activation after DNA DSBs involves the MRN complex because compromised MRN function can affect ATM localization to damage-induced foci or show decreased ATM activity and a reduction in phosphorylation of ATM substrates (25, 41, 72, 163). In addition to ATM and MRN, other key effectors of the DSB response include histone H2AX, 53BP1, mediator of damage checkpoint 1 (MDC1), and BRCA1 (170). After DNA damage, these factors rapidly mobilize to the sites of DNA DSBs and initiate and participate in an ATM-dependent signaling cascade that leads to the resolution of the break through DNA repair, or cell death after excessive DNA damage, often through p53-mediated apoptosis. Collectively, these proteins function as key regulators of the DNA damage response, and a clear interdependency exists among them as inactivation of any renders cells hypersensitive to DSBs (72, 126, 141). In collaboration with the MRN complex and Artemis, ATM participates in the repair of a subset of DSBs via NHEJ (131). ATM and the MRN complex have also been implicated in DSBR via HR (75, 109).

MRN COMPLEX, ATAXIA TELANGIECTASIA-LIKE DISORDER, AND NIJMEGEN BREAKAGE SYNDROME

Hypomorphic mutations of MRE11 lead to ataxia telangiectasia-like disorder (A-TLD) and mutations of NBS1 result in Nijmegen breakage syndrome (NBS) (24, 146). Although A-T, A-TLD, and NBS share some common features, they are distinguished by different neuropathology. A-TLD individuals are characterized by neurodegeneration, albeit less severe than A-T, whereas NBS individuals are microcephalic without neurodegeneration (152). However, outside of the nervous system there is substantial similarity between A-T and NBS that includes radiosensitivity, immune deficiency, and cancer proneness (particularly lymphomas). Although A-TLD individuals share some features with NBS and A-T, the small number and the lack of extensive clinical characterization of these individuals make a comprehensive comparison difficult. For example, minor immune deficiency is present in A-TLD, but it is much less striking than in A-T and NBS (152). Furthermore, cancer predisposition is a feature of A-T and NBS, but this has not been observed in A-TLD; a mouse model of A-TLD is also not cancer prone, whereas mice deficient for ATM or NBS1 are cancer prone (156, 174).

The interrelationship of these diseases results from the requirement of the MRN complex for normal ATM activation after DSBs. Although the precise details of ATM activation by the MRN complex have yet to be fully elucidated, a zinc-hook structure present in the flexible coiled-coil region of Rad50 is critical for the function of this complex (61, 176). The binding of MRN to DNA induces a conformational change leading to the extension of the RAD50 coiled coils to form an intercomplex tether between distinct MRN complexes bound to opposing DNA strands (108, 175). The ability of the MRN complex to tether broken DNA ends provides a physical basis for the recruitment of critical signaling kinases such as ATM. For example, after DNA damage, NBS1 accumulation at DSBs is required for localization of activated ATM at these DSBs (72, 95). In terms of disease, it will be important to understand the relative impact of the mutations of NBS1 and MRE11 on MRN function during the response to DNA damage. The microcephaly characteristic of NBS compared to the neurodegeneration of A-TLD suggests that MRN-ATM function may be subject to other regulatory mechanisms in the nervous system. In this regard, it is noteworthy that microcephaly has been linked to an increasing number of genes that encode proteins likely important for the response to DNA damage (90, 182) (see above). As discussed above, the MRN complex participates with ATM in the repair of a subset of DSBs by NHEJ (131), and has also been linked to DSBR via HR (75, 109).

BRCA1

Individuals with heterozygous *BRCA1* mutations exhibit a greatly increased risk of breast and ovarian cancer (105). BRCA1 interacts with many different proteins, and BRCA1-deficient cells exhibit increased chromosomal aberrations in response to ionizing radiation (IR) and are hypersensitive to various genotoxins, including cisplatin, MMS, IR, and topoisomerase poisons. The role(s) of BRCA1 that account for its tumor suppressor function remain unclear, partly because BRCA1 regulates many different processes including transcription (106, 145), RNA processing (74), differentiation and apoptosis (100, 140, 155), X chromosome inactivation (51, 161), telomere maintenance (181), and chromatin remodeling (15, 64).

BRCA1 is also a component of the DNA damage signaling/DNA repair machinery and is important for G1, S, and G2/M cell cycle checkpoint control (40). Among the earliest events at DSBs is recruitment of the MRN complex and ATM activation (25, 81, 82, 163), leading to phosphorylation of H2AX on Ser129 (19, 48, 133). Phosphorylated H2AX, denoted γH2AX, is required for accumulation of numerous DNA damage response proteins into subnuclear foci at DSBs, including BRCA1, Rad51, MDC1, 53BP1, and components of the MRN complex, although not necessarily for their initial recruitment (12, 26, 122, 147). BRCA1 is required for ATM/ATR-mediated phosphorylation of several transducer/effector proteins including p53, c-Jun, CtIP, Chk2, and NBS1, and interacts with the MRN nuclease complex (50, 193). BRCA1 is also a target for phosphorylation by ATM and ATR (33, 83, 159), and phosphorylation by ATM is important for BRCA1 to maintain resistance to ionizing radiation (IR) (33).

BRCA1 also interacts with the HR factors BRCA2 (28) and BRIP1/BACH1, a helicase mutated in Fanconi anemia (FA) (see below) (23, 87, 91). A link between BRCA1 and FA is also suggested by the interaction of BRCA1 with FANCA (49) and FANCD2 (52, 151), and by the delay in appearance of crosslink-induced BRCA1 foci in FA cells (127). BRIP1/BACH1 mutations may account for early-onset breast cancer in some patients that lack BRCA1 or BRCA2 mutations (22). Moreover, BRCA1 is required for accumulation of the HR protein Rad51 into subnuclear foci following IR or cisplatin damage (14), and, along with NBS1, for ATM accumulation and phosphorylation of SMC1, another protein implicated in DSBR (71, 72). BRCA1 assembles with BARD1 in a E3 ubiquitin ligase complex that promotes protein ubiquitination (7, 59, 145, 179). BARD1 is also mutated in some sporadic breast, ovarian, and uterine cancers (154, 178), and BRCA1/BARD1 heterodimer is important for DSBR by HR and possibly single-strand annealing (29, 113, 142, 173). There are conflicting reports concerning whether or not BRCA1 contributes to NHEJ. BRCA1 has been reported not to be required for NHEJ, and by promoting HR may actually suppress certain types of NHEJ events, perhaps through a mechanism regulated by Chk2 (104, 142, 167). However, other studies have observed a requirement for BRCA1 for NHEJ, and it has been suggested that BRCA1 may promote the fidelity of end-joining reactions (9, 34, 191, 192).

BRCA2

BRCA2 mutation results in predisposition to breast and ovarian cancer and is also implicated in prostate cancer (80, 160, 177). The clearest role for BRCA2 following DNA damage is during HR-mediated repair of DSBs induced by genotoxins or arising at collapsed replication forks (38, 93, 114, 162, 189). BRCA2 does not appear to be required for the repair of DSBs by NHEJ (167, 180). The primary role of BRCA2 during HR appears to be to interact with Rad51 and to regulate the formation of Rad51 nucleoprotein filaments (38, 123, 185, 186, 188, 189). Binding of RAD51 to BRCA2 is regulated by CDK-mediated phosphorylation, enabling regulation of this process during the cell cycle and following DNA damage (47). The Rad51 mediator role of BRCA2 is facilitated by several additional interactions, including the proteins BCCIP (94) and DSS1 (56, 77, 101), and BRCA2 function may also be regulated by EMSY (65) and USP11 (138). BRCA2 is also a component of BRCC, an E3 ligase complex that contains BRCA1 and RAD51 and that ubiquitinates p53 (45). As described below, BRCA2 is intimately related to FA because BRCA2 interacts with FANCG (66, 68) and ubiquitinated FANCD2 (67, 169), and was recently identified as the gene mutated in FANCD1 patients (63).

FANCONI ANEMIA

FA results from mutation in 1 of at least 11 genes and is associated with developmental defects, progressive bone marrow failure, and cancer susceptibility. The molecular defects that result in these phenotypes are not fully elucidated. With the notable exception of FANCD1/BRCA2, most FANC proteins form a complex that is required for monoubiquitination and recruitment of FANCD2 into chromatin (52). However, core components of the FA pathway, such as FANCC, FANCG, and FANCL, may also have roles independent of FANCD2 monoubiquitination (102). FA cells are hypersensitive to crosslinking agents, most likely due to a defect in the repair of DSBs arising during crosslink excision (43, 118, 149, 171, 184). FA proteins may operate during HR in conjunction with BRCA1, BRCA2, and Rad51. This is suggested by the observations that FANCD1 is in fact BRCA2 and that FANCD2 associates and/or interacts with BRCA1, BRCA2, and Rad51 (42, 52, 63, 67, 151). A connection between FA and HR is further suggested by interactions between FANCC and FANCD1/BRCA2 (73), and between FANCG and both FANCD1/BRCA2 and the Rad51 mediator protein XRCC3 (66, 68). FA proteins also interact with the MRN complex, perhaps promoting DSBR mediated by HR or NHEJ and/or cell cycle checkpoint control (44, 115, 127). It is possible that the role of FA at DSBs is restricted to those arising during DNA replication (4, 62, 143), or that FA proteins control pathway choice during HR, perhaps favoring a specific type of gene conversion or single-strand annealing (116, 183, 187).

CONCLUSIONS AND FUTURE ISSUES

In the preceding sections we briefly summarized syndromes resulting from defective repair of DNA strand breaks. Although the spectrum of molecular defects in these diseases is broad, in many cases common outcomes are observed. For example, radiosensitivity, immunodeficiency, and predisposition to lymphomas or other cancers are common phenotypes in checkpoint/DSBR-defective disorders (RS-SCID, Lig4 syndrome, XLF disorder, A-T, NBS, BRCA1, BRCA2, and FA). In contrast, disorders with defects in SSBR appear to selectively target the nervous system, resulting in cerebellar ataxia and neurodegeneration (AOA1, SCAN1). There are plausible explanations for some of these differences. For example, it is easy to see how checkpoint control and defective removal of DSBs might be crucial for suppressing genetic instability and cancer, and why defects in NHEJ commonly result in lymphoma and immunodeficiency [because of their role in V(D)J recombination]. Similarly, it is not too difficult to speculate why unrepaired SSBs might pose a particular threat to long-lived postmitotic neurons (see above for details), which lack the ability to remove such lesions using

HR-mediated DSBR during DNA replication. However, there are significant overlaps between the pathology associated with each class of disease, and in some cases striking variability in the pathology associated within each class of disease, that are not readily explained. For example, why does the DSB/checkpoint-related disease A-T resemble the SSB-related diseases AOA1 and SCAN1 in neurological phenotype, yet differ from A-TLD (another DSB/checkpoint-related disease) in predisposition to cancer? Similarly, why do SCAN1 and AOA1 result in cerebellar ataxia, yet only AOA1 results in oculomotor apraxia? And what aspects of MRN function reflect the different phenotypes between A-TLD and NBS? Clearly, relating the molecular defects observed in DNA strand break repair–associated diseases to tissue-specific outcomes remains a challenging priority.

SUMMARY POINTS

1. Defects in the repair of or response to chromosomal DNA strand breaks result in a range of genetic diseases.
2. Diseases primarily associated with defects in the repair of single-strand breaks exhibit a pathology that is restricted to cerebellar ataxia and neurodegeneration.
3. Diseases primarily associated with defects in the repair/response to double-strand breaks exhibit a more diverse pathology and can include radiosensitivity, immunodeficiency, developmental abnormalities, microcephaly, and cancer predisposition.

LITERATURE CITED

1. Ahel I, Rass U, El-Khamisy SF, Katyal S, Clements PM, et al. 2006. The neurodegenerative disease protein aprataxin resolves abortive DNA ligation intermediates. *Nature* 443:713–16
2. Ahnesorg P, Smith P, Jackson SP. 2006. XLF interacts with the XRCC4-DNA ligase IV complex to promote DNA nonhomologous end-joining. *Cell* 124:301–13
3. Alderton GK, Galbiati L, Griffith E, Surinya KH, Neitzel H, et al. 2006. Regulation of mitotic entry by microcephalin and its overlap with ATR signaling. *Nat. Cell Biol.* 8:725–33
4. Andreassen PR, D'Andrea AD, Taniguchi T. 2004. ATR couples FANCD2 monoubiquitination to the DNA-damage response. *Genes Dev.* 18:1958–63
5. Angele S, Lauge A, Fernet M, Moullan N, Beauvais P, et al. 2003. Phenotypic cellular characterization of an ataxia telangiectasia patient carrying a causal homozygous missense mutation. *Hum. Mutat.* 21:169–70
6. Audebert M, Salles B, Weinfeld M, Calsou P. 2006. Involvement of polynucleotide kinase in a poly(ADP-ribose) polymerase-1-dependent DNA double-strand breaks rejoining pathway. *J. Mol. Biol.* 356:257–65
7. Baer R, Ludwig T. 2002. The BRCA1/BARD1 heterodimer, a tumor suppressor complex with ubiquitin E3 ligase activity. *Curr. Opin. Genet. Dev.* 12:86–91
8. Bakkenist CJ, Kastan MB. 2003. DNA damage activates ATM through intermolecular autophosphorylation and dimer dissociation. *Nature* 421:499–506
9. Baldeyron C, Jacquemin E, Smith J, Jacquemont C, De Oliveira I, et al. 2002. A single mutated BRCA1 allele leads to impaired fidelity of double strand break end-joining. *Oncogene* 21:1401–10

10. Barnes DE, Tomkinson AE, Lehmann AR, Webster AD, Lindahl T. 1992. Mutations in the DNA ligase I gene of an individual with immunodeficiencies and cellular hypersensitivity to DNA-damaging agents. *Cell* 69:495–503
11. Barzilai A, Rotman G, Shiloh Y. 2002. ATM deficiency and oxidative stress: a new dimension of defective response to DNA damage. *DNA Repair (Amst.)* 1:3–25
12. Bassing CH, Chua KF, Sekiguchi J, Suh H, Whitlow SR, et al. 2002. Increased ionizing radiation sensitivity and genomic instability in the absence of histone H2AX. *Proc. Natl. Acad. Sci. USA* 99:8173–78
13. Ben-Omran TI, Cerosaletti K, Concannon P, Weitzman S, Nezarati MM. 2005. A patient with mutations in DNA ligase IV: clinical features and overlap with Nijmegen breakage syndrome. *Am. J. Med. Genet. A* 137:283–87
14. Bhattacharyya A, Ear US, Koller BH, Weichselbaum RR, Bishop DK. 2000. The breast cancer susceptibility gene *BRCA1* is required for subnuclear assembly of Rad51 and survival following treatment with the DNA cross-linking agent cisplatin. *J. Biol. Chem.* 275:23899–903
15. Bochar DA, Wang L, Beniya H, Kinev A, Xue Y, et al. 2000. BRCA1 is associated with a human SWI/SNF-related complex: linking chromatin remodeling to breast cancer. *Cell* 102:257–65
16. Brenner C. 2002. Hint, Fhit, and GalT: function, structure, evolution, and mechanism of three branches of the histidine triad superfamily of nucleotide hydrolases and transferases. *Biochemistry* 41:9003–14
17. Buck D, Malivert L, de Chasseval R, Barraud A, Fondaneche MC, et al. 2006. Cernunnos, a novel nonhomologous end-joining factor, is mutated in human immunodeficiency with microcephaly. *Cell* 124:287–99
18. Budman J, Chu G. 2005. Processing of DNA for nonhomologous end-joining by cell-free extract. *EMBO J.* 24:849–60
19. Burma S, Chen BP, Murphy M, Kurimasa A, Chen DJ. 2001. ATM phosphorylates histone H2AX in response to DNA double-strand breaks. *J. Biol. Chem.* 276:42462–67
20. Caldecott KW. 2003. DNA single-strand break repair and spinocerebellar ataxia. *Cell* 112:7–10
21. Caldecott KW. 2003. XRCC1 and DNA strand break repair. *DNA Repair. (Amst.)* 2:955–69
22. Cantor S, Drapkin R, Zhang F, Lin Y, Han J, et al. 2004. The BRCA1-associated protein BACH1 is a DNA helicase targeted by clinically relevant inactivating mutations. *Proc. Natl. Acad. Sci. USA* 101:2357–62
23. Cantor SB, Bell DW, Ganesan S, Kass EM, Drapkin R, et al. 2001. BACH1, a novel helicase-like protein, interacts directly with BRCA1 and contributes to its DNA repair function. *Cell* 105:149–60
24. Carney JP, Maser RS, Olivares H, Davis EM, Le Beau M, et al. 1998. The hMre11/hRad50 protein complex and Nijmegen breakage syndrome: linkage of double-strand break repair to the cellular DNA damage response. *Cell* 93:477–86
25. Carson CT, Schwartz RA, Stracker TH, Lilley CE, Lee DV, Weitzman MD. 2003. The Mre11 complex is required for ATM activation and the G2/M checkpoint. *EMBO J.* 22:6610–20
26. Celeste A, Fernandez-Capetillo O, Kruhlak MJ, Pilch DR, Staudt DW, et al. 2003. Histone H2AX phosphorylation is dispensable for the initial recognition of DNA breaks. *Nat. Cell Biol.* 5:675–79

27. Chappell C, Hanakahi LA, Karimi-Busheri F, Weinfeld M, West SC. 2002. Involvement of human polynucleotide kinase in double-strand break repair by nonhomologous end joining. *EMBO J.* 21:2827–32
28. Chen J, Silver DP, Walpita D, Cantor SB, Gazdar AF, et al. 1998. Stable interaction between the products of the BRCA1 and BRCA2 tumor suppressor genes in mitotic and meiotic cells. *Mol. Cell* 2:317–28
29. Chen JJ, Silver D, Cantor S, Livingston DM, Scully R. 1999. BRCA1, BRCA2, and Rad51 operate in a common DNA damage response pathway. *Cancer Res.* 59:1752s–56s
30. Chun HH, Gatti RA. 2004. Ataxia-telangiectasia, an evolving phenotype. *DNA Repair (Amst.)* 3:1187–96
31. Clements PM, Breslin C, Deeks ED, Byrd PJ, Ju L, et al. 2004. The ataxia-oculomotor apraxia 1 gene product has a role distinct from ATM and interacts with the DNA strand break repair proteins XRCC1 and XRCC4. *DNA Repair (Amst.)* 3:1493–502
32. Collis SJ, DeWeese TL, Jeggo PA, Parker AR. 2004. The life and death of DNA-PK. *Oncogene* 24:949–61
33. Cortez D, Wang Y, Qin J, Elledge SJ. 1999. Requirement of ATM-dependent phosphorylation of brca1 in the DNA damage response to double-strand breaks. *Science* 286:1162–66
34. Coupier I, Baldeyron C, Rousseau A, Mosseri V, Pages-Berhouet S, et al. 2004. Fidelity of DNA double-strand break repair in heterozygous cell lines harboring BRCA1 missense mutations. *Oncogene* 23:914–19
35. D'Amours D, Desnoyers S, D'Silva I, Poirier GG. 1999. Poly(ADP-ribosyl)ation reactions in the regulation of nuclear functions. *Biochem. J.* 342(Pt 2):249–68
36. D'Amours D, Jackson SP. 2002. The MRE11 complex: at the crossroads of DNA repair and checkpoint signaling. *Nat. Rev. Mol. Cell Biol.* 3:317–27
37. Date H, Onodera O, Tanaka H, Iwabuchi K, Uekawa K, et al. 2001. Early-onset ataxia with ocular motor apraxia and hypoalbuminemia is caused by mutations in a new HIT superfamily gene. *Nat. Genet.* 29:184–88
38. Davies AA, Masson JY, McIlwraith MJ, Stasiak AZ, Stasiak A, et al. 2001. Role of BRCA2 in control of the RAD51 recombination and DNA repair protein. *Mol. Cell* 7:273–82
39. Deng C, Brown JA, You D, Brown JM. 2005. Multiple endonucleases function to repair covalent topoisomerase I complexes in *Saccharomyces cerevisiae*. *Genetics* 170:591–600
40. Deng CX. 2006. BRCA1: cell cycle checkpoint, genetic instability, DNA damage response and cancer evolution. *Nucleic Acids Res.* 34:1416–26
41. Difilippantonio S, Celeste A, Fernandez-Capetillo O, Chen HT, Reina San Martin B, et al. 2005. Role of Nbs1 in the activation of the Atm kinase revealed in humanized mouse models. *Nat. Cell Biol.* 7:675–85
42. Digweed M, Rothe S, Demuth I, Scholz R, Schindler D, et al. 2002. Attenuation of the formation of DNA-repair foci containing RAD51 in Fanconi anaemia. *Carcinogenesis* 23:1121–26
43. Donahue SL, Campbell C. 2002. A DNA double strand break repair defect in Fanconi anemia fibroblasts. *J. Biol. Chem.* 277:46243–47
44. Donahue SL, Campbell C. 2004. A Rad50-dependent pathway of DNA repair is deficient in Fanconi anemia fibroblasts. *Nucleic Acids Res.* 32:3248–57
45. Dong Y, Hakimi MA, Chen X, Kumaraswamy E, Cooch NS, et al. 2003. Regulation of BRCC, a holoenzyme complex containing BRCA1 and BRCA2, by a signalosome-like subunit and its role in DNA repair. *Mol. Cell* 12:1087–99

46. El-Khamisy SF, Saifi GM, Weinfeld M, Johansson F, Helleday T, et al. 2005. Defective DNA single-strand break repair in spinocerebellar ataxia with axonal neuropathy-1. *Nature* 434:108–13
47. Esashi F, Christ N, Gannon J, Liu Y, Hunt T, et al. 2005. CDK-dependent phosphorylation of BRCA2 as a regulatory mechanism for recombinational repair. *Nature* 434:598–604
48. Fernandez-Capetillo O, Chen HT, Celeste A, Ward I, Romanienko PJ, et al. 2002. DNA damage-induced G2-M checkpoint activation by histone H2AX and 53BP1. *Nat. Cell Biol.* 4:993–97
49. Folias A, Matkovic M, Bruun D, Reid S, Hejna J, et al. 2002. BRCA1 interacts directly with the Fanconi anemia protein FANCA. *Hum. Mol. Genet.* 11:2591–97
50. Foray N, Marot D, Gabriel A, Randrianarison V, Carr AM, et al. 2003. A subset of ATM- and ATR-dependent phosphorylation events requires the BRCA1 protein. *EMBO J.* 22:2860–71
51. Ganesan S, Silver DP, Greenberg RA, Avni D, Drapkin R, et al. 2002. BRCA1 supports XIST RNA concentration on the inactive X chromosome. *Cell* 111:393–405
52. Garcia-Higuera I, Taniguchi T, Ganesan S, Meyn MS, Timmers C, et al. 2001. Interaction of the Fanconi anemia proteins and BRCA1 in a common pathway. *Mol. Cell* 7:249–62
53. Gatti RA, Becker-Catania S, Chun HH, Sun X, Mitui M, et al. 2001. The pathogenesis of ataxia-telangiectasia. Learning from a Rosetta Stone. *Clin. Rev. Allergy Immunol.* 20:87–108
54. Girard PM, Kysela B, Harer CJ, Doherty AJ, Jeggo PA. 2004. Analysis of DNA ligase IV mutations found in LIG4 syndrome patients: the impact of two linked polymorphisms. *Hum. Mol. Genet.* 13:2369–76
55. Goodarzi AA, Yu Y, Riballo E, Douglas P, Walker SA, et al. 2006. DNA-PK autophosphorylation facilitates Artemis endonuclease activity. *EMBO J.* 25:3880–89
56. Gudmundsdottir K, Lord CJ, Witt E, Tutt AN, Ashworth A. 2004. DSS1 is required for RAD51 focus formation and genomic stability in mammalian cells. *EMBO Rep.* 5:989–93
57. Gumy-Pause F, Wacker P, Sappino AP. 2004. ATM gene and lymphoid malignancies. *Leukemia* 18:238–42
58. Haber D. 2000. BRCA1: an emerging role in the cellular response to DNA damage. *Lancet* 355:2090–91
59. Hashizume R, Fukuda M, Maeda I, Nishikawa H, Oyake D, et al. 2001. The RING heterodimer BRCA1-BARD1 is a ubiquitin ligase inactivated by a breast cancer-derived mutation. *J. Biol. Chem.* 276:14537–40
60. Hentges P, Ahnesorg P, Pitcher RS, Bruce CK, Kysela B, et al. 2006. Evolutionary and functional conservation of the DNA non homologous end joining protein, XLF/Cernunnos. *J. Biol. Chem.* 281:37517–26
61. Hopfner KP, Craig L, Moncalian G, Zinkel RA, Usui T, et al. 2002. The Rad50 zinc-hook is a structure joining Mre11 complexes in DNA recombination and repair. *Nature* 418:562–66
62. Howlett NG, Taniguchi T, Durkin SG, D'Andrea AD, Glover TW. 2005. The Fanconi anemia pathway is required for the DNA replication stress response and for the regulation of common fragile site stability. *Hum. Mol. Genet.* 14:693–701
63. Howlett NG, Taniguchi T, Olson S, Cox B, Waisfisz Q, et al. 2002. Biallelic inactivation of BRCA2 in Fanconi anemia. *Science* 297:606–9
64. Hu YF, Hao ZL, Li R. 1999. Chromatin remodeling and activation of chromosomal DNA replication by an acidic transcriptional activation domain from BRCA1. *Genes Dev.* 13:637–42

65. Hughes-Davies L, Huntsman D, Ruas M, Fuks F, Bye J, et al. 2003. EMSY links the BRCA2 pathway to sporadic breast and ovarian cancer. *Cell* 115:523–35
66. Hussain S, Wilson JB, Blom E, Thompson LH, Sung P, et al. 2006. Tetratricopeptide-motif-mediated interaction of FANCG with recombination proteins XRCC3 and BRCA2. *DNA Repair (Amst.)* 5:629–40
67. Hussain S, Wilson JB, Medhurst AL, Hejna J, Witt E, et al. 2004. Direct interaction of FANCD2 with BRCA2 in DNA damage response pathways. *Hum. Mol. Genet.* 13:1241–48
68. Hussain S, Witt E, Huber PA, Medhurst AL, Ashworth A, Mathew CG. 2003. Direct interaction of the Fanconi anaemia protein FANCG with BRCA2/FANCD1. *Hum. Mol. Genet.* 12:2503–10
69. Karmakar P, Snowden CM, Ramsden DA, Bohr VA. 2002. Ku heterodimer binds to both ends of the Werner protein and functional interaction occurs at the Werner N-terminus. *Nucleic Acids Res.* 30:3583–91
70. Kastan MB, Bartek J. 2004. Cell-cycle checkpoints and cancer. *Nature* 432:316–23
71. Kim ST, Xu B, Kastan MB. 2002. Involvement of the cohesin protein, Smc1, in Atm-dependent and independent responses to DNA damage. *Genes Dev.* 16:560–70
72. Kitagawa R, Bakkenist CJ, McKinnon PJ, Kastan MB. 2004. Phosphorylation of SMC1 is a critical downstream event in the ATM-NBS1-BRCA1 pathway. *Genes Dev.* 18:1423–38
73. Kitao H, Yamamoto K, Matsushita N, Ohzeki M, Ishiai M, Takata M. 2006. Functional interplay between BRCA2/FancD1 and FancC in DNA repair. *J. Biol. Chem.* 281:21312–20
74. Kleiman FE, Manley JL. 1999. Functional interaction of BRCA1-associated BARD1 with polyadenylation factor CstF-50. *Science* 285:1576–79
75. Kobayashi J. 2004. Molecular mechanism of the recruitment of NBS1/hMRE11/hRAD50 complex to DNA double-strand breaks: NBS1 binds to gamma-H2AX through FHA/BRCT domain. *J. Radiat. Res. (Tokyo)* 45:473–78
76. Koch CA, Agyei R, Galicia S, Metalnikov P, O'Donnell P, et al. 2004. Xrcc4 physically links DNA end processing by polynucleotide kinase to DNA ligation by DNA ligase IV. *EMBO J.* 23:3874–85
77. Kojic M, Yang H, Kostrub CF, Pavletich NP, Holloman WK. 2003. The BRCA2-interacting protein DSS1 is vital for DNA repair, recombination, and genome stability in *Ustilago maydis*. *Mol. Cell* 12:1043–49
78. Kozlov SV, Graham ME, Peng C, Chen P, Robinson PJ, Lavin MF. 2006. Involvement of novel autophosphorylation sites in ATM activation. *EMBO J.* 25:3504–14
79. Kuzminov A. 2001. Single-strand interruptions in replicating chromosomes cause double-strand breaks. *Proc. Natl. Acad. Sci. USA* 98:8241–46
80. Lancaster JM, Wooster R, Mangion J, Phelan CM, Cochran C, et al. 1996. BRCA2 mutations in primary breast and ovarian cancers. *Nat. Genet.* 13:238–40
81. Lee JH, Paull TT. 2004. Direct activation of the ATM protein kinase by the Mre11/Rad50/Nbs1 complex. *Science* 304:93–96
82. Lee JH, Paull TT. 2005. ATM activation by DNA double-strand breaks through the Mre11-Rad50-Nbs1 complex. *Science* 308:551–54
83. Lee JS, Collins KM, Brown AL, Lee CH, Chung JH. 2000. hCds1-mediated phosphorylation of BRCA1 regulates the DNA damage response. *Nature* 404:201–4
84. Lee Y, Barnes DE, Lindahl T, McKinnon PJ. 2000. Defective neurogenesis resulting from DNA ligase IV deficiency requires Atm. *Genes Dev.* 14:2576–80

85. Lees-Miller SP, Meek K. 2003. Repair of DNA double strand breaks by nonhomologous end joining. *Biochimie* 85:1161–73
86. Lehmann AR, Willis AE, Broughton BC, James MR, Steingrimsdottir H, et al. 1988. Relation between the human fibroblast strain 46BR and cell lines representative of Bloom's syndrome. *Cancer Res.* 48:6343–47
87. Levran O, Attwooll C, Henry RT, Milton KL, Neveling K, et al. 2005. The BRCA1-interacting helicase BRIP1 is deficient in Fanconi anemia. *Nat. Genet.* 37:931–33
88. Lieber MR, Ma Y, Pannicke U, Schwarz K. 2003. Mechanism and regulation of human nonhomologous DNA end-joining. *Nat. Rev. Mol. Cell Biol.* 4:712–20
89. Lin SY, Rai R, Li K, Xu ZX, Elledge SJ. 2005. BRIT1/MCPH1 is a DNA damage responsive protein that regulates the Brca1-Chk1 pathway, implicating checkpoint dysfunction in microcephaly. *Proc. Natl. Acad. Sci. USA* 102:15105–9
90. Deleted in proof
91. Litman R, Peng M, Jin Z, Zhang F, Zhang J, et al. 2005. BACH1 is critical for homologous recombination and appears to be the Fanconi anemia gene product FANCJ. *Cancer Cell* 8:255–65
92. Liu C, Pouliot JJ, Nash HA. 2002. Repair of topoisomerase I covalent complexes in the absence of the tyrosyl-DNA phosphodiesterase Tdp1. *Proc. Natl. Acad. Sci. USA* 99:14970–75
93. Lomonosov M, Anand S, Sangrithi M, Davies R, Venkitaraman AR. 2003. Stabilization of stalled DNA replication forks by the BRCA2 breast cancer susceptibility protein. *Genes Dev.* 17:3017–22
94. Lu H, Guo X, Meng X, Liu J, Allen C, et al. 2005. The BRCA2-interacting protein BCCIP functions in RAD51 and BRCA2 focus formation and homologous recombinational repair. *Mol. Cell Biol.* 25:1949–57
95. Lukas C, Falck J, Bartkova J, Bartek J, Lukas J. 2003. Distinct spatiotemporal dynamics of mammalian checkpoint regulators induced by DNA damage. *Nat. Cell Biol.* 5:255–60
96. Luo H, Chan DW, Yang T, Rodriguez M, Chen BP, et al. 2004. A new XRCC1-containing complex and its role in cellular survival of methyl methanesulfonate treatment. *Mol. Cell Biol.* 24:8356–65
97. Ma Y, Lu H, Tippin B, Goodman MF, Shimazaki N, et al. 2004. A biochemically defined system for mammalian nonhomologous DNA end joining. *Mol. Cell* 16:701–13
98. Ma Y, Pannicke U, Schwarz K, Lieber MR. 2002. Hairpin opening and overhang processing by an artemis/DNA-dependent protein kinase complex in nonhomologous end joining and V(D)J recombination. *Cell* 108:781–94
99. Mackenney VJ, Barnes DE, Lindahl T. 1997. Specific function of DNA ligase I in simian virus 40 DNA replication by human cell-free extracts is mediated by the amino-terminal noncatalytic domain. *J. Biol. Chem.* 272:11550–56
100. Mak TW, Hakem A, McPherson JP, Shehabeldin A, Zablocki E, et al. 2000. Brca1 required for T cell lineage development but not TCR loci rearrangement. *Nat. Immunol.* 1:77–82
101. Marston NJ, Richards WJ, Hughes D, Bertwistle D, Marshall CJ, Ashworth A. 1999. Interaction between the product of the breast cancer susceptibility gene *BRCA2* and DSS1, a protein functionally conserved from yeast to mammals. *Mol. Cell Biol.* 19:4633–42
102. Matsushita N, Kitao H, Ishiai M, Nagashima N, Hirano S, et al. 2005. A FancD2-monoubiquitin fusion reveals hidden functions of Fanconi anemia core complex in DNA repair. *Mol. Cell* 19:841–47
103. McKinnon PJ. 2004. ATM and ataxia telangiectasia. *EMBO Rep.* 5:772–76

104. Merel P, Prieur A, Pfeiffer P, Delattre O. 2002. Absence of major defects in nonhomologous DNA end joining in human breast cancer cell lines. *Oncogene* 21:5654–59
105. Miki Y, Swensen J, Shattuck-Eidens D, Futreal PA, Harshman K, et al. 1994. A strong candidate for the breast and ovarian cancer susceptibility gene *BRCA1*. *Science* 266:66–71
106. Monteiro AN. 2000. BRCA1: exploring the links to transcription. *Trends Biochem. Sci.* 25:469–74
107. Moreira MC, Barbot C, Tachi N, Kozuka N, Uchida E, et al. 2001. The gene mutated in ataxia-ocular apraxia 1 encodes the new HIT/Zn-finger protein aprataxin. *Nat. Genet.* 29:189–93
108. Moreno-Herrero F, de Jager M, Dekker NH, Kanaar R, Wyman C, Dekker C. 2005. Mesoscale conformational changes in the DNA-repair complex Rad50/Mre11/Nbs1 upon binding DNA. *Nature* 437:440–43
109. Morrison C, Sonoda E, Takao N, Shinohara A, Yamamoto K, Takeda S. 2000. The controlling role of ATM in homologous recombinational repair of DNA damage. *EMBO J.* 19:463–71
110. Moshous D, Callebaut I, de Chasseval R, Corneo B, Cavazzana-Calvo M, et al. 2001. Artemis, a novel DNA double-strand break repair/V(D)J recombination protein, is mutated in human severe combined immune deficiency. *Cell* 105:177–86
111. Moshous D, Pannetier C, Chasseval Rd R, Deist Fl F, Cavazzana-Calvo M, et al. 2003. Partial T and B lymphocyte immunodeficiency and predisposition to lymphoma in patients with hypomorphic mutations in Artemis. *J. Clin. Invest.* 111:381–87
112. Motoyama N, Naka K. 2004. DNA damage tumor suppressor genes and genomic instability. *Curr. Opin. Genet. Dev.* 14:11–16
113. Moynahan ME, Chiu JW, Koller BH, Jasin M. 1999. Brca1 controls homology-directed DNA repair. *Mol. Cell* 4:511–18
114. Moynahan ME, Pierce AJ, Jasin M. 2001. BRCA2 is required for homology-directed repair of chromosomal breaks. *Mol. Cell* 7:263–72
115. Nakanishi K, Taniguchi T, Ranganathan V, New HV, Moreau LA, et al. 2002. Interaction of FANCD2 and NBS1 in the DNA damage response. *Nat. Cell Biol.* 4:913–20
116. Nakanishi K, Yang YG, Pierce AJ, Taniguchi T, Digweed M, et al. 2005. Human Fanconi anemia monoubiquitination pathway promotes homologous DNA repair. *Proc. Natl. Acad. Sci. USA* 102:1110–15
117. Nick McElhinny SA, Snowden CM, McCarville J, Ramsden DA. 2000. Ku recruits the XRCC4-ligase IV complex to DNA ends. *Mol. Cell Biol.* 20:2996–3003
118. Niedernhofer LJ, Lalai AS, Hoeijmakers JH. 2005. Fanconi anemia (cross)linked to DNA repair. *Cell* 123:1191–98
119. Nocentini S. 1995. Comet assay analysis of repair of DNA strand breaks in normal and deficient human cells exposed to radiations and chemicals. Evidence for a repair pathway specificity of DNA ligation. *Radiat. Res.* 144:170–80
120. O'Driscoll M, Cerosaletti KM, Girard PM, Dai Y, Stumm M, et al. 2001. DNA ligase IV mutations identified in patients exhibiting developmental delay and immunodeficiency. *Mol. Cell* 8:1175–85
121. O'Driscoll M, Gennery AR, Seidel J, Concannon P, Jeggo PA. 2004. An overview of three disorders associated with genetic instability: LIG4 syndrome, RS-SCID and ATR-Seckel syndrome. *DNA Repair* 3:1227–35
122. Paull TT, Rogakou EP, Yamazaki V, Kirchgessner CU, Gellert M, Bonner WM. 2000. A critical role for histone H2AX in recruitment of repair factors to nuclear foci after DNA damage. *Curr. Biol.* 10:886–95

123. Pellegrini L, Yu DS, Lo T, Anand S, Lee M, et al. 2002. Insights into DNA recombination from the structure of a RAD51-BRCA2 complex. *Nature* 420:287–93
124. Pellegrini M, Celeste A, Difilippantonio S, Guo R, Wang W, et al. 2006. Autophosphorylation at serine 1987 is dispensable for murine Atm activation in vivo. *Nature* 443:222–25
125. Petrini JH, Stracker TH. 2003. The cellular response to DNA double-strand breaks: defining the sensors and mediators. *Trends Cell Biol.* 13:458–62
126. Petrini JH, Stracker TH. 2003. The cellular response to DNA double-strand breaks: defining the sensors and mediators. *Trends Cell Biol.* 13:458–62
127. Pichierri P, Averbeck D, Rosselli F. 2002. DNA cross-link-dependent RAD50/MRE11/NBS1 subnuclear assembly requires the Fanconi anemia C protein. *Hum. Mol. Genet.* 11:2531–46
128. Plo I, Liao ZY, Barcelo JM, Kohlhagen G, Caldecott KW, et al. 2003. Association of XRCC1 and tyrosyl DNA phosphodiesterase (Tdp1) for the repair of topoisomerase I-mediated DNA lesions. *DNA Repair (Amst.)* 2:1087–100
129. Pouliot JJ, Robertson CA, Nash HA. 2001. Pathways for repair of topoisomerase I covalent complexes in *Saccharomyces cerevisiae*. *Genes Cells* 6:677–87
130. Ramsden DA, Gellert M. 1998. Ku protein stimulates DNA end joining by mammalian DNA ligases: a direct role for Ku in repair of DNA double-strand breaks. *EMBO J.* 17:609–14
131. Riballo E, Kuhne M, Rief N, Doherty A, Smith GC, et al. 2004. A pathway of double-strand break rejoining dependent upon ATM, Artemis, and proteins locating to gamma-H2AX foci. *Mol. Cell* 16:715–24
132. Deleted in proof
133. Rogakou EP, Pilch DR, Orr AH, Ivanova VS, Bonner WM. 1998. DNA double-stranded breaks induce histone H2AX phosphorylation on serine 139. *J. Biol. Chem.* 273:5858–68
134. Rolig RL, McKinnon PJ. 2000. Linking DNA damage and neurodegeneration. *Trends Neurosci.* 23:417–24
135. Rooney S, Alt FW, Lombard D, Whitlow S, Eckersdorff M, et al. 2003. Defective DNA repair and increased genomic instability in Artemis-deficient murine cells. *J. Exp. Med.* 197:553–65
136. Rouse J, Jackson SP. 2002. Interfaces between the detection, signaling, and repair of DNA damage. *Science* 297:547–51
137. Sano Y, Date H, Igarashi S, Onodera O, Oyake M, et al. 2004. Aprataxin, the causative protein for EAOH is a nuclear protein with a potential role as a DNA repair protein. *Ann. Neurol.* 55:241–49
138. Schoenfeld AR, Apgar S, Dolios G, Wang R, Aaronson SA. 2004. BRCA2 is ubiquitinated in vivo and interacts with USP11, a deubiquitinating enzyme that exhibits prosurvival function in the cellular response to DNA damage. *Mol. Cell Biol.* 24:7444–55
139. Sekiguchi J, Ferguson DO, Chen HT, Yang EM, Earle J, et al. 2001. Genetic interactions between ATM and the nonhomologous end-joining factors in genomic stability and development. *Proc. Natl. Acad. Sci. USA* 98:3243–48
140. Shao N, Chai YL, Shyam E, Reddy P, Rao VN. 1996. Induction of apoptosis by the tumor suppressor protein BRCA1. *Oncogene* 13:1–7
141. Shiloh Y. 2003. ATM and related protein kinases: safeguarding genome integrity. *Nat. Rev. Cancer* 3:155–68
142. Snouwaert JN, Gowen LC, Latour AM, Mohn AR, Xiao A, et al. 1999. BRCA1 deficient embryonic stem cells display a decreased homologous recombination frequency and an

increased frequency of nonhomologous recombination that is corrected by expression of a brca1 transgene. *Oncogene* 18:7900–7
143. Sobeck A, Stone S, Costanzo V, de Graaf B, Reuter T, et al. 2006. Fanconi anemia proteins are required to prevent accumulation of replication-associated DNA double-strand breaks. *Mol. Cell Biol.* 26:425–37
144. Somia NV, Jessop JK, Melton DW. 1993. Phenotypic correction of a human cell line (46BR) with aberrant DNA ligase I activity. *Mutat. Res.* 294:51–58
145. Starita LM, Parvin JD. 2003. The multiple nuclear functions of BRCA1: transcription, ubiquitination and DNA repair. *Curr. Opin. Cell Biol.* 15:345–50
146. Stewart GS, Maser RS, Stankovic T, Bressan DA, Kaplan MI, et al. 1999. The DNA double-strand break repair gene hMRE11 is mutated in individuals with an ataxia-telangiectasia-like disorder. *Cell* 99:577–87
147. Stewart GS, Wang B, Bignell CR, Taylor AM, Elledge SJ. 2003. MDC1 is a mediator of the mammalian DNA damage checkpoint. *Nature* 421:961–66
148. Sun Y, Jiang X, Chen S, Fernandes N, Price BD. 2005. A role for the Tip60 histone acetyltransferase in the acetylation and activation of ATM. *Proc. Natl. Acad. Sci. USA* 102:13182–87
149. Sun Y, Moses RE. 1991. Reactivation of psoralen-reacted plasmid DNA in Fanconi anemia, xeroderma pigmentosum, and normal human fibroblast cells. *Somat. Cell Mol. Genet.* 17:229–38
150. Takashima H, Boerkoel CF, John J, Saifi GM, Salih MA, et al. 2002. Mutation of TDP1, encoding a topoisomerase I-dependent DNA damage repair enzyme, in spinocerebellar ataxia with axonal neuropathy. *Nat. Genet.* 32:267–72
151. Taniguchi T, Garcia-Higuera I, Andreassen PR, Gregory RC, Grompe M, D'Andrea AD. 2002. S-phase-specific interaction of the Fanconi anemia protein, FANCD2, with BRCA1 and RAD51. *Blood* 100:2414–20
152. Taylor AM, Groom A, Byrd PJ. 2004. Ataxia-telangiectasia-like disorder (ATLD)-its clinical presentation and molecular basis. *DNA Repair (Amst.)* 3:1219–25
153. Teo IA, Broughton BC, Day RS, James MR, Karran P, et al. 1983. A biochemical defect in the repair of alkylated DNA in cells from an immunodeficient patient (46BR). *Carcinogenesis* 4:559–64
154. Thai TH, Du F, Tsan JT, Jin Y, Phung A, et al. 1998. Mutations in the BRCA1-associated RING domain (BARD1) gene in primary breast, ovarian and uterine cancers. *Hum. Mol. Genet.* 7:195–202
155. Thangaraju M, Kaufmann SH, Couch FJ. 2000. BRCA1 facilitates stress-induced apoptosis in breast and ovarian cancer cell lines. *J. Biol. Chem.* 275:33487–96
156. Theunissen JW, Kaplan MI, Hunt PA, Williams BR, Ferguson DO, et al. 2003. Checkpoint failure and chromosomal instability without lymphomagenesis in Mre11(ATLD1/ATLD1) mice. *Mol. Cell* 12:1511–23
157. Thompson LH, West MG. 2000. XRCC1 keeps DNA from getting stranded. *Mutat. Res.* 459:1–18
158. Thorstenson YR, Roxas A, Kroiss R, Jenkins MA, Yu KM, et al. 2003. Contributions of ATM mutations to familial breast and ovarian cancer. *Cancer Res.* 63:3325–33
159. Tibbetts RS, Cortez D, Brumbaugh KM, Scully R, Livingston D, et al. 2000. Functional interactions between BRCA1 and the checkpoint kinase ATR during genotoxic stress. *Genes Dev.* 14:2989–3002
160. Tischkowitz M, Eeles R. 2003. Mutations in BRCA1 and BRCA2 and predisposition to prostate cancer. *Lancet* 362:80

161. Turner JM, Mahadevaiah SK, Fernandez-Capetillo O, Nussenzweig A, Xu X, et al. 2005. Silencing of unsynapsed meiotic chromosomes in the mouse. *Nat. Genet.* 37:41–47
162. Tutt A, Bertwistle D, Valentine J, Gabriel A, Swift S, et al. 2001. Mutation in Brca2 stimulates error-prone homology-directed repair of DNA double-strand breaks occurring between repeated sequences. *EMBO J.* 20:4704–16
163. Uziel T, Lerenthal Y, Moyal L, Andegeko Y, Mittelman L, Shiloh Y. 2003. Requirement of the MRN complex for ATM activation by DNA damage. *EMBO J.* 22:5612–21
164. van den Bosch M, Bree RT, Lowndes NF. 2003. The MRN complex: coordinating and mediating the response to broken chromosomes. *EMBO Rep.* 4:844–49
165. Vance JR, Wilson TE. 2002. Yeast Tdp1 and Rad1-Rad10 function as redundant pathways for repairing Top1 replicative damage. *Proc. Natl. Acad. Sci. USA* 99:13669–74
166. Vance JR, Wilson TE. 2002. Yeast Tdp1 and Rad1-Rad10 function as redundant pathways for repairing Top1 replicative damage. *Proc. Natl. Acad. Sci. USA* 99:13669–74
167. Wang H, Zeng ZC, Bui TA, DiBiase SJ, Qin W, et al. 2001. Nonhomologous end-joining of ionizing radiation-induced DNA double-stranded breaks in human tumor cells deficient in BRCA1 or BRCA2. *Cancer Res.* 61:270–77
168. Wang J, Pluth JM, Cooper PK, Cowan MJ, Chen DJ, Yannone SM. 2005. Artemis deficiency confers a DNA double-strand break repair defect and Artemis phosphorylation status is altered by DNA damage and cell cycle progression. *DNA Repair (Amst.)* 4:556–70
169. Wang X, Andreassen PR, D'Andrea AD. 2004. Functional interaction of monoubiquitinated FANCD2 and BRCA2/FANCD1 in chromatin. *Mol. Cell Biol.* 24:5850–62
170. Ward JM, Devor-Henneman DE. 2004. Mouse models of human familial cancer syndromes. *Toxicol. Pathol.* 32(Suppl. 1):90–98
171. Weksberg R, Buchwald M, Sargent P, Thompson MW, Siminovitch L. 1979. Specific cellular defects in patients with Fanconi anemia. *J. Cell Physiol.* 101:311–23
172. West SC. 2003. Molecular views of recombination proteins and their control. *Nat. Rev. Mol. Cell Biol.* 4:435–45
173. Westermark UK, Reyngold M, Olshen AB, Baer R, Jasin M, Moynahan ME. 2003. BARD1 participates with BRCA1 in homology-directed repair of chromosome breaks. *Mol. Cell Biol.* 23:7926–36
174. Williams BR, Mirzoeva OK, Morgan WF, Lin J, Dunnick W, Petrini JH. 2002. A murine model of Nijmegen breakage syndrome. *Curr. Biol.* 12:648–53
175. Williams RS, Bernstein N, Lee MS, Rakovszky ML, Cui D, et al. 2005. Structural basis for phosphorylation-dependent signaling in the DNA-damage response. *Biochem. Cell Biol.* 83:721–27
176. Wiltzius JJ, Hohl M, Fleming JC, Petrini JH. 2005. The Rad50 hook domain is a critical determinant of Mre11 complex functions. *Nat. Struct. Mol. Biol.* 12:403–7
177. Wooster R, Bignell G, Lancaster J, Swift S, Seal S, et al. 1995. Identification of the breast cancer susceptibility gene *BRCA2*. *Nature* 378:789–92
178. Wu JY, Vlastos AT, Pelte MF, Caligo MA, Bianco A, et al. 2006. Aberrant expression of BARD1 in breast and ovarian cancers with poor prognosis. *Int. J. Cancer* 118:1215–26
179. Wu LC, Wang ZW, Tsan JT, Spillman MA, Phung A, et al. 1996. Identification of a RING protein that can interact in vivo with the BRCA1 gene product. *Nat. Genet.* 14:430–40
180. Xia F, Taghian DG, DeFrank JS, Zeng ZC, Willers H, et al. 2001. Deficiency of human BRCA2 leads to impaired homologous recombination but maintains normal nonhomologous end joining. *Proc. Natl. Acad. Sci. USA* 98:8644–49
181. Xiong J, Fan S, Meng Q, Schramm L, Wang C, et al. 2003. BRCA1 inhibition of telomerase activity in cultured cells. *Mol. Cell Biol.* 23:8668–90

182. Xu X, Lee J, Stern DF. 2004. Microcephalin is a DNA damage response protein involved in regulation of CHK1 and BRCA1. *J. Biol. Chem.* 279:34091–94
183. Yamamoto K, Hirano S, Ishiai M, Morishima K, Kitao H, et al. 2005. Fanconi anemia protein FANCD2 promotes immunoglobulin gene conversion and DNA repair through a mechanism related to homologous recombination. *Mol. Cell Biol.* 25:34–43
184. Yamamoto K, Ishiai M, Matsushita N, Arakawa H, Lamerdin JE, et al. 2003. Fanconi anemia FANCG protein in mitigating radiation- and enzyme-induced DNA double-strand breaks by homologous recombination in vertebrate cells. *Mol. Cell Biol.* 23:5421–30
185. Yang H, Jeffrey PD, Miller J, Kinnucan E, Sun Y, et al. 2002. BRCA2 function in DNA binding and recombination from a BRCA2-DSS1-ssDNA structure. *Science* 297:1837–48
186. Yang H, Li Q, Fan J, Holloman WK, Pavletich NP. 2005. The BRCA2 homologue Brh2 nucleates RAD51 filament formation at a dsDNA-ssDNA junction. *Nature* 433:653–57
187. Yang YG, Herceg Z, Nakanishi K, Demuth I, Piccoli C, et al. 2005. The Fanconi anemia group A protein modulates homologous repair of DNA double-strand breaks in mammalian cells. *Carcinogenesis* 26:1731–40
188. Yu DS, Sonoda E, Takeda S, Huang CL, Pellegrini L, et al. 2003. Dynamic control of Rad51 recombinase by self-association and interaction with BRCA2. *Mol. Cell* 12:1029–41
189. Yuan SS, Lee SY, Chen G, Song M, Tomlinson GE, Lee EY. 1999. BRCA2 is required for ionizing radiation-induced assembly of Rad51 complex in vivo. *Cancer Res.* 59:3547–51
190. Zhang X, Succi J, Feng Z, Prithivirajsingh S, Story MD, Legerski RJ. 2004. Artemis is a phosphorylation target of ATM and ATR and is involved in the G2/M DNA damage checkpoint response. *Mol. Cell Biol.* 24:9207–20
191. Zhong Q, Boyer TG, Chen PL, Lee WH. 2002. Deficient nonhomologous end-joining activity in cell-free extracts from Brca1-null fibroblasts. *Cancer Res.* 62:3966–70
192. Zhong Q, Chen CF, Chen PL, Lee WH. 2002. BRCA1 facilitates microhomology-mediated end joining of DNA double strand breaks. *J. Biol. Chem.* 277:28641–47
193. Zhong Q, Chen CF, Li S, Chen Y, Wang CC, et al. 1999. Association of BRCA1 with the hRad50-hMre11-p95 complex and the DNA damage response. *Science* 285:747–50
194. Zhou W, Doetsch PW. 1993. Effects of abasic sites and DNA single-strand breaks on prokaryotic RNA polymerases. *Proc. Natl. Acad. Sci. USA* 90:6601–5
195. Zhou W, Doetsch PW. 1994. Transcription bypass or blockage at single-strand breaks on the DNA template strand: effect of different 3′ and 5′ flanking groups on the T7 RNA polymerase elongation complex. *Biochemistry* 33:14926–34
196. Zhou W, Reines D, Doetsch PW. 1995. T7 RNA polymerase bypass of large gaps on the template strand reveals a critical role of the nontemplate strand in elongation. *Cell* 82:577–85

The Genetic Lexicon of Dyslexia

Silvia Paracchini, Thomas Scerri, and Anthony P. Monaco

Wellcome Trust Centre for Human Genetics, University of Oxford, Oxford OX3 7BN, United Kingdom; email: silviap@well.ox.ac.uk, clicker@well.ox.ac.uk, anthony.monaco@well.ox.ac.uk

Annu. Rev. Genomics Hum. Genet. 2007. 8:57–79

First published online as a Review in Advance on April 19, 2007

The *Annual Review of Genomics and Human Genetics* is online at genom.annualreviews.org

This article's doi:
10.1146/annurev.genom.8.080706.092312

1527-8204/07/0922-0057$20.00

Key Words

reading disability, association studies, gene expression, epistasis, cognition, neuronal migration

Abstract

Reading abilities are acquired only through specific teaching and training. A significant proportion of children fail to achieve these skills despite normal intellectual abilities and an appropriate opportunity to learn. Difficulty in learning to read is attributable to specific dysfunctions of the brain, which so far remain poorly understood. However, it is recognized that the neurological basis for dyslexia, or reading disability, is caused in large part by genetic factors. Linkage studies have successfully identified several regions of the human genome that are likely to harbor susceptibility genes for dyslexia. In the past few years there have been exciting advances with the identification of four candidate genes located within three of these linked chromosome regions: *DYX1C1* on chromosome 15, *ROBO1* on chromosome 3, and *KIAA0319* and *DCDC2* on chromosome 6. Functional studies of these genes are offering new insights about the biological mechanisms underlying the development of dyslexia and, in general, of cognition.

RD: reading disability

Transparent language: a language where the letters, or groups of letters, are uniquely mapped to a single speech sound of the spoken language. For example, the "gh" combination will always make the same sound in Italian while in English it can produce the sounds "g" as in "ghost" or "ff" as in "laugh"

Phoneme: the smallest phonetic unit of speech sound; combinations of phonemes make words

Grapheme: all of the letters and letter combinations that represent a phoneme

DYSLEXIA PREVALENCE

Reading disability (RD) is defined as an unexpected difficulty in reading and spelling that cannot be explained by other obvious causes (49). RD is one of the most prevalent neurobehavioral disorders among school-age children, with a reported frequency ranging from 5% to 10% (73, 86) and even up to 17.5% (84). This wide range is the result of different factors, including clinical features such as variation in RD definition and diagnostic criteria, but also other aspects, such as native language (24). The frequency of RD is higher in nontransparent languages (such as English) compared with quasitransparent languages (such as Italian), where diagnosis relies almost exclusively on reading speed (59).

Various studies have suggested a three to four times higher prevalence of dyslexia in males than females (32, 101), whereas other studies have reported a gender ratio very close to one (20, 48, 86). In some cases the excess of males could be explained by a referral bias, with more boys referred to clinics (30, 101).

RD is usually diagnosed during the first years at school, but early signs can manifest in language-related tasks in preschool children (28). Despite the flourishing of commercial remediation programs, most dyslexics continue to have reading difficulties throughout adulthood, affecting both their academic achievements and their social and emotional life (27).

THE NEUROLOGICAL BASIS OF DYSLEXIA

The etiology of RD remains unknown and different theories have been proposed to explain the underlying neurological and cognitive causes (49, 78). Each theory is supported by alternative sets of empirical evidence, mainly represented by different subgroups of RD patients, which show specific characteristics. Supporters of the phonological theory attribute the causes of the reading impairment exclusively to a cognitive deficit specific to the representation and processing of speech sounds (89, 90). While learning to read, children acquire the ability to break down words into the constituent phonemes, which are graphically represented by letters in the graphemes (**Table 1**). The learning of the grapheme-phoneme correspondence, and as a consequence the complete reading process, is compromised if the sounds are poorly represented, stored, or retrieved (9, 80). Conversely, although there is a general agreement in recognizing the phonological deficit in dyslexic individuals, this is often considered a secondary effect of a much broader impairment. In fact, three other theories (78) have connected the causes of reading impairment to auditory processing (96), visual processing (61), and motor control (68) deficits. Although each of the three theories is supported by the identification of these deficits in individuals affected by RD that cannot be explained by the phonological theory, critics emphasize the fact that the deficits are detected only in subsets

Table 1 Terminology used to describe units of written and spoken language

Item	Examples						
Pictures							
Words	Book			Scarf			
Graphemes	B	OO	K	S	C	AR	F
Phonemes	/b/	/oo/	/k/	/s/	/k/	/ahr/	/f/

of the dyslexic population (78). The magnocellular theory is an attempt to integrate all these observations, postulating that failures in the magnocellular pathway will affect sensory modalities (visual, auditory, and tactile) as well as cerebellum regulation, leading to a motor control deficit (92, 93). However, the magnocellular theory cannot explain the absence of sensory motor disorders in a significant proportion of cases (80).

The existence of all these categories of RD individuals demonstrates how complex and variable the dyslexic phenotype can be and how difficult it is to give a precise definition of dyslexia. The picture is further complicated by the co-occurrence of RD with other disorders, in particular attention-deficit/hyperactivity disorder (ADHD) (107) and specific language impairment (SLI) (7). In both cases the major underlying question is whether RD is a separate disorder or a manifestation of an extended common deficit. And, from a genetic point of view, are the same genes involved in the development of RD, SLI, and ADHD or are different genes responsible for each specific disorder?

PHENOTYPE ASSESSMENT FOR GENETIC STUDIES

A crucial aspect of genetic studies is the appropriate definition of the phenotypes to gather the optimal sample. Ideally, individuals for whom RD could be traced back to specific causes (presence of specific neurological deficits, social or educational deprivation) should be excluded, while all those cases where there is a predicted genetic component (RD history in the family) should be preferentially recruited. It is also important to obtain a homogeneous sample ascertained through very well defined criteria. But which exact criteria can be used to define such a complex phenotype?

RD represents the lower tail of a normal distribution of reading ability found in the general population (85). A threshold is usually fixed to establish a point on this distribution below which an individual is considered dyslexic. A common way to fix such a threshold is to use the discrepancy between the observed reading ability and that expected on the basis of general intelligence (IQ) (98). A reading ability score two or more standard deviations (SDs) below the expected score is often adopted for sample ascertainment. Normally, a minimum IQ score is also fixed, so that the individuals selected will typically have reading problems despite a high IQ. Another method is based on discrepancy between actual reading skills and expectations based on chronological age.

Once appropriate ascertainment criteria have been established, it is important to select the appropriate phenotypes to be measured for the subsequent molecular genetic studies. Most genetic studies now use a battery of psychometric tests aimed at measuring a range of reading-related cognitive skills (15, 35, 43, 46, 76) (**Table 2**). The outcome of these tests is measured as a continuous variable, reflecting individual performance along the complete spectrum of reading abilities. They are therefore suitable for genetic analysis through the use of quantitative approaches. These reading measures are often highly correlated, although affected individuals can show notably variable performance profiles across different tests (13, 38, 63, 71).

Unfortunately, there is not universal agreement on which precise ascertainment criteria and psychometric tests should be applied, and different research groups typically use a specific selection of them, often dependent on the language of the population under study. In some cases, researchers prefer to perform genetic analyses according to categorical modalities (affected/not affected) rather than quantitative genetics, despite using psychometric tests to ascertain their samples (82, 99).

The lack of uniform measures makes it difficult to compare studies and perform comprehensive meta-analyses across different samples.

Magnocellular pathway: pathway responsible for the detection of objects and their boundaries as well as the perception of stimulus change including motion

ADHD: attention deficit hyperactivity disorder

SLI: specific language impairment

Table 2 Reading-related measures to dissect cognition of reading

Measures	Description	Example
Single-word reading	Individuals are asked to read aloud unrelated words of increasing difficulty until some error threshold is reached. This is the most commonly used test to diagnose dyslexia.	Cat Chair Mushroom Television Dictionary Laboratory
Spelling	Test the ability to spell real words of increasing difficulty and various types. This test is also used to diagnose dyslexia.	
Phonological decoding	Ability to convert printed letter strings into speech sounds according to specific phonetic rules. Usually tests involve the reading of pseudowords.	Siglop Dorkit Pamdin
Orthographic coding	Test the ability to recognize a word as a unit without fragmenting it. It is usually tested by reading irregular words that cannot be decoded by phonetic rules alone.	Yacht Salmon
	In the "forced choice" test individuals are asked to identify real words from pseudowords that would produce the same sounds.	Rain or rane? Fite or fight?
Homonym choice	Similar to the forced choice orthographic coding test, except that the two test words are both real words pronounced the same but with different meanings.	Seven days constitutes a: Week or weak?
Phonological awareness	Ability to recognize and manipulate the smallest components of words, the phonemes. Individuals are usually asked to move phonemes around within the same word or swap them between words. Also known as "spoonerism."	spoon & dog  doon & spog
Rapid automatic naming	Ability to rapidly retrieve the names of visually presented stimuli (numbers, colors, objects). This skill is associated with reading fluency and is related to language ability.	

THE GENETIC COMPONENTS OF DYSLEXIA

Heritability: proportion of the phenotypic variance among individuals that can be attributed to genotypic differences

Quantitative trait locus (QTL): genes involved in determining individual differences for any quantitative measured trait, the extreme ends of which might be classified as a disorder

RD is familial and heritable, with family history being one of the most important risk factors (23, 75). The incidence of RD among siblings of affected individuals has been consistently reported to be approximately 40% (31, 44, 75, 102). Familial clustering is a good indication that genetic factors are involved in the etiology of the disorder but may also indicate the effect of shared environmental factors. Twin studies have been crucial to differentiate the relative contribution of genetics and environment. A large study of twins affected by RD has shown that the concordance rate in monozygotic twins, who carry the same genetic background, was 68% compared to 38% in dizygotic twins, who are genetically no more similar than nontwin siblings (20). These values suggest that genetic factors contribute significantly to RD, but cannot completely explain the causes of this disorder, which is likely the result of multiple genetic and environmental interactions.

Transmission of RD in family studies (74) and regression analysis of twin data (22) not only confirms the high heritability of RD but also shows that it is a genetically heterogeneous condition that usually is not inherited as a Mendelian trait. Linkage studies confirm this heterogeneity, identifying several genomic regions that may carry susceptibility quantitative trait loci (QTLs) for RD (33). Despite the use of different approaches, including differing proband ascertainment criteria, family structures, genotyping technologies, analysis software and algorithms, and phenotypic measurements, there is a remarkable consensus in the genomic locations of the QTLs when compared with other complex traits. In fact, there is strong evidence for nine relatively well defined loci harboring RD susceptibility genes. A detailed compilation of the linked regions from various studies was recently published (108). Here we will focus on the genetic studies concerning chromosomes 15, 3, and 6 because these are the regions in which candidate genes have so far been proposed.

CHROMOSOME 15

The First Candidate Gene

The first RD susceptibility QTL to be implicated was on chromosome 15 (88). Initially, it was reported that a RD locus resided near the centromere of chromosome 15 (88), but this finding could not be replicated (6, 35, 46, 82). A locus in the middle of the long arm of chromosome 15 (15q15.2-15q21.2) was found in a number of subsequent studies (46, 62, 67, 82). Moreover, two Finnish families with members affected by RD have been reported to carry two different balanced translocations, t(2;15)(p13;q22) and t(2;15)(q11;q21), involving the 15q21-22 region (70). One of the translocations occurred in four family members, but only one individual had evidence for dyslexia. The other translocation cosegregated with reading problems in all four affected family members (a father and three children). The breakpoint of this translocation was refined by Taipale et al. (95), and was found to occur in an interval encompassing exons 8 and 9 of a novel gene at 15q21. This gene is now known as either *EKN1* or *DYX1C1*.

SNP: single nucleotide polymorphism

Sequencing of the *DYX1C1* cDNA from 20 unrelated dyslexic individuals of Finnish descent revealed 8 single nucleotide polymorphisms (SNPs), which were then genotyped in a further 35 dyslexics and 113 controls (**Table 3**) (95). Association analysis of these SNPs revealed a higher frequency of the alleles -3A and 1249T in the dyslexia group, with P-values of 0.006 and 0.02, respectively. In a second set of 54 dyslexics and 82 controls association analysis yielded P-values of 0.02 and 0.1 for -3A and 1249T,

Table 3 Association studies of *DYX1C1*

Reference	Proband's disorder	Country of origin	Most significantly reported P-values[a] for individual SNPs or haplotypes within *DYX1C1*		
			-3G > A	1249G > T	-3G > A:1249G > T
Taipale et al. (95)	Dyslexia	Finland	0.002 (A)	0.006 (T)	0.015 (A:T)
Scerri et al. (81)[b]	Dyslexia	U.K.	n/s	0.0076 (G)	0.0140 (G:G) 0.0182 (G:T)
Wigg et al. (105)	Dyslexia	Canada[c]	0.021 (G)	n/s	0.026 (G:G)
Cope et al. (17)	Dyslexia	U.K.	n/s	n/s	n/s
Marino et al. (62)	Dyslexia	Italy	n/s	n/s	n/s
Meng et al. (65)	Dyslexia	U.S.[c]	n/s	n/s	n/t
Bellini et al. (4)	Dyslexia	Italy	n/s	n/s	n/t
Ylisaukko-Oja et al. (110)	Autism	Finland	n/s	n/s	n/s
Wigg et al. (106)	ADHD	Canada[c]	n/s	n/s	n/t

[a]Significant P-values (<0.05) for association to poor traits.
[b]Analysis of severe RD subgroup (134 families).
[c]Majority likely to be European Caucasian and first language English.
Abbreviations: n/s = not significant, n/t = not tested.

respectively. Haplotype analysis of these two SNPs revealed significant association of the -3A:1249T haplotype with RD. The -3G > A transversion lies within a putative transcription factor binding site and the 1249G > T transversion results in a truncated protein product with the loss of 4 amino acids. It has been postulated that these two SNPs are functionally relevant to RD susceptibility (95), although additional studies failed to confirm the association between these alleles and RD (see below).

Little is currently known about *DYX1C1*. The gene consists of 10 exons, and encodes for a 420 amino acid protein that localizes to the cell nucleus (95). Expression of the gene occurs in numerous tissues including the brain, although data on expression within specific areas of the brain are limited. *Dyx1c1*, the mouse ortholog of *DYX1C1*, encodes a 420 amino acid protein that has at least 79% sequence identity to the human form. Both human and mice forms of the protein contain three tetratricopeptide repeat (TPR) domains at their C terminus (95) that generally mediate protein-protein interactions (8). It has been shown through yeast two-hybrid screening that the mammalian U-box C-terminal Hsp-interacting protein (CHIP) binds to DYX1C1 in the mouse (52). The interaction was mediated by the heat shock protein 70 (hsp70) chaperone protein. Due to the structural similarity between Dyx1c1 and XAP2, another TPR domain-containing protein that binds to CHIP, it has been postulated that DYX1C1 may behave similarly to XAP2 in inhibiting the ubiquitin ligase activity of CHIP (52). It has been suggested that impairment in these proteins might lead to increased inhibition of ubiquitin ligase activity, resulting in protein accumulation, as occurs in several neurodegenerative diseases (Alzheimer's, Parkinson's, and Huntington's) (52), although this seems unlikely to be the cause of RD.

Taipale et al. (95) reported that the rate of coding divergence was approximately three times greater in *DYX1C1* compared with *FOXP2*, a gene implicated in speech and language (58). However, it has been noted (34) that this report is misleading because they did not consider the local mutation rate (i.e., of the noncoding variants) of *DYX1C1* when making their calculations. Furthermore, in another study that analyzed 103 protein-coding genes, including *DYX1C1*, from human, chimpanzee, gorilla, and orangutan for human-specific enhancements of amino acid changes compared with the apes, *DYX1C1* did not show any significant signal (56).

Attempts at Replicating *DYX1C1* Associations

A flurry of subsequent studies tried to replicate the initial association of *DYX1C1* with RD (95) (**Table 3**). The analysis of *DYX1C1* variants has been conducted in several populations by testing their association with numerous reading-related traits (whether categorical or quantitative), but thus far neither of the alleles (-3A or 1249T) nor the haplotype -3A:1249T have achieved significance for replication (**Table 3**) (4, 17, 62, 65, 81, 105). In fact, a couple of groups have found association to poor reading performance with the alternative alleles (-3G and 1249G) (81, 105). One of these studies also found association to a previously untested SNP (rs11629841), and haplotypes including this SNP (105), but again these failed to replicate in a subsequent independent study (17).

There are several points that may explain the lack of consistency in all these studies. First, it should be noted that for the family in which dyslexia cosegregated with the t(2;15)(q11;q21) translocation, one of the children that had poor reading skills also had below normal intelligence (95); hence, RD might have been one aspect of a broader cognitive deficit. And second, the design of the original case-control study was suboptimal. The case and control samples in the study that found associations with the -3A and 1249T alleles originated from the same families, rather than being unrelated. Despite all these negative replication studies, genetic variation in

DYX1C1 could still contribute to RD. It remains possible that different causal variants within this gene exist in the populations studied, and that the different languages or variations in linkage disequilibrium (LD) could explain inconsistent findings.

The Finnish population, in which the initial study was performed, is a genetic isolate (1) that was established around 10,000 years ago, with a limited number of founders, that has since gone through several bottlenecks, the most recent occurring in the early eighteenth century (18). In such populations, regions of LD will generally extend further, thus making them ideally suited for low-resolution mapping. Hence a causal genetic variant in the Finnish population could be in significant LD with a particular genetic background (e.g., the -3A:1249T haplotype) some distance away. Testing more heterogeneous populations will not necessarily pick up this same association signal as the LD throughout the region would be weaker.

Autism and ADHD, two neurodevelopmental disorders with complex genetic etiologies, have also been linked to 15q (2, 97). Autistic individuals have severe social and language impairments, but sometimes possess advanced reading skills (100). Individuals with ADHD are characterized by inattention, overactivity, and impulsiveness, and are at greater risk of developing dyslexia (107). Accordingly, *DYX1C1* variants are now being genotyped in samples with each of these disorders. So far, neither a Finnish autistic case-control study (110) nor an ADHD family-based study (106) have found evidence for association with any single SNP within *DYX1C1*. However, analysis of a haplotypic marker comprising six SNPs across *DYX1C1* revealed a significant biased transmission (P-value = 0.009) of one haplotype with a 41% frequency for ADHD as a categorical trait (106). Subsequent quantitative association analysis produced significant results for the biased transmission of the same haplotype for several ADHD traits, but not for the reading-related traits that had been measured in this ADHD sample. Therefore, testing *DYX1C1* variants in further ADHD samples seems warranted.

In summary, the lack of consistent replications of the original association found between *DYX1C1* and RD strongly suggests that the two variants (-3A and 1249T) do not contribute significantly to RD. Therefore, it is not possible to claim that *DYX1C1* is the gene responsible for the QTL observed at 15q21. It may well be the causal gene for the phenotype observed in the family with the translocation t(2;15)(q11;q21), but its role in influencing reading ability in the general population seems unlikely.

Linkage disequilibrium (LD): two loci are said to be in strong linkage disequilibrium if there has been little (or no) ancestral recombination between them

CHROMOSOME 3: A STORY FOR A UNIQUE FAMILY

The *DYX5* locus on chromosome 3p12-q13 has not attracted as much attention as the locus on chromosome 15. Few studies have looked at this region. Linkage was first reported in a large four-generation Finnish family with 21 severely dyslexic individuals. Dyslexia segregated in an apparently autosomal dominant fashion in this family, suggesting a mutation in a single gene (69). Through a genome-wide scan approach, a 20-cM-long haplotype that spanned the centromere of chromosome 3 was found to be shared by 19 of the 21 dyslexic individuals. Other evidence for linkage at the DYX5 locus came from the genome-wide scans performed by Fisher et al. (2002) in two independent samples of U.K. and U.S. families. Although the U.K. samples showed some sign of linkage at 3p13, the most significant signal came from the U.S. sample, where the linked region was on the other arm of the chromosome (3q13). The *DYX5* region has also shown linkage to speech sound disorder, which shares phonological deficits with RD (91) A candidate gene was identified within the critical interval when an individual with dyslexia was found to have a translocation t(3;8)(p12:q11) that includes the *DYX5* region (50). The translocation breakpoint on chromosome 3 disrupted the first intron of

ROBO1, orthologous to the *Drosophila roundabout* gene.

Hannula-Jouppi et al. (50) returned to the original Finnish family and observed that the 19 dyslexic individuals carried a rare haplotype spanning *ROBO1*, which was not detected in the other samples examined, including both controls and the remaining family members. Gene expression analysis conducted in lymphocytes from four affected members of the family showed that the transcription of *ROBO1* from this haplotype was absent or attenuated. They concluded that dyslexia may be caused by a reduction in the gene expression of *ROBO1*.

ROBO1 is widely expressed in the brain (**Figure 1**), and its role in brain development makes it an attractive candidate for RD. *Robo* in *Drosophila* encodes a transmembrane receptor that controls axon guidance across

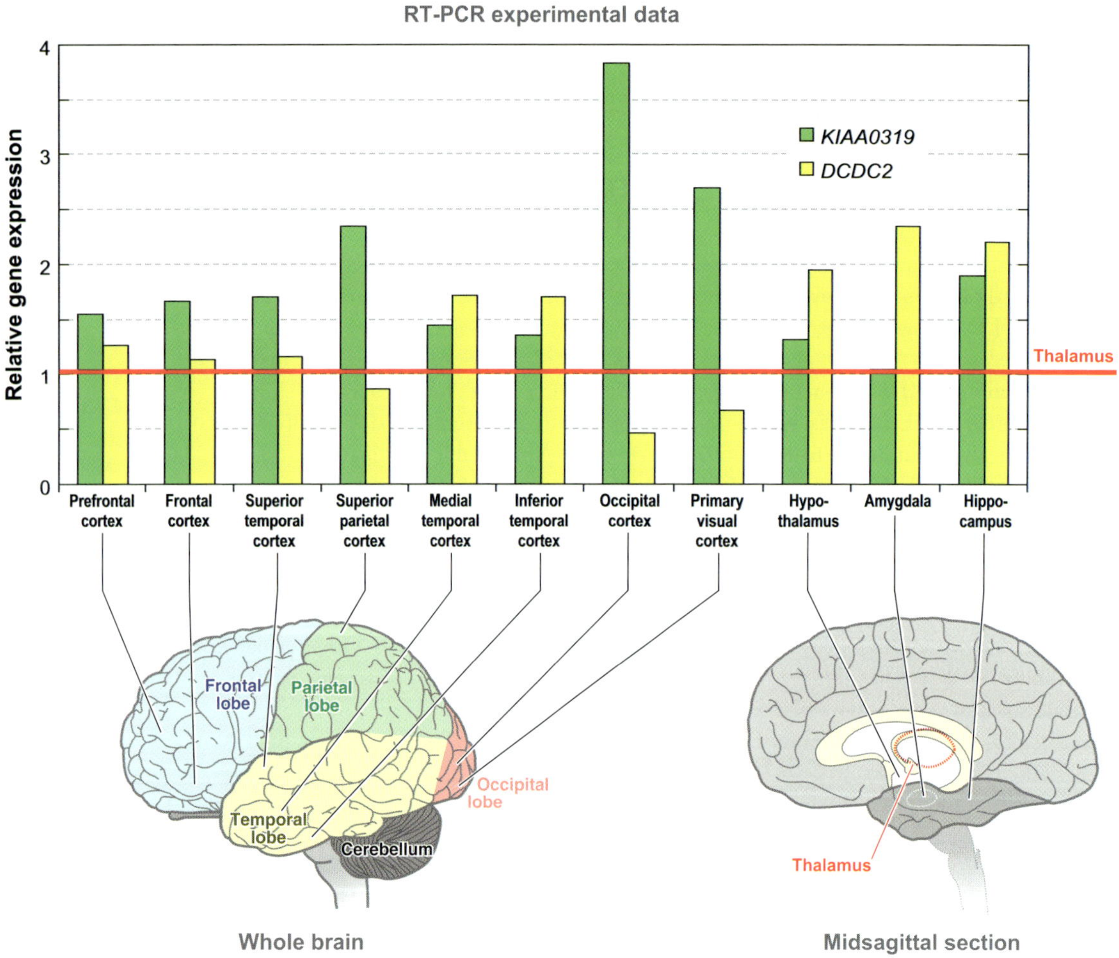

Figure 1

Candidate gene expression patterns in the brain. The relative expression levels of different genes in various tissues have been determined utilizing real-time PCR (RT-PCR) and microarrays. The first graph shows the results of a RT-PCR experiment conducted by Meng et al. (66), which illustrates the expression levels of *KIAA0319* and *DCDC2* in various brain regions relative to the thalamus, and the second graph represents data taken from a publicly available microarray experiment (94) showing the relative expression levels of *KIAA0319* and *ROBO1* in various brain regions relative to the thalamus.

the midline (55). However, several issues related to this study remain unresolved. First, the contribution of the translocation to the development of dyslexia is questionable because one family member had severe dyslexia but did not carry the translocation (50). Second, the expression analysis was conducted on cDNAs derived from lymphocytes of just four dyslexic individuals, leaving open the following questions: whether the results are representative of (*a*) the other affected members of the family and (*b*) the expression level of *ROBO1* in the brain (34). Moreover, the reduction in the expression level of the allele associated with dyslexia was extremely variable across the four individuals, making it difficult to explain a uniformly severe phenotype. It is important that these findings should be replicated in other samples. Although it is possible that the effect of *ROBO1* may be restricted to this unusual family, it will be interesting to see if any association can be detected in the samples that showed linkage to chromosome 3 (35, 91). The identification of the functional variant(s) responsible for reducing the expression of *ROBO1* would provide evidence in support of a causal link to dyslexia.

CHROMOSOME 6

Two Candidates for One Locus

The chromosome 6 story is the most intriguing so far. Positive linkage signals on the short arm of chromosome 6 have been consistently reported in independent studies. The focused attention of several groups on this locus has resulted in the identification of

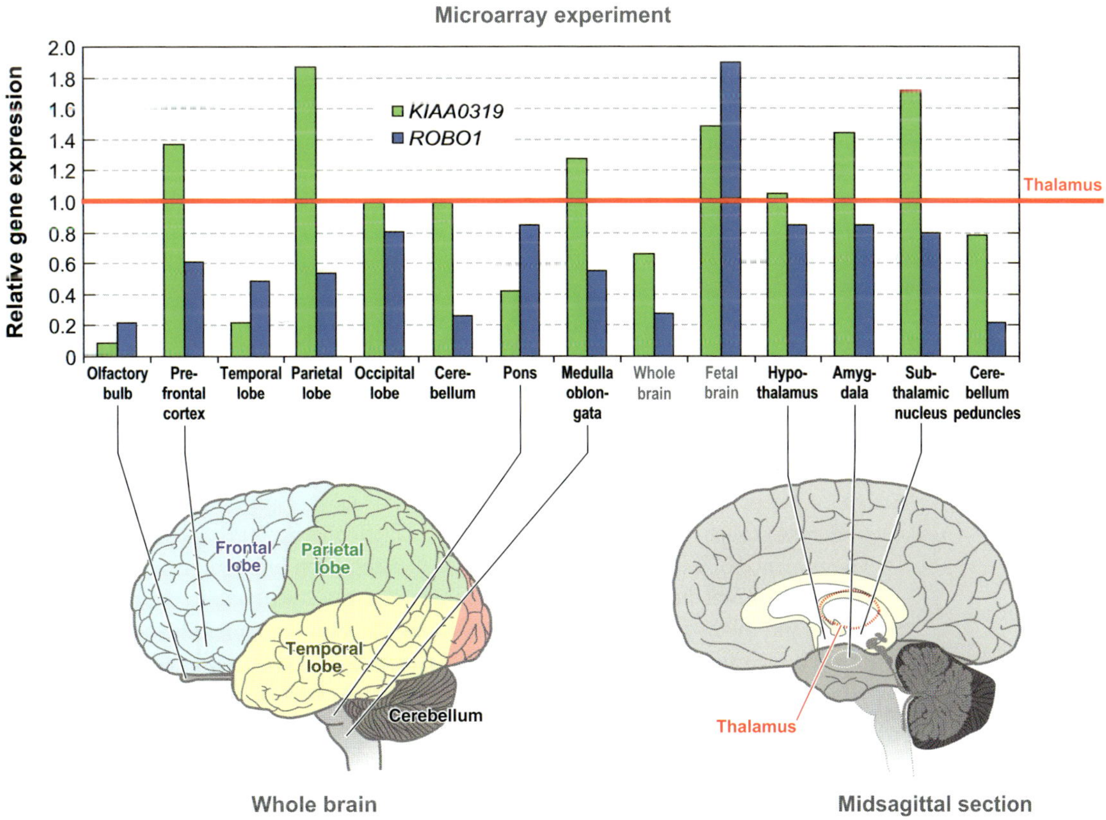

Figure 1

(*Continued*)

Table 4 Different criteria employed for the association studies of the chromosome 6 locus

Reference	Gene associated	Country of origin	Phenotype definition	Sample composition	Severity selection criteria	Sample after selection
Deffenbacher et al. (19)	*DCDC2*	U.S.[b]	Quantitative	349 nuclear families[a]	A sib with severe score in at least one trait	114 nuclear families
	KIAA0319					
Francks et al. (39)	*KIAA0319*	U.K.	Quantitative	264 nuclear families	A sib with severe score on a composite measure of traits mostly contributing to the linkage	126 nuclear families
	KIAA0319	U.S.[b]		159 nuclear families		124 nuclear families
Cope et al. (16)	*KIAA0319*	U.K.	Categorical	223 cases/273 controls		
		U.K.		143 trios		
Meng et al. (66)	*DCDC2*	U.S.[b]	Quantitative	153 nuclear families		
Schumacher et al. (83)	*DCDC2*	Germany	Categorical	137 trios	Probands with severe spelling disabilities	47 trios
		Germany		239 trios		64 trios
Harold et al. (51)	*KIAA0319*	U.K.	Quantitative	Sample as Francks et al. (39)		
		U.K.	Categorical	350 cases/273 controls[c]		

[a]Used for linkage analysis.

[b]Samples derived from the CLDRC collection.

[c]Extended case-control sample from Cope et al. (16).

not one, but two candidate RD susceptibility genes: *KIAA0319* and *DCDC2*. The two genes share many similarities: They are physically close, are both expressed in the brain (**Figure 1**), show similar functions, and have each been independently identified by at least two different groups (**Table 4**).

The presence of a susceptibility QTL on chromosome 6p was first reported by Smith et al. (87), and then replicated by other studies using both categorical (46, 47, 99) and quantitative (11, 12, 36, 43, 54) approaches (**Figure 2**). There have been a few negative reports as well (29, 76, 82), which may reflect the genetic heterogeneity of the dyslexic phenotype. Although well defined, the 6p interval is more than 16 Mb in length and contains several hundred genes, making the search for the dyslexia candidate challenging. However, various association studies not only succeeded in detecting positive signals at this locus, but actually confined them to within the *DCDC2* and *KIAA0319* genes, which are less than 200 kb apart (**Figure 2**).

Deffenbacher et al. (19) found evidence of association in both genes in a sample of dizygotic twin families collected through the Colorado Learning Disabilities Research Center (CLDRC) (21). Their linkage analysis suggested that both orthographic and phonological skills are influenced by this QTL. Since the linkage was stronger in families with the most severe phenotypic scores, they limited their quantitative association analysis to families with the most severely affected individuals. They genotyped 31 SNPs that

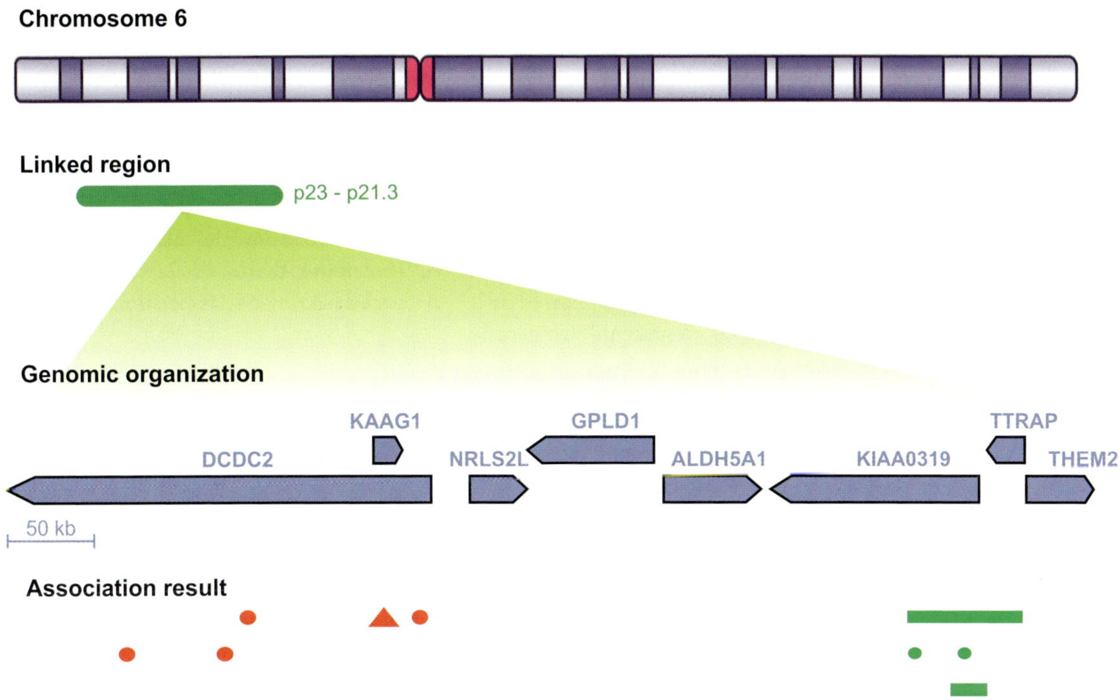

Figure 2

Results of association studies at the chromosome 6 locus. Linkage studies have identified a consensus region at chromosome 6p23-21.3 that has been implicated in harboring reading disability (RD) susceptibility variants. Association studies have detected positive signals within the *DCDC2* and *KIAA0319* genes, which are located within 200 kb of each other. The dots, triangle, and bars represent the SNPs, deletion, and genomic region, respectively, that produced the most significant association with RD. The markers from the studies of Meng et al. (66) and Schumacher et al. (83) are shown in red, and the markers from the studies of Francks et al. (39), Cope et al. (16), and Harold et al. (51) are in green.

spanned 10 genes, and found that 8 SNPs within *DCDC2* yielded association across different phenotypes and with different analytical approaches (P-values in the range of 0.001–0.05). Compelling association was also detected in the *KIAA0319/TTRAP/THEM2* cluster (P-values around 0.01). In a subsequent study, Francks et al. (39) performed association analysis on brain-expressed genes within the 6p locus in a sample of 89 families from the United Kingdom. The strongest signal (P = 0.0004) in this study was in the *KIAA0319/TTRAP/THEM2* gene cluster. The association was replicated in independent collections of families from the United Kingdom and United States, the latter selected from the CLDRC collection. The association was consistent across different reading measures and got stronger in subsamples selected for severity. A specific haplotype (tagged by the three SNPs: rs4504469, rs2038137, and rs2143340) spanning a 77-kb region of LD across the three genes was associated with dyslexia in both the U.K. and U.S. families. However, the strong LD in the region prevented refinement of the association by further genetic mapping, and the failure to detect any disruptive mutations left open the question as to which of the three genes was implicated in the QTL effect.

Cope et al. (16) also found strong association within this region in a completely independent U.K. sample when using a categorical definition of dyslexia. The association was detected with a classical case-control study design and confirmed via transmission disequilibrium testing. Notably, one of the strongest associations was detected with the same allele of the rs4504469 *KIAA0319* coding SNP, which defined the risk haplotype identified by Francks et al. (39).

With the *KIAA0319/TTRAP/THEM2* gene cluster being associated with RD in three independent samples, it appeared as if the search for the *DYX2* susceptibility gene(s) was drawing to an end. However, two studies brought *DCDC2* back into the limelight.

Meng et al. (66) reported association with *DCDC2* in another set of families derived from the CLDRC collection. In a previous study, this research group found a positive association signal ($P = 0.0007$) between the orthographic choice measure and JA04, a microsatellite marker located in the first exon of *KIAA0319*. They genotyped 149 SNPs across the 1.5 Mb surrounding JA04 in 153 families (54). The most significant quantitative association peak ($P = 0.0003$) was between a SNP located in intron 6 of *DCDC2* and the discriminant score phenotype (a weighted composite measure of reading and spelling subtests). No association was detected with orthographic choice or any other phenotype. They also detected an intronic deletion of 2445 bp in 10 families. The deletion removes a 168-bp region containing a microsatellite. When the analysis was conducted by pooling the deletion with rare microsatellite alleles, significant association ($P = 0.00002$) was detected with the homonym choice phenotype. However, neither the deletion itself nor the individual microsatellite alleles showed evidence of association by themselves. Hence, it is possible that this association is driven by the rare microsatellite alleles in a few individuals that happen to have extreme scores in a very specific phenotype. Therefore, the functional role of the deletion is currently based solely on this association.

Further support for *DCDC2* as a candidate gene comes from a study by Schumacher et al. (83). Using a categorical definition of dyslexia based on spelling abilities, they found association with markers within the *DCDC2* gene in two independent trio samples of German descent, most significantly with a two-marker haplotype in intron 7. The association was strongest in the most severely affected patients of both samples. The intronic deletion was not tested, but they analyzed the three SNPs defining the *KIAA0319* risk haplotype without detecting any association.

KIAA0319 or DCDC2? Taken together, these reports lead to the question of which gene, *KIAA0319* or *DCDC2*, is responsible for the QTL effect on chromosome 6p? It is still not possible to exclude either gene as a candidate for RD susceptibility, and there are good reasons to believe that both genes might contribute to the QTL effect. Because the two genes are located so close together, it is reasonable to argue that the associations detected in both genes are simply the signal of a single functional mutation, not yet identified. However, the lack of significant LD across and between the two genes (51) makes it unlikely. On the other hand, the coexistence of two candidate genes at the same locus might explain why the 6p region has been so consistently linked in many different samples, and is perfectly compatible with a model for a multifactorial disorder.

So why is there a lack of uniformity across different studies? There are several explanations accounting for these apparently inconsistent results. The most obvious is that the various studies have been conducted using heterogeneous criteria, both in terms of sample selection and analytical strategy, which can be critical in determining different outcomes (**Table 4**). Because of these differences, it is difficult to directly compare these findings. A plausible explanation is that different subgroups of dyslexic patients are determined by the effect of one or the other gene. If, for example, the *DCDC2* gene more specifically affects the process underlying the spelling phenotype, it would explain why *DCDC2* is associated with dyslexia in the German sample, which was selected for being impaired in spelling ability. In this respect the Colorado sample is emblematic. Three different studies have used this sample and three different outcomes were shown; one study detected association with *KIAA0319* (39), one with *DCDC2* (66), and one with both genes (19). How is this possible? All the Colorado families were recruited using the same criteria, but despite a significant sample overlap, the exact subset of families used by each of the three studies differed. Meng et al. (66) found association with *DCDC2* in a sample of 153 families, whereas the other two studies obtained their most significant results from subgroups of families selected for severity of phenotype. Francks et al. (39) detected association in the *KIAA0319* gene using 124 families with the most negative scores in a composite measure of single-word reading and phonological decoding, because these measures showed the highest linkage peaks in the Colorado sample. Deffenbacher et al. (19) employed 114 families that had at least one child with a severe score in at least one of the phenotypes and found suggestive association with both genes. This picture not only supports the idea that both genes are possible candidates, but also shows how the analysis is sensitive to subtle differences in sample composition, especially if the sample sizes are relatively small.

Another factor that makes a direct comparison across the various studies difficult is the use of different sets of SNPs employed by each research group. To address this issue, a collaborative effort (51) has analyzed the same selection of *KIAA0319* and *DCDC2* markers in both the U.K. samples previously employed by Francks et al. (39) and Cope et al. (16). They analyzed all the markers that had previously showed association within *DCDC2*, including the intronic deletion (66) and the two-marker haplotype (83). In addition, a new set of *KIAA0319* markers were analyzed (51). None of the *DCDC2* markers showed significant association, whereas some of the newly genotyped SNPs within the *KIAA0319* gene showed the strongest association so far detected (P-values < 0.0001). Interestingly, the most significant association signals were clustered within a few kilobases around the first *KIAA0319* exon. The data suggest that *KIAA0319* is involved in the development of RD in both of the U.K. samples. In particular, it suggests the presence of a functional mutation located within the regulatory sequences near the 5′ end of *KIAA0319*. Although *DCDC2* does not seem to have a major influence on RD in these samples, its role in the etiology of dyslexia cannot be excluded.

GENE EXPRESSION AS A KEY MECHANISM

A definitive answer to resolve the debate between the two genes and to understand how a genetic background can affect reading abilities would come from the identification of the causal mutations underlying the associations detected within *KIAA0319* and *DCDC2*. So far no functional effects of any of the associated markers have been found. However, some progress has been made for the *KIAA0319* gene. Paracchini et al. (72) hypothesized that the haplotype previously found to be associated with dyslexia (39) would affect gene expression. They tested this hypothesis in both lymphoblastoid and neuroblastoma cell lines heterozygous for the risk haplotype by comparing the level of transcription generated from the chromosome carrying the risk haplotype versus chromosomes carrying other haplotypes. The experimental strategy consisted of quantitatively measuring the relative abundance of the two alleles of SNPs located either within the transcript or close to the regulatory regions of a specific gene via mass spectrometry. A lower level of the alleles on the risk haplotype was consistently observed for all the SNPs tested within the *KIAA0319* gene in all the selected cell lines. These results suggest that the effect of the risk haplotype could be explained by the presence of one or more variants that reduce the expression of the *KIAA0319* gene, in agreement with the association results by Harold et al. (51).

As we previously discussed, a reduced gene expression was also associated with the risk haplotype for the *ROBO1* gene (50). Even if those findings are still waiting further replication, they support the idea that altered regulation of gene expression could be a common theme for genes involved in dyslexia.

RD is a complex trait, expected to be the result of the concomitant effect of multiple genetic and environmental factors. Each of these factors is anticipated to contribute a small effect to the overall trait, thus creating a continuum of abilities within the general population. A susceptibility genetic variant will be relevant to a dysfunction depending on the epistatic architecture of the genome. A subtle factor such as variation in the level of a specific protein, if combined with other elements, could be enough to play a role in the development of a complex trait (57). It has already been shown that allelic variation influencing gene expression is a common feature of the human genome (60, 109), and this variation could contribute significantly to human variability. Therefore, finding that a specific allele associated with RD is also regulating the expression of a specific gene supports this model.

NEURONAL MIGRATION: THE PATH FROM GENES TO COGNITION

The candidate genes discussed so far have been proposed on the basis of genetic mapping approaches and statistical analysis. But how can these genes be relevant to the development of RD? In other words, what do these genes do? The answer will not only be the key to understanding the biological mechanism leading to dyslexia, but it will also provide essential tools to penetrate the labyrinth of cognition (34). There is evidence that all four candidates are required during the development of the brain, and they all seem to take part in pathways involved in axon growth and neuronal migration (41).

Neuronal migration is a key step in the development of the neocortex, resulting in the organization of neurons in six specialized laminar layers (53). Defects in neuronal migration are the cause of several human syndromes with symptoms of epilepsy and mental retardation (5, 64). Interestingly, abnormalities in neuronal migration have been implicated in dyslexia by anatomical studies conducted by Galaburda and coworkers in postmortem brains (40, 42). The cerebral cortex derived from individuals diagnosed with RD consistently showed the presence of misplaced groups of cells (heterotopias) resulting from abnormal neuronal migration, mainly localized in the left hemisphere surrounding the sylvian fissure, a region traditionally associated with language. More than 20 years later, these findings still remain restricted to the eight brains analyzed in the original studies because a similar investigation has not been repeated. However, another connection between abnormal neuronal migration and reading impairment has emerged from behavioral studies (14) of patients affected by periventricular nodular heterotopia (PNH), a neuronal migration defect caused by mutation in the filamin A (*FLNA*) gene (37). Patients with PNH have been shown to have deficits in reading skills despite normal intelligence (14).

ROBO1 is widely expressed in the developing nervous system, and the role of its homologs in both invertebrates and vertebrates is well established (55). It encodes a guidance receptor that is essential to direct the outgrowth of both axon and dendrite during the formation of the neuronal network. The extracellular chemorepellent *Slit* acts as a ligand for *Robo*. The *Slit-Robo* guidance system controls axon growth across the commissures of the brain and spinal cord (77), and in vitro studies have shown that the Slit family also acts as a repellent in neuronal migration (104).

The exact functions of the other three candidates (*DCDC2*, *KIAA0319*, and *DYX1C1*) have yet to be elucidated, but they have all been shown to be involved in glia-guided neuronal migration during the formation of the cortex (66, 72, 103) (**Figure 3**). To test for a possible role of these candidates in neuronal

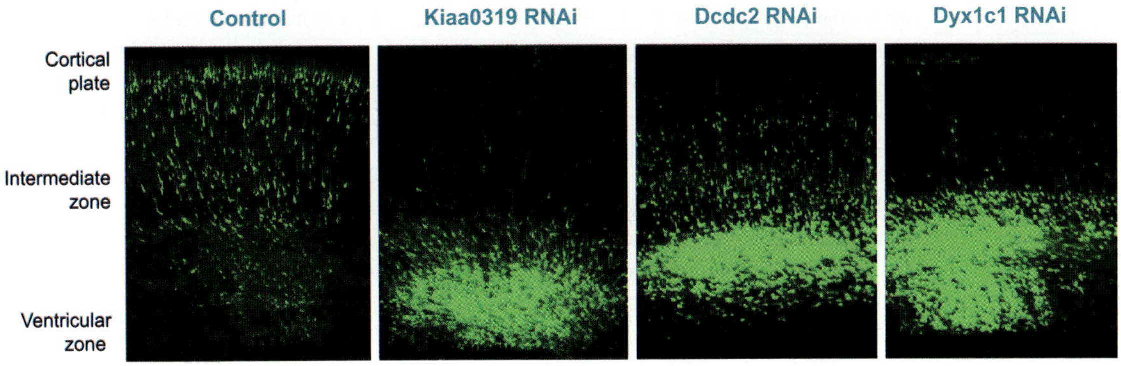

Figure 3

RNA interference (RNAi) of *Kiaa0319*, *Dcdc2*, and *Dyx1c1* disrupts migration in the developing rat neocortex. Images of sections of embryonic rat neocortex are shown four days after the electroporation of short hairpin RNA (shRNA) vectors and an EGFP vector, as described by Bai et al. (3). In the control experiment, where cells were transfected with a neutral shRNA, most of the neurons migrated well away from the ventricular zone and many reach the cortical plate. Neurons transfected with shRNA vectors against *Kiaa0319*, *Dcdc2*, and *Dyx1c1* migrate abnormally and are arrested in the ventricular and intermediate zones. Credit: Joe LoTurco, University of Connecticut.

migration, they have been tested by in utero RNA interference (RNAi) (3) in the developing cerebral neocortex of rats. With this technique, it is possible to inhibit the activity of a selected gene only in those neurons that are preparing to migrate toward their final destinations in the multilayer structure of the cortex. A limited number of genes has been described so far to direct neuronal migration, and surprisingly, these three RD candidates are somehow involved in this crucial event for the correct development of the cortex.

DCDC2 contains two doublecortin peptide domains and belongs to the protein family of DCX, which has a well-recognized role in neuronal migration (45). DCX is a cytoplasmic protein that regulates the organization of microtubules during neuronal migration. Mutations in *DCX*, mainly located within the doublecortin peptide domains, cause double-cortex syndrome and lissencephaly in humans, two neuronal migration defects (25). Local loss of function of *Dcdc2* induced by RNAi results in abnormal migration, which is characterized by a reduced progression of the distance covered by the migrating neurons (66) (**Figure 3**).

KIAA0319 is a membrane protein with a large extracellular domain, characterized by five polycystic kidney disease (PKD) domains, originally described in the PKD1 protein (10) and implicated in cell adhesion functions. The RNAi of rat *Kiaa0319* resulted in a significant reduction in the distance migrated by neurons, which mostly remained within the ventricular proliferative zone (72) (**Figure 3**). The observed phenotype suggests that the fault in the migration process could be caused by a defective association between the migrating neurons and the radial glia fibers. It is therefore possible that the PKD domains of KIAA0319 are responsible for mediating the proper adhesion between the neurons and the radial glia. An involvement of the *KIAA0319* gene in neuronal migration was also supported by in situ hybridization data, which showed that *KIAA0319* has a very distinct spatial-temporal pattern of expression in fetal mouse and human brains. In particular, *KIAA0319* is expressed in the developing neocortex when neuronal migration occurs (72).

The structure of DYX1C1 does not relate to cell adhesion properties or cytoskeletal

RNA interference (RNAi): the introduction of an RNA molecule into a cell to cause the degradation of the complementary cellular mRNA, leading to the knockdown of a specific gene activity

Double-cortex syndrome: a neuronal migration disorder characterized by the appearance of an extra layer of neurons under the normal cortex

Lissencephaly: a neuronal migration disorder where neurons fail to get organized in a six-layered structure, resulting in a smooth brain surface

Epistasis: the interaction between two or more genes to control a single phenotype

maintenance. However, RNAi studies have shown that *DYX1C1* is required for neuronal migration. Specifically, RNAi against DYX1C1 delays neurons in the deep layer, preventing them from reaching the cortical plate (103) (**Figure 3**). Moreover, many of the delayed neurons eventually reach an abnormal placement or produce heterotopias that resemble those observed in the postmortem dyslexic brains (42).

But, how can general neuronal migration genes be linked specifically to dyslexia rather than show association with a more extensive phenotype? We might expect to detect their effect on neuronal migration only in the region of the cortex controlling reading pathways. However, even if the exact expression patterns of the RD candidates genes in the human brain remain to be elucidated, there is no evidence to suggest that the expression of these genes is restricted to specific cortical areas or limited to embryogenesis (**Figure 1**). On the other hand, some genes have variable expression patterns in different cortical areas (26), and it is possible that their epistatic interactions with the RD candidates are at the basis of dyslexia (79). Epistasis could also explain co-morbidity with other cognitive disorders. While some gene-gene interactions may cause dyslexia, other different but partially overlapping interactions could influence different cognitive functions. Thus, it seems worthwhile testing the RD candidates that show co-morbidity with dyslexia for other cognitive disorders, such as SLI and ADHD.

CONCLUSION: THE ROLE OF GENETICS IN UNCOVERING THE BIOLOGY OF DYSLEXIA

The four candidate genes for RD represent an important advance not only for the genetics of dyslexia but also for the field of neuroscience in general. The involvement of the four candidates in brain development and their possible role in neuronal migration provide a hint to unravel the cognitive functions of the brain. While no functional mutations have been identified for any of the four candidate genes, it is encouraging that two of the candidates (*KIAA0319* and *DCDC2*) are supported by compelling genetic evidence in independent samples (16, 39, 65, 83). Identifying the causal genetic variants will be one of the ultimate goals of the newly formed NeuroDys consortium (**http://www.neurodys.com**). NeuroDys is a multidisciplinary project, grouping together 13 research groups from 10 European countries. It will have access to nearly 4000 children with RD for genetic testing, overcoming the recurrent limitation of the lack of power due to small sample size. Moreover, the analysis will be performed with homogenous criteria across all the samples from different countries, thus eliminating the discrepancies introduced by separate approaches.

The isolation of the RD functional genetic variants will mark only the beginning of understanding the link between genetics and cognitive functions. It will be interesting to see whether different genetic backgrounds, characterized by the combinations of various risk alleles, translate into distinguishable phenotypes and may partially explain the variation in the severity of impairment.

Another issue is whether the genetic profile at these loci correlates with abnormalities in brain structure or function. Neuroimaging studies of individuals carrying different risk variants could reveal how genetics contributes in shaping the brain and in influencing its activity. Moreover, these alleles might contribute to differentiating an individual's response toward external exposures, providing an important tool to study gene-environment interactions in the etiology of RD. NeuroDys has the potential to address these issues, thanks to the ample variety of expertise of the participants, which ranges from genetics to cognitive sciences to brain imaging.

The answers to these questions will provide the tools necessary to disclose the neural pathways that contribute to reading, and will ultimately improve the diagnoses and treatment of dyslexia by taking into consideration the individuality of each patient. Even if the answers are still far off, it looks like genetics has begun to unravel the complex picture of reading ability, and is playing a crucial role in the understanding of dyslexia.

SUMMARY POINTS

1. The *DYX1C1*, *DCDC2*, *KIAA0319*, and *ROBO1* genes, proposed as RD susceptibility candidates, seem to be implicated in brain development. In particular, *DCDC2*, *KIAA0319*, and *DYX1C1* have been shown to be involved in neuronal migration.
2. Both candidates (*DCDC2* and *KIAA0319*) at the chromosome 6 locus were found to be associated with RD in at least two independent samples.
3. Association with *DYX1C1*, the first proposed candidate, has not been convincingly replicated.
4. No functional mutations have yet been identified for any of the four candidates.
5. Alteration in gene expression has been proposed as a possible mechanism disrupting the correct functions of two candidates (*ROBO1* and *KIAA0319*).
6. No single gene, but rather the epistasis of multiple factors, is expected to play a major role in the development of RD.

FUTURE ISSUES

1. How can genes expressed in different regions of the brain and involved in a general mechanism such as neuronal migration specifically affect a distinct cognitive function?
2. Do specific risk alleles correlate with particular brain changes or subgroups of RD phenotypes?
3. Are the RD candidates relevant to other cognitive disorders?

ACKNOWLEDGMENTS

S.P., T.S., and A.P.M. are supported by the Wellcome Trust. We wish to thank Javier Alegre for help with the annotation of **Figure 1**, Joe LoTurco for providing images of the RNAi experiments, and Dianne Newbury and Simon Fisher for making useful suggestions to the manuscript.

LITERATURE CITED

1. Arcos-Burgos M, Muenke M. 2002. Genetics of population isolates. *Clin. Genet.* 61:233–47
2. Bacchelli E, Maestrini E. 2006. Autism spectrum disorders: molecular genetic advances. *Am. J. Med. Genet. C Semin. Med. Genet.* 142:13–23
3. **Bai J, Ramos RL, Ackman JB, Thomas AM, Lee RV, LoTurco JJ. 2003. RNAi reveals doublecortin is required for radial migration in rat neocortex. *Nat. Neurosci.* 6:1277–83**

3. Describes the same technical procedure used to show that the *KIAA0319*, *DCDC2*, and *DYX1C1* candidates are involved in neuronal migration.

4. Bellini G, Bravaccio C, Calamoneri F, Donatella Cocuzza M, Fiorillo P, et al. 2005. No evidence for association between dyslexia and DYX1C1 functional variants in a group of children and adolescents from Southern Italy. *J. Mol. Neurosci.* 27:311–14
5. Bielas S, Higginbotham H, Koizumi H, Tanaka T, Gleeson JG. 2004. Cortical neuronal migration mutants suggest separate but intersecting pathways. *Annu. Rev. Cell Dev. Biol.* 20:593–618
6. Bisgaard ML, Eiberg H, Moller N, Niebuhr E, Mohr J. 1987. Dyslexia and chromosome 15 heteromorphism: negative lod score in a Danish material. *Clin. Genet.* 32:118–19
7. Bishop DV, Snowling MJ. 2004. Developmental dyslexia and specific language impairment: same or different? *Psychol. Bull.* 130:858–86
8. Blatch GL, Lassle M. 1999. The tetratricopeptide repeat: a structural motif mediating protein-protein interactions. *Bioessays* 21:932–39
9. Brady SA, Shankweiler DP. 1991. *Phonological Processes in Literacy*. Hillsdale, New Jersey: Lawrence Erlbaum
10. Bycroft M, Bateman A, Clarke J, Hamill SJ, Sandford R, et al. 1999. The structure of a PKD domain from polycystin-1: implications for polycystic kidney disease. *EMBO J.* 18:297–305
11. Cardon LR, Smith SD, Fulker DW, Kimberling WJ, Pennington BF, DeFries JC. 1994. Quantitative trait locus for reading disability on chromosome 6. *Science* 266:276–79
12. Cardon LR, Smith SD, Fulker DW, Kimberling WJ, Pennington BF, DeFries JC. 1995. Quantitative trait locus for reading disability: correction. *Science* 268:1553
13. Castles A, Coltheart M. 1993. Varieties of developmental dyslexia. *Cognition* 47:149–80
14. Chang BS, Ly J, Appignani B, Bodell A, Apse KA, et al. 2005. Reading impairment in the neuronal migration disorder of periventricular nodular heterotopia. *Neurology* 64:799–803
15. Chapman NH, Igo RP, Thomson JB, Matsushita M, Brkanac Z, et al. 2004. Linkage analyses of four regions previously implicated in dyslexia: confirmation of a locus on chromosome 15q. *Am. J. Med. Genet. B Neuropsychiatr. Genet.* 131:67–75
16. Cope N, Harold D, Hill G, Moskvina V, Stevenson J, et al. 2005. Strong evidence that KIAA0319 on chromosome 6p is a susceptibility gene for developmental dyslexia. *Am. J. Hum. Genet.* 76:581–91
17. Cope NA, Hill G, van den Bree M, Harold D, Moskvina V, et al. 2005. No support for association between dyslexia susceptibility 1 candidate 1 and developmental dyslexia. *Mol. Psychiat.* 10:237–38
18. de la Chapelle A, Wright FA. 1998. Linkage disequilibrium mapping in isolated populations: the example of Finland revisited. *Proc. Natl. Acad. Sci. USA* 95:12416–23
19. Deffenbacher KE, Kenyon JB, Hoover DM, Olson RK, Pennington BF, et al. 2004. Refinement of the 6p21.3 quantitative trait locus influencing dyslexia: linkage and association analyses. *Hum. Genet.* 115:128–38
20. DeFries JC, Alarcón M. 1996. Genetics of specific reading disability. *Ment. Retard. Dev. Disabil. Res. Rev.* 2:39– 47
21. DeFries JC, Filipek PA, Fulker DW, Olson RK, Pennington BF, et al. 1997. Colorado learning disabilities research center. *Learn. Disabil. Q.* 8:7–19
22. DeFries JC, Fulker DW. 1985. Multiple regression analysis of twin data. *Behav. Genet.* 15:467–73
23. DeFries JC, Singer SM, Foch TT, Lewitter FI. 1978. Familial nature of reading disability. *Br. J. Psychiat.* 132:361–67
24. Demonet JF, Taylor MJ, Chaix Y. 2004. Developmental dyslexia. *Lancet* 363:1451–60

25. des Portes V, Francis F, Pinard JM, Desguerre I, Moutard ML, et al. 1998. Doublecortin is the major gene causing X-linked subcortical laminar heterotopia (SCLH). *Hum. Mol. Genet.* 7:1063–70
26. Donoghue MJ, Rakic P. 1999. Molecular gradients and compartments in the embryonic primate cerebral cortex. *Cereb. Cortex* 9:586–600
27. Eden GF, Moats L. 2002. The role of neuroscience in the remediation of students with dyslexia. *Nat. Neurosci.* 5(Suppl.):1080–84
28. Felton RH, Naylor CE, Wood FB. 1990. Neuropsychological profile of adult dyslexics. *Brain Lang.* 39:485–97
29. Field LL, Kaplan BJ. 1998. Absence of linkage of phonological coding dyslexia to chromosome 6p23-p21.3 in a large family data set. *Am. J. Hum. Genet.* 63:1448–56
30. Finucci JM, Childs B. 1981. Are there really more dyslexic boys than girls? In *Sex Differences in Dyslexia*, ed. A Ansara, N Geshwind, AM Galaburda, M Albert, M Gartrell, pp. 1–9: Towson, MD: Orton Dyslexia Soc.
31. Finucci JM, Guthrie JT, Childs AL, Abbey H, Childs B. 1976. The genetics of specific reading disability. *Ann. Hum. Genet.* 40:1–23
32. Finucci JM, Isaacs SD, Whitehouse CC, Childs B. 1983. Classification of spelling errors and their relationship to reading ability, sex, grade placement, and intelligence. *Brain Lang.* 20:340–55
33. Fisher SE, DeFries JC. 2002. Developmental dyslexia: genetic dissection of a complex cognitive trait. *Nat. Rev. Neurosci.* 3:767–80
34. Fisher SE, Francks C. 2006. Genes, cognition and dyslexia: learning to read the genome. *Trends Cogn. Sci.* 10:250–57
35. Fisher SE, Francks C, Marlow AJ, MacPhie IL, Newbury DF, et al. 2002. Independent genome-wide scans identify a chromosome 18 quantitative-trait locus influencing dyslexia. *Nat. Genet.* 30:86–91
36. Fisher SE, Marlow AJ, Lamb J, Maestrini E, Williams DF, et al. 1999. A quantitative-trait locus on chromosome 6p influences different aspects of developmental dyslexia. *Am. J. Hum. Genet.* 64:146–56
37. Fox JW, Lamperti ED, Eksioglu YZ, Hong SE, Feng Y, et al. 1998. Mutations in filamin 1 prevent migration of cerebral cortical neurons in human periventricular heterotopia. *Neuron* 21:1315–25
38. Francks C, MacPhie IL, Monaco AP. 2002. The genetic basis of dyslexia. *Lancet Neurol.* 1:483–90
39. Francks C, Paracchini S, Smith SD, Richardson AJ, Scerri TS, et al. 2004. A 77-kb region of chromosome 6p22.2 is associated with dyslexia in families from the United Kingdom and from the United States. *Am. J. Hum. Genet.* 75:1046–58
40. Galaburda AM, Kemper TL. 1979. Cytoarchitectonic abnormalities in developmental dyslexia: a case study. *Ann. Neurol.* 6:94–100
41. Galaburda AM, LoTurco J, Ramus F, Fitch RH, Rosen GD. 2006. From genes to behavior in developmental dyslexia. *Nat. Neurosci.* 9:1213–17
42. Galaburda AM, Sherman GF, Rosen GD, Aboitiz F, Geschwind N. 1985. Developmental dyslexia: four consecutive patients with cortical anomalies. *Ann. Neurol.* 18:222–33
43. Gayan J, Smith SD, Cherny SS, Cardon LR, Fulker DW, et al. 1999. Quantitative-trait locus for specific language and reading deficits on chromosome 6p. *Am. J. Hum. Genet.* 64:157–64
44. Gilger JW, Pennington BF, DeFries JC. 1991. Risk for reading disability as a function of parental history in three family studies. *Read. Writ.* 3:205–17

33. A comprehensive analysis of different genetic approaches to identify dyslexia susceptibility loci, with particular emphasis on the application of QTL mapping methods.

35. The first QTL-based genome-wide linkage scans for dyslexia, strongly implicating a chromosome 18 locus and detecting other interesting loci.

42. Anatomical study of postmortem brains linking neuronal migration to dyslexia for the first time.

45. Gleeson JG, Lin PT, Flanagan LA, Walsh CA. 1999. Doublecortin is a microtubule-associated protein and is expressed widely by migrating neurons. *Neuron* 23:257–71
46. Grigorenko EL, Wood FB, Meyer MS, Hart LA, Speed WC, et al. 1997. Susceptibility loci for distinct components of developmental dyslexia on chromosomes 6 and 15. *Am. J. Hum. Genet.* 60:27–39
47. Grigorenko EL, Wood FB, Meyer MS, Pauls DL. 2000. Chromosome 6p influences on different dyslexia-related cognitive processes: further confirmation. *Am. J. Hum. Genet.* 66:715–23
48. Guerin DW, Griffin JR, Gottfried AW, Christenson GN. 1993. Dyslexic subtypes and severity levels: Are there gender differences? *Optom. Vis. Sci.* 70:348–51
49. **Habib M. 2000. The neurological basis of developmental dyslexia: an overview and working hypothesis. *Brain* 123(Pt.12):2373–99**

49. Discussion of multidisciplinary studies into the biology of dyslexia.

50. Hannula-Jouppi K, Kaminen-Ahola N, Taipale M, Eklund R, Nopola-Hemmi J, et al. 2005. The axon guidance receptor gene ROBO1 is a candidate gene for developmental dyslexia. *PLOS Genet.* 1:e50
51. Harold D, Paracchini S, Scerri T, Dennis M, Cope N, et al. 2006. Further evidence that the KIAA0319 gene confers susceptibility to developmental dyslexia. *Mol. Psychiat.* 11:1085–91
52. Hatakeyama S, Matsumoto M, Yada M, Nakayama KI. 2004. Interaction of U-box-type ubiquitin-protein ligases (E3s) with molecular chaperones. *Genes Cells* 9:533–48
53. Hatten ME. 1999. Central nervous system neuronal migration. *Annu. Rev. Neurosci.* 22:511–39
54. Kaplan DE, Gayan J, Ahn J, Won TW, Pauls D, et al. 2002. Evidence for linkage and association with reading disability on 6p21.3–22. *Am. J. Hum. Genet.* 70:1287–98
55. Kidd T, Brose K, Mitchell KJ, Fetter RD, Tessier-Lavigne M, et al. 1998. Roundabout controls axon crossing of the CNS midline and defines a novel subfamily of evolutionarily conserved guidance receptors. *Cell* 92:205–15
56. Kitano T, Liu YH, Ueda S, Saitou N. 2004. Human-specific amino acid changes found in 103 protein-coding genes. *Mol. Biol. Evol.* 21:936–44
57. Knight JC. 2005. Regulatory polymorphisms underlying complex disease traits. *J. Mol. Med.* 83:97–109
58. Lai CS, Gerrelli D, Monaco AP, Fisher SE, Copp AJ. 2003. FOXP2 expression during brain development coincides with adult sites of pathology in a severe speech and language disorder. *Brain* 126:2455–62
59. Lindgren SD, De Renzi E, Richman LC. 1985. Cross-national comparisons of developmental dyslexia in Italy and the United States. *Child Dev.* 56:1404–17
60. Lo HS, Wang Z, Hu Y, Yang HH, Gere S, et al. 2003. Allelic variation in gene expression is common in the human genome. *Genome Res.* 13:1855–62
61. Lovegrove WJ, Bowling A, Badcock D, Blackwood M. 1980. Specific reading disability: differences in contrast sensitivity as a function of spatial frequency. *Science* 210:439–40
62. Marino C, Giorda R, Vanzin L, Nobile M, Lorusso ML, et al. 2004. A locus on 15q15–15qter influences dyslexia: further support from a transmission/disequilibrium study in an Italian speaking population. *J. Med. Genet.* 41:42–46
63. Marlow AJ, Fisher SE, Richardson AJ, Francks C, Talcott JB, et al. 2001. Investigation of quantitative measures related to reading disability in a large sample of sib-pairs from the UK. *Behav. Genet.* 31:219–30
64. McManus MF, Golden JA. 2005. Neuronal migration in developmental disorders. *J. Child. Neurol.* 20:280–86

65. Meng H, Hager K, Held M, Page GP, Olson RK, et al. 2005. TDT-association analysis of EKN1 and dyslexia in a Colorado twin cohort. *Hum. Genet.* 118:87–90
66. Meng H, Smith SD, Hager K, Held M, Liu J, et al. 2005. DCDC2 is associated with reading disability and modulates neuronal development in the brain. *Proc. Natl. Acad. Sci. USA.* 102:17053–58
67. Morris DW, Robinson L, Turic D, Duke M, Webb V, et al. 2000. Family-based association mapping provides evidence for a gene for reading disability on chromosome 15q. *Hum. Mol. Genet.* 9:843–48
68. Nicolson R, Fawcett AJ, Dean P. 2001. Dyslexia, development and the cerebellum. *Trends Neurosci.* 24:515–16
69. Nopola-Hemmi J, Myllyluoma B, Haltia T, Taipale M, Ollikainen V, et al. 2001. A dominant gene for developmental dyslexia on chromosome 3. *J. Med. Genet.* 38:658–64
70. Nopola-Hemmi J, Taipale M, Haltia T, Lehesjoki AE, Voutilainen A, Kere J. 2000. Two translocations of chromosome 15q associated with dyslexia. *J. Med. Genet.* 37:771–75
71. Olson RK, Forsberg H, Wise B. 1994. Genes, environment, and the development of orthographic skills. In *The Varieties of Orthographic Knowledge, 1: Theoretical and Developmental Issues*, ed. VW Berninger, pp. 27–71. Dordrecht, The Netherlands: Kluwer Academic
72. **Paracchini S, Thomas A, Castro S, Lai C, Murugan P, et al. 2006. The chromosome 6p22 haplotype associated with dyslexia reduces the expression of KIAA0319, a novel gene involved in neuronal migration. *Hum. Mol. Genet.* 15:1659–66**

72. Shows a link between a genetic background and a possible mechanism underlying dyslexia susceptibility for the first time.

73. Pennington BF. 1990. The genetics of dyslexia. *J. Child. Psychol. Psychiat.* 31:193–201
74. Pennington BF, Gilger JW, Pauls D, Smith SA, Smith SD, DeFries JC. 1991. Evidence for major gene transmission of developmental dyslexia. *JAMA* 266:1527–34
75. Pennington BF, Gilger JW. 1996. How is dyslexia transmitted? In *Developmental Dyslexia. Neural, Cognitive and Genetic Mechanisms*, ed. CH Chase, GD Rosen, GF Sherman, pp. 41–61. Baltimore, MD: York Press
76. Petryshen TL, Kaplan BJ, Liu MF, Field LL. 2000. Absence of significant linkage between phonological coding dyslexia and chromosome 6p23–21.3, as determined by use of quantitative-trait methods: confirmation of qualitative analyses. *Am. J. Hum. Genet.* 66:708–14
77. Rajagopalan S, Nicolas E, Vivancos V, Berger J, Dickson BJ. 2000. Crossing the midline: roles and regulation of Robo receptors. *Neuron* 28:767–77
78. Ramus F. 2003. Developmental dyslexia: specific phonological deficit or general sensorimotor dysfunction? *Curr. Opin. Neurobiol.* 13:212–18
79. Ramus F. 2006. Genes, brain, and cognition: a roadmap for the cognitive scientist. *Cognition* 101:247–69
80. Ramus F, Rosen S, Dakin SC, Day BL, Castellote JM, et al. 2003. Theories of developmental dyslexia: insights from a multiple case study of dyslexic adults. *Brain* 126:841–65
81. Scerri TS, Fisher SE, Francks C, MacPhie IL, Paracchini S, et al. 2004. Putative functional alleles of DYX1C1 are not associated with dyslexia susceptibility in a large sample of sibling pairs from the UK. *J. Med. Genet.* 41:853–57
82. Schulte-Korne G, Grimm T, Nothen MM, Muller-Myhsok B, Cichon S, et al. 1998. Evidence for linkage of spelling disability to chromosome 15. *Am. J. Hum. Genet.* 63:279–82
83. Schumacher J, Anthoni H, Dahdouh F, Konig IR, Hillmer AM, et al. 2006. Strong genetic evidence of DCDC2 as a susceptibility gene for dyslexia. *Am. J. Hum. Genet.* 78:52–62
84. Shaywitz SE. 1998. Dyslexia. *N. Engl. J. Med.* 338:307–12

85. Shaywitz SE, Escobar MD, Shaywitz BA, Fletcher JM, Makuch R. 1992. Evidence that dyslexia may represent the lower tail of a normal distribution of reading ability. *N. Engl. J. Med.* 326:145–50
86. Shaywitz SE, Shaywitz BA, Fletcher JM, Escobar MD. 1990. Prevalence of reading disability in boys and girls. Results of the Connecticut Longitudinal Study. *JAMA* 264:998–1002
87. Smith SD, Kimberling WJ, Pennington BF. 1991. Screening for multiple genes influencing dyslexia. *Read. Writ. Interdiscip. J.* 3:285–98
88. Smith SD, Kimberling WJ, Pennington BF, Lubs HA. 1983. Specific reading disability: identification of an inherited form through linkage analysis. *Science* 219:1345–47
89. Snowling MJ. 1981. Phonemic deficits in developmental dyslexia. *Psychol. Res.* 43:219–34
90. Stanovich KE. 1988. Explaining the differences between the dyslexic and the garden-variety poor reader: the phonological-core variable-difference model. *J. Learn Disabil.* 21:590–604
91. Stein CM, Schick JH, Gerry Taylor H, Shriberg LD, Millard C, et al. 2004. Pleiotropic effects of a chromosome 3 locus on speech-sound disorder and reading. *Am. J. Hum. Genet.* 74:283–97
92. Stein J, Talcott J, Witton C. 2001. The sensorimotorbasis of developmental dyslexia. In *Dyslexia: Theory and Good Practice*, ed. AJ Fawcett, pp. 65–88. London: Whurr
93. Stein J, Walsh V. 1997. To see but not to read: the magnocellular theory of dyslexia. *Trends Neurosci.* 20:147–52
94. Su AI, Wiltshire T, Batalov S, Lapp H, Ching KA, et al. 2004. A gene atlas of the mouse and human protein-encoding transcriptomes. *Proc. Natl. Acad. Sci. USA* 101:6062–67
95. Taipale M, Kaminen N, Nopola-Hemmi J, Haltia T, Myllyluoma B, et al. 2003. A candidate gene for developmental dyslexia encodes a nuclear tetratricopeptide repeat domain protein dynamically regulated in brain. *Proc. Natl. Acad. Sci. USA* 100:11553–58

95. *DYX1C1* is the first dyslexia susceptibility gene to be reported. These findings remain controversial due to the inconsistent results obtained in follow-up studies.

96. Tallal P. 1980. Auditory temporal perception, phonics, and reading disabilities in children. *Brain Lang.* 9:182–98
97. Thapar A, O'Donovan M, Owen MJ. 2005. The genetics of attention deficit hyperactivity disorder. *Hum. Mol. Genet.* 14(Spec. No. 2): R275–82
98. Thomson ME. 1982. The assessment of children with specific reading difficulties (dyslexia) using the British Ability Scales. *Br. J. Psychol.* 73:461–78
99. Turic D, Robinson L, Duke M, Morris DW, Webb V, et al. 2003. Linkage disequilibrium mapping provides further evidence of a gene for reading disability on chromosome 6p21.3–22. *Mol. Psychiat.* 8:176–85
100. Turkeltaub PE, Flowers DL, Verbalis A, Miranda M, Gareau L, Eden GF. 2004. The neural basis of hyperlexic reading: an FMRI case study. *Neuron* 41:11–25
101. Vogel SA. 1990. Gender differences in intelligence, language, visual-motor abilities, and academic achievement in students with learning disabilities: a review of the literature. *J. Learn. Disabil.* 23:44–52
102. Vogler GP, DeFries JC, Decker SN. 1985. Family history as an indicator of risk for reading disability. *J. Learn. Disabil.* 18:419–21
103. Wang Y, Paramasivam M, Thomas A, Bai J, Kaminen-Ahola N, et al. 2006. DYX1C1 functions in neuronal migration in developing neocortex. *Neuroscience.* 143:515–22
104. Ward M, McCann C, DeWulf M, Wu JY, Rao Y. 2003. Distinguishing between directional guidance and motility regulation in neuronal migration. *J. Neurosci.* 23:5170–77
105. Wigg KG, Couto JM, Feng Y, Anderson B, Cate-Carter TD, et al. 2004. Support for EKN1 as the susceptibility locus for dyslexia on 15q21. *Mol. Psychiat.* 9:1111–21

106. Wigg KG, Couto JM, Feng Y, Crosbie J, Anderson B, et al. 2005. Investigation of the relationship of attention deficit hyperactivity disorder to the EKN1 gene on chromosome 15q21. *Scient. Stud. Read.* 9:261–83
107. Willcutt EG, Pennington BF, Olson RK, Chhabildas N, Hulslander J. 2005. Neuropsychological analyses of comorbidity between reading disability and attention deficit hyperactivity disorder: in search of the common deficit. *Dev. Neuropsychol.* 27:35–78
108. Williams J, O'Donovan MC. 2006. The genetics of developmental dyslexia. *Eur. J. Hum. Genet.* 14:681–89
109. Yan H, Yuan W, Velculescu VE, Vogelstein B, Kinzler KW. 2002. Allelic variation in human gene expression. *Science* 297:1143
110. Ylisaukko-Oja T, Peyrard-Janvid M, Lindgren CM, Rehnstrom K, Vanhala R, et al. 2005. Family-based association study of DYX1C1 variants in autism. *Eur. J. Hum. Genet.* 13:127–30

RELATED RESOURCES

Eckert M. 2004. Neuroanatomical markers for dyslexia: a review of dyslexia structural imaging studies. *Neuroscientist* 10:362–71

Nadarajah B, Parnavelas JG. 2002. Modes of neuronal migration in the developing cerebral cortex. *Nat. Rev. Neurosci.* 3:423–32

Newbury DF, Bishop DV, Monaco AP. 2005. Genetic influences on language impairment and phonological short-term memory. *Trends Cogn. Sci.* 9:528–34

Pennington BF. 2006. From single to multiple deficit models of developmental disorders. *Cognition* 101:385–413

Ramus F. 2004. Neurobiology of dyslexia: a reinterpretation of the data. *Trends Neurosci.* 27:720–26

National Center for Biotechnology Information. *Gene Expression Omnibus.* **http://www.ncbi.nlm.nih.gov/projects/geo/**

NeuroDys. *European Dyslexia Research.* **http://www.neurodys.com/**

Applications of RNA Interference in Mammalian Systems*

Scott E. Martin and Natasha J. Caplen

Gene Silencing Section, Genetics Branch, Center for Cancer Research, National Cancer Institute, National Institutes of Health, Bethesda, Maryland 20892; email: martinsc@mail.nih.gov, ncaplen@mail.nih.gov

Annu. Rev. Genomics Hum. Genet. 2007. 8:81–108

First published online as a Review in Advance on May 3, 2007.

The *Annual Review of Genomics and Human Genetics* is online at genom.annualreviews.org

This article's doi:
10.1146/annurev.genom.8.080706.092424

1527-8204/07/0922-0081$20.00

*The U.S. Government has the right to retain a nonexclusive, royalty-free license in and to any copyright covering this paper.

Key Words

siRNA, shRNA, screen, gene expression, gene function, gene silencing, RNAi

Abstract

RNA interference (RNAi) can mediate the long- or short-term silencing of gene expression at the DNA, RNA, and/or protein level. Although several triggers of RNAi have been identified, the best characterized of these are small interfering RNAs (siRNAs), which can decrease gene expression through mRNA transcript cleavage, and endogenous microRNAs (miRNAs), which primarily inhibit protein translation. An improved understanding of RNAi has provided new, powerful tools for conducting functional studies in a gene-specific manner. In various applications, RNAi has been used to create model systems, to identify novel molecular targets, to study gene function in a genome-wide fashion, and to create new avenues for clinical therapeutics. Here, we review many of the ongoing applications of RNAi in mammalian and human systems, and discuss how advances in our knowledge of the RNAi machinery have enhanced the use of these technologies.

INTRODUCTION

RNA interference (RNAi): a gene-silencing mechanism induced by double-stranded RNA; can silence gene expression at both a transcriptional and post-transcriptional level

TGS: transcriptional gene silencing

RNA-induced silencing complex (RISC): central component of the RNAi pathway. RNAi effectors guide RISC to nucleic acids with at least partial complementarity

microRNAs (miRNAs): endogenous RNA species associated with RNAi that are generated through the processing of hairpin-containing transcripts

The awarding of the 2006 Nobel Prize in medicine for the discovery of RNA interference (RNAi) (**http://nobelprize.org/nobel_prizes/medicine/laureates/2006/**) clearly illustrates the importance of this field. First identified in *Caenorhabditis elegans* (31), and subsequently found to explain previous observations in plants and fungi (for reviews see 97, 141), RNAi is a gene-silencing mechanism that is induced by double-stranded RNA (dsRNA). Since its initial discovery, RNAi has been established in a number of additional organisms including *Drosophila melanogaster* (60, 90) and mammalian cells (15, 27). An increasingly diverse set of biological processes has been associated with RNAi. These processes include transcriptional gene silencing (TGS), antiviral responses, and, most relevant for this discussion, the post-transcriptional regulation of endogenous gene expression. Methods that exploit RNAi have been rapidly developed. Although primarily used for gene function analysis, ranging in scale from the individual gene to genome-wide, RNAi-based technologies have been used in a number of contexts including clinical applications.

The initial descriptions of RNAi focused on the post-transcriptional suppression of target genes mediated by the introduction of homologous dsRNA [over $\approx$100 nucleotides (nts)] into model organisms. Subsequently, these dsRNAs were found to be processed into smaller 21–23 nt dsRNAs, termed small interfering RNAs (siRNAs), with 3′ dinucleotide overhangs generated by the RNase III endoribonuclease Dicer (4, 28). siRNAs were found to be active independent of processing from larger dsRNAs (28). As the immune response precludes the use of long dsRNAs in mammalian cells, it was not until this discovery that RNAi could be identified in these systems (15, 27).

siRNAs direct the cleavage of targeted mRNAs. Cleavage is mediated by a single strand of the siRNA duplex, termed the guide strand, after incorporation into a ribonucleoprotein complex known as the RNA-induced silencing complex (RISC). RISC contains Argonaute proteins. This family of proteins is highly diverse, but all members are characterized by the presence of two domains, the Piwi-Argonaute-Zwille (PAZ) and PIWI domains (for review see 106). The PAZ domain specifically recognizes the characteristic 3′ termini of processed effectors and the PIWI domain adopts an RNase H-like structure that can catalyze the enzymatic cleavage of RNA. There are eight known Argonaute proteins in humans, but of these only Argonaute 2 (Ago2) has been found to generate cleavage-competent RISC (77). In addition to target cleavage, Ago2 is also responsible for guide strand selection. This occurs through the asymmetric unwinding of duplex RNAs, whereby the guide strand is preferentially retained within RISC and the other strand, termed the passenger strand, is degraded (85, 112).

Although the introduction of exogenous siRNAs results in the RISC-dependent cleavage of target transcripts, the documented occurrence of endogenous cleavage complexes is not common in mammalian cells. Rather, it is another species of small RNA, termed microRNAs (miRNAs), that uses the innate RNAi machinery. (For a detailed discussion and key references related to miRNAs, see the Mendell review in this volume.) miRNAs interact with transcripts possessing partial complementarity, primarily within target 3′ untranslated regions (UTRs). miRNAs were originally identified as a species of small RNA ($\approx$22 nt) that regulates genes required for development in the nematode *C. elegans*. Known as small temporal RNAs, these were the first examples of a large number of small endogenous RNAs that can regulate gene expression (see **http://microrna.sanger.ac.uk/** for a database of all miRNAs). miRNAs are generated through the processing of genomically encoded primary miRNA transcripts (pri-miRNAs) by a multisubunit complex that at its core consists of the RNase III endoribonuclease Drosha and, in mammalian cells, the DGCR8 protein. The processing of primary miRNA transcripts yields hairpin structures known as precursor miRNAs (pre-miRNAs). Following export to the cytoplasm via Exportin5 (Exp5), pre-miRNAs are processed by Dicer to produce mature miRNAs that incorporate into miRNA ribonucleoprotein complexes (miRNPs). These complexes are similar, if not identical, to RISC. Early studies suggested that mismatches between miRNAs and target transcripts resulted in translational repression without corresponding loss of mRNA. However, more recent studies have shown that miRNAs can induce reductions in target mRNA levels. Furthermore, miRNAs have been shown to colocalize with their target transcripts in sites of mRNA degradation known as cytoplasmic processing bodies, or P-bodies, where degradation through a process of cap removal followed by 5′–3′ exonuclease activity may occur. It has been estimated that miRNAs may regulate at least 30% of protein-encoding genes. Consequently, miRNAs are emerging as a fundamental regulatory unit in human biology. They have been linked to a number of important processes including development, differentiation, and even cancer. Understanding the role of miRNAs in human disease will be a major area of study in the coming years and may enable the development of new therapeutic approaches (see Sidebar: microRNAs).

Although originally thought to be restricted to post-transcriptional gene silencing within the cytoplasm of mammalian cells, RNAi has since been found to occur within the nucleus and to regulate a number of additional processes (**Figure 1**). For example, nuclear-localized RNAs can be degraded through RNAi (116). Furthermore, RNAi can induce TGS. Studies have shown that RNAi-mediated TGS can occur through DNA methylation and/or histone modification through the targeting of promoter regions (17, 58, 92, 137, 146). Others have found that TGS can occur independent of these DNA modifications (54). RNAi can also regulate selfish genetic elements. For example, endogenous siRNAs generated from within the 5′ UTR of the LINE-1 (L1) element can regulate L1 retrotransposons. Also consistent with a role for RNAi in the regulation of selfish genetic elements, *Dicer* knockout murine embryos exhibit an increased abundance of certain endogenous retroviruses (for review see 47).

In addition to miRNAs and siRNAs, other innate RNAi effectors have been identified. One class of these is the Piwi-interacting RNAs (piRNAs). piRNAs seem to be uniquely expressed in the mammalian germline, particularly in the testes (1, 34, 38, 145). The functional role of piRNAs is currently unclear, but a role in spermatogenesis is likely. Furthermore, a number of other small RNAs associated with RNAi have been identified in different species, including trans-activating siRNAs (tasiRNAs), studied in plants and nematodes, and small scan RNAs (ScnRNAs), found in *Tetrahymena* (for review see 62). To date, it

microRNAs

The analysis of microRNA (miRNA) expression profiles has indicated a relationship between miRNA expression and various biological processes including cancer (for review see 13). For example, the overexpression of hsa-miR-155 has been reported in Hodgkin's and some Burkitt's lymphomas (64, 65, 88) and is also associated with poor outcome for patients with lung cancer (152). Conversely, the downregulation of two genomically adjacent miRNAs, hsa-miR-15 and hsa-miR-16, has been reported in chronic lymphocytic leukemia (14). In addition to these types of correlations, a number of studies have identified the potential for miRNAs to act as oncogenes or tumor suppressors. For example, the expression of hsa-miR-155 can induce splegnomelema in transgenic mice (21). Additionally, the expression of miR-15 and miR-16 can induce apoptosis in leukemia cells (18). Notably, single-stranded oligoribonucleotides with 2′O-methyl modifications can inhibit the activity of complementary miRNAs (50). Thus, the use of these inhibitors has been suggested as a possible therapeutic approach.

Exp5: Exportin5 protein

UTR: untranslated region

RNAi effector: a small RNA that can direct RNAi against complementary targets (e.g., siRNAs or shRNAs)

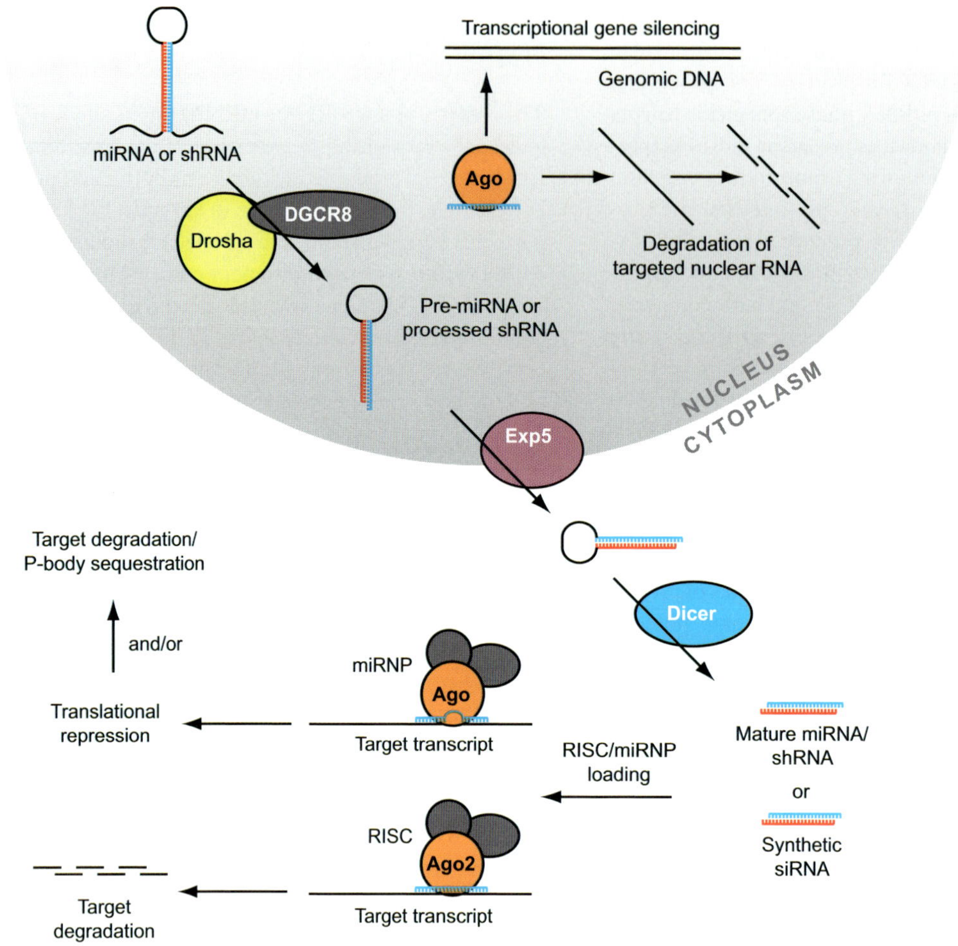

Figure 1

Simplified schematic of the RNAi pathway in mammalian cells. Only processes mentioned in the text are illustrated.

is unclear whether these types of small noncoding RNAs are also present in mammalian systems. However, it is apparent that RNAi-associated small RNAs can have an enormous effect on a variety of biological processes (see **Figure 1** for an overview of RNAi in mammalian cells).

INDUCING RNAi IN MAMMALIAN CELLS

In most cases, the aim of RNAi-based experiments is the sequence-dependent cleavage and reduction of protein-encoding mRNAs. Although most studies have focused on the RNAi analysis of these targets, any RNA species can be targeted (for example, noncoding RNA transcripts or viral RNAs). Only a limited number of mammalian cell types can tolerate RNAi induced by large, exogenous dsRNAs (e.g., embryonic stem cells). Thus, it is usually necessary to use one of two broad categories of RNAi effector molecules in mammalian systems. These include siRNA duplexes, formed through the annealing of two independent RNA strands, or single-stranded RNA molecules that contain a dsRNA domain, termed short-hairpin RNAs (shRNAs). In both cases, RNAi effectors are designed to possess full complementarity with target transcripts, thereby resulting in their cleavage.

RNAi Effectors Used for Biological Analysis in Mammalian Cells

siRNAs can be generated through the annealing of synthetic oligonucleotides. Most synthetic siRNAs consist of 19 perfectly matched complementary ribonucleotides and 3′ dinucleotide overhangs that, for ease of synthesis, often consist of deoxyribonucleotides. Synthetic siRNAs are available from a number of commercial vendors. More rarely, siRNAs are generated by a number of other methods including in vitro transcription, plasmid-based tandem or convergent expression cassettes, polymerase chain reaction (PCR) or the endonuclease digestion of large dsRNAs that produce pools of siRNAs. The introduction of synthetic siRNAs into cultured mammalian cells usually uses standard physico-chemical transfection methods, such as those based on cationic lipids, cationic polymers, or electroporation. Empirical testing is required to determine the most efficacious transfection conditions for any given cell system. Once well-optimized, transfected siRNAs can yield a substantial decrease in the steady-state levels of target mRNAs for ~24–120 h.

As opposed to direct transfection, shRNAs are usually expressed from plasmids or viral-based expression vectors. shRNAs are designed to mimic miRNA precursors. Consequently, they are processed by the endogenous RNAi machinery and loaded into RISC. A number of different shRNA expression systems have been described. Variations include differences in promoter-terminator combinations, linker sequences, flanking sequences, duplex length, and regulatory elements that can be used for spatial and/or temporal-specific expression. Additionally, selection markers, used to generate stable cell lines, and unique sequence elements, used to identify active shRNAs among larger populations (discussed below), have been employed. As opposed to siRNAs, the stable expression of shRNAs allows for a nontransient reduction of targeted mRNAs. Thus, the choice of RNAi effector (siRNA or shRNA) depends on the question under investigation. No matter the choice, it is always important to include negative control effector molecules in any RNAi-based experiment (reasons discussed in detail below). These controls, which are commercially available, incorporate sequences with minimal complementarity to any endogenous transcript.

small interfering RNAs (siRNAs): annealed RNA duplexes of ~21 nucleotides in length that can direct RNAi; generally synthetic in origin

short-hairpin RNAs (shRNAs): hairpin-containing transcripts designed to mimic miRNA precursors; intracellularly processed shRNAs can direct RNAi

Maximizing Gene Silencing by RNAi

In addition to improving conditions for their cellular introduction and/or expression and subsequent processing, many studies have been directed toward maximizing activity as a function of effector sequence (61, 121). For example, an understanding of any bias related to guide strand selection has obvious implications for design. Analysis has revealed that the strand most easily unwound from its 5′ end is preferentially incorporated into RISC (61, 115). Thus, effector design incorporates such bias to encourage selection of intended guide strands. Studies have indicated other positional biases. For example, high thermodynamic stability is preferred between nucleotides 5–10 of the guide strand (61, 115). Furthermore, empirical comparisons between large sets of effective and ineffective siRNAs have led to the development of algorithms that assist in the generation of active siRNAs. These types of design tools are incorporated into the production of commercial siRNAs and are also publicly available. Sequences generated by these tools merely have an increased probability of mediating RNAi. Only experimentation will establish the activity of any given RNAi effector. Of note, an increasing number of validated sequences are available from commercial sources and are being characterized and collated by the scientific community (**http://www.ncbi.nlm.nih.gov/projects/genome/RNAi/**).

off-target effects: unintended effects arising through the application of RNAi; can occur in a sequence-dependent or -independent manner

Considerations of the target are also important for maximizing RNAi. As RNAi effectors are designed according to reported reference sequences, any discrepancies between those and the actual target sequences within systems under study, for example as a consequence of single nucleotide polymorphisms (SNPs), may prevent efficient RNAi (84a). However, the influence of sequence discrepancies may be less than predicted owing to the fact that RISC can sometimes tolerate mismatches within targets, especially those distal from the cleavage site (26, 84). Despite possibly interfering with RNAi, sequence aberrations can potentially be used to selectively target mutated transcripts associated with disease. This approach has been applied in a number of contexts including the targeting of cancer-specific mutations, the targeting of a single-base mutation associated with the dominant genetic disorder spinocerebellar ataxia, and, most recently, for the silencing of mutant β-globin as an approach toward treating sickle cell anemia (9, 26, 89). In addition to potential sequence discrepancies, one should also ensure that RNAi effectors target all known transcript variants of genes under study (84a). Inevitably, even with an increased understanding of RNAi, effectors invariably exhibit a spectrum of activity. Thus, it may be prudent to obtain more than one effector against targets under investigation.

OFF-TARGET EFFECTS ASSOCIATED WITH RNAi

The ability of RNAi effectors to elicit specific downregulation of intended targets while minimizing or controlling for unintended effects, termed off-target effects, is critical for the meaningful application of RNAi. Off-target effects are known to arise from a variety of mechanisms, which include both sequence-independent and sequence-dependent processes. Sequence-independent effects, or nonspecific effects, generally involve those relating to transfection conditions (e.g., lipid transfection reagents), inhibition of endogenous miRNA activity, or stimulation of pathways associated with the immune response. Sequence-dependent effects primarily concern the unintentional silencing of targets sharing partial complementarity with RNAi effector molecules through miRNA-like interactions, but also include receptor-mediated immune stimulation through the recognition of certain nucleotide motifs. As discussed below, there are a number of approaches toward controlling for both types of off-target effects (summarized in **Table 1**).

Nonspecific Effects: Disruption of Endogenous RNAi

Nonspecific effects resulting from the inhibition of endogenous miRNA activity appear to depend on saturation of Exp5 (37, 154).

Table 1 List of off-target effects encountered in mammalian systems and steps to minimize their impact

Off-target effect	Type	Steps to mitigate
Saturation of the endogenous RNAi machinery	Sequence-independent	Use effectors at lowest possible concentration
		Use negative control effectors for comparison
Immune response	Sequence-independent	Use effectors at lowest possible concentration
Immune response	Sequence-dependent	Avoid known stimulatory motifs
		Use chemically modified effectors
		Use multiple effectors to confirm phenotypes
Silencing of unintended targets through partial complementarity	Sequence-dependent	Use effectors at lowest possible concentration
		Use multiple effectors to confirm phenotypes

For example, the shRNA-mediated inhibition of miRNA activity is mitigated by the overexpression of Exp5 (37, 154). Also consistent with Exp5 as a saturatable component of RNAi, its overexpression, but not the overexpression of other RNAi components, enhances the activity of both miRNAs and shRNAs (37, 154). As siRNAs do not require export from the nucleus, their activity would not be expected to depend on Exp5. However, some studies have found that siRNA-mediated RNAi is dependent on Exp5, where Exp5 prevents entry, and subsequent dilution, of siRNAs into non-nucleolar areas of the nucleus (102). This is still controversial, as other studies have found no relationship between Exp5 and siRNA activity (37, 154). At the very least, it is clear that the activity of endogenous miRNAs can be disrupted by the overexpression of shRNAs. The consequences of this not only manifest in cell culture, but also in vivo, as Grimm and colleagues (37) have shown that a high percentage of shRNAs can cause lethality in mice regardless of shRNA target, or even the presence of a target. Moreover, this toxicity correlated with high shRNA expression (37). Findings that RNAi effectors can saturate the endogenous machinery emphasize the importance of using RNAi effectors at the lowest possible effective concentrations. Additionally, the use of negative control siRNAs or shRNAs is paramount for the proper interpretation of results. In the case of shRNAs, it does not seem adequate to simply use an empty vector control, as this does not control for shRNA-mediated inhibition of the endogenous miRNA machinery.

Nonspecific Effects: The Immune Response

Despite early perceptions that siRNAs of less than 30 nts would avoid the immunostimulatory activity exhibited by larger RNA molecules (15, 27), it has since been shown that siRNAs can activate the immune response in a sequence-independent, concentration-dependent manner (57, 109, 114, 128). For example, 21-nt siRNAs have been shown to induce an interferon response in human glioblastoma T98G cells through a process dependent on the activation of the dsRNA-dependent protein kinase (PKR), and at least partially dependent on siRNA concentration (128). Similarly, both externally delivered siRNAs and shRNAs were found to induce an interferon response in HEK293 and HaCaT keratinocyte cell lines (57). Additional studies in HEK293 cells found this response to be primarily dependent on Toll-like receptor 3 (TLR3) (57). Importantly, the induction of an interferon response is cell-line dependent, with long siRNAs of 27 nt unable to activate a response in certain cell lines, including HeLa cells (114). The expression of shRNAs can also induce an interferon response (8). As with siRNAs, shRNA-mediated activation also appears to be concentration dependent (8). Thus, similar to saturation of the endogenous RNAi machinery, the use of lowest effective concentrations and negative control RNAi effectors are necessary to control for stimulation of interferon-type responses. Additionally, chemical modifications that help prevent the activation of PKR have been described (110).

Sequence-Dependent Effects: The Immune Response

siRNAs can also induce an immune response through sequence-dependent effects, particularly when it is part of a lipid or polycation complex in vivo. More specifically, certain nucleotide motifs, especially GU-rich sequences, can induce interferon-α (IFN-α), interleukin-6 (IL-6), and tumor necrosis factor-α (TNF-α), probably through activation of TLRs. For example, a subset of liposome-encapsulated siRNAs was found to induce a substantial, dose-dependent IFN-α response in mice (56). These siRNAs also stimulated an immune response in human peripheral blood mononuclear cells and isolated plasmacytoid dendritic cells. The stimulatory siRNAs were found to share

UGUGU motifs that were presumably recognized by endosomal TLR7 and/or TLR8 (56). Similarly, siRNAs were found to stimulate IFN-α production in human plasmacytoid dendritic cells through a GUCCUUCAA motif (48). In this case, experiments confirmed that stimulation was dependent on recognition by TLR7. Because nonimmune cells do not express detectable TLR7 or TLR8, sequence-dependent immune stimulation is not thought to influence experiments conducted in commonly used cell lines (83), but sequence-dependent stimulation is clearly an important issue regarding in vivo applications.

Sequence-Dependent Effects: Shared Complementarity Between Effectors and Nontargeted mRNAs

Although specific siRNA nucleotide motifs can activate an immune response, the primary source of sequence-dependent off-target effects originates from partial complementarity between RNAi effectors and off-target transcripts. Such interactions are similar to those exhibited by endogenous miRNAs, which usually share complementarity between nucleotides within their 5′ ends and regions within target 3′ UTRs (70). In fact, much like miRNA targets, off-targeted transcripts are enriched in those containing complementarity between their 3′ UTRs and hexamer (nts 2–7) and heptamer (nts 2–8) sequences within 5′ ends of RNAi effectors (5, 52). Some studies have found these effects to be nontitratable, with dose responses mirroring that of on-target transcripts (51). Others have found these effects to be concentration-dependent, whereby the use of low siRNA concentrations can significantly mitigate off-target interactions (122). Importantly, most detailed studies of off-target effects are conducted using gene expression analysis. However, since miRNAs can impede translation in a manner disproportionate with alterations in target mRNA levels (24, 25), the magnitude of off-target effects may be underestimated.

Sequence-dependent off-target effects can have functional consequences. For example, different siRNAs targeting the same gene can exhibit varying effects on the mRNA and protein levels of key cellular genes, independent of on-target silencing (120). Accordingly, a high percentage of siRNAs can induce a toxic phenotype. For example, 51 of 176 randomly selected siRNAs directed against either firefly luciferase or human *DBI* reduced the viability of HeLa cells by more than 25%, a trend that was reproducible in different cell lines (30). From a practical perspective, off-target effects can have a profound effect on experimental results. For example, Lin and colleagues determined that the top three "hits" from a siRNA-based screen for targets affecting the hypoxia-related HIF-1 pathway resulted from off-target effects (76). For two of these three "hits," activity could be traced to interactions within the 3′ UTR of *HIF-1A* itself. Of note, off-target effects not only affect experiments conducted in mammalian systems, but can also influence studies in *Drosophila* (68, 81).

There are a number of ways to control for, and help minimize, sequence-dependent off-target effects. Many of these relate to siRNA design features. For example, the use of asymmetric design, which helps to minimize the loading of passenger strands into RISC, thereby reducing associated off-target effects, and the use of siRNAs designed to avoid homology with untargeted transcripts. Both of these considerations are typically incorporated into the design of commercially available siRNAs. Increased stringency may be gained through the development of new algorithms that include emphasis on avoiding complementarity between siRNAs and untargeted 3′ UTRs (5). Chemical modifications that reduce sequence-dependent off-target effects have also been described. For example, the incorporation of 2′-O-methyl groups within the first two 5′ nucleotides of siRNA passenger strands reduces passenger strand-mediated activity (52). Similarly, a 2′-O-methyl ribosyl substitution at position 2 of the guide strand can significantly reduce sequence-dependent off-target effects (52). Modified siRNAs exhibiting reduced off-target effects are commercially available. As with sequence-independent effects, experimental conditions should be optimized to use the lowest effective dose of the RNAi effector.

Despite all of these considerations, the occurrence of sequence-dependent off-target effects may be unavoidable. Consequently, efforts should be made to help validate hits. All RNAi-derived phenotypes should be confirmed with additional RNAi effectors against the same target. Moreover, the downregulation of target mRNA and protein levels should be characterized and correlated with the observed effects. For example, the inactivity of a follow-up siRNA does not necessarily imply that the activity of the first resulted from off-target effects, especially if the second siRNA is unable to downregulate target levels. Conversely, a phenotype induced by only a fraction of siRNAs directed against the same target, despite equivalent silencing by all siRNAs, would be suspicious. Overall, it is difficult to prescribe the number of independent RNAi effectors necessary for target validation, but it would certainly require at least two. No matter how many independent RNAi effectors are tested, it could be possible that observed phenotypes result from cooperative effects between target-specific downregulation and nonspecific effects. Even a rescue experiment using a target construct resistant to RNAi could not control for such scenarios. Thus, confirmation of phenotypes under different experimental conditions (e.g., the use of a different lipid reagent or the use of an siRNA to confirm an shRNA-derived phenotype) may help to eliminate some of these possibilities. Additionally, RNAi-independent methods, such as the chemical inhibition of identified targets, should be used to corroborate phenotypes where possible.

STUDIES USING RNAi

The types of studies using siRNAs or shRNAs have been highly diverse and are being performed on an increasingly routine basis. Most reports describe the knockdown of one or a limited number of genes directly relevant to the biological question under investigation. Other, slightly broader studies have used RNAi to identify causal genes within disease-associated genomic regions (74). Analogously, RNAi is being used to validate statistical relationships between gene expression and phenotypes and to investigate compound modes of action. For example, RNAi has been used to identify both biomarkers associated with and genes responsible for compound activity (78, 79). However, in contrast to these focused applications, it is the development of large-scale RNAi analysis that has the greatest potential to impact the discovery of novel gene function.

RNAi-Based Screens

The generation of whole genome sequences for human and an increasing number of model organisms has presented biologists with an extremely valuable resource. To use this information fully, it is essential to develop complementary analytical approaches that can be performed on a similar scale. Studies of DNA copy number by comparative genomic hybridization, mRNA expression by microarray, and the profiling of the proteome by various methods have all been adapted for large-scale analysis. Now the use of RNAi-based methods is enabling corresponding functional analysis to be conducted on a similar scale. Genome-wide RNAi analysis was first developed for the model organisms *C. elegans* (see **http://www.wormbase.org/db/searches/rnai_search** to search for *C. elegans* RNAi phenotypes) and *Drosophila* (see **http://flyrnai.org/cgi-bin/RNAi_screens.pl** for details of a number of large-scale screens). Recently, analogous RNAi screens were described in human cells.

Several different strategies for large-scale RNAi-based screening in mammalian cells have been developed (see **Figure 2** for an overview). Perhaps the most straightforward

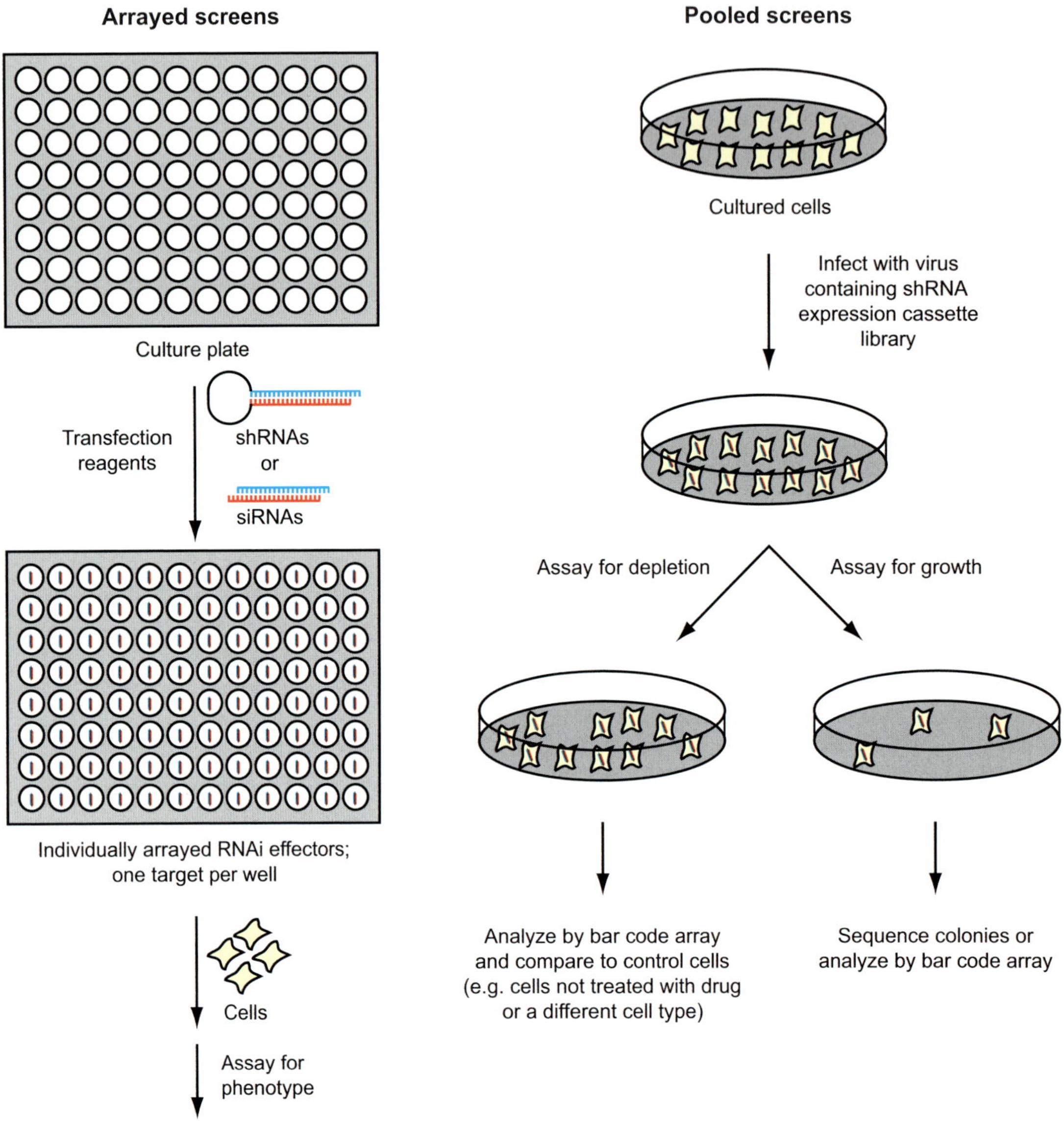

Figure 2

The different formats for conducting RNAi-based screens. Arrayed screens (*left*) are usually conducted in standard tissue culture plates where one gene target is assayed per well. This type of format is used for both small interfering RNA (siRNA)- and short hairpin RNA (shRNA)-based screens. As illustrated, RNAi effectors can be precomplexed with transfection reagents followed by the addition of cells. In most screens, effectors are evaluated in at least duplicate and usually a number of sequence-independent effectors are evaluated per target. Pooled screens (*right*) are usually conducted by infecting cells with an shRNA library in a "one pot" format. Active shRNAs are identified through sequence analysis or by bar code array.

involves the use of arrayed synthetic siRNA libraries. These libraries are available from a number of commercial vendors and in a variety of formats, ranging from a genome-wide scale to smaller customized subsets. A number of studies using these types of resources have been described. Some of these involve the elegant integration of siRNA screens with image-based high content analysis. For example, a screen of siRNAs directed against the human kinome identified regulators of clathrin and caveolae-mediated endocytosis by imaging the internalization of vesicular stomatis virus and simian virus 40, respectively (107). Similarly, a more focused screen, restricted to known endocytosis-related genes, identified those associated with endocytosis of the dopamine transporter (130). Recently, a fully automated process was used for screening siRNAs with a time-lapse imaging system in human cells (98).

Synthetic siRNA-based screens are also used to identify novel molecular targets, both in the context of stand-alone targets and those augmenting the activity of existing chemotherapeutics. For example, a screen of ~5000 genes found several associated with migration in highly mobile SKOV-3 cells (20). The activity of one identified target, MAP4K4, was mediated through the c-Jun N-terminal kinase, leading to the rational use of a c-Jun N-terminal kinase small molecule inhibitor to mitigate SKOV-3 motility (20). Additionally, a screen of human kinases and phosphatases identified those contributing to the circumvention of apoptosis in cancer cells (82). Several novel regulators were identified, including PINK1 kinase, whose inhibition was shown to sensitize HeLa and BT474 cells to taxol (82). Analogously, a screen identified targets that enhanced TRAIL-induced apoptosis in HeLa cells (2). This type of strategy provides an important avenue toward identifying rational drug combinations, which may result in potent therapies that avoid high doses, and associated side effects, of any one agent.

siRNA-based screens are usually conducted in standard plate-based formats (e.g., 96 or 384 well plates). However, the potential adaptation of RNAi screening to a microarray format has been explored. Several "proof of principle studies" have been conducted, whereby small RNAs are precomplexed with a carrier and arrayed on slides to enable the localized transfection of overlaid cells (29, 95, 127). Standard image-based systems can be used to assess resulting phenotypes. One potential advantage of this approach is that it may facilitate the generation of a more standardized format for use by the scientific community, thereby enabling more extensive cross-comparison of data.

Another major approach to large-scale RNAi-based screens is the use of shRNA libraries. A number of these systems have been described (3, 91, 99, 104, 126). Generally, these systems incorporate shRNAs into selectable viral vector plasmid backbones, with expression driven by RNA polymerase III promoters. Both retroviral and lentiviral systems have been described. Lentiviral systems are advantageous for certain applications, as they are able to infect primary and nondividing cells. Most shRNAs contained within these constructs are based on hairpin structures intended to mimic pre-miRNAs. Furthermore, they incorporate many of the same design rules as those used for synthetic siRNAs. Recently, a library based on the mimicry of primary, rather than precursor, miRNAs, was described (126). Such shRNAs generate upward of 12 times more mature small RNAs than those based on the mimicry of pre-miRNAs (126).

shRNA libraries have been screened in both arrayed and pooled formats (**Figure 2**). Inherently, a pooled format has a significant throughput advantage. However, pooling requires the ability to subsequently identify active shRNAs within treated cell populations. In some cases, sequence analysis can be used. This is especially possible when screening for selectable phenotypes that yield a growth advantage, resulting in the generation

bar code: the shRNA sequence or a unique sequence incorporated into shRNA expression cassettes that is used to identify individual shRNAs within pooled screens of positive clones. For example, a screen using a system engineered to exhibit innate p53-dependent growth arrest identified five novel inhibitors of this process (3). Similarly, novel tumor suppressors were identified using shRNAs that transformed human primary cells (66). Despite the successful use of this strategy, there are a number of limitations. For example, using this type of approach it is difficult to screen for shRNAs that inhibit growth, thereby leading to the depletion of active shRNAs from the larger, inactive population. Furthermore, even in systems engineered to assay for growth advantages, high hit rates may make sequencing-based identification impractical. To remedy these issues, many shRNA expression systems employ an identification strategy based on the use of bar codes. Bar codes are unique nucleotide sequences that are used to help identify enriched or depleted shRNAs within treated cell populations. Bar codes can be assayed by DNA microarray analysis. Different bar code strategies have been employed. In one approach, the shRNA sequences are used as bar code identifiers (3). In a different approach, unique 60-nt bar codes are incorporated into shRNA expression cassettes (99, 104, 126). Several studies have demonstrated the utility of bar codes. In an elegant example, this strategy was used to identify molecular targets unique to a subset of diffuse large B-cell lymphoma (99). This was accomplished by using an inducible shRNA expression system, whereby bar code analysis of an induced versus noninduced cell population enabled the identification of shRNAs exclusively depleted from the induced set. As with siRNA-based screens, bar code screens can be used to probe the mode of action of anticancer agents. For example, a bar code strategy found that DNA-damage induced signaling enhances the activity of a small molecule MDM2 inhibitor, suggesting that optimal efficacy may be observed in cells with wild-type p53 and activated DNA-damage signaling (10). The successful application of pooled shRNA screens is promising, as it may make large-scale RNAi screens more accessible to laboratories lacking the extensive liquid-handling capabilities required for large arrayed screens.

In addition to synthetic siRNA and shRNA libraries, the generation of RNAi libraries from cDNAs has also been described. For example, *Escherichia coli* RNase III has been used to generate siRNAs from cDNA-derived dsRNA (63, 153). The screening of a library prepared by this strategy was able to identify novel regulators of cell division in HeLa cells (63). In another cDNA-based approach, expression systems are produced through a process partially dependent on the use of *Mme*1 restriction enzyme (80, 123, 125). This enzyme cleaves 20 nt from its recognition site, making it very amenable to the generation of siRNAs. As all of these systems are based on random generation from cDNAs, the generated effectors do not incorporate rational design features, possibly resulting in a relatively higher frequency of deleterious events, such as off-target effects. Despite the development of cDNA-based strategies, the uses of both siRNA and shRNA libraries are clearly the more abundant methodologies employed in large-scale screens.

Regardless of strategy, a key promise of large-scale RNAi functional analysis is its potential impact on molecular target discovery and drug development. This is well illustrated by an shRNA-based screen of deubiquitinating enzymes that identified cylindromatosis (CYLD) as a novel suppressor of NF-κB activation (11). Inactivating mutations in CYLD are linked to a rare genetic disorder known as familial cylindromatosis, which predisposes individuals to the development of certain skin tumors. As the activation of NF-κB is antiapoptotic, it was hypothesized that loss of CYLD in cylindromatosis results in aberrant growth due to inhibition of apoptosis. Consequently, the inhibition of NF-κB was hypothesized as a potential treatment for this disorder (11). The clinical impact of this finding was quickly realized with the performance of a trial assessing the topical application of the NF-κB inhibitor salicylic acid (103). Of 12 treated lesions, 2 lesions showed complete remission whereas another 8 exhibited some response to treatment (103). Future RNAi-based studies will hopefully be able to link more diseases with existing therapeutics.

THE IN VIVO APPLICATION OF RNAi-BASED TECHNOLOGIES

RNAi has enormous potential for the treatment of many genetic and acquired diseases. For example, RNAi could potentially be used to reduce the levels of toxic gain-of-function proteins, trigger cytotoxicity within tumors, or block viral replication. The use of RNAi-based therapeutics is especially appealing as RNAi can be used to modulate the expression of proteins not normally accessible by more traditional pharmaceutical approaches. For example, nondruggable targets lacking ligand-binding domains or proteins sharing high degrees of structural homology that are difficult to target as individuals are all accessible by RNAi.

The in vivo application of RNAi was described within a year of the first cell culture experiments, with reports describing the transient inhibition of transgenes within the livers of mice. This was accomplished through high-pressure tail vein injection of both siRNAs and shRNAs (71, 86). Subsequent in vivo studies have focused on the improved delivery and efficacy of RNAi effectors. These efforts have used the experience gained through two decades of developing ribozyme and antisense-based therapeutics and the gene therapy field as a whole. Currently, most in vivo studies using synthetic siRNAs use lipid-based carriers with or without modification of the siRNA itself, whereas most shRNA-based studies employ the standard viral vector expression systems used in traditional gene therapy. As evidenced by ongoing clinical trials, significant progress has been made in the field of in vivo RNAi (for an overview of RNAi therapeutic strategies see **Figure 3**).

In Vivo Application of Synthetic siRNAs

The in vivo delivery of synthetic siRNAs must account for the need to ensure resistance to exonuclease digestion, the maintenance of duplex stability, good pharmacokinetics, and the minimization of nonspecific immunological responses. Accordingly, a number of siRNA chemical modifications that address these issues have been examined. Many of these modifications are analogous to those incorporated in RNase H-dependent antisense oligonucleotides. A common modification to improve stability is the use of a partial phosphorothioate backbone, particularly within the 3′ overhangs of both siRNA strands. Furthermore, the inclusion of 2′-O-methyl dinucleotides at the 3′ end of the antisense strand has also been shown to improve stability. As mentioned, avoiding immune stimulation is also critical. The selection of sequences that avoid GU-rich sequences and/or modification with 2′-O-Me nucleotides or locked nucleic acids (LNAs) have all been shown to inhibit stimulation of the immune system without concomitant loss of efficacy (48, 55). Chemical modifications have also been engineered to improve cellular uptake. For example, cholesterol-conjugated siRNAs, corresponding to the *ApoB* gene, have been delivered into the livers of mice as a potential strategy for the treatment of familial cholesterolemia and, possibly, for the broader treatment of atherosclerosis (131). These conjugates were found to induce a significant decrease in both liver *ApoB* mRNA and plasma ApoB protein levels, as well as downstream lipoprotein and cholesterol levels. These effects were much greater than those observed with nonconjugated analogs. Unfortunately, the quantity of material necessary for efficient silencing was incompatible with scale-up to larger preclinical models, thus follow-up studies in nonhuman primates

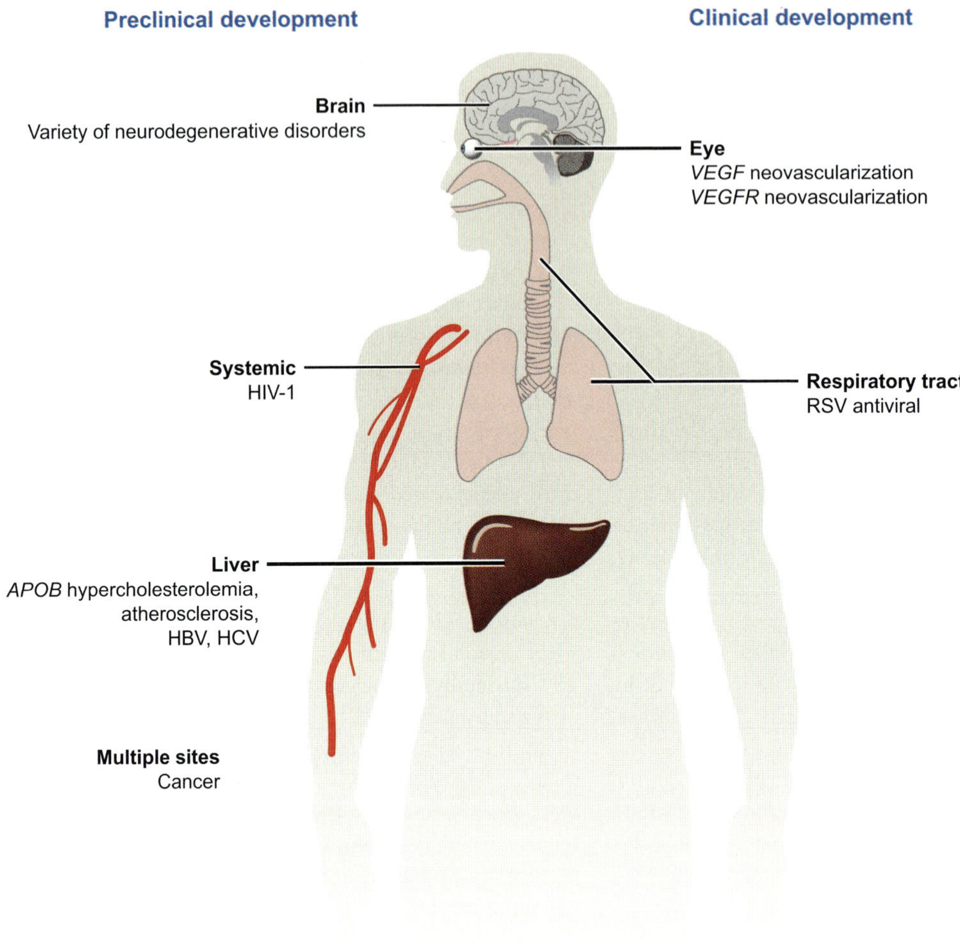

Figure 3

An illustration of the different in vivo applications of RNAi. A number of disorders have been targeted through RNAi in preclinical models (*left*). Many issues regarding the in vivo application of RNAi are analogous to those previously encountered in the field of gene therapy. Despite these complications, a number of RNAi-based therapeutics are currently in clinical development (*right*).

used a different delivery strategy (liposomal carriers, discussed below) (158).

In addition to directly modifying siRNAs for improved characteristics, carrier molecules also have the potential to protect siRNAs from the extracellular environment and improve intracellular delivery. A wide variety of polymer- or lipid-based delivery systems have been described. For example, cationic polyethylenimines have been used for siRNA transfection in vivo, including delivery to lung and xenografts following subcutaneous, intraperitoneal, and intrathecal administration (39, 136). A large number of different liposome-based carriers have also been developed for the in vivo delivery of siRNAs (for review see 73). One such system that has been relatively well characterized uses lipid-polyethylene glycol (PEG) mixtures to encapsulate siRNAs. This delivery system has been used for the systemic delivery of *APOB*-targeted siRNAs into the livers of nonhuman primates, causing a significant reduction in both *APOB* mRNA and protein levels. Furthermore, a relatively sustained (11-day) reduction in low-density lipoprotein was observed in animals receiving the highest dose of siRNA lipid (2.5 mg/kg). Similarly, this lipid-encapsulated siRNA system has been used in studies directed toward inhibiting viral infections (33, 94).

In Vivo Application of shRNAs

The first studies applying shRNAs in vivo used plasmid DNA (71); however, most subsequent studies have focused on the use of viral vectors. The choice of viral delivery system usually depends on the cell type under investigation and on the need for short- or long-term shRNA expression. For example, adenoviral (AV) and herpes simplex viral vector systems have been primarily used for short-term expression, while adeno-associated viral (AAV) vectors and the integrating viral vector systems based on retroviruses (RVs) and lentiviruses (LVs) have usually been used for long-term expression or for applications in nondividing cells.

An important adaptation of RNAi has come from the ability to stably express shRNAs in blastocytes or embryonic stem cells, from which transgenic animals can be generated. Initial "proof of concept" experiments used shRNAs to target overexpressed marker genes (e.g., green fluorescence protein) in transgenic animals. These studies used either direct injection (41) or lentiviral transduction of early embryos (41, 138). Subsequent studies have demonstrated the feasibility of targeting endogenous genes within embryonic stem (ES) cells (16, 69, 118). These models broadly mimic the phenotype of traditional knockout mice. Consequently, the constitutive expression of an shRNA for the generation of an RNAi-based transgenic is only compatible with genes that do not compromise animal viability. To circumvent embryonic lethality, shRNA-based conditional expression systems have been developed. These include Cre-Lox-based systems whereby the shRNA is flanked by LoxP sites that prevent shRNA expression (32, 142). Tissue-specific or temporal-specific shRNA expression can then be achieved by crossing shRNA transgenic mice with *Cre* recombinase expressing mice (101, 142). Although Cre-lox RNAi-based systems are irreversible, reversible expression, predominantly using doxycycline-based control systems, has been described (133, 149). RNAi transgenics have also been used in animals not normally amenable to traditional homologous recombination techniques, including rat (22) and goat. In the case of goat, a RNAi transgenic was generated through somatic cell nuclear transfer from a LV-transduced goat fibroblast stably expressing an shRNA corresponding to the prion protein (35). While the development of RNAi transgenics was initially hailed with great excitement, its broader use has not been adapted as quickly as may have been anticipated. This may be due to difficulties in obtaining efficient lentiviral transfection of embryos or ES cells, difficulties in generating ES clones that stably express shRNAs or problems associated with variations in knockdown efficiency. The recent adaptation of the more conventional pronuclear injection procedure may enable wider use of RNAi in the development of transgenics (108).

shRNAs have also been used in xenograft tumor models, particularly in mice. One of the first examples of this was targeting an activated mutant of *K-RAS* found in the pancreatic carcinoma cell line CAPAN-1 using a mutant-specific shRNA expressed from a retroviral vector (9). In contrast to control cells, the *KRAS*-targeted cells failed to form tumors in athymic mice, demonstrating the ability of RNAi-mediated silencing to suppress tumor formation in vivo. Another interesting example involves targeting the tumor suppressor *TP53* in mouse *Eμ-Myc* hematopoietic stem cells. By using shRNAs that mediated different degrees of *TP53* message reduction, the percentage of mice developing lymphoma could be varied as a function of TP53 protein levels (42). More recent variants of this method include the use of an inducible expression system that can be activated upon xenograft tumor formation, potentially generating a better clinical model for the identification and validation of anticancer molecular targets (59, 72).

The Development of RNAi-Based Therapeutics: Organ and Disease-Based Studies

The eye and neovascularization. The first clinical trial of an RNAi effector, a synthetic siRNA, was initiated just three years after the identification of RNAi in mammalian cells. Initially tested in mouse (113), the aim of these trials is to decrease expression of the vascular endothelial growth factor (VEGF) in order to suppress abnormal blood vessel development in the eyes of patients with age-related macular degeneration (AMD) and, more recently, diabetic retinopathy (**http://www.acuitypharma.com/**). A similar clinical trial in patients with AMD is being conducted with a siRNA targeting the VEGF receptor (**http://www.sirna.com/**).

The brain and neurological disorders. Using traditional transgenic approaches, it has been difficult to generate models of neurological disorders, including Parkinson's disease (PD). Thus, RNAi has been widely pursued as a new approach for developing models of neurological disorders. Due to the need for long-term suppression of gene expression and the need to treat nondividing cells, most brain-based studies have focused on the use of shRNA expression vectors derived from AAVs (44, 45, 117, 150) or LVs (36, 111). However, adenoviral systems (151), herpes-simplex systems (46), and synthetic siRNAs have also been used (134, 135). Examples of using RNAi to generate neurological or behavioral models include the introduction of an AAV shRNA against tyrosine hydroxylase into mid-brain dopamine neurons as a model for PD, as well as the AAV expression of an shRNA corresponding to the Leptin receptor as a model of feeding behavior (44, 45).

RNAi has also been explored for the potential treatment of neurodegenerative disease, particularly those associated with a dominant genetic inheritance. Many neurodegenerative disorders are associated with the expression of an aberrant protein that may be inappropriately aggregated, deposited, sequestrated, or mislocalized. Numerous groups have hypothesized that the induction of RNAi directed against these proteins may modulate disease progression. Associated studies have included the use of shRNAs to target the Huntingtin protein of Huntington's disease in mouse models (40, 117), the use of *DYT1* for the treatment of dystonia (36), and the targeting of *ataxin-1* to treat spinocerebellar ataxia (150). In addition, studies targeting the superoxide dismutase gene have been conducted in models of amyotropic lateral sclerosis (111).

The liver, metabolic disorders, and hepatitis (nonviral and viral). A number of RNAi-based strategies have been proposed to treat various liver diseases, including inherited metabolic diseases such as hypercholesterolemia (131, 158). However, most liver studies have focused on using RNAi to prevent hepatitis. One of the first of these studies investigated the feasibility of preventing liver damage caused by Fas-mediated apoptosis in a mouse model of acute hepatitis (129). Using a synthetic siRNA against *FAS*, which was introduced by hydrodynamic tail vein injection, a significant reduction in liver cytotoxicity and substantial effects on liver pathology and survival were seen as compared with controls. In a similar study, the RNAi knockdown of *Caspase 8* was used to prevent acute liver failure (155). However, the main focus of RNAi liver studies has been the development of anti-Hepatitis B virus (HBV) or anti-Hepatitis C virus (HCV) strategies. Antiviral strategies using RNAi-based approaches fall into three broad categories: those targeting (*a*) a host gene required for viral infection (e.g., a cell surface receptor), (*b*) the viral genome itself, or (*c*) a viral transcript encoding a protein essential for virus replication. One initial study regarding HBV involved plasmids expressing shRNAs against pregenomic viral RNA and in some cases viral mRNAs. Using a plasmid DNA-based replicating HBV model system, this study showed that HBV-targeting shRNAs decreased HBV DNA replication and the expression of two viral proteins (87). Chemically modified synthetic siRNAs encapsulated in liposomes have also been used to target HBV. As with the shRNA-based system, siRNAs were shown to reduce both HBV DNA and HBV protein levels in a similar HBV model system (94). Importantly, the use of an adenoviral vector expressing HBV shRNAs in an HBV-transgenic mouse model suggests that RNAi can substantially reduce HBV gene expression and replication in an established HBV infection (140). Unfortunately, no robust in vivo model of HCV infection exists, but experiments in cell culture have shown that both siRNAs and shRNAs targeting either the HCV genome or different viral transcripts can significantly reduce viral replicon replication (67, 143, 148).

Other antiviral RNAi therapies. HIV-1 has been another important target for the development of RNAi-based therapeutics. A number of different anti-HIV strategies have been explored. These include targeting HIV-1 cell surface receptors (for example, CD4) as well as strategies that block viral integration or replication. For example, RNAi has been directed against the HIV-1 *gag* gene that encodes the major structural proteins of HIV-1 (100). Other targets have included the HIV-1 long terminal repeat, HIV-1 accessory proteins, and the regulatory proteins *tat* and *rev* (19, 53). Current studies are addressing the need to obtain long-term expression of anti-HIV-1 RNAi effectors (for review see 93).

Other antiviral RNAi studies have focused on using RNAi against the negative-RNA strand virus respiratory syncytial virus (RSV), which has a significant impact on children and the elderly. A number of promising anti-RSV strategies have been described. For example, the intranasal administration of a synthetic siRNA directed against RSV reduced viral titer in lung when administered before or at the time of viral infection. A less profound effect was observed when administered several days after infection (6). Additionally, an shRNA corresponding to a different RSV mRNA was also able to decrease viral titer in lung. However, as with the siRNA, minimal activity was seen when the RNAi effector was administered post-infection (157). Notably, a siRNA-based clinical trial targeting RSV has been initiated (**http://www.alnylam.com/**).

In addition to HIV-1 and RSV, RNAi-based therapeutics have been developed for several other viruses including Ebola (33), Herpes simplex virus 2 (105), and SARS coronavirus (75). No matter the target, an important consideration for any RNAi-based antiviral is sequence variations between different strains or isolates of the same virus. Additionally, viruses have the potential to generate RNAi-resistant mutations. For example, the long-term culturing of HIV-1 in cell lines expressing an anti-HIV-1 shRNA can result in high levels of sequence alterations (mutations and deletions) that render the virus resistant (7, 23, 147). Using multiple shRNAs targeting independent sequences may help to mitigate this effect, but sequence adaptation/variation represents a significant hurdle in the use of antiviral RNAi therapeutics.

Cancer. There are many events in cancer that could be directly targeted by RNAi-based therapeutics. Especially amenable to RNAi are those cancers driven by a single genetic factor—for example, the oncogenic fusion transcript EWS-FLI1 that is associated with Ewing's sarcoma. Using a nonviral delivery method based on transferrin receptor targeting, an siRNA corresponding to the *EWS-FLI1* fusion transcript exhibited significant antitumor effects in a mouse model of Ewing's sarcoma (49). RNAi has also been used to target aberrant genes within cancer-related signaling pathways. For example, RNAi has been directed against activated *BRAF*, a mutant commonly associated with melanoma. In one study, a human melanoma cell line heterozygous for this mutation was transduced with either a lentiviral vector expressing a mutant-specific shRNA, an shRNA corresponding to the wild-type *BRAF* sequence, or a control shRNA. These cells were then used to generate xenografts in different cohorts of mice (132). Mice receiving either of the *BRAF* shRNAs exhibited smaller tumors than control mice, with the mutant-specific variant showing the least amount of tumor growth. In another study, an inducible *BRAF* shRNA was able to inhibit tumor growth in vivo and shrink pre-existing melanoma xenografts (43). RNAi has been used for the in vivo targeting of a number of other genes integral to signaling cascades. Some of these include *K-RAS* (study described above), *EGFR*, and *c-Met* (9, 119, 156).

RNAi has also been used to target genes that are differentially expressed in cancer. For example, Survivin is minimally expressed in normal cells, but frequently expressed in many tumors where it is thought to alter cell cycle regulation and suppress apoptosis. Several in vivo cancer model systems, including glioma, esophageal carcinoma, and rhabdomyosarcoma xenografts, have been used to assess the effect of *Survivin* downregulation (12, 139, 144). In all cases, a substantial inhibition of tumor growth was observed following RNAi against *Survivin*. In an analogous example, the induction of RNAi against telomerase, another gene preferentially expressed in tumor cells, showed substantial inhibition of tumor growth in xenografts of cervical and bladder cancers (96, 159). In one of these studies, RNAi enhanced the sensitivity of tumor cells to existing chemotherapeutic agents (96). The use of RNAi to identify these types of "synthetic lethal" combinations will likely be a very important field of study in coming years. Related to this is the potential for RNAi to inhibit proteins associated with chemotherapeutic resistance (for example, MDR1). The feasibility of such an approach was demonstrated by showing that an shRNA targeting *MDR1* was able to mitigate drug resistance in vivo (124). No matter the approach, it is likely that at least some RNAi-based clinical trials will be developed over the next few years, particularly for cancers where there are few existing treatment options.

CONCLUSIONS

The discovery of RNAi has dramatically changed our understanding of how gene expression is modulated. Furthermore, the ability to harness this mechanism through the use of artificial triggers has become an invaluable tool for the scientific community. RNAi has not only been applied to understand the functions of a limited number of genes in focused studies, but has been employed in larger, genome-wide applications. The information provided by these studies has significantly enhanced our ability to identify new disease-related molecular targets and rationally improve therapeutic development. Furthermore, RNAi-based technologies have, in themselves, the potential for clinical application. While there has been rapid progress in moving RNAi-based technologies into the clinic, a substantial number of issues still remain. Some of these issues are similar to those encountered by the gene therapy and antisense fields. Specifically, ensuring efficient delivery of nucleic acids with minimal stimulation of host responses is absolutely critical. Other issues relate to the possibility of overwhelming the endogenous RNAi machinery and the potential downregulation of nontargeted transcripts. However, the enormous potential of RNAi makes overcoming these obstacles a worthwhile endeavor.

SUMMARY POINTS

1. RNAi is a conserved endogenous mechanism that can mediate gene silencing at a transcriptional and post-transcriptional level in mammalian cells. Artificial triggers of RNAi have been rapidly developed to modulate gene expression. Applications of RNAi have ranged from the single gene to genome-wide scales.
2. RNAi effectors can induce unintended effects, termed off-target effects. Off-target effects can result in the misinterpretation of RNAi-induced phenotypes. There are a number of steps that should be taken to minimize the impact of off-target effects.
3. RNAi has been used for large-scale functional screens in human cells. A number of elegant screening strategies have been described. These screens have yielded a wealth of information, ranging from a better understanding of biological processes to the identification of novel anticancer molecular targets.
4. A variety of RNAi-based strategies have been successfully developed to target genes in vivo. Although most studies have been conducted in preclinical models, there are currently a number of clinical trials involving RNAi-based therapeutics.

DISCLOSURE STATEMENT

NC collaborates with Qiagen Inc., Germantown, MD, on research investigating aspects of the use of synthetic siRNAs to mediate RNAi.

ACKNOWLEDGMENTS

We thank previous and current members of the Gene Silencing Section, Genetics Branch for useful discussion and comments. The research conducted by the Gene Silencing Section, Genetics Branch, CCR, NCI, NIH is supported by the Intramural Research Program (Center for Cancer Research, National Cancer Institute) of the National Institutes of Health.

LITERATURE CITED

1. Aravin A, Gaidatzis D, Pfeffer S, Lagos-Quintana M, Landgraf P, et al. 2006. A novel class of small RNAs bind to MILI protein in mouse testes. *Nature* 442:203–7
2. Aza-Blanc P, Cooper CL, Wagner K, Batalov S, Deveraux QL, Cooke MP. 2003. Identification of modulators of TRAIL-induced apoptosis via RNAi-based phenotypic screening. *Mol. Cell.* 12:627–37
3. Berns K, Hijmans EM, Mullenders J, Brummelkamp TR, Velds A, et al. 2004. A large-scale RNAi screen in human cells identifies new components of the p53 pathway. *Nature* 428:431–37
4. Bernstein E, Caudy AA, Hammond SM, Hannon GJ. 2001. Role for a bidentate ribonuclease in the initiation step of RNA interference. *Nature* 409:363–66
5. Birmingham A, Anderson EM, Reynolds A, Ilsley-Tyree D, Leake D, et al. 2006. 3′ UTR seed matches, but not overall identity, are associated with RNAi off-targets. *Nat. Methods* 3:199–204

6. Bitko V, Musiyenko A, Shulyayeva O, Barik S. 2005. Inhibition of respiratory viruses by nasally administered siRNA. *Nat. Med.* 11:50–55
7. Boden D, Pusch O, Lee F, Tucker L, Ramratnam B. 2003. Human immunodeficiency virus type 1 escape from RNA interference. *J. Virol.* 77:11531–35
8. Bridge AJ, Pebernard S, Ducraux A, Nicoulaz AL, Iggo R. 2003. Induction of an interferon response by RNAi vectors in mammalian cells. *Nat. Genet.* 34:263–64
9. Brummelkamp TR, Bernards R, Agami R. 2002. Stable suppression of tumorigenicity by virus-mediated RNA interference. *Cancer Cell* 2:243–47
10. Brummelkamp TR, Fabius AW, Mullenders J, Madiredjo M, Velds A, et al. 2006. An shRNA barcode screen provides insight into cancer cell vulnerability to MDM2 inhibitors. *Nat. Chem. Biol.* 2:202–6
11. Brummelkamp TR, Nijman SM, Dirac AM, Bernards R. 2003. Loss of the cylindromatosis tumor suppressor inhibits apoptosis by activating NF-kappaB. *Nature* 424:797–801
12. Caldas H, Holloway MP, Hall BM, Qualman SJ, Altura RA. 2006. Survivin-directed RNA interference cocktail is a potent suppressor of tumor growth in vivo. *J. Med. Genet.* 43:119–28
13. Calin GA, Croce CM. 2006. MicroRNA-cancer connection: the beginning of a new tale. *Cancer Res.* 66:7390–94
14. Calin GA, Dumitru CD, Shimizu M, Bichi R, Zupo S, et al. 2002. Frequent deletions and down-regulation of micro-RNA genes miR15 and miR16 at 13q14 in chronic lymphocytic leukemia. *Proc. Natl. Acad. Sci. USA* 99:15524–29
15. Caplen NJ, Parrish S, Imani F, Fire A, Morgan RA. 2001. Specific inhibition of gene expression by small double-stranded RNAs in invertebrate and vertebrate systems. *Proc. Natl. Acad. Sci. USA* 98:9742–47
16. Carmell MA, Zhang L, Conklin DS, Hannon GJ, Rosenquist TA. 2003. Germline transmission of RNAi in mice. *Nat. Struct. Biol.* 10:91–92
17. Castanotto D, Tommasi S, Li M, Li H, Yanow S, et al. 2005. Short hairpin RNA-directed cytosine (CpG) methylation of the RASSF1A gene promoter in HeLa cells. *Mol. Ther.* 12:179–83
18. Cimmino A, Calin GA, Fabbri M, Iorio MV, Ferracin M, et al. 2005. miR-15 and miR-16 induce apoptosis by targeting BCL2. *Proc. Natl. Acad. Sci. USA* 102:13944–49
19. Coburn GA, Cullen BR. 2002. Potent and specific inhibition of human immunodeficiency virus type 1 replication by RNA interference. *J. Virol.* 76:9225–31
20. Collins CS, Hong J, Sapinoso L, Zhou Y, Liu Z, et al. 2006. A small interfering RNA screen for modulators of tumor cell motility identifies MAP4K4 as a promigratory kinase. *Proc. Natl. Acad. Sci. USA* 103:3775–80
21. Costinean S, Zanesi N, Pekarsky Y, Tili E, Volinia S, et al. 2006. Pre-B cell proliferation and lymphoblastic leukemia/high-grade lymphoma in E(mu)-miR155 transgenic mice. *Proc. Natl. Acad. Sci. USA* 103:7024–29
22. Dann CT, Alvarado AL, Hammer RE, Garbers DL. 2006. Heritable and stable gene knockdown in rats. *Proc. Natl. Acad. Sci. USA* 103:11246–51
23. Das AT, Brummelkamp TR, Westerhout EM, Vink M, Madiredjo M, et al. 2004. Human immunodeficiency virus type 1 escapes from RNA interference-mediated inhibition. *J. Virol.* 78:2601–5
24. Doench JG, Petersen CP, Sharp PA. 2003. siRNAs can function as miRNAs. *Genes Dev.* 17:438–42
25. Doench JG, Sharp PA. 2004. Specificity of microRNA target selection in translational repression. *Genes Dev.* 18:504–11

26. Dykxhoorn DM, Schlehuber LD, London IM, Lieberman J. 2006. Determinants of specific RNA interference-mediated silencing of human beta-globin alleles differing by a single nucleotide polymorphism. *Proc. Natl. Acad. Sci. USA* 103:5953–58
27. Elbashir SM, Harborth J, Lendeckel W, Yalcin A, Weber K, Tuschl T. 2001. Duplexes of 21-nucleotide RNAs mediate RNA interference in cultured mammalian cells. *Nature* 411:494–98
28. Elbashir SM, Lendeckel W, Tuschl T. 2001. RNA interference is mediated by 21- and 22-nucleotide RNAs. *Genes Dev.* 15:188–200
29. Erfle H, Simpson JC, Bastiaens PI, Pepperkok R. 2004. siRNA cell arrays for high-content screening microscopy. *Biotechniques* 37:454–8, 60, 62
30. Fedorov Y, Anderson EM, Birmingham A, Reynolds A, Karpilow J, et al. 2006. Off-target effects by siRNA can induce toxic phenotype. *RNA* 12:1188–96
31. Fire A, Xu S, Montgomery MK, Kostas SA, Driver SE, Mello CC. 1998. Potent and specific genetic interference by double-stranded RNA in *Caenorhabditis elegans*. *Nature* 391:806–11
32. Fritsch L, Martinez LA, Sekhri R, Naguibneva I, Gerard M, et al. 2004. Conditional gene knock-down by CRE-dependent short interfering RNAs. *EMBO Rep.* 5:178–82
33. Geisbert TW, Hensley LE, Kagan E, Yu EZ, Geisbert JB, et al. 2006. Postexposure protection of guinea pigs against a lethal ebola virus challenge is conferred by RNA interference. *J. Infect. Dis.* 193:1650–57
34. Girard A, Sachidanandam R, Hannon GJ, Carmell MA. 2006. A germline-specific class of small RNAs binds mammalian Piwi proteins. *Nature* 442:199–202
35. Golding MC, Long CR, Carmell MA, Hannon GJ, Westhusin ME. 2006. Suppression of prion protein in livestock by RNA interference. *Proc. Natl. Acad. Sci. USA* 103:5285–90
36. Gonzalez-Alegre P, Bode N, Davidson BL, Paulson HL. 2005. Silencing primary dystonia: lentiviral-mediated RNA interference therapy for DYT1 dystonia. *J. Neurosci.* 25:10502–9
37. Grimm D, Streetz KL, Jopling CL, Storm TA, Pandey K, et al. 2006. Fatality in mice due to oversaturation of cellular microRNA/short hairpin RNA pathways. *Nature* 441:537–41
38. Grivna ST, Beyret E, Wang Z, Lin H. 2006. A novel class of small RNAs in mouse spermatogenic cells. *Genes Dev.* 20:1709–14
39. Grzelinski M, Urban-Klein B, Martens T, Lamszus K, Bakowsky U, et al. 2006. RNA interference-mediated gene silencing of pleiotrophin through polyethylenimine-complexed small interfering RNAs in vivo exerts antitumoral effects in glioblastoma xenografts. *Hum. Gene Ther.* 17:751–66
40. Harper SQ, Staber PD, He X, Eliason SL, Martins IH, et al. 2005. RNA interference improves motor and neuropathological abnormalities in a Huntington's disease mouse model. *Proc. Natl. Acad. Sci. USA* 102:5820–25
41. Hasuwa H, Kaseda K, Einarsdottir T, Okabe M. 2002. Small interfering RNA and gene silencing in transgenic mice and rats. *FEBS Lett.* 532:227–30
42. Hemann MT, Fridman JS, Zilfou JT, Hernando E, Paddison PJ, et al. 2003. An epi-allelic series of p53 hypomorphs created by stable RNAi produces distinct tumor phenotypes in vivo. *Nat. Genet.* 33:396–400
43. Hoeflich KP, Gray DC, Eby MT, Tien JY, Wong L, et al. 2006. Oncogenic BRAF is required for tumor growth and maintenance in melanoma models. *Cancer Res.* 66:999–1006
44. Hommel JD, Sears RM, Georgescu D, Simmons DL, DiLeone RJ. 2003. Local gene knockdown in the brain using viral-mediated RNA interference. *Nat. Med.* 9:1539–44

45. Hommel JD, Trinko R, Sears RM, Georgescu D, Liu ZW, et al. 2006. Leptin receptor signaling in midbrain dopamine neurons regulates feeding. *Neuron* 51:801–10
46. Hong CS, Goins WF, Goss JR, Burton EA, Glorioso JC. 2006. Herpes simplex virus RNAi and neprilysin gene transfer vectors reduce accumulation of Alzheimer's disease-related amyloid-beta peptide in vivo. *Gene Ther.* 13:1068–79
47. Horman SR, Svoboda P, Prak ET. 2006. The potential regulation of l1 mobility by RNA interference. *J. Biomed. Biotechnol.* 2006:32713
48. Hornung V, Guenthner-Biller M, Bourquin C, Ablasser A, Schlee M, et al. 2005. Sequence-specific potent induction of IFN-alpha by short interfering RNA in plasmacytoid dendritic cells through TLR7. *Nat. Med.* 11:263–70
49. Hu-Lieskovan S, Heidel JD, Bartlett DW, Davis ME, Triche TJ. 2005. Sequence-specific knockdown of EWS-FLI1 by targeted, nonviral delivery of small interfering RNA inhibits tumor growth in a murine model of metastatic Ewing's sarcoma. *Cancer Res.* 65:8984–92
50. Hutvagner G, Simard MJ, Mello CC, Zamore PD. 2004. Sequence-specific inhibition of small RNA function. *PLoS Biol.* 2:E98
51. Jackson AL, Bartz SR, Schelter J, Kobayashi SV, Burchard J, et al. 2003. Expression profiling reveals off-target gene regulation by RNAi. *Nat. Biotechnol.* 21:635–37
52. Jackson AL, Burchard J, Leake D, Reynolds A, Schelter J, et al. 2006. Position-specific chemical modification of siRNAs reduces "off-target" transcript silencing. *RNA* 12:1197–205
53. Jacque JM, Triques K, Stevenson M. 2002. Modulation of HIV-1 replication by RNA interference. *Nature* 418:435–38
54. Janowski BA, Huffman KE, Schwartz JC, Ram R, Nordsell R, et al. 2006. Involvement of AGO1 and AGO2 in mammalian transcriptional silencing. *Nat. Struct. Mol. Biol.* 13:787–92
55. Judge AD, Bola G, Lee AC, MacLachlan I. 2006. Design of noninflammatory synthetic siRNA mediating potent gene silencing in vivo. *Mol. Ther.* 13:494–505
56. Judge AD, Sood V, Shaw JR, Fang D, McClintock K, MacLachlan I. 2005. Sequence-dependent stimulation of the mammalian innate immune response by synthetic siRNA. *Nat. Biotechnol.* 23:457–62
57. Kariko K, Bhuyan P, Capodici J, Weissman D. 2004. Small interfering RNAs mediate sequence-independent gene suppression and induce immune activation by signaling through toll-like receptor 3. *J. Immunol.* 172:6545–49
58. Kawasaki H, Taira K. 2004. Induction of DNA methylation and gene silencing by short interfering RNAs in human cells. *Nature* 431:211–17
59. Ke N, Zhou D, Chatterton JE, Liu G, Chionis J, et al. 2006. A new inducible RNAi xenograft model for assessing the staged tumor response to mTOR silencing. *Exp. Cell Res.* 312:2726–34
60. Kennerdell JR, Carthew RW. 1998. Use of dsRNA-mediated genetic interference to demonstrate that frizzled and frizzled 2 act in the wingless pathway. *Cell* 95:1017–26
61. Khvorova A, Reynolds A, Jayasena SD. 2003. Functional siRNAs and miRNAs exhibit strand bias. *Cell* 115:209–16
62. Kim VN. 2006. Small RNAs just got bigger: piwi-interacting RNAs (piRNAs) in mammalian testes. *Genes Dev.* 20:1993–97
63. Kittler R, Heninger AK, Franke K, Habermann B, Buchholz F. 2005. Production of endoribonuclease-prepared short interfering RNAs for gene silencing in mammalian cells. *Nat. Methods* 2:779–84

64. Kluiver J, Haralambieva E, de Jong D, Blokzijl T, Jacobs S, et al. 2006. Lack of BIC and microRNA miR-155 expression in primary cases of Burkitt lymphoma. *Genes Chromosomes Cancer* 45:147–53
65. Kluiver J, Poppema S, de Jong D, Blokzijl T, Harms G, et al. 2005. BIC and miR-155 are highly expressed in Hodgkin, primary mediastinal and diffuse large B cell lymphomas. *J. Pathol.* 207:243–49
66. Kolfschoten IG, van Leeuwen B, Berns K, Mullenders J, Beijersbergen RL, et al. 2005. A genetic screen identifies PITX1 as a suppressor of RAS activity and tumorigenicity. *Cell* 121:849–58
67. Kronke J, Kittler R, Buchholz F, Windisch MP, Pietschmann T, et al. 2004. Alternative approaches for efficient inhibition of hepatitis C virus RNA replication by small interfering RNAs. *J. Virol.* 78:3436–46
68. Kulkarni MM, Booker M, Silver SJ, Friedman A, Hong P, et al. 2006. Evidence of off-target effects associated with long dsRNAs in *Drosophila melanogaster* cell-based assays. *Nat. Methods* 3:833–38
69. Kunath T, Gish G, Lickert H, Jones N, Pawson T, Rossant J. 2003. Transgenic RNA interference in ES cell-derived embryos recapitulates a genetic null phenotype. *Nat. Biotechnol.* 21:559–61
70. Lewis BP, Burge CB, Bartel DP. 2005. Conserved seed pairing, often flanked by adenosines, indicates that thousands of human genes are microRNA targets. *Cell* 120:15–20
71. Lewis DL, Hagstrom JE, Loomis AG, Wolff JA, Herweijer H. 2002. Efficient delivery of siRNA for inhibition of gene expression in postnatal mice. *Nat. Genet.* 32:107–8
72. Li L, Lin X, Staver M, Shoemaker A, Semizarov D, et al. 2005. Evaluating hypoxia-inducible factor-1alpha as a cancer therapeutic target via inducible RNA interference in vivo. *Cancer Res.* 65:7249–58
73. Li SD, Huang L. 2006. Gene therapy progress and prospects: nonviral gene therapy by systemic delivery. *Gene Ther.* 13:1313–19
74. Li T, Chang CY, Jin DY, Lin PJ, Khvorova A, Stafford DW. 2004. Identification of the gene for vitamin K epoxide reductase. *Nature* 427:541–44
75. Lim LP, Lau NC, Garrett-Engele P, Grimson A, Schelter JM, et al. 2005. Microarray analysis shows that some microRNAs downregulate large numbers of target mRNAs. *Nature* 433:769–73
76. Lin X, Ruan X, Anderson MG, McDowell JA, Kroeger PE, et al. 2005. siRNA-mediated off-target gene silencing triggered by a 7 nt complementation. *Nucleic Acids Res.* 33:4527–35
77. Liu J, Carmell MA, Rivas FV, Marsden CG, Thomson JM, et al. 2004. Argonaute2 is the catalytic engine of mammalian RNAi. *Science* 305:1437–41
78. Lorenzi PL, Reinhold WC, Rudelius M, Gunsior M, Shankavaram U, et al. 2006. Asparagine synthetase as a causal, predictive biomarker for L-asparaginase activity in ovarian cancer cells. *Mol. Cancer Ther.* 5:2613–23.
79. Ludwig JA, Szakacs G, Martin SE, Chu BF, Cardarelli C, et al. 2006. Selective toxicity of NSC73306 in MDR1-positive cells as a new strategy to circumvent multidrug resistance in cancer. *Cancer Res.* 66:4808–15
80. Luo B, Heard AD, Lodish HF. 2004. Small interfering RNA production by enzymatic engineering of DNA (SPEED). *Proc. Natl. Acad. Sci. USA* 101:5494–99
81. Ma Y, Creanga A, Lum L, Beachy PA. 2006. Prevalence of off-target effects in *Drosophila* RNA interference screens. *Nature* 443:359–63

82. MacKeigan JP, Murphy LO, Blenis J. 2005. Sensitized RNAi screen of human kinases and phosphatases identifies new regulators of apoptosis and chemoresistance. *Nat. Cell Biol.* 7:591–600
83. Marques JT, Devosse T, Wang D, Zamanian-Daryoush M, Serbinowski P, et al. 2006. A structural basis for discriminating between self and nonself double-stranded RNAs in mammalian cells. *Nat. Biotechnol* 24:559–65
84. Martin SE, Caplen NJ. 2006. Mismatched siRNAs downregulate mRNAs as a function of target site location. *FEBS Lett.* 580:3694–98
84a. Martin SE, Jones TL, Thomas CL, Lorenzi PL, Nguyen DA, et al. 2007. Multiplexing siRNAs to compress RNAi-based screen size in human cells. *Nucleic Acids Res.* doi:10.1093/nar/gkm141
85. Matranga C, Tomari Y, Shin C, Bartel DP, Zamore PD. 2005. Passenger-strand cleavage facilitates assembly of siRNA into Ago2-containing RNAi enzyme complexes. *Cell* 123:607–20
86. McCaffrey AP, Meuse L, Pham TT, Conklin DS, Hannon GJ, Kay MA. 2002. RNA interference in adult mice. *Nature* 418:38–39
87. McCaffrey AP, Nakai H, Pandey K, Huang Z, Salazar FH, et al. 2003. Inhibition of hepatitis B virus in mice by RNA interference. *Nat. Biotechnol.* 21:639–44
88. Metzler M, Wilda M, Busch K, Viehmann S, Borkhardt A. 2004. High expression of precursor microRNA-155/BIC RNA in children with Burkitt lymphoma. *Genes Chromosomes Cancer* 39:167–69
89. Miller VM, Xia H, Marrs GL, Gouvion CM, Lee G, et al. 2003. Allele-specific silencing of dominant disease genes. *Proc. Natl. Acad. Sci. USA* 100:7195–200
90. Misquitta L, Paterson BM. 1999. Targeted disruption of gene function in *Drosophila* by RNA interference (RNA-i): a role for nautilus in embryonic somatic muscle formation. *Proc. Natl. Acad. Sci. USA* 96:1451–56
91. Moffat J, Grueneberg DA, Yang X, Kim SY, Kloepfer AM, et al. 2006. A lentiviral RNAi library for human and mouse genes applied to an arrayed viral high-content screen. *Cell* 124:1283–98
92. Morris KV, Chan SW, Jacobsen SE, Looney DJ. 2004. Small interfering RNA-induced transcriptional gene silencing in human cells. *Science* 305:1289–92
93. Morris KV, Rossi JJ. 2006. Lentivirus-mediated RNA interference therapy for human immunodeficiency virus type 1 infection. *Hum. Gene Ther.* 17:479–86
94. Morrissey DV, Lockridge JA, Shaw L, Blanchard K, Jensen K, et al. 2005. Potent and persistent in vivo anti-HBV activity of chemically modified siRNAs. *Nat. Biotechnol.* 23:1002–7
95. Mousses S, Caplen NJ, Cornelison R, Weaver D, Basik M, et al. 2003. RNAi microarray analysis in cultured mammalian cells. *Genome Res.* 13:2341–47
96. Nakamura M, Masutomi K, Kyo S, Hashimoto M, Maida Y, et al. 2005. Efficient inhibition of human telomerase reverse transcriptase expression by RNA interference sensitizes cancer cells to ionizing radiation and chemotherapy. *Hum. Gene Ther.* 16:859–68
97. Nakayashiki H, Kadotani N, Mayama S. 2006. Evolution and diversification of RNA silencing proteins in fungi. *J. Mol. Evol.* 63:127–35
98. Neumann B, Held M, Liebel U, Erfle H, Rogers P, et al. 2006. High-throughput RNAi screening by time-lapse imaging of live human cells. *Nat. Methods* 3:385–90
99. Ngo VN, Davis RE, Lamy L, Yu X, Zhao H, et al. 2006. A loss-of-function RNA interference screen for molecular targets in cancer. *Nature* 441:106–10

100. Novina CD, Murray MF, Dykxhoorn DM, Beresford PJ, Riess J, et al. 2002. siRNA-directed inhibition of HIV-1 infection. *Nat. Med.* 8:681–86
101. Oberdoerffer P, Kanellopoulou C, Heissmeyer V, Paeper C, Borowski C, et al. 2005. Efficiency of RNA interference in the mouse hematopoietic system varies between cell types and developmental stages. *Mol. Cell. Biol.* 25:3896–905
102. Ohrt T, Merkle D, Birkenfeld K, Echeverri CJ, Schwille P. 2006. In situ fluorescence analysis demonstrates active siRNA exclusion from the nucleus by Exportin 5. *Nucleic Acids Res.* 34:1369–80
103. Oosterkamp HM, Neering H, Nijman SM, Dirac AM, Mooi WJ, et al. 2006. An evaluation of the efficacy of topical application of salicylic acid for the treatment of familial cylindromatosis. *Br. J. Dermatol.* 155:182–85
104. Paddison PJ, Silva JM, Conklin DS, Schlabach M, Li M, et al. 2004. A resource for large-scale RNA-interference-based screens in mammals. *Nature* 428:427–31
105. Palliser D, Chowdhury D, Wang QY, Lee SJ, Bronson RT, et al. 2006. An siRNA-based microbicide protects mice from lethal herpes simplex virus 2 infection. *Nature* 439:89–94
106. Parker JS, Barford D. 2006. Argonaute: a scaffold for the function of short regulatory RNAs. *Trends Biochem. Sci.* 31:622–30.
107. Pelkmans L, Fava E, Grabner H, Hannus M, Habermann B, et al. 2005. Genome-wide analysis of human kinases in clathrin- and caveolae/raft-mediated endocytosis. *Nature* 436:78–86
108. Peng S, York JP, Zhang P. 2006. A transgenic approach for RNA interference-based genetic screening in mice. *Proc. Natl. Acad. Sci. USA* 103:2252–56
109. Persengiev SP, Zhu X, Green MR. 2004. Nonspecific, concentration-dependent stimulation and repression of mammalian gene expression by small interfering RNAs (siRNAs). *RNA* 10:12–18
110. Puthenveetil S, Whitby L, Ren J, Kelnar K, Krebs JF, Beal PA. 2006. Controlling activation of the RNA-dependent protein kinase by siRNAs using site-specific chemical modification. *Nucleic Acids Res.* 34:4900–11
111. Ralph GS, Radcliffe PA, Day DM, Carthy JM, Leroux MA, et al. 2005. Silencing mutant SOD1 using RNAi protects against neurodegeneration and extends survival in an ALS model. *Nat. Med.* 11:429–33
112. Rand TA, Petersen S, Du F, Wang X. 2005. Argonaute2 cleaves the antiguide strand of siRNA during RISC activation. *Cell* 123:621–29
113. Reich SJ, Fosnot J, Kuroki A, Tang W, Yang X, et al. 2003. Small interfering RNA (siRNA) targeting VEGF effectively inhibits ocular neovascularization in a mouse model. *Mol. Vis.* 9:210–16
114. Reynolds A, Anderson EM, Vermeulen A, Fedorov Y, Robinson K, et al. 2006. Induction of the interferon response by siRNA is cell type- and duplex length-dependent. *RNA* 12:988–93
115. Reynolds A, Leake D, Boese Q, Scaringe S, Marshall WS, Khvorova A. 2004. Rational siRNA design for RNA interference. *Nat. Biotechnol.* 22:326–30
116. Robb GB, Brown KM, Khurana J, Rana TM. 2005. Specific and potent RNAi in the nucleus of human cells. *Nat. Struct. Mol. Biol.* 12:133–37
117. Rodriguez-Lebron E, Denovan-Wright EM, Nash K, Lewin AS, Mandel RJ. 2005. Intrastriatal rAAV-mediated delivery of antihuntingtin shRNAs induces partial reversal of disease progression in R6/1 Huntington's disease transgenic mice. *Mol. Ther.* 12:618–33

118. Rubinson DA, Dillon CP, Kwiatkowski AV, Sievers C, Yang L, et al. 2003. A lentivirus-based system to functionally silence genes in primary mammalian cells, stem cells and transgenic mice by RNA interference. *Nat. Genet.* 33:401–6
119. Saydam O, Glauser DL, Heid I, Turkeri G, Hilbe M, et al. 2005. Herpes simplex virus 1 amplicon vector-mediated siRNA targeting epidermal growth factor receptor inhibits growth of human glioma cells in vivo. *Mol. Ther.* 12:803–12
120. Scacheri PC, Rozenblatt-Rosen O, Caplen NJ, Wolfsberg TG, Umayam L, et al. 2004. Short interfering RNAs can induce unexpected and divergent changes in the levels of untargeted proteins in mammalian cells. *Proc. Natl. Acad. Sci. USA* 101:1892–97
121. Schwarz DS, Hutvagner G, Du T, Xu Z, Aronin N, Zamore PD. 2003. Asymmetry in the assembly of the RNAi enzyme complex. *Cell* 115:199–208
122. Semizarov D, Frost L, Sarthy A, Kroeger P, Halbert DN, Fesik SW. 2003. Specificity of short interfering RNA determined through gene expression signatures. *Proc. Natl. Acad. Sci. USA* 100:6347–52
123. Sen G, Wehrman TS, Myers JW, Blau HM. 2004. Restriction enzyme-generated siRNA (REGS) vectors and libraries. *Nat. Genet.* 36:183–89
124. Shi Z, Liang YJ, Chen ZS, Wang XW, Wang XH, et al. 2006. Reversal of MDR1/P-glycoprotein-mediated multidrug resistance by vector-based RNA interference in vitro and in vivo. *Cancer Biol. Ther.* 5:39–47
125. Shirane D, Sugao K, Namiki S, Tanabe M, Iino M, Hirose K. 2004. Enzymatic production of RNAi libraries from cDNAs. *Nat. Genet.* 36:190–96
126. Silva JM, Li MZ, Chang K, Ge W, Golding MC, et al. 2005. Second-generation shRNA libraries covering the mouse and human genomes. *Nat. Genet.* 37:1281–88
127. Silva JM, Mizuno H, Brady A, Lucito R, Hannon GJ. 2004. RNA interference microarrays: high-throughput loss-of-function genetics in mammalian cells. *Proc. Natl. Acad. Sci. USA* 101:6548–52
128. Sledz CA, Holko M, de Veer MJ, Silverman RH, Williams BR. 2003. Activation of the interferon system by short-interfering RNAs. *Nat. Cell Biol.* 5:834–39
129. Song E, Lee SK, Wang J, Ince N, Ouyang N, et al. 2003. RNA interference targeting Fas protects mice from fulminant hepatitis. *Nat. Med.* 9:347–51
130. Sorkina T, Miranda M, Dionne KR, Hoover BR, Zahniser NR, Sorkin A. 2006. RNA interference screen reveals an essential role of Nedd4–2 in dopamine transporter ubiquitination and endocytosis. *J. Neurosci.* 26:8195–205
131. Soutschek J, Akinc A, Bramlage B, Charisse K, Constien R, et al. 2004. Therapeutic silencing of an endogenous gene by systemic administration of modified siRNAs. *Nature* 432:173–78
132. Sumimoto H, Miyagishi M, Miyoshi H, Yamagata S, Shimizu A, et al. 2004. Inhibition of growth and invasive ability of melanoma by inactivation of mutated BRAF with lentivirus-mediated RNA interference. *Oncogene* 23:6031–39
133. Szulc J, Wiznerowicz M, Sauvain MO, Trono D, Aebischer P. 2006. A versatile tool for conditional gene expression and knockdown. *Nat. Methods* 3:109–16
134. Thakker DR, Natt F, Husken D, Maier R, Muller M, et al. 2004. Neurochemical and behavioral consequences of widespread gene knockdown in the adult mouse brain by using nonviral RNA interference. *Proc. Natl. Acad. Sci. USA* 101:17270–75
135. Thakker DR, Natt F, Husken D, van der Putten H, Maier R, et al. 2005. siRNA-mediated knockdown of the serotonin transporter in the adult mouse brain. *Mol. Psychiatry* 10:782–89.

136. Thomas M, Lu JJ, Ge Q, Zhang C, Chen J, Klibanov AM. 2005. Full deacylation of polyethylenimine dramatically boosts its gene delivery efficiency and specificity to mouse lung. *Proc. Natl. Acad. Sci. USA* 102:5679–84
137. Ting AH, Schuebel KE, Herman JG, Baylin SB. 2005. Short double-stranded RNA induces transcriptional gene silencing in human cancer cells in the absence of DNA methylation. *Nat. Genet.* 37:906–10
138. Tiscornia G, Singer O, Ikawa M, Verma IM. 2003. A general method for gene knockdown in mice by using lentiviral vectors expressing small interfering RNA. *Proc. Natl. Acad. Sci. USA* 100:1844–48
139. Uchida H, Tanaka T, Sasaki K, Kato K, Dehari H, et al. 2004. Adenovirus-mediated transfer of siRNA against survivin induced apoptosis and attenuated tumor cell growth in vitro and in vivo. *Mol. Ther.* 10:162–71
140. Uprichard SL, Boyd B, Althage A, Chisari FV. 2005. Clearance of hepatitis B virus from the liver of transgenic mice by short hairpin RNAs. *Proc. Natl. Acad. Sci. USA* 102:773–78
141. Vaucheret H. 2006. Post-transcriptional small RNA pathways in plants: mechanisms and regulations. *Genes Dev.* 20:759–71
142. Ventura A, Meissner A, Dillon CP, McManus M, Sharp PA, et al. 2004. Cre-lox-regulated conditional RNA interference from transgenes. *Proc. Natl. Acad. Sci. USA* 101:10380–85
143. Wang Q, Contag CH, Ilves H, Johnston BH, Kaspar RL. 2005. Small hairpin RNAs efficiently inhibit hepatitis C IRES-mediated gene expression in human tissue culture cells and a mouse model. *Mol. Ther.* 12:562–68
144. Wang Y, Zhu H, Quan L, Zhou C, Bai J, et al. 2005. Downregulation of survivin by RNAi inhibits the growth of esophageal carcinoma cells. *Cancer Biol. Ther.* 4:974–78
145. Watanabe T, Takeda A, Tsukiyama T, Mise K, Okuno T, et al. 2006. Identification and characterization of two novel classes of small RNAs in the mouse germline: retrotransposon-derived siRNAs in oocytes and germline small RNAs in testes. *Genes Dev.* 20:1732–43
146. Weinberg MS, Villeneuve LM, Ehsani A, Amarzguioui M, Aagaard L, et al. 2006. The antisense strand of small interfering RNAs directs histone methylation and transcriptional gene silencing in human cells. *RNA* 12:256–62
147. Westerhout EM, Ooms M, Vink M, Das AT, Berkhout B. 2005. HIV-1 can escape from RNA interference by evolving an alternative structure in its RNA genome. *Nucleic Acids Res.* 33:796–804
148. Wilson JA, Jayasena S, Khvorova A, Sabatinos S, Rodrigue-Gervais IG, et al. 2003. RNA interference blocks gene expression and RNA synthesis from hepatitis C replicons propagated in human liver cells. *Proc. Natl. Acad. Sci. USA* 100:2783–88
149. Wiznerowicz M, Szulc J, Trono D. 2006. Tuning silence: conditional systems for RNA interference. *Nat. Methods* 3:682–88
150. Xia H, Mao Q, Eliason SL, Harper SQ, Martins IH, et al. 2004. RNAi suppresses polyglutamine-induced neurodegeneration in a model of spinocerebellar ataxia. *Nat. Med.* 10:816–20
151. Xia H, Mao Q, Paulson HL, Davidson BL. 2002. siRNA-mediated gene silencing in vitro and in vivo. *Nat. Biotechnol.* 20:1006–10
152. Yanaihara N, Caplen N, Bowman E, Seike M, Kumamoto K, et al. 2006. Unique microRNA molecular profiles in lung cancer diagnosis and prognosis. *Cancer Cell* 9:189–98
153. Yang D, Buchholz F, Huang Z, Goga A, Chen CY, et al. 2002. Short RNA duplexes produced by hydrolysis with *Escherichia coli* RNase III mediate effective RNA interference in mammalian cells. *Proc. Natl. Acad. Sci. USA* 99:9942–47

154. Yi R, Doehle BP, Qin Y, Macara IG, Cullen BR. 2005. Overexpression of exportin 5 enhances RNA interference mediated by short hairpin RNAs and microRNAs. *RNA* 11:220–26
155. Zender L, Hutker S, Liedtke C, Tillmann HL, Zender S, et al. 2003. Caspase 8 small interfering RNA prevents acute liver failure in mice. *Proc. Natl. Acad. Sci. USA* 100:7797–802
156. Zhang SZ, Pan FY, Xu JF, Yuan J, Guo SY, et al. 2005. Knockdown of c-Met by adenovirus-delivered small interfering RNA inhibits hepatocellular carcinoma growth in vitro and in vivo. *Mol. Cancer Ther.* 4:1577–84
157. Zhang W, Yang H, Kong X, Mohapatra S, San Juan-Vergara H, et al. 2005. Inhibition of respiratory syncytial virus infection with intranasal siRNA nanoparticles targeting the viral NS1 gene. *Nat. Med.* 11: 56–62
158. Zimmermann TS, Lee AC, Akinc A, Bramlage B, Bumcrot D, et al. 2006. RNAi-mediated gene silencing in nonhuman primates. *Nature* 441:111–14
159. Zou L, Zhang P, Luo C, Tu Z. 2006. shRNA-targeted hTERT suppress cell proliferation of bladder cancer by inhibiting telomerase activity. *Cancer Chemother. Pharmacol.* 57:328–34

The Pathophysiology of Fragile X Syndrome

Olga Penagarikano,[1] Jennifer G. Mulle,[1] and Stephen T. Warren[1,2,3]

[1]Department of Human Genetics, [2]Department of Biochemistry and [3]Department of Pediatrics, Emory University School of Medicine, Atlanta, Georgia 30322; email: openaga@genetics.emory.edu, jmulle@genetics.emory.edu, swarren@genetics.emory.edu

Annu. Rev. Genomics Hum. Genet. 2007. 8:109–29

First published online as a Review in Advance on May 3, 2007.

The *Annual Review of Genomics and Human Genetics* is online at genom.annualreviews.org

This article's doi:
10.1146/annurev.genom.8.080706.092249

1527-8204/07/0922-0109$20.00

Key Words

mental retardation, trinucleotide repeat expansion, FMR1, FMRP, synaptic plasticity, autism

Abstract

Fragile X syndrome is the most common form of inherited mental retardation. The disorder is mainly caused by the expansion of the trinucleotide sequence CGG located in the 5′ UTR of the *FMR1* gene on the X chromosome. The abnormal expansion of this triplet leads to hypermethylation and consequent silencing of the *FMR1* gene. Thus, the absence of the encoded protein (FMRP) is the basis for the phenotype. FMRP is a selective RNA-binding protein that associates with polyribosomes and acts as a negative regulator of translation. FMRP appears to play an important role in synaptic plasticity by regulating the synthesis of proteins encoded by certain mRNAs localized in the dendrite. An advancing understanding of the pathophysiology of this disorder has led to promising strategies for pharmacologic interventions.

FX: fragile X

Fragile sites: chromosomal regions that appear as nonrandom chromatin discontinuities at metaphase during cell division

FXS: fragile X syndrome

INTRODUCTION

In 1911, Thomas H. Morgan first pointed out sex-linked inheritance of genetic traits in *Drosophila melanogaster* when he noted that the phenotype "white-eyed" affected only males. Morgan suggested that there could be a recessive genetic trait affecting only males if the gene responsible for it was located on the X chromosome, in the region not common with the Y chromosome. Twenty-five years after Morgan described sex-linked inheritance in flies, human studies were noting a higher incidence of mental retardation in males compared with females (90), and the first pedigrees showing mental retardation associated to the X chromosome were published (79). However, the idea of human intelligence being affected by X-linked genes was not fully appreciated until 1974, when Lehrke published that 25–50% of mental retardation may be attributed to X-chromosome mutations (73). Although this estimation was later discovered to be too high, his work established the concept of X-linked mental retardation (XLMR). Today, approximately 2% of the general population has an intelligence quotient (IQ) below what is considered normal (IQ < 70) and 15–20% of this is believed to be due to XLMR, with 202 conditions described, of which, to date, 45 have the responsible gene cloned (22). Although each of these X-linked MR loci contributes a relatively small percentage of human XLMR, one locus, that responsible for fragile X syndrome, contributes the largest fraction.

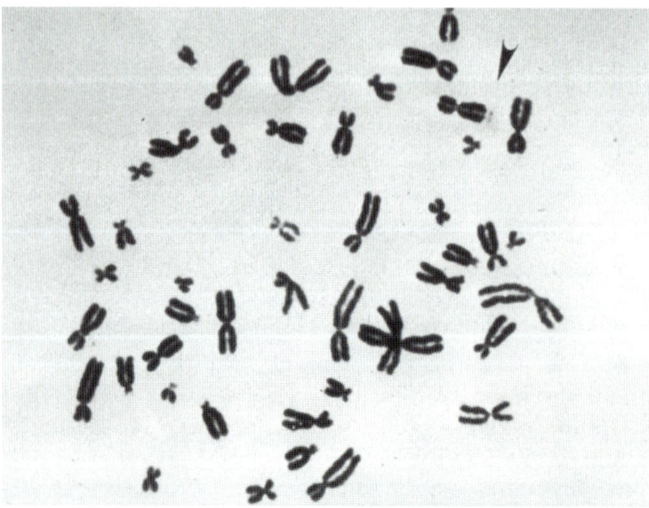

Figure 1

The "marker X" chromosome. Metaphase chromosomes showing the peculiar constriction at the end of the long arm of chromosome X that is characteristic in fragile X (FX) individuals.

FRAGILE X SYNDROME

One of the first large families where mental retardation was transmitted in an X-linked fashion was described by Martin & Bell in 1943 (79). In 1969, Lubs detected an unusual secondary constriction at the end of the long arm of the X chromosome in four mentally retarded males and three obligate carrier females in a single family (76) (**Figure 1**). However, this observation remained a mystery for more than a decade until Sutherland showed that specific culture conditions are needed for the cytogenetic expression of the "marker X," enabling the recognition of additional families (105). Fragile sites are named with the term FRA, designating fragility, followed by the chromosome where they are located and a letter indicating the order in which they were found (14). Thus, the fragile site associated to the fragile X chromosome, now localized specifically to band Xq27.3 (53), was named FRAXA, as it was the first fragile site described on chromosome X. Richards & Sutherland (93) later analyzed the original pedigree described by Martin and Bell and showed that the affected males did indeed have fragile X syndrome. Consequently, the descriptors Martin-Bell syndrome or marker X syndrome have now been largely superseded by the designation fragile X syndrome (FXS).

FRAGILE X PHENOTYPE

Initially, the fragile or marker X chromosome was not clearly associated with any other phenotype but mental retardation. However, once the laboratory conditions necessary for the reproducible expression of the cytogenetic fragile site were established, a number of affected families were detected and the associated phenotype emerged.

Mental retardation is the most prominent phenotype, with IQ values typically between 20 and 70 (41). The cognitive dysfunction particularly affects short-term memory for complex information, visuospatial skills, and speech. A delay in speech is common and often the first symptom that brings the child to medical attention. Some patients show hyperactivity, hypersensitivity to sensorial stimuli, and attention deficit, and between 15–50% of affected individuals show some autistic behavior such as poor visual contact, tactile defensiveness, and repetitive behaviors (26).

Individuals affected by FXS show a subtle, yet characteristic, facial morphology of macrocephaly, with a long narrow face and prominent forehead, jaw, and ears (23) (**Figure 2**). Macroorchidism, or enlarged testicles, are commonly seen in postpubescent male patients. Other physical features such as the presence of a high arched palate, hypotonia (51), flat feet, increased joint laxity, and mitral valve prolapse (74) have also been reported. These findings are collectively believed to be due to a poorly understood connective tissue abnormality that has received little research attention (58).

Several investigators have also reported a neuroendocrine dysfunction in FXS. Although poorly understood at the molecular level, an unusual overgrowth pattern has been attributed to this and includes an increased birth weight, macrocephaly, and increased stature, particularly in childhood. Some individuals affected by FXS have also been diagnosed as Sotos (115) or Prader-Willi syndromes (32) due to their striking physical similarity. Endocrine studies of these patients have reported hypothalamus dysfunction (55).

Figure 2

The fragile X phenotype. Note the characteristic facial features of the syndrome: long, narrow face, prominent forehead, jaw, and ears.

UNUSUAL INHERITANCE PATTERN

As soon as families showing the FX chromosome were identified, it became apparent that the inheritance pattern was not consistent with a recessive X-linked condition, as previously thought. The main conflicting data were the existence of nonaffected male carriers and affected female carriers (84). Sherman et al. (99, 100) conducted extensive pedigree analyses and reported that 20% of males carrying the mutated gene were not affected. These individuals were termed normal transmitting males (NTMs). In addition, 30% of carrier females showed some form of mental

Macroorchidism: testicular volume >30 ml, which is principally developed during the postpubertal period

NTM: normal transmitting male

FMR1: fragile X mental retardation 1

Microsatellite: repetitive DNA sequence consisting of 1–4 base repeat units

Dynamic mutation: A repetitive DNA sequence whose increasing length, beyond a certain threshold, confers an increasing inter-generational instability

impairment. They also noticed that the risk of inheriting the syndrome depended on the position of an individual within the pedigree: The mothers of NTM showed three times less risk of having affected sons than their daughters. This phenomenon is known as the Sherman paradox and was not resolved until 1991, when the gene responsible for the syndrome was identified and, at the same time, a new mutational mechanism revealing the particular inheritance model: trinucleotide repeat expansion.

A NEW CONCEPT OF MUTATION

According to Mendel's work in 1860, the inheritance of genetic traits is determined by units of discrete nature, later named genes, which are transmitted intact and independently to the next generation. Almost 100 years later Benzer showed that genes could suffer mutations, but, once produced, these mutations would be transmitted intact to the next generation. This concept is still valid for most genes; however, since 1991, when the gene responsible for FXS was cloned, it has been revealed that specific repetitive sequences in some genes are unstable and can mutate serially upon transmission.

The Fragile X Mutation

The existence of the Xq27.3 cytogenetic fragile site associated with the syndrome was a target for subsequent molecular studies. Pedigree analysis localized both the causal locus and the fragile site to a 22-cM region on the X chromosome (88), and further studies revealed a number of linked markers that reduced the interval to 1–2 Mb and strengthened the localization of the disease locus to the fragile site (57, 95, 106, 107). Warren et al. (117) described somatic cell hybrids carrying the human fragile X chromosome with rodent chromosome translocations at the FX site. Heitz et al. (54) identified a yeast artificial chromosome (YAC) containing two markers known to flank the fragile site hybrid break points and showed that this YAC contains a CpG island aberrantly methylated in fragile X patients. Furthermore, these abnormally methylated fragments were unstable in affected individuals and increased in size when transmitted through a pedigree (88). In 1991, Verkerk et al. (114) used these somatic cell hybrids and identified a gene, expressed in the brain, that showed a tandem repeat of the CGG trinucleotide in its 5′ region, which was expanded in affected individuals and was coincident with the fragile site. The gene was named Fragile X Mental Retardation 1 (FMR1), assuming that it was the first of a number of yet-identified genes associated with the FX site. We now appreciate that the *FMR1* gene resides in a gene desert, and this single gene is responsible for the syndrome.

Microsatellite sequences, such as the one responsible for FXS, are very frequent in the human genome, being present every ~2 kb and representing approximately 90 Mb of the human genome (62). Most of them show relatively high levels of polymorphism, but are usually stable upon transmission. The variability in the repeat length represents a slight instability in the sequence that usually appears as an expansion or contraction of one or two repeats (12). These small changes in repeat length usually remain in the "normal" range; therefore, most microsatellites are not associated with any disease. However, trinucleotide repeats can show a much greater instability. In particular, when the transmitting allele is beyond a certain threshold size, it generates a new lengthened allele when transmitted that is often a large expansion that leads to genetic disorders. This new and unexpected multistep mutational mechanism is called dynamic mutation (94). FXS, along with Kennedy disease [spinal-bulbar muscular atrophy (SBMA)], was the first to be identified among the disorders associated with trinucleotide repeat expansion. To date, at least 16 disorders are known to be caused by trinucleotide repeat expansion (28). The trinucleotide repeat expansion can lie within the coding region of the gene, as in Kennedy disease or other polyglutamine diseases, where the expanded CAG triplet, which codes for glutamine, is included in the encoded protein and the new mutant protein gains a novel deleterious function. Alternatively, the triplet expansion can be located in a noncoding region, as in FXS, and, depending on the triplet and on the location (within an intron or within a nontranslated region of an exon), the pathogenic mechanism is different, usually leading to a diminished expression or even silencing of the gene.

Trinucleotide Repeat Expansion

Since the molecular basis of this new mutational mechanism was described, special efforts have been made to explain the phenomenon of expansion. The proposed models include unequal exchange of genetic material during meiosis or mitosis due to an incorrect pairing of the chromatids along the repetitive sequence and gene conversion. Although these mechanisms could be the cause of some of the expansions observed in trinucleotide repeats, the initial event leading to the instability of a normal allele thought to be slipped-strand mispairing during DNA replication (70). Richards & Sutherland (93) suggested that the slippage would occur specifically in the Okazaki fragments. Okazaki fragments formed by long or uninterrupted repetitive sequences will not be anchored by unique sequences at their ends, showing a risk of suffering slippage. If slippage occurs, DNA repairing enzymes could add nucleotides to the template strand to make it complementary to the newly formed strand, generating an expansion. In support of this, the absence of a functional flap endonuclease 1 (FEN-1), which is necessary for correct processing of Okazaki fragments, has dramatic effects on repeat instability. Indeed, FEN-1 activity is diminished when the flap length is longer than 11 repeats (72). If the flap is not correctly processed, it will lead to an expansion of the repeat. Initially, a gradual accumulation of one or a few repeats per generation would occur; massive expansions could take place by some continuous slippage due to a block in the replication process generated by alternative structures formed by the trinucleotide repeat sequences (89). Because slipped-strand mispairing during DNA replication is the most accepted mechanism of trinucleotide repeat expansion, it is natural to think that the mismatch repair system (MRS) would play an important role. However, to date no clear data exist regarding its role in triplet repeat instability.

Okazaki fragment: short fragment of DNA created on the lagging strand during DNA replication due to the intrinsic 5′–3′ polarity of DNA polymerase

Flap endonuclease 1 (FEN-1): removes 5′ flaps from Okazaki fragments. Flaps containing repeats could acquire different DNA conformations that enable them to escape FEN-1 cleavage

THE *FMR1* GENE

The *FMR1* gene is 38 kb in length and contains 17 exons (**Figure 3**). The 4.4-kb full-length mRNA codes for a protein with a maximum length of 632 amino acids and a molecular mass of 80 kDa, although different transcripts can be produced by alternative splicing (7). The translation start is located 69 base pairs (bp) downstream the repetitive region and the repeat lays in the 5′ untranslated region (UTR) within the first exon of the gene (7). In addition, the gene shows a CpG island located 250 bp upstream from the repeat (38). The repeat length is polymorphic and ranges from 6–54 CGG in normal individuals (44). Female carriers and NTMs show an expansion between 55 and 200 repeats, termed premutation. When the expansion is more than 200 repeats, it is termed full mutation. As a result of the expansion, the repeat, the upstream CpG island, and the surrounding sequence become hypermethylated and the gene silenced. Thus, the absence of the encoded protein is the basis for the phenotype (91). Silencing the expanded *FMR1* allele is a unique example of a local epigenetic change driven by variation in the gene sequence.

Methylation of the *FMR1* gene acts both directly, by inhibiting the binding of transcription factors, and indirectly, by inducing chromatin condensation, which, in turn, prevents the transcription machinery from binding. The transcriptionally inactive status of the *FMR1* gene shows four occupied

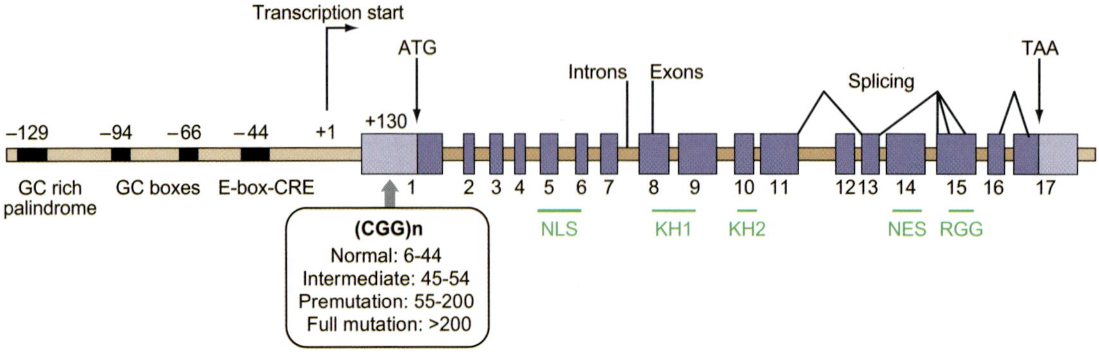

Figure 3

Schematic representation of the *fragile X mental retardation 1* (*FMR1*) gene. The *FMR1* gene shows 17 exons that undergo alternative splicing. The CGG repeat is located within the untranslated region of the first exon. Expansion over 200 repeats leads to methylation of the promoter and transcriptional silencing. The location of the principal domains of the normally encoded protein is also indicated.

transcription factor-binding sites in its promoter, including a palindrome, two GC-like boxes, and an overlapping E-box-cAMP response element (CRE) site. Studies on the proteins that bind to these sites have identified the factors upstream stimulatory factor 1 (USF1) and USF2, nuclear respiratory factor 1 (NRF1) and NRF2, specificity protein 1 (Sp1) and cAMP response element binding protein (CREB) as the most actively involved in *FMR1* transcriptional activity (69, 103). In addition, the 5′ region of the *FMR1* gene is normally associated with histone proteins H3 and H4 acetylated in their lysine residues, but this acetylation is reduced in FX cells (25). Changes in histone methylation have also been described, with histone 3 showing methylated lysine 4 and unmethylated lysine 9 in normal cells but the opposite methylation pattern in FX cells (24). Attempts to pharmacologically reactivate the gene have shown a synergistic effect using both DNA methylation and histone deacetylation inhibitors (20); however, the connection between histone modifications and DNA methylation remains unclear. It was recently shown that the transcriptionally active versus inactive state of the *FMR1* gene display different broader chromatin conformations. Chromatin segments located throughout a 50-kb domain around the promoter interact less frequently when *FMR1* is expressed compared with when the gene is repressed. Therefore, the expression-correlated change in conformation affects a significantly larger domain than the one marked by histone modifications (46). However, the cause-and-effect relationship among promoter activation, local histone modifications, and broader changes in chromatin remains to be determined.

All data suggest that the transcriptional repression of the *FMR1* gene occurs when both the CpG island and the CGG repeat become hypermethylated. It is assumed that methylation and consequent inactivation of the gene occur after fertilization during embryogenesis and that the protein is expressed during the first stages of embryonic development (75, 119). However, this early expression does not seem to play an important biological role as individuals affected as consequence of deletions in the *FMR1* gene, and therefore not expressing the protein at any point during their development, do not show any phenotypic differences from individuals affected owing to gene methylation (45, 120).

Factors Influencing CGG Repeat Expansion

With FXS, as in every dynamic mutation, there is a multistep process of expansion that takes place over many generations. Normal repeats (6–54 CGG) are usually transmitted stably, but sometimes increase (or decrease) by a few repeats, creating a pool of at-risk alleles that can eventually become premutations. Premutations (55–200 CGG) tend to expand when maternally transmitted, as male spermatogenesis is unable to maintain full-mutation alleles (78). The probability of expansion depends on the repeat length. Alleles larger than 90–100 CGG repeats show an almost 100% risk of expansion to full mutation (>200 CGG) in the next generation (86). Smaller alleles, although unstable, usually do not expand to a full mutation in one generation. To date, the smallest allele that has suffered that kind of change has 59 repeats (87). In addition, sequencing analysis of normal *FMR1* alleles revealed that the CGG repeat is not pure, but presents an AGG interruption every 9–10 repeat units, and the most common allele is $(CGG)_{9-10}AGG(CGG)_9AGG(CGG)_9$ (71). However, in FX families the premutated alleles usually have one or no interruptions. Loss or lack of one of these interruptions is an important determinant for repeat instability (37). The 59 repeat allele that expanded to a full mutation in one generation had no interruptions, supporting their important role. The fact that most of the premutation alleles bigger than 90 repeats expand to full mutation in one generation, and assuming that the maximum number of AGG interruptions that they show is probably 2, and that they are located every 9–10 CGG, suggests that approximately 70 pure repeats are needed for the allele to be fully expanded. Because the most likely mechanism of expansion is slipped-strand mispairing during DNA replication in the Okazaki fragments, which range among 25–300 nucleotides, and 70 repeats make 210 nucleotides, there is a high probability that this fragment is not anchored by unique flanking sequences, which supports the idea of strand slippage.

Analysis of polymorphic markers within and near the *FMR1* gene shows that certain haplotypes are preferably associated with the FX mutation. This observation is denoted as linkage disequilibrium and suggests the existence of a pool of founder chromosomes that lead to the FX chromosomes (71). Thus, apparently nonrelated individuals could have common ancestors, and the origin of the mutation could be the same. Different mutational routes that correspond to the different haplotypes observed in FX individuals may exist. They typically involve the generation of alleles with longer tracts of pure CGG repeats either by loss of AGG interruptions or by gradual increase in the repeat length.

Linkage disequilibrium: nonrandom association of alleles at two or more linked loci

It is still not clear when the repeat expansion occurs. First, the observation that affected individuals are mosaics, showing both premutation and full-mutation alleles or full-mutation alleles of different sizes, together with the fact that affected males show only premutations in their sperm (42) led to the assumption of a postzygotic model, where the zygote harbors a premutation and the expansion occurs mitotically during early embriogenesis. It was later pointed out that there was no correlation between the repeat length and the degree of mosaicism and that no mechanism that distinguishes a maternal transmission, suffering expansion, from a paternal transmission, not suffering expansion, was known. Therefore, the mosaicism observed in patients would likely be due to the contraction of somatic alleles rather than to the different expansion of maternal premutated alleles during gametogenesis, and expansion would probably be prezygotic (83). The prezygotic model of expansion is supported by the observation of full-mutation alleles in oocytes and in fetal spermatogonia, but only premutations in testes at a later stage, suggesting that the expansion must occur before germinal differentiation (78).

Animal models will be essential to determine the timing and the mechanism of both the expansion and the methylation and inactivation seen in humans. However, although some expansions were recently reported in a mouse containing a human $(CGG)_{98}$ repeat

POF: premature ovarian failure

FXTAS: fragile X tremor ataxia syndrome

(15) (even occasionally reaching the full-mutation range), to date there are no reports of massive expansions occurring in the mouse at the same frequencies as are observed in human pedigrees. In addition, mice with long repeats (~230 CGG) show an absence of abnormal methylation, indicating that the FX full mutation has not yet been mimicked in mice.

Fragile X Syndrome Without CGG Expansion

Because the cause of the phenotype associated with FXS is the lack of the protein encoded by the *FMR1* gene, it is reasonable to think that other mutations apart from the CGG expansion could lead to the disease. Actually, more than 15 deletions affecting part of or the entire gene, and ranging from 1.6 kb to 13 Mb, have been described (45, 52, 56, 63, 81, 109, 120). Besides deletions, point mutations that lead to the phenotype due to the production of a nonfunctional protein have also been described (31, 77). It is important to consider that most diagnostic testing for FXS is limited to the examination of the CGG repeat and largely ignores conventional mutations. However, because the full mutation leads to a functional null allele, it is likely that there are many nonsense and missense mutations of *FMR1* that are being missed.

FRAGILE X PREMUTATION

For many years, male and female premutation carriers were considered asymptomatic. This view was gradually challenged by reports of different premutation-associated clinical phenotypes over the past decade. The prevalence of premutation alleles >54 repeats in the general population is estimated at 1/259 females and 1/813 males (96). As early as 1986, Fryns (43) noted an increased rate of twinning among female premutation carriers, suggesting the possibility of ovarian dysfunction. A few years later, Cronister et al. (27) noticed that women carrying the FX mutation who were cognitively normal had a higher prevalence of premature ovarian failure. The mean age of menopause in the general population is 51. It is assumed that it occurs as a result of the reduction of the initial oocyte pool below a determined level. Premature ovarian failure (POF) is defined as menopause before 40 years of age. Both genetic and environmental causes are known (97), but 60% of the cases have no obvious cause. When premutation carriers could be distinguished from mutation carriers, it was shown that only the former showed this higher prevalence (16–24%), whereas the latter showed the same risk as noncarrier relatives and the general population (~1%) (3). In addition, an increased level of follicle-stimulating hormone (FSH), which is as an indicator of ovarian aging, was found in premutation carriers (104). Because only a proportion of women who carry the premutation have ovarian dysfunction, it is important to identify the risk factors that lead to clinical variability. The most obvious factors are those related to the *FMR1* gene. In a recent report, Sullivan et al. (104) found a significant influence of repeat size on the risk for ovarian failure, with increasing prevalence of POF and decreasing age at menopause correlating with an increasing repeat size up to 100 repeats.

More recently, an emerging group of premutation carriers showing neurological problems was identified. Hagerman et al. (50) identified five male patients with clinical features of intention tremor and cerebellar ataxia at 50–60 years of age. The disorder was designated as fragile X tremor ataxia syndrome (FXTAS). Prior to onset, these patients had normal cognitive ability. It was later shown that this phenotype is not exclusive to males, as earlier suggested, but that females can also be affected, although they show a later onset, likely due to the degree of inactivation of the affected X chromosome (49). These patients show a general reduction in brain size together with cerebellar damage, severe loss of Purkinje cells, spongiosis of the white matter, and axon dystrophy. In addition, eosinophilic intranuclear inclusions, mainly formed by RNA and proteins, in neurons and astrocytes throughout the brain were found in postmortem samples obtained from symptomatic male premutation carriers (47). These types of granules have also been found in other neurodegenerative diseases caused by trinuclotide repeat expansion, such as polyglutamine diseases and myotonic dystrophy.

The existence of phenotypes associated exclusively with the FX premutation suggests that a molecular cause other than the reduction in FMRP levels must exist. mRNA levels increased up to tenfold in premutation carriers, despite the reduction of FMRP expression (110). Transcriptional activity seems to increase in response to a decreased translation rate (40). A dominant RNA gain-of-function mechanism has been proposed as the molecular mechanism underlying FXTAS. The elevated number of *FMR1* transcripts present in premutation carriers might result in sequestration of cellular proteins and formation of inclusions. A mouse model for FXTAS that displays similar neuronal intranuclear inclusions was recently generated (112). Further research on the development and composition of the inclusions will help scientists to better understand the mechanism of RNA-mediated neurodegeneration. Ovarian dysfunction could also result from a toxic effect on the follicles of an increased amount of *FMR1* RNA, leading to increased follicular atresia rates rather than a reduction of the initial oocyte pool size. Future studies on ovarian tissue from premutation carriers will be necessary to investigate this hypothesis.

Fragile X Mental Retardation Protein

Understanding the pathogenesis of FXS involves studying its cause and, thus, the function of the molecule absent in affected individuals: FMRP. The protein encoded by the *FMR1* gene is expressed widely, particularly in the brain and testes (33), which is consistent with the primary characteristics of the syndrome, mental retardation, and macroorchidism. As a result of alternative splicing, the *FMR1* gene can generate at least 12 different proteins between 67–80 kDa (33, 101). Their expression does not seem to be tissue specific, as different tissues show the same expression pattern (113). Alternative splicing affects exons 12, 14, 15, and 17. In general, the most common isoforms lack exon 12, and isoforms lacking exon 14 show the lowest expression (102). FMRP is predominantly localized in the cytoplasm; however, it was shown that isoforms lacking exon 14 were localized to the nucleus, consistent with the identification of a nuclear exportation signal (NES) encoded by this exon. FMRP also contains a nuclear localization signal (NLS) in exons 1–5, indicating that FMRP shuttles between the nucleus and cytoplasm. This has been corroborated by electron microscopy (39).

FMRP: fragile X mental retardation protein

FMRP shows sequence motifs characteristic of RNA-binding proteins. It has two heterogeneus nuclear ribonucleoprotein K Homology (KH) domains in exons 8 and 10, an Arginine-Glycine-Glycine (RGG) box in exon 15 (7), and an RNA-binding domain in the amino-terminal region of the protein (2). In vitro it binds to approximately 4% of the mRNAs in fetal human brain, including its own message (7). FMRP has also been shown to associate with polyribosomes in an RNA-dependent manner via messenger ribonucleoprotein particles (mRNP) of about 660 kDa, which contain several other proteins including its two autosomal homologs fragile X related protein 1 (FXR1P) and FXR2P encoded by autosomal genes localized to the chromosomal regions 3q28 and 17p13.1, respectively (123). FXR1P and FXR2P show a 86% and 70% level of similarity with FMRP, respectively, and show the same principal functional domains as FMRP: NLS and NES, the KH domains, the RGG box, and a homodimerization domain. They are mainly localized to the cytoplasm associated with polysomes as part of the same protein complex. They also shuttle between the nucleus and cytoplasm, and FXR2P shows a nucleolar localization signal (NoS) and shuttles between the

Synaptic plasticity: ability of the synapse to change in strength or change in transmembrane potential

cytoplasm and the nucleolus (108). In addition to FMRP homologs, other proteins that interact with FMRP or that are present in the same complex have been identified. However, it has not been determined whether all complexes are the same or if there are differences in the associated proteins. If they were different, each complex could be affected by the absence of FMRP in a different way. These proteins include nuclear FMRP interacting protein (NUFIP), cytoplasmic FMRP interacting protein 1 (CYFIP1) and CYFIP2, an 82-kD FMRP interacting protein (82-FIP), nucleolin, and Y-Box factor 1/p50 (YB1/p50) (10, 11, 17, 18, 98).

Posttranslational modifications of FMRP have also been reported. FMRP was recently shown to be normally methylated in the RGG box, mainly in arginine 544, and this methylation seems to regulate its protein-protein and protein-RNA interactions (35). FMRP also shows a cluster of phosphorylated residues on the N-terminal side. The primary site of phosphorylation is at serine 499 and it triggers phosphorylation of nearby residues when the residue is phosphorylated. This phosphorylation does not affect the rate of degradation of FMRP or the species of RNA with which FMRP is associated; rather, it has been observed that the phosphorylated form of FMRP is associated with apparently stalled polyribosomes whereas the unphosphorylated form is associated with actively translating polyribosomes, suggesting that phosphorylation could regulate FMRP function (19).

Identification of FMRP mRNA ligands in vivo will lead to essential information about the functional roles of FMRP. Several works have focused on the search for mRNAs bound to the protein (16, 20, 29, 82). These experiments led to the identification of mRNAs that encode for proteins with an important role for neuronal function, synaptic plasticity, and neuronal maturation. Efforts are being made to elucidate the way that FMRP regulates its target mRNAs. On one hand, FMRP could recognize its targets directly. Darnell et al. (29) showed that the RGG box binds to an RNA loop structure named G quartet, where four guanines are stabilized by Hoogsteen-type hydrogen bonds in a planar conformation. Most of the mRNA targets identified for FMRP in the above-mentioned studies contain a potential G-quartet structure (**Figure 4**). Chen et al. (20) identified another set of RNA ligands containing U-rich sequences with 5–23 bases of repeating U pentamers. In addition, the KH2 domain has been shown to bind a complex RNA tertiary structure named loop-loop pseudoknot or "kissing complex" (30). The authors showed that RNAs with this structure can compete off the association of FMRP with polyribosomes, suggesting that this structure could be crucial for FMRP translational regulation. On the other hand, an indirect or more complex interaction between FMRP and its target mRNAs could occur. Recent data have shown the association of FMRP with the microRNA pathway, which suggests a new way that FMRP could modulate translation. MicroRNAs are short (21 nucleotides) noncoding regulatory RNAs that modulate gene expression by blocking translation of partially homolog mRNAs. MicroRNAs are processed from longer RNAs by an RNAse III enzyme named Dicer and are loaded into an RNA-induced silencing complex (RISC) protein complex, which mediates the interaction between microRNAs and their mRNA targets. FMRP associates with both microRNAs and proteins from the RISC complex as well as with Dicer activity, suggesting that it could also be involved in processing microRNA precursors (66). The "kissing complex" may represent the interaction between an mRNA and a microRNA or other noncoding RNA. FMRP was reported to bind to the small nonmessenger cytoplasmic RNA BC1, which simultaneously associates to mRNA targets known to be regulated by FMRP (122). BC1 has been suggested to mediate the formation of a translational inhibition complex by association to a polyA-binding protein and eIF4A (116). However, BC1 is found in lighter polysome fractions than FMRP,

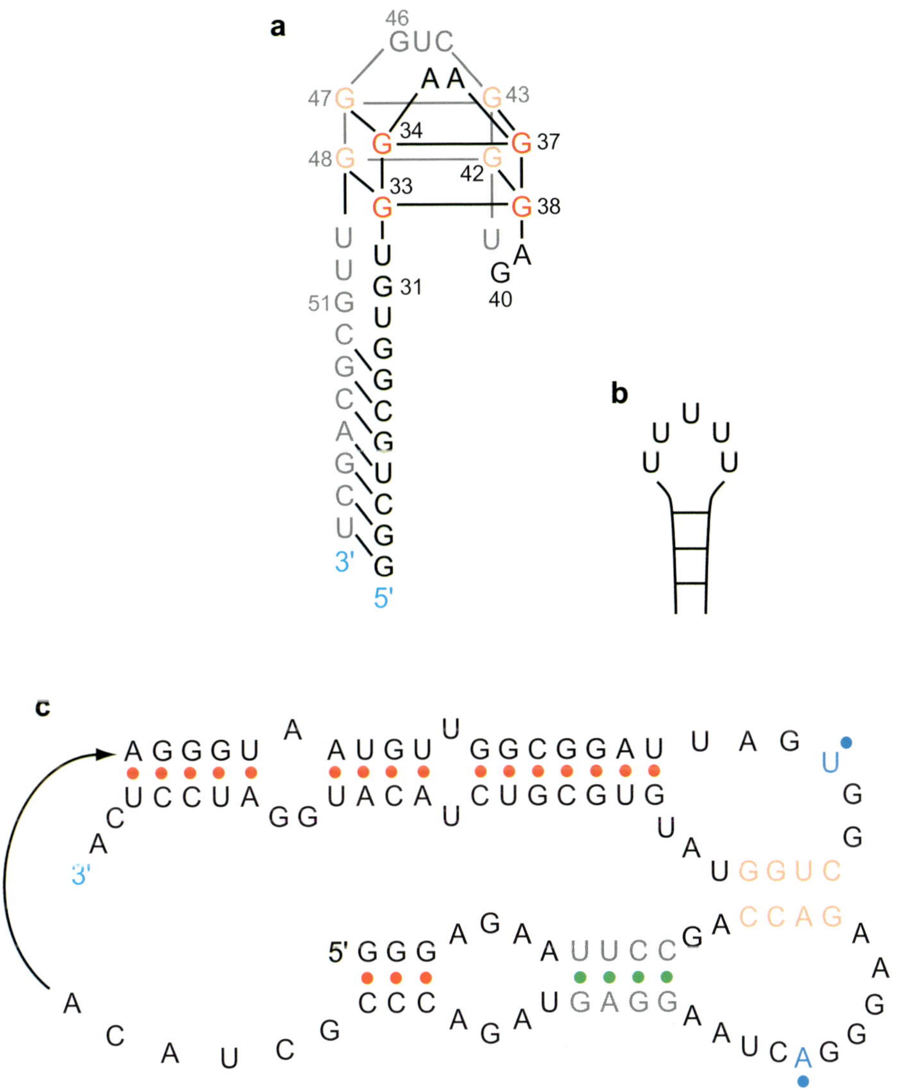

Figure 4

RNA structures recognized by fragile X mental retardation protein (FMRP). (*a*) G quartet, (*b*) U pentamer, (*c*) kissing complex.

making their direct interaction unlikely (68). Recent data support the interaction of FMRP with the RISC complex in regulating synaptic protein synthesis. FMRP interacts with the calcium-calmodulin-dependent kinase II (CaMKII) mRNA, from which dendritic local protein synthesis is required for memory formation (48). Translation of this mRNA is inhibited by RISC and relieved by neuronal stimulation, associating the miRNA pathway with the control of local protein synthesis in neurons (8).

NEURONAL-MOLECULAR CONNECTION

FMRP expression is widely seen during mouse embryogenesis. During development, the expression level diminishes in some tissues, leading to a more specific expression pattern (9). In humans, embryonic studies have shown FMRP expression in the central nervous system of a 9-week-old fetus whereas in a 25-week-old fetus FMRP expression was restricted to neurons (1). Autopsies performed in individuals affected with FXS have

Dendritic spine: small membranous extrusion that protrudes from a dendrite (branched projections of a neuron) and forms one half of a synapse

Synapse: specialized junction through which cells of the nervous system signal to one another from axons to dendrites

LTP: long-term potentiation

LTD: long-term depression

made the associated brain abnormalities apparent. They show more immature (longer and thinner) dendritic spines, where most of the synapses occur, and a higher density of them, suggesting a possible misregulation in their development and elimination (65). An abnormal development of dendritic spines in FMR1 knockout mice has also been reported (85). During development, synapses are often produced in numbers that exceed those that will ultimately survive. Maturation and pruning are therefore important to establish the final synaptic pattern. Abnormalities in the dendritic spines have also been associated with nonsyndromic mental retardation as well as with Down, Rett, Coffin-Lowry, and Rubinstein-Taybi syndromes (34). One of the neurological problems in FXS likely associated with this spine morphology is the presence of epileptic seizures because the characteristic spine morphology could cause a greater neuronal excitation (64). Specific proteins and biochemical pathways have been associated with spine morphogenesis; many of these pathways ultimately lead to actin rearrangement in the spine cytoskeleton. The local synthesis of FMRP can potentially influence other proteins or signaling cascades involved in morphogenesis such as Rac1, MAP1B, CaMKII, calbindin, alpha-glucocorticoid receptor, and cadherins (48). Thus, FXS is considered a synaptic disease.

FMRP is an RNA-binding protein associated with polyribosomes, suggesting that it has a role in modulating protein synthesis in neurons, where it is largely expressed. In neurons, protein synthesis in both cell bodies and dendritic spines is important for synaptic plasticity, and certain mRNAs are translated in response to a specific synaptic activation. Synaptic plasticity is essential for memory and learning processes and involves long-term potentiation (LTP) and long-term depression (LTD), which are associated with synapse creation and elimination in addition to synaptic transmission. Depending on the brain region, synaptic plasticity is expressed pre- and/or postsynaptically. At the postsynaptic level, LTD can be triggered by the activation of postsynaptic N-methyl-D-aspartate (NMDA) receptors or group I metabotropic glutamate receptors (mGluRs). Both types of LTD do not occlude each other, but result in a decrease of surface expression of alpha-amino-3-hydroxy-5-methylisoxazole-4-propionic acid (AMPA) receptors (61). The activation of group I mGluRs (mGluR1 and mGluR5) was reported to stimulate the synthesis of FMRP in synapses (118) and in neuronal lysate (59). *Fmr1* knockout mice (36) showed enhanced mGluR-LTD in the hippocampal Schaffer collateral synapses of the CA1 area (60) and in the cerebellar parallel fiber to Purkinje cell synapses (67). NMDAR-LTD in *Fmr1* knockout mice remained the same as that in the wild-type control, indicating that the phenotype was specific to the mGluR-dependent form of synaptic plasticity. In addition, protein synthesis dependent on mGluR activation was enhanced in the absence of FMRP (60). These data led to the hypothesis that FMRP normally represses further protein synthesis (13). It was recently shown that the rapid increase in FMRP synthesis associated with the stimulation that can cause mGluR-LTD is transient and followed by a rapid decrease in FMRP back to baseline levels, suggesting that FMRP is dynamically regulated during mGluR-LTD, with rapid synthesis followed by rapid degradation in hippocampal slices and homogenate. The authors also showed that the ubiquitin-proteasome degradation pathway played a role in FMRP regulation after mGluR stimulation because FMRP was polyubiquitinated in the same temporal window where its levels begin to decline (59). Therefore, it is reasonable to propose that the rapid degradation of FMRP might be required to permit its target mRNAs to be translated and consolidate mGluR-LTD. In addition to local protein synthesis at synapses, FMRP appears to be involved in the mRNA transport to the synaptic locations. Stimulation of mGluRs increases the transport of

FMRP-containing granules to the dendrites (5, 6). Together, these data suggest a dual role for FMRP: delivery and release of mRNAs for activity-dependent translation.

TREATMENT

Useful guidelines for health supervision of FX children were published by the American Academy of Pediatrics (4) and include advice for both physical and behavioral components of the syndrome. The therapy currently used in FX individuals is designed specifically for each patient and is focused on his/her specific behavior symptoms. Unfortunately, there is not yet a treatment to compensate for the absence of FMRP. Attempts have been made to reactivate the silenced *FMR1* gene and therefore restore the production of the protein (21, 92); however, reactivation processes are too general and genes other than *FMR1* may be reactivated. In addition, toxicity of such approaches is very high. Regardless, reactivation of a full mutation would still produce a transcript with a long repeat that renders the message inefficiently translated. The possibility of gene therapy is currently not possible due to difficulties in the delivery of the normal gene into neurons and its restoration of the phenotype.

More recently, the understanding of the function of FMRP and the consequence of its absence has led to potential therapeutic approaches. Because FMRP is a translational suppressor and mGluR 5 activation of the excitatory pathway stimulates local protein synthesis, the absence of FMRP leads to the apparent overstimulation of translation by mGluR5 signaling (13). This imbalance, which has fundamental consequences with regard to synaptic plasticity, may be countered by mGluR5 antagonists (**Figure 5**). One such antagonist is MPEP. Remarkably, MPEP was recently shown to be highly efficient in rescuing phenotypes associated with FMRP loss in animal models for FXS. In the *Drosophila* model, McBride et al. (80) showed that MPEP can rescue memory deficits in courtship behavior and mushroom body defects observed in mutant flies. Similarly, Yan et al. (121) showed that FMR1 knockout mice treated with MPEP exhibit a reduction in the sensitivity to audiogenic seizures and respond better to the open field test, with a reduced tendency to mainly be in the center of the field. Using zebrafish as a model for the study of human FXS, Tucker et al. (111) recently reported that MPEP was able to rescue the axonal branching defect observed in these animals. These data indicate that the interaction between mGluR signaling and FMR1 function is responsible for some of the symptoms associated with FXS. Thus, drugs targeting mGLuR5 or downstream signaling may be of therapeutic potential in FXS. It is anticipated that clinical trials will begin within the next two years. FXS is instructive regarding genetic disease in general. Once given the "toe hold" of the identification of the responsible gene, basic scientific investigation into the normal and abnormal function of encoded proteins can lead to sufficient molecular understanding to suggest traditional pharmacologic interventions for even the most seemingly intractable disorders.

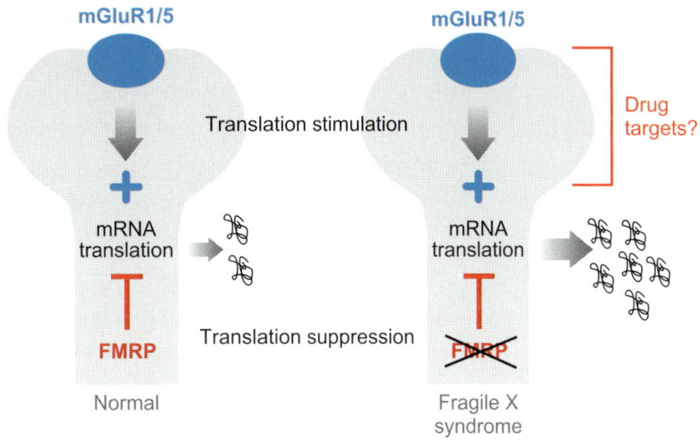

Figure 5

Role of fragile X mental retardation protein (FMRP) in modulating synaptic plasticity. The activation of group I metabotropic receptors (mGluR) stimulates translation of specific mRNAs at synapses. FMRP normally acts as a translational repressor regulating such expression; however, in the absence of FMRP, overexpression of such messages is found.

SUMMARY POINTS

1. FXS is the most common form of inherited mental retardation, has striking behavioral overlap with autism, and may be considered the best-understood form of autism.
2. FXS is caused by the functional absence of FMRP encoded by the *FMR1* X-linked gene, usually due to silencing of the gene caused by expansion of the trinucleotide repeat CGG located in its 5′ UTR.
3. FMRP is a selective RNA-binding protein associated with polyribosomes and expressed in neurons including the dendritic spine, where it has been shown to regulate translation of mRNAs important for synaptic plasticity and neuronal maturation.
4. How FMRP plays its role in translational regulation is still unclear and involves direct interaction with mRNAs as well as interaction with the microRNA pathway.

FUTURE ISSUES TO BE RESOLVED

1. Animal models for both expansion of the CGG repeat and silencing of the *FMR1* gene will be essential to determine the timing and mechanism of repeat expansion and subsequent epigenetic changes in *FMR1*.
2. The molecular details of how FMRP regulates mRNA translation need to be elucidated as does the regulation of FMRP itself.
3. A better understanding is needed of the role of FMRP in local protein synthesis in the dendrite, its normal role in synaptic plasticity, and the neuronal consequence of its absence.
4. Pharmacological approaches have been identified that may compensate for the loss of FMRP, and therapeutic approaches now need to be developed to improve the quality of life of patients with fragile X syndrome.

DISCLOSURE STATEMENT

SW serves as Chair of the Scientific Advisory Committee for Seaside Therapeutics, Inc.

LITERATURE CITED

1. Abitbol M, Menini C, Delezoide AL, Rhyner T, Vekemans M, Mallet J. 1993. Nucleus basalis magnocellularis and hippocampus are the major sites of FMR-1 expression in the human fetal brain. *Nat. Genet.* 4(2):147–55
2. Adinolfi S, Ramos A, Martin SR, Dal Piaz F, Pucci P, et al. 2003. The N-terminus of the Fragile X mental retardation protein contains a novel domain involved in dimerization and RNA binding. *Biochemistry* 42(35):10437–44
3. Allingham-Hawkins DJ, Babul-Hirji R, Chitayat D, Holden JJ, Yang KT, et al. 1999. Fragile X premutation is a significant risk factor for premature ovarian failure: the International Collaborative POF in Fragile X study—preliminary data. *Am. J. Med. Genet.* 83(4):322–25

4. American Academy of Pediatrics Committee on Genetics. 1996. Health supervision for children with Fragile X syndrome *Pediatrics.* 98(3 Pt 1):467–72
5. Antar LN, Dictenberg JB, Plociniak M, Afroz R, Bassell GJ. 2005. Localization of FMRP associated mRNA granules and requirement of microtubules for activity-dependent trafficking in hippocampal neurons. *Genes Brain Behav.* 4:350–59
6. Antar LN, Afroz R, Dictenberg JB, Carroll RC, Bassell GJ. 2004. Metabotropic glutamate receptor activation regulates Fragile X mental retardation protein and FMR1 mRNA localization differentially in dendrites and at synapses. *J. Neurosci.* 24(11):2648–55
7. **Ashley CTJ, Wilkinson KD, Reines D, Warren ST. 1993. FMR1 protein: conserved RNP family domains and selective RNA binding. *Science* 262(5133):563–66**

7. Key report showing the RNA-binding properties of FMRP and giving the first insight into its molecular function.

8. Ashraf SI, McLoon AL, Sclarsic SM, Kunes S. 2006. Synaptic protein synthesis associated with memory is regulated by the RISC pathway in Drosophila. *Cell* 124:191–205
9. Bakker CE, de Diego Otero Y, Bontekoe C, Raghoe P, Luteijn T, et al. 2000. Immunocytochemical and biochemical characterization of FMRP, FXR1P, and FXR2P in the mouse. *Exp. Cell. Res.* 258(1):162–70
10. Bardoni B, Castets M, Huot ME, Schenck A, Adinolfi S, et al. 2003. 82-FIP, a novel FMRP (fragile X mental retardation protein) interacting protein, shows a cell cycle-dependent intracellular localization. *Hum. Mol. Genet.* 12(14):1689–98
11. Bardoni B, Schenck A, Mandel JL. 1999. A novel RNA-binding nuclear protein that interacts with the fragile X mental retardation (FMR1) protein. *Hum. Mol. Genet.* 8(13):2557–66
12. Barker GC. 2002. Microsatellite DNA: a tool for population genetic analysis. *Trans. R. Soc. Trop. Med. Hyg.* 96(Suppl. 1):21–24
13. **Bear MF, Huber KM, Warren ST. 2004. The mGlu theory of Fragile X mental retardation. *Trends Neurosci.* 27(7):370–77**

13. An overview of the role of FMRP in synaptic plasticity where the authors present a theory that can account for diverse neurological and psychiatric aspects of FXS.

14. Berger R, Bloomfield CD, Sutherland GR. 1985. Report of the committee on chromosome rearrangements in neoplasia and on fragile sites. *Cytogenet. Cell. Genet.* 40(1–4):490–535
15. Brouwer JR, Mientjes EJ, Bakker CE, Niewenhuizen IM, Severijnen LA, et al. 2006. Elevated Fmr1 mRNA levels and reduced protein expression in a mouse model with an unmethylated Fragile X full mutation. *Exp. Cell Res.* 313(2):244–53
16. **Brown V, Jin P, Ceman S, Darnell JC, O'Donell WT, et al. 2001. Microarray identification of FMRP-associated brain mRNAs and altered mRNA translational profiles in Fragile X syndrome. *Cell* 107(4):477–87**

16. The first identification of candidate mRNAs regulated by FMRP in mouse brain.

17. Ceman S, Brown V, Warren ST. 1999. Isolation of an FMRP-associated messenger ribonucleoprotein particle and identification of nucleolin and the fragile X-related proteins as components of the complex. *Cell. Biol.* 19:7925–32
18. Ceman S, Nelson R, Warren ST. 2000. Identification of mouse YB1/p50 as a component of the FMRP-associated mRNP particle. *Biochem. Biophys. Res. Commun.* 279(3):904–8
19. Ceman S, O'Donnell WT, Reed M, Patton S, Pohl J, Warren ST. 2003. Phosphorylation influences the translation state of FMRP associated polyribosomes. *Hum. Mol. Genet.* 12:3295–305
20. Chen L, Yun SW, Seto J, Liu W, Toth M. 2003. The Fragile X mental retardation protein binds and regulates a novel class of mRNAs containing U rich target sequences. *Neuroscience* 120(4):1005–17
21. Chiurazzi P, Pomponi MG, Pietrobono R, Bakker CE, Neri G, et al. 1999. Synergistic effect of histone hyperacetylation and DNA demethylation in the reactivation of the *FMR1* gene. *Hum. Mol. Genet.* 8(12):2317–23

22. Chiurazzi P, Tabolacci E, Neri G. 2004. X-linked mental retardation (XLMR): from clinical conditions to cloned genes. *Crit. Rev. Clin. Lab. Sci.* 41(2):117–58
23. Chudley AE, Hagerman RJ. 1987. Fragile X syndrome. *J. Pediatr.* 110(6):821–31
24. Coffee B, Zhang F, Ceman S, Warren ST, Reines D. 2002. Histone modifications depict an aberrantly heterochromatinized *FMR1* gene in Fragile X syndrome. *Am. J. Hum. Genet.* 71(4):923–32
25. Coffee B, Zhang F, Warren ST, Reines D. 1999. Acetylated histones are associated with *FMR1* in normal but not fragile X-syndrome cells. *Nat. Genet.* 22(1):98–101
26. Cornish K, Sudhalter V, Turk J. 2004. Attention and language in fragile X. *Ment. Ret. Dev. Dis.* 10:11–16
27. Cronister A, Schreiner R, Wittenberger M, Amiri K, Harris K, et al. 1991. Heterozygous fragile X female: historical, physical, cognitive, and cytogenetic features. *Am. J. Med. Genet.* 38(2–3):269–74
28. **Cummings CJ, Zoghbi HY. 2000. Trinucleotide repeats: mechanisms and pathophysiology. *Annu. Rev. Genomics Hum. Genet.* 1:281–328**

28. The latest review on the growing number of diseases known to be caused by trinucleotide repeat expansions since the discovery of the causative mutation of FXS.

29. Darnell JC, Jensen KB, Jin P, Brown V, Warren ST, et al. 2001. Fragile X mental retardation protein targets G quartet mRNAs important for neuronal function. *Cell* 107(4):489–99
30. Darnell JC, Fraser CE, Mostovetsky O, Stefani G, Jones TA, et al. 2005. Kissing complex RNAs mediate interaction between the Fragile X mental retardation protein KH2 domain and brain polyribosomes. *Genes Dev.* 19:903–18
31. de Boulle K, Verkerk AJMH, Reyniers E, Vits L, Hendrickx J, et al. 1993. A point mutation in the FMR-1 gene associated with the Fragile X mental retardation. *Nat. Genet.* 3:31–35
32. de Vries BB, Fryns JP, Butler MG, Canziani F, Wesby-van Swaay E, et al. 1993. Clinical and molecular studies in fragile X patients with a Prader-Willi-like phenotype. *J. Med. Genet.* 30(9):761–66
33. Devys D, Lutz Y, Rouyer N, Bellocq JP, Mandel JL. 1993. The FMR-1 protein is cytoplasmic, most abundant in neurons and appears normal in carriers of a Fragile X premutation. *Nat. Genet.* 4:335–40
34. Dierssen M, Ramakers GJ. 2006. Dendritic pathology in mental retardation: from molecular genetics to neurobiology. *Genes Brain Behav.* 5(Suppl. 2):48–60
35. Dolzhanskaya N, Merz G, Aletta JM, Denman RB. 2006. Methylation regulates the intracellular protein-protein and protein-RNA interactions of FMRP. *J. Cell Sci.* 119:1933–46
36. Dutch-Belgian Fragile X Consortium. 1994. Fmr1 knockout mice: a model to study fragile X mental retardation. *Cell* 78:23–33
37. Eichler EE, Holden JJ, Popovich BW, Reiss AL, Snow K, et al. 1994. Length of uninterrupted CGG repeats determines instability in the FMR 1 gene. *Nat. Genet.* 8(1):88–94
38. Eichler EE, Richards S, Gibbs RA, Nelson DL. 1993. Fine structure of the human *FMR1* gene. *Hum. Mol. Genet.* 2(8):1147–53
39. **Feng Y, Gutekunst C, Eberhart DE, Yi H, Warren ST, et al. 1997. Fragile X mental retardation protein: nucleocytoplasmic shuttling and association with somatodendritic ribosomes. *J. Neurosci.* 17(5):1539–47**

39. Places FMRP in the postsynaptic compartment implicating it in local protein synthesis.

40. Feng Y, Zhang F, Lokey LK, Chastain JL, Lakkis L, et al. 1995. Translational suppression by trinucleotide repeat expansion at FMR1. *Science* 268(5211):731–34
41. Fisch GS, Simensen RJ, Schroer RJ. 2002. Longitudinal changes in cognitive and adaptive behavior scores in children and adolescents with the Fragile X mutation or autism. *J. Autism Dev. Disord.* 32(2):107–14

42. Fisch GS, Snow K, Thibodeau SN, Chalifaux M, Holden JJ, et al. 1995. The Fragile X premutation in carriers and its effect on mutation size in offspring. *Am. J. Hum. Genet.* 56(5):1147–55
43. Fryns JP. 1984. The fragile X syndrome. A study of 83 families. *Clin. Genet.* 26(6):497–528
44. Fu Y, Kuhl DPA, Pizzuti A, Pieretti M, Sutcliffe JS, et al. 1991. Variation of the CGG repeat at the Fragile X site results in genetic instability: resolution of the Sherman paradox. *Cell* 67:1047–58
45. Gedeon AK, Baker E, Robinson H, Partington MW, Gross B, et al. 1992. Fragile X syndrome without CCG amplification has an *FMR1* deletion. *Nat. Genet.* 1(5):341–44
46. Gheldof N, Tabuchi TM, Dekker J. 2006. The active FMR1 promoter is associated with a large domain of altered chromatin conformation with embedded local histone modifications. *Proc. Natl. Acad. Sci. USA* 103(33):12463–68
47. Greco CM, Hagerman RJ, Tassone F, Chudley AE, Del Bigio MR, et al. 2002. Neuronal intranuclear inclusions in a new cerebellar tremor/ataxia syndrome among fragile X carriers. *Brain* 125(Pt. 8):1760–71
48. Grossman AW, Aldridge GM, Weiler IJ, Greenough WT. 2006. Local protein synthesis and spine morphogenesis: Fragile X syndrome and beyond . *J. Neurosci.* 26(27):7151–55
49. Hagerman PJ, Greco CM, Hagerman RJ. 2003. A cerebellar tremor/ataxia syndrome among fragile X premutation carriers. *Cytogenet. Genome Res.* 100(1–4):206–12
50. Hagerman RJ, Hull CE, Safanda JF, Carpenter I, Staley LW, et al. 1994. High functioning Fragile X males: demonstration of an unmethylated fully expanded FMR-1 mutation associated with protein expression. *Am. J. Med. Genet.* 51(4):298–308
51. Hagerman RJ, McBogg P, Hagerman PJ. 1983. The Fragile X syndrome: history, diagnosis, and treatment. *J. Dev. Behav. Pediatr.* 4(2):122–30
52. Hammond LS, Macias MM, Tarleton JC, Pai GS. 1997. Fragile X syndrome and deletions in *FMR1*: new case and review of the literature. *Am. J. Med. Genet.* 72:430–34
53. Harrison CJ, Jack EM, Allen TD, Harris R. 1983. The fragile X: a scanning electron microscope study. *J. Med. Genet.* 20(4):280–85
54. Heitz D, Rousseau F, Devis D, Saccone S, Abderrahim H, et al. 1991. Isolation of sequences that span the fragile X and identification of a fragile X related CpG island. *Science* 251:1236–39
55. Hessl D, Rivera SM, Reiss AL. 2004. The neuroanatomy and neuroendocrinology of Fragile X syndrome. *Ment. Ret. Dev. Dis.* 10:17–24
56. Hirst M, Grewal P, Flannery A, Slatter R, Maher E, et al. 1995. Two new cases of *FMR1* deletion associated with mental impairment. *Am. J. Hum. Genet.* 56(1):67–74
57. Hirst MC, Rack K, Nakahori Y, Roche A, Bell MV, et al. 1991. A YAC contig across the fragile X site defines the region of fragility. *Nucleic Acids Res.* 19(12):3283–88
58. Hjalgrim H, Fisher Hansen B, Brondum-Nielsen K, Nolting D, Kjaer I. 2000. Aspects of skeletal development in Fragile X syndrome fetuses. *Am. J. Med. Genet.* 95(2):123–29
59. Hou L, Antion MD, Hu D, Spencer CM, Paylor R, Klann E. 2006. Dynamic translational and proteasomal regulation of fragile X mental retardation protein controls mGluR-dependent long-term depression. *Neuron* 51(4):441–54
60. Huber KM, Gallagher SM, Warren ST, Bear MF. 2002. Altered synaptic plasticity in a mouse model of Fragile X mental retardation. *Proc. Nat. Acad. Sci. USA* 99(11):7746–50
61. Huber KM, Kayser MS, Bear MF. 2000. Role for rapid dendritic protein synthesis in hippocampal mGluR-dependent long-term depression. *Science* 288(5469):1254–57
62. IHGSC (International Human Genome Sequencing Consortium). 2001. Initial sequencing analysis of the human genome. *Nature* 409:860–921

63. Imbert G, Feng Y, Nelson DL, Warren ST, Mandel JL. 1998. *FMR1* and mutations in Fragile X syndrome: molecular biology, biochemistry, and genetics. In *Genetic Instabilities and Hereditary Neurological Diseases*, ed. RD Wells, ST Warren, 3:27–53. San Diego: Academic
64. Incorpora G, Sorge G, Sorge A, Pavone L. 2002. Epilepsy in Fragile X syndrome. *Brain. Dev.* 24(8):766–69
65. Irwin SA, Patel B, Idupulapati M, Harris JB, Crisostomo RA, et al. 2001. Abnormal dendritic spine characteristics in the temporal and visual cortices of patients with fragile-X syndrome: a quantitative examination. *Am. J. Med. Genet.* 98(2):161–67
66. **Jin P, Zarnescu DC, Ceman S, Nakamoto M, Mowrey J, et al. 2004. Biochemical and genetic interaction between the Fragile X mental retardation protein and the microRNA pathway. *Nat. Neurosci.* 7(2):113–17**

66. First report showing the in vivo interaction between mammalian FMRP and both microRNAs and the components of the microRNA pathway.

67. Koekkoek SK, Yamaguchi K, Milojkovic BA, Dortland BR, Ruigrok TJ, et al. 2005. Deletion of FMR1 in Purkinje cells enhances parallel fiber LTD, enlarges spines, and attenuates cerebellar eyelid conditioning in Fragile X syndrome. *Neuron* 47(3):339–52
68. Krichevsky AM, Kosik KS. 2001. Neuronal RNA granules: a link between RNA localization and stimulation-dependent translation. *Neuron* 32(4):683–96
69. Kumari D, Gabrielian A, Wheeler D, Usdin K. 2005. The roles of Sp1, Sp3, USF1/USF2 and NRF-1 in the regulation and three-dimensional structure of the Fragile X mental retardation gene promoter. *Biochem. J.* 386(Pt. 2):297–303
70. Kunkel TA. 1993. Nucleotide repeats. Slippery DNA and diseases. *Nature* 365(6443):207–8
71. Kunst CB, Warren ST. 1994. Cryptic and polar variation of the Fragile X repeat could result in predisposing normal alleles. *Cell* 17:853–61
72. Lee S, Park MS. 2002. Human FEN-1 can process the 5′-flap DNA of CTG/CAG triplet repeat derived from human genetic diseases by length and sequence dependent manner. *Exp. Mol. Med.* 34(4):313–17
73. Lehrke RG. 1974. X-linked mental retardation and verbal disability. *Birth Defects Orig. Artic. Ser.* 10(1):1–100
74. Loehr JP, Synhorst DP, Wolfe RR, Hagerman RJ. 1986. Aortic root dilatation and mitral valve prolapse in the Fragile X syndrome. *Am. J. Med. Genet.* 23(1–2):189–94
75. Losekoot M, Hoogendoorn E, Olmer R, Jansen CC, Oosterwijk JC, et al. 1997. Prenatal diagnosis of the Fragile X syndrome: loss of mutation owing to a double recombinant or gene conversion event at the *FMR1* locus. *J. Med. Genet.* 34(11):924–26
76. Lubs HA. 1969. A marker X chromosome. *Am. J. Hum. Genet.* 21(3):231–44
77. Lugenbeel KA, Peier AM, Carson NL, Chudley AE, Nelson DL. 1995. Intragenic loss of function mutations demonstrate the primary role of *FMR1* in Fragile X syndrome. *Nat. Genet.* 10(4):483–85
78. Malter H, Iber JC, Willemsen R, Graaff E, Tarleton JC, et al. 1997. Characterization of the full Fragile X syndrome mutation in fetal gametes. *Nat. Genet.* 15:165–69
79. Martin J, Bell J. 1943. A pedigree of mental defect showing sex linkage. *Arch. Neurol. Psychiat.* 6:154–57
80. McBride SM, Choi CH, Wang Y, Liebelt D, Braunstein E, et al. 2005. Pharmacological rescue of synaptic plasticity, courtship behaviour and mushroom body defects in a Drosophila model for fragile X syndrome. *Neuron* 45(5):753–64
81. Meijer H, de Graaff E, Merckx DM, Jongbloed RJ, de Die-Smulders CE, et al. 1994. A deletion of 1.6 kb proximal to the CGG repeat of the *FMR1* gene causes the clinical phenotype of the Fragile X Syndrome. *Hum. Mol. Genet.* 3(4):615–20

82. Miyashiro KY, Beckel-Mitchener A, Purk TP, Becker KG, Barret T, et al. 2003. RNA cargoes associating with FMRP reveal deficits in cellular functioning in *FMR1* null mice. *Neuron* 37(3):417–31
83. Moutou C, Vincent MC, Biancalana V, Mandel JL. 1997. Transition from premutation to full mutation in Fragile X syndrome is likely to be prezygotic. *Hum. Mol. Genet.* 6(7):971–79
84. Nielsen KB, Tommerup N, Poulsen H, Mikkelsen M. 1981. X-linked mental retardation with fragile X. A pedigree showing transmission by apparently unaffected males and partial expression in female carriers. *Hum. Genet.* 59(1):23–25
85. Nimchinsky EA, Oberlander AM, Svoboda K. 2001. Abnormal development of dendritic spines in FMR1 knock-out mice. *J. Neurosci.* 21(14):5139–46
86. **Nolin S, Brown WT, Glicksman A, Houck GE, Gargano AD, et al. 2003. Expansion of the Fragile X CGG repeat in females with premutation or intermediate alleles. *Am. J. Hum. Genet.* 72:454–64**

86. Collaborative study of 13 laboratories in 8 countries was established to examine four issues concerning FMR1 CGG-repeat instability.

87. Nolin SL, Lewis FA, Ye LL, Houck GE, Glicksman AE, et al. 1996. Familial transmission of the FMR1 CGG repeat. *Am. J. Med. Genet.* 59:1252–61
88. Oberle I, Rousseau F, Heitz D, Kretz C, Devys D, et al. 1991. Instability of a 550 base pair DNA segment and abnormal methylation in fragile X syndrome. *Science* 252:1097–102
89. Pearson CE, Edamura KN, Cleary JD. 2005. Repeat instability: mechanisms of dynamic mutations. *Nat. Rev. Genet.* 6(10):729–42
90. Penrose LS. 1938. A clinical and genetic study of 1,280 cases of mental defect. In *Special Report Series, Medical Research Council 299*. London: Her Majesty's Stationary Office
91. Pieretti M, Zhang F, Fu YH, Warren ST, Oostra BA, et al. 1991. Absence of expression of the *FMR1* gene in Fragile X syndrome. *Cell* 66:817–22
92. Pietrobono R, Pomponi MG, Tabolacci E, Oostra BA, Chiurazzi P, et al. 2002. Quantitative analysis of DNA demethylation and transcriptional reactivation of the *FMR1* gene in Fragile X cells treated with 5-azadeoxycytidine. *Nucl. Acids Res.* 30(14):3278–85
93. Richards RI, Sutherland GR. 1994. Simple repeat DNA is not replicated simply. *Nat. Genet.* 6(2):114–16
94. Richards RI, Sutherland GR. 1997. Dynamic mutation: possible mechanisms and significance in human disease. *Trends Biochem. Sci.* 22(11):432–36
95. Rousseau F, Heitz D, Biancalana V, Blumenfeld S, Kretz C, et al. 1991. Direct diagnosis by DNA analysis of the Fragile X syndrome of mental retardation. *N. Engl. J. Med.* 325(24):1673–81
96. Rousseau F, Rouillard P, Morel ML, Khandjian EW, Morgan K. 1995. Prevalence of carriers of premutation size alleles of the FMR1 gene and implications for the population genetics of the fragile X syndrome. *Am. J. Hum. Genet.* 57:1006–18
97. Santoro N. 2003. Mechanisms of premature ovarian failure. *Ann. Endocrinol. (Paris)* 64(2):87–92
98. Schenck A, Bardoni B, Moro A, Bagni C, Mandel JL. 2001. A highly conserved protein family interacting with the fragile X mental retardation protein (FMRP) and displaying selective interactions with FMRP-related proteins FXR1P and FXR2P. *Proc. Natl. Acad. Sci. USA* 98(15):8844–49
99. Sherman SL, Jacobs PA, Morton NE, Froster-Iskenius U, Howard-Peebles PN, et al. 1985. Further segregation analysis of the Fragile X syndrome with special reference to transmitting males. *Hum. Genet.* 69(4):289–99
100. Sherman SL, Morton NE, Jacobs PA, Turner G. 1984. The marker (X) syndrome: a cytogenetic and genetic analysis. *Ann. Hum. Genet.* 48(Pt. 1):21–37

101. Siomi H, Siomi MC, Nussbaum RL, Dreyfuss G. 1993. The protein product of the fragile X gene, FMR1, has characteristics of an RNA-binding protein. *Cell* 74(2):291–98
102. Sittler A, Devys D, Weber C, Mandel JL. 1996. Alternative splicing of exon 14 determines nuclear or cytoplasmic localisation of *FMR1* protein isoforms. *Hum. Mol. Genet.* 5(1):95–102
103. Smith KT, Nicholls RD, Reines D. 2006. The gene encoding the fragile X RNA-binding protein is controlled by nuclear respiratory factor 2 and the CREB family of transcription factors. *Nucl. Acids Res.* 34(4):1205–15
104. Sullivan AK, Marcus M, Epstein MP, Allen EG, Anido AE, et al. 2005. Association of FMR1 repeat size with ovarian dysfunction. *Hum. Reprod.* 20(2):402–12
105. Sutherland GR. 1977. Fragile sites on human chromosomes: demonstration of their dependence on the type of tissue culture medium. *Science* 197(4300):265–66
106. Suthers GK, Callen DF, Hyland VJ, Kozman HM, Baker E, et al. 1989. A new DNA marker tightly linked to the fragile X locus (FRAXA). *Science* 246(4935):1298–300
107. Suthers GK, Hyland VJ, Callen DF, Oberle I, Rocchi M, et al. 1990. Physical mapping of new DNA probes near the fragile X mutation (FRAXA) by using a panel of cell lines. *Am. J. Hum. Genet.* 47(2):187–95
108. Tamanini F, Kirkpatrick LL, Schonkeren J, van Unen L, Bontekoe C, et al. 2000. The fragile X-related proteins FXR1P and FXR2P contain a functional nucleolar-targeting signal equivalent to the HIV-1 regulatory proteins. *Hum. Mol. Genet.* 9(10):1487–93
109. Tarleton J, Richie R, Schwartz C, Rao K, Aylsworth AS, et al. 1993. An extensive de novo deletion removing *FMR1* in a patient with mental retardation and the Fragile X syndrome phenotype. *Hum. Mol. Genet.* 2(11):1973–74
110. Tassone F, Hagerman RJ, Taylor AK, Mills JB, Harris SW, et al. 2000. Clinical involvement and protein expression in individuals with FMR1 premutation. *Am. J. Med. Genet.* 91(2):144–52
111. Tucker B, Richards RI, Lardelli M. 2006. Contribution of mGluR and Fmr1 functional pathways to neurite morphogenesis, craniofacial development and fragile X syndrome. *Hum. Mol. Genet.* 15(23):3446–58
112. Van Dam D, Errijgers V, Kooy RF, Willemsen R, Mientjes E, et al. 2005. Cognitive decline, neuromotor and behavioural disturbances in a mouse model for fragile-X-associated tremor/ataxia syndrome (FXTAS). *Behav. Brain Res.* 162(2):233–39
113. Verkerk AJ, de Graaff E, De Boulle K, Eichler EE, Konecki DS, et al. 1993. Alternative splicing in the fragile X gene *FMR1*. *Hum. Mol. Genet.* 2:399–404
114. Verkerk AJMH, Pieretti M, Sutcliffe JS, Fu YH, Kuhl DPA, et al. 1991. Identification of a gene (FMR-1) containing a CGG repeat coincident with a breakpoint cluster region exhibiting length variation in Fragile X syndrome. *Cell* 65:905–14

114. Reports the identification of the gene responsible for FXS.

115. Verloes A, Sacre JP, Geubelle F. 1987. Sotos syndrome and fragile X chromosomes. *Lancet* 2(8554):329
116. Wang H, Iacoangeli A, Popp S, Muslimov IA, Imataka H, et al. 2002. Dendritic BC1 RNA: functional role in regulation of translation initiation. *J. Neurosci.* 22(23):10232–41
117. Warren ST, Zhang F, Licameli GR, Peters GF. 1987. The fragile X site in somatic cell hybrids: an approach for molecular cloning of fragile sites. *Science* 237(4813):420–23
118. Weiler IJ, Irwin SA, Klintsova AY, Spencer CM, Brazelton AD, et al. 1997. Fragile X mental retardation protein is translated near synapses in response to neurotransmitter activation. *Proc. Natl. Acad. Sci. USA* 94(10):5395–400

118. First evidence of the potential role of FMRP in modulating synaptic plasticity.

119. Willemsen R, Oosterwijk JC, Los FJ, Galjaard H, Oostra BA. 1996. Prenatal diagnosis of Fragile X syndrome. *Lancet* 348(9032):967–68

120. Wohrle D, Kotzot D, Hirst MC, Manca A, Korn B, et al. 1992. A microdeletion of less than 250 kb, including the proximal part of the FMR-I gene and the fragile-X site, in a male with the clinical phenotype of fragile-X syndrome. *Am. J. Hum. Genet.* (51)2:299–306
121. Yan QJ, Rammal M, Tranfaglia M, Bauchwitz RP. 2005. Suppression of two major Fragile X syndrome mouse model phenotypes by the mGluR5 antagonist MPEP. *Neuropharmacology* 49(7):1053–66
122. Zalfa F, Giorgi M, Primerano B, Moro A, Di Penta A, et al. 2003. The Fragile X syndrome protein FMRP associates with BC1 RNA and regulates the translation of specific mRNAs at synapses. *Cell* 112(3):317–27
123. Zhang Y, O'Connor JP, Siomi MC, Srinivasan S, Dutra A, et al. 1995. The Fragile X mental retardation syndrome protein interacts with novel homologs Fragile X R1 and Fragile X R2. *EMBO J.* 14(21):5358–66

RELATED RESOURCES

Bilen J, Bonini NM. 2005. *Drosophila* as a model for human neurodegenerative disease. *Annu. Rev. Genet.* 39:153–71

Cao X, Yeo G, Muotri AR, Kuwabara T, Gauge FH. 2006. Noncoding RNAs in the mammalian central nervous system. *Annu. Rev. Neurosci.* 29:77–103

Matnya A, Kusher SA, Silva A. 2002. Genetic approaches to molecular and cellular cognition: a focus on LTP and learning and memory. *Annu. Rev. Genet.* 36:687–720

Nimchinsky EA, Sabatini BL, Svoboda K. 2002. Structure and function of dendritic spines. *Annu. Rev. Physiol.* 64:313–53

Steward O, Schuman EM. 2001. Protein synthesis at synaptic sites on dendrites. *Annu. Rev. Neurosci.* 24:299–325

Mapping, Fine Mapping, and Molecular Dissection of Quantitative Trait Loci in Domestic Animals

Michel Georges

Unit of Animal Genomics, GIGA-R and Faculty of Veterinary Medicine, University of Liège, 4000-Liège, Belgium

Annu. Rev. Genomics Hum. Genet. 2007. 8:131–62

First published online as a Review in Advance on May 3, 2007.

The *Annual Review of Genomics and Human Genetics* is online at genom.annualreviews.org

This article's doi: 10.1146/annurev.genom.8.080706.092408

1527-8204/07/0922-0131$20.00

Key Words

positional cloning, genetical genomics, genomic selection

Abstract

Artificial selection has created myriad breeds of domestic animals, each characterized by unique phenotypes pertaining to behavior, morphology, physiology, and disease. Most domestic animal populations share features with isolated founder populations, making them well suited for positional cloning. Genome sequences are now available for most domestic species, and with them a panoply of tools including high-density single-nucleotide polymorphism panels. As a result, domestic animal populations are becoming invaluable resources for studying the molecular architecture of complex traits and of adaptation. Here we review recent progress and issues in the positional identification of genes underlying complex traits in domestic animals. As many phenotypes studied in animals are quantitative, we focus on mapping, fine mapping, and cloning of quantitative trait loci.

DOMESTIC ANIMAL POPULATIONS: A TREASURE TROVE OF PHENOTYPIC DIVERSITY

Animal Breeding: A Phenotype-Driven Mutagenesis Screen of Unprecedented Scale

Forward genetics remains one of the most powerful approaches to elucidate gene function. This justifies the implementation of large-scale phenotype-driven mutagenesis screens in the mouse to fill the so-called phenotype gap (10). Male mice are mutagenized, generating of the order of 2500 mutations per gamete (~100-fold higher than the spontaneous mutation rate), and mutant phenotypes are recovered in the offspring either in dominant or in, considerably more demanding, recessive screens (52).

Animal breeders have been conducting similar phenotype-driven screens in multiple species for the past ~10000 years, exploiting naturally occurring mutations that either predate or occurred after domestication. During this time, breeders identified animals exhibiting innumerable desirable attributes pertaining to behavior, morphology, physiology, disease resistance, etc. Attentive breeders can pick up even the most subtle features that would likely go unnoticed in present-day screens performed in mice. Ongoing selection, serially cumulating the effects of many mutations over time, ultimately led to drastic phenotypic divergence and the wealth of phenotypic variation nowadays displayed by domestic animals. One example of the power of artificial selection is the awesome phenotypic variation exhibited by the ~400 dog breeds accrued over a few thousand years, which probably exceeds the level of phenotypic variation observed among wild-type canids accrued over more than 10 million years (58). The performances of livestock species have increased remarkably over the past 50 years. For example, in the United States, average milk yield per cow has nearly doubled in 40 years. Most of this progress is genetic and results from breeding designs rooted in advanced quantitative genetics that have generated highly specialized breeds (22).

The emergence of molecular genetics in the 1980s spurred efforts to identify the chromosomal regions, genes, and mutations underlying phenotypic variation in domestic animals. The first experiments were primarily conducted in livestock species, targeting agronomically important traits and motivated by the desire for more effective marker-assisted selection (MAS). These applied objectives remain a major driver for most ongoing efforts in animal genomics, but it is becoming increasingly apparent that studying the genomes of domestic animals has immense value beyond agronomic applications. The availability of the genomic sequences of domestic animals will contribute to understanding genome organization and function via comparative genomics (63). Moreover, identifying the molecular determinants of their tremendous phenotypic diversity has the potential to significantly contribute to the phenotype-genotype map and to understanding adaptation (2).

Domestic Animals: Collections of Isolated Founder Populations

Genetically isolated populations, either of small constant size or having expanded after a bottleneck, are a coveted commodity in human genetics (79). A number of factors might facilitate positional cloning in genetic isolates when compared with the general population. These include (*a*) more extensive linkage disequilibrium (LD) and the possibility to apply "drift mapping," (*b*) reduced allelic heterogeneity due to founder effects and the possibility to map through the identification of shared haplotypes, (*c*) reduced locus heterogeneity and, thus, genetic background noise, and (*d*) reduced environmental noise. Positional identification of single-gene disorders has proven effective in genetically isolated populations. However, whether the advantages of genetic isolates extend to the analysis of complex traits remains a matter of debate (89, 97).

Domestic animal populations share several features in common with genetically isolated human populations. It is an old practice of animal breeders to mate related individuals to generate pure-bred animals that satisfy specific standards. Myriad breeds were generated that way, made official more recently by the creation of breed registries or studbooks. The latter can be "closed" (i.e., an animal can only be registered if both its sire and dam are), thus enforcing the genetic isolation, or "open," allowing for a certain degree of migration as long as immigrants comply with the standards of the breed. The number of registered animals per breed varies depending on its popularity. However, and as a result of the extensive use of popular sires, especially in species where artificial insemination (AI) is common practice, the effective population size is often limited, even for the most populous breeds. For example, there are several tens of millions of Holstein-Friesian (HF) dairy cattle in the world, yet their effective population size is estimated to be in the hundreds at most (8). The founder effects linked to the extensive use of specific sires cause bursts of Mendelian disorders. For instance, in the 1990s in the United States ~14% of HF AI bulls were carriers of bovine leukocyte adhesion deficiency (BLAD) and 0.2% of calves were affected by this lethal disease (91). More recently, ~25% of HF AI bulls were carriers of complex vertebral malformation (CVM) (99). Both defects are thought to have been disseminated in the population by one of the founder sires of the modern HF breed: Carlin Ivanhoe Bell. In other domestic species, most breeds are likewise characterized by their burden of inherited defects. In November 2006, Online Mendelian Inheritance in Animals (**http://omia.angis.org.au/**) listed 2539 distinct phenotypes reported in animals, most of which were inherited disorders.

As expected from the population structure, LD extends over much longer chromosome segments in domestic animals when compared with the general human population (28, 58, 64, 75, 95), and this feature is increasingly exploited to map and clone genes and mutations underlying genetic defects in animals (see below).

To a large extent, the jury is still out on how useful domestic animal populations will be for dissecting complex traits, but the combination of long-range LD combined with reduction in allelic and locus heterogeneity should offer favorable grounds. This review focuses on strategies, issues, and results related to the identification of genes underlying complex traits in animals. As most studies in animals have dealt with continuously distributed quantitative traits, the discussion mainly addresses quantitative trait loci (QTL) mapping, fine mapping, and cloning.

QUANTITATIVE TRAIT LOCI MAPPING IN DOMESTIC ANIMALS

Both Experimental and Commercial Pedigrees are Utilized to Map Quantitative Trait Loci in Domestic Animals

The ability to mate animals at will has allowed animal breeders to generate experimental crosses de novo for the purpose of gene mapping. Typically, F1 animals are generated by crossing breeds that are highly divergent for the traits of interest, and the F1 animals are then either intercrossed (F2) or backcrossed (BC) to one of the parental lines. Although the BC design may provide more detection and resolution power when pursuing specific QTL (e.g., QTL with dominance and BC to the recessive line), in most circumstances the F2 design is preferred because of its polyvalence (18). In animal breeding, the choice between an F2 or a BC is most often dictated by practical considerations. A BC pedigree only requires the generation of a few F1 males that can then be mated to available pure-bred females. An F2, however, requires the additional generation of sufficient numbers of F1 females.

Many such experimental crosses have been generated over the years in cattle, sheep, deer, pigs, poultry, and fish. Experimental crosses that deserve special attention are those that were established between domestic strains and their wild-type progenitors. Such F2 pedigrees were generated between European Wild Boar and domesticated Large White pigs (1), as well as between red junglefowl and White Leghorn chickens (51). The aim of these pedigree resources is to identify the genes that underlie the spectacular phenotypic divergence that has accrued as a result of ~10,000 years of artificial selection.

The generation of experimental F2 and BC crosses counting the hundreds of offspring needed to have reasonable power to map QTL is a costly experiment. Moreover, mapping QTL is only the first step in the quest for underlying genes. It is well established that, when working with crosses between inbred strains of mice, for example, fine mapping QTL to chromosome intervals of a size that is compatible with positional cloning requires the generation of thousands of additional progeny, whether through the generation of advanced intercross lines (AILs) or other designs (18). In domestic animals, fine mapping QTL via this road does not seem like an attractive proposition. Finally, F2 and BC designs target the genes that underlie the phenotypic difference between breeds, but most of the ongoing breeding programs exploit the genetic variation that exists within breeds. For example, breeding companies working with dairy cattle are more interested in identifying the genes that differentiate the top dairy bulls from their average contemporaries than in identifying the genes that explain why HF dairy cattle produce more milk than do Angus beef cattle.

For these reasons, several groups have turned to commercial pedigrees to map QTL: the pedigrees are readily available, fine-mapping approaches based on population-wide LD can be implemented, and the targeted QTL are still segregating within the populations of interest. Despite the concern that QTL with large effects might not segregate any longer in commercial populations as a result of selection, this proposition has been particularly coveted in cattle. Indeed, as a result of extensive use of AI, some cattle populations are characterized by amazing "harems" or paternal half-sib pedigrees, which are ideally suited for QTL mapping. For instance, in dairy cattle, bulls that have more than 10,000 milking daughters are commonplace, and some have had more than 200,000. Moreover, in many countries, phenotypic data for the most relevant traits are systematically recorded on hundreds of thousands of animals. These are processed in specialized centers to dissect the environmental and genetic variance components (VCs) yielding precise estimates of heritability as well as "breeding values" (i.e., estimates of the additive genetic predisposition of an animal accounting for known genealogy and corrected for nongenetic effects) for all individuals. In dairy cattle, the use of large sets of paternal half-sisters is one classic design to map QTL and is referred to as the "daughter design." A more popular design is the "granddaughter design," which consists of large sets of paternal half-brothers. This may seem paradoxical as milk production can only be measured in females. In the granddaughter design, the phenotypes used for QTL mapping are the males' breeding values estimated from the performances of their respective daughters by progeny testing. Progeny tests are typically based on at least 50 daughters, and the bulls' breeding value estimates are thus more accurate predictors of genetic merit than the cows' breeding values estimated from their own performances and pedigree data. As a result, the granddaughter design requires 2.5–3 times fewer genotyped animals than does the daughter design (31, 105) (**Figure 1**).

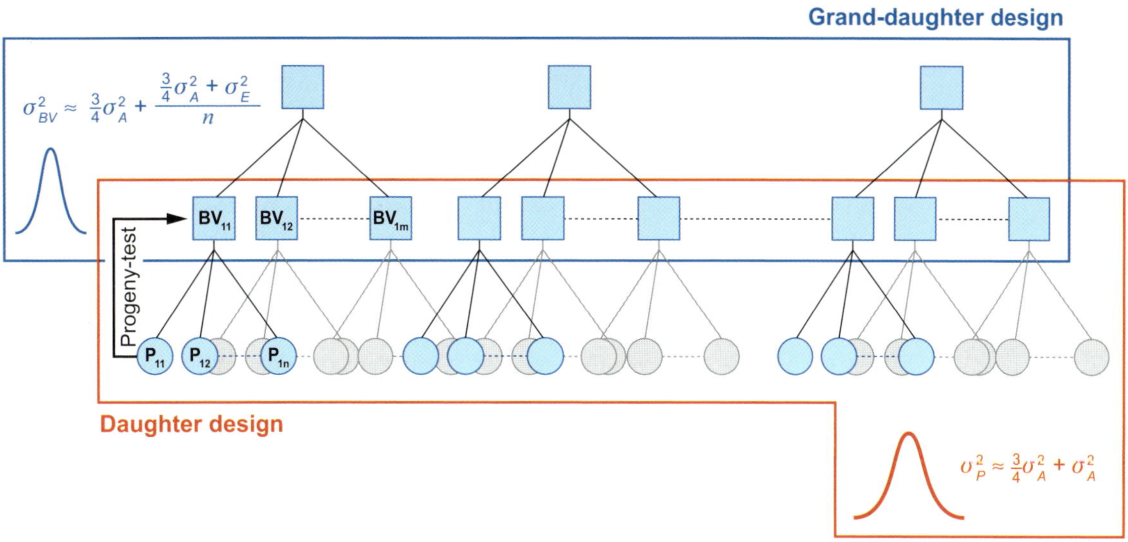

Figure 1

The "granddaughter design" (GDD) is based on a series of paternal half-brother pedigrees, typically 10 to 20 sire groups with 10 to 100 sons each. The phenotypes used for quantitative trait loci (QTL) mapping are the sons' breeding values (BVs) estimated by progeny testing from the phenotypes (P) of at least 50 daughters each. The within-half-brother family variance of BVs is approximately $\sigma^2_{BV} \approx {}^3/_4\sigma^2_A + ({}^3/_4\sigma^2_A + \sigma^2_E)/n$, where σ^2_A is the additive genetic variance, σ^2_E is the nongenetic variance, and n is the number of daughters per son. The "daughter design" (DD) is based on a series of paternal half-daughter pedigrees. The number of daughters per sire is typically much larger than the number of sons per sire in the GDD. The phenotypes used for QTL mapping are the daughters' phenotypes. The within-half-daughter family variance of Ps is approximately $\sigma^2_P \approx {}^3/_4\sigma^2_A + \sigma^2_E$. As the progeny test is typically based on ~100 daughters (= n), σ^2_P is larger than σ^2_{BV} and consequently the sample size needed to detect a QTL that is larger for the DD than for the GDD (e.g., 31).

The Vast Majority of Quantitative Trait Loci Have Been Mapped Using Simple Statistical Methods

To verify whether a given chromosome region harbors a QTL, offspring are sorted in a limited number of genotype classes using polymorphic markers, and the phenotypic means of the different classes are compared. This procedure is repeated sequentially across the genome to identify the chromosomal regions that have a significant effect on the phenotype.

So far, virtually all QTL mapping studies in domestic animals have been conducted with panels of 100–300 microsatellite markers covering the genome, corresponding to an average distance between markers of ~5–20 centiMorgans (cM). With the recent development of genome wide panels of single nucleotide polymorphisms (SNPs) for most domestic species (58, 110), and cost-effective array-based genotyping, SNPs are rapidly becoming the preferred markers for QTL mapping studies in animals.

When working with experimental crosses, the offspring are typically sorted into two (BC: 11 and 12) or three (F2: 11, 12, and 22) genotype classes at each tested position. This can be achieved unambiguously only at marker positions. In a procedure called interval mapping, one also tests marker intervals (55). At these intermarker positions, and as a result of recombination, one does not know with certainty to which class a given individual belongs. This issue is circumvented by computing the corresponding probabilities (e.g., P_{11}, P_{12}, and P_{22} in an F2) conditional on flanking marker genotypes, which can then be used in maximum likelihood (ML) or linear regression (LR) approaches to test for the presence of a QTL. For example, when using the popular LR in an F2 (40), one expresses the phenotype of individual i as $y_i = \mu + (p_{i,11} - p_{i,22})a + p_{i,12}d + \varepsilon_i$, where μ corresponds to the midpoint between the means of classes 11 and 22, $2a$ to the phenotypic difference between the means of classes 11 and 22 (i.e., twice the allele substitution effect), d to the difference between μ and the mean of class 12 (i.e., the dominance deviation), and ε_i to the residual error of individual i. One then identifies the values of μ, a, and d that minimize the sum of squares of residual errors (SSE) over all individuals. The presence of a QTL with significant effect on the phenotype at a given marker position can then be tested knowing that

$$\left(\frac{n-3}{2}\right)\frac{SSR}{SSE}$$

follows a $F_{2,n-3}$ distribution. The sum of squares regression (SSR) measures the variability in the phenotypes due to the segregation of the tested position (**Figure 2**).

As previously mentioned, most studies performed in commercial populations have exploited the "harem" component of the data, i.e., the linkage information contained in the segregation of paternal chromosomes. The principles of QTL mapping in half-sib pedigrees are similar to the above. Animals are first sorted by paternal half-sib pedigree and then, within each half-sib pedigree, according to the homolog (*1* or *2*) inherited from the father at the tested chromosome position. Using paternal half-sib pedigrees precludes exploration of the X chromosome at least via linkage analysis (86). If using LR approaches (54), the phenotype of offspring i of sire j is expressed as $y_{ij} = \mu_j + (p_{i,1} - p_{i,2})a_j + \varepsilon_{ij}$, where μ_j is the phenotypic mean of the offspring of sire j, $p_{i,1}$ and $p_{i,2}$ are the probabilities that offspring i has inherited homolog 1 (respectively 2) at the tested position conditional on flanking marker genotypes, a_j is the 1-to-2 allele substitution effect for sire j, and ε_{ij} is the residual error of offspring i of sire j. Least square values of μ_j and a_j are estimated separately for each family by LR as one cannot assume that all sires will be heterozygous for the QTL and sires can be heterozygous for different QTL alleles; half-sib pedigrees are indeed sampled at random in an outbred population. The presence of a QTL with a significant effect on the phenotype at a given marker position can then be tested using the across-pedigree statistic

$$\frac{\sum_{j=1}^{k} SSR_j/k}{\sum_{j=1}^{k} SSE_j/(n-2k)},$$

where n is the total number of offspring and k is the number of half-sib families. This ratio is distributed as an F statistic under the null hypothesis of no QTL.

Note that by restricting focus to the half-sib component of the pedigree, the previous approach, although robust, potentially ignores a considerable amount of useful linkage information. To access this supplementary information, there have been substantial efforts to develop effective full-pedigree analysis methods (43). Animal breeding has a long-standing tradition of using mixed models and thus it is not surprising that VC methods dominate full-pedigree QTL analysis in animal genetics. Typically, when testing for the presence of a QTL at a given map position, a distinct additive allelic effect is estimated for each chromosome in the sample. Therefore, one has to estimate $2n$ effects (in addition to all other effects in the model) if studying a sample of n individuals. This is possible in a random effect setting by imposing a covariance structure on the solutions. The covariance between the effects of any pair of chromosomes in the sample is assumed to be proportional to the probability that the pair is identical by descent (IBD) at the considered map position. IBD probabilities are computed for all pairs of chromosomes conditional on marker data. This is fairly easily achieved (recursively) with complete marker genotypes, but becomes less

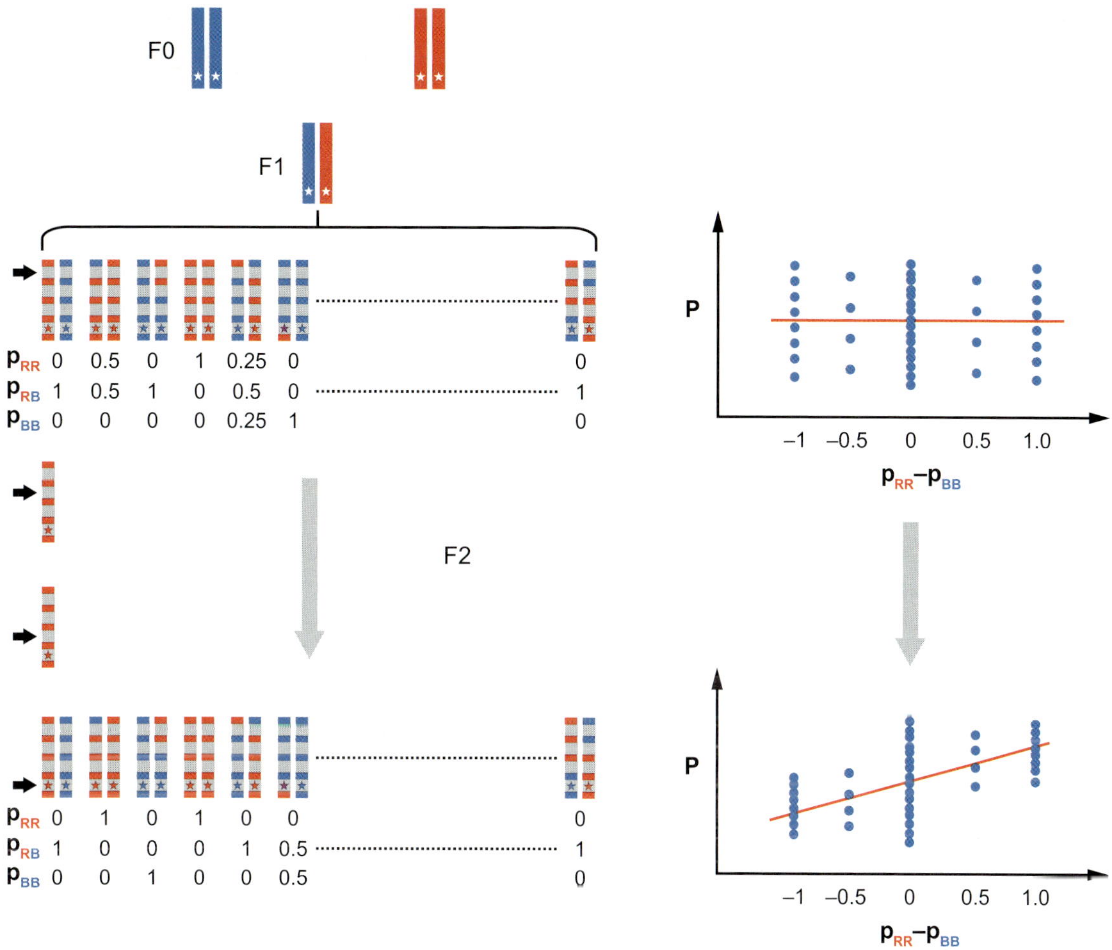

Figure 2

Principles of quantitative trait loci (QTL) interval mapping using linear regression (LR) illustrated for an F2 cross. An F2 population is generated by intercrossing "blue" and "red" parental strains differing for a phenotype of interest. The F2 population is genotyped with a battery of genetic markers covering the genome at regular intervals of ~10 centiMorgans (cM), shown as colored bars on the chromosomes of the F2 individuals. Marker intervals are "interrogated" successively (*black arrows*) for the presence of a QTL. For each interval, and for each F2 individual, one computes the probability that the individual is homozygous "red-red" (p_{RR}), heterozygous "red-blue" (p_{RB}), or homozygous "blue-blue" (p_{BB}), using the observable genotypes at flanking marker loci. The additive effect of a given interval on the phenotype is estimated by regressing the phenotypes on $p_{RR}-p_{BB}$, as shown in the panels on the right. In the absence of a QTL in the tested interval (e.g., interval 1), the regression coefficient does not deviate significantly from 0. In the presence of a QTL in the corresponding interval (shown by the star in interval 4), the regression coefficient may deviate significantly from 0.

trivial with missing genotypes, requiring the use of complex Monte Carlo Markov Chain (MCMC) methods. VCs and individual chromosome effects can then be estimated using this so-called G-matrix via restricted maximum likelihood (REML) methods. In addition to the haplotype effect described above, one would typically include an individual animal effect in the model to correct for stratification, i.e., residual polygenic coancestry that might be correlated with the coancestry at the tested locus (61). VC QTL analysis methods are evolving rapidly (6). So far, their impact in detecting QTL has been relatively limited: No QTL has been detected using VC approaches that would not have been uncovered using the basic approaches described above. However, VC methods will likely make a major contribution when performing combined linkage and LD analyses (see below).

The statistical significance of the QTL signal obtained at a given map position is typically measured by its probability of occurrence under the null hypothesis of "no QTL." This probability has to be adjusted for the number of tests performed, including the analysis of multiple chromosome positions. This is conveniently achieved using permutation testing (26): Phenotypes are shuffled among individuals and genome scans are performed on thousands of such permutated data sets. The values of the test statistics obtained with the real data are compared with the distribution of the highest test statistics obtained for each permutated data set to determine their significance level. A QTL is considered significant if its signal is only encountered once per 20 genome scans with permutated data sets (corresponding to a genome-wide p-value of 0.05). A QTL is considered suggestive if its signal is encountered, on average, once per genome scan with permutated data set (corresponding to a genome-wide p-value of $1\text{-}e^{1} = 0.63$). An alternative approach to evaluating the statistical significance of a given QTL is its false discovery rate (FDR) or expected proportion of false positives among the QTL exceeding the corresponding signal (93, 106). The FDR extracts information from the distribution of p-values over all performed tests. For example, a tenfold excess (10%) of tests significant at the 1% level suggests that 10% of them are false positives.

When detecting a QTL, it is good to have an idea of the accuracy of its predicted position. Confidence intervals (CIs) for the QTL location can be conveniently estimated by bootstrapping methods (102). Data sets are generated in silico by sampling with replacement from the real data set; thus, some individuals will be randomly represented multiple times whereas others will be omitted. The variation of most likely QTL positions across such bootstrap samples is a measure of the accuracy of the QTL location. A CI for the QTL can be defined as the smallest chromosome segment encompassing 95% of the most likely QTL positions across bootstrap samples.

DNA Pooling Combined with Large Paternal Half-Sib Pedigrees Provides Extreme Quantitative Trait Loci Detection Power

The QTL mapping approaches described above can be seen as sorting individuals by genotype and looking for differences between the phenotypic means of the groups. Alternatively, one can sort the individuals by phenotype and look for differences between the genotype frequencies of the groups. When dealing with quantitative traits, individuals are sorted in a discrete number of classes ranging from the low to the high tail of the distribution. Typically, only the classes corresponding to extreme high and low phenotypes are selected for further analysis (i.e., selective genotyping), as these animals provide most of the mapping power (55).

Considerable cost savings can be achieved by genotyping not individual animals, but rather pools of DNA to which all animals contribute equally (i.e. bulk segregant analysis or selective DNA pooling) using genotyping methods that allow for precise quantitation of marker allele frequencies (59).

The common occurrence of large paternal half-sister groups (i.e., thousands to tens of thousands of sisters) combined with the ability to extract DNA from milk pools have made this proposition particularly effective in dairy cattle populations. Using this approach,

QTL affecting milk protein percentage with FDR <0.10 were found on all but two of the 29 bovine autosomes, jointly accounting for a large fraction of the trait variance in the population (71).

The QTL effects estimated from selective genotyping experiments suffer a severe upward bias, requiring either a mathematical correction and/or a re-evaluation in an unselected population. One major limitation of selective genotyping is that the selected population is ideally suited only for analyzing the selected phenotype, although formulas exist that enable unbiased estimation of QTL effects for correlated traits (9).

Innumerable Quantitative Trait Loci Affecting Virtually All Analyzed Traits Have Been Mapped in Domestic Animals

Using the approaches described above, large numbers of QTL influencing a panoply of phenotypes have been identified in livestock. In November 2006, the **http://www.animalgenome.org/bioinfo/** site compiled data on 630 QTL influencing 89 traits in cattle, 1287 QTL influencing 246 traits in pig, and 657 QTL influencing 112 traits in poultry.

Hayes & Goddard (41) combined the results of four QTL mapping experiments targeting growth, carcass, and meat quality in pigs (F2 designs), and three experiments targeting milk yield and composition in cattle (granddaughter designs), in an attempt to gain insight into the molecular architecture of the corresponding quantitative traits. The predicted distribution of QTL effects was consistent with many genes of small effect and few of large effect, consistent with classical population genetics theory (77). Fifty to 100 QTL were estimated to influence the corresponding traits, and ~25% of the leading QTL explained 90% of the genetic variance.

It is worth remembering, however, that the utilized mapping methods cannot effectively distinguish lod score profiles generated by one or a limited number of QTL with moderate to large effect from those generated by a large number of evenly spread QTL with small effect (the "infinitesimal model") (21, 101). Thus, these predictions are circumspect.

About Imprinted Quantitative Trait Loci and Other Parent-of-Origin Effects

Approximately 100 mammalian genes are known to undergo parental imprinting (**http://www.mgu.har.mrc.ac.uk/research/imprinting/**). Close to half of these are preferentially expressed from the paternal allele, and the remainder from the maternal allele. Imprinted genes tend to map to a limited number (~15) of clusters in the genome (80). Imprinting is thought to primarily affect genes that regulate the transfer of maternal resources to offspring in polygamous species (i.e., the parental conflict hypothesis) (108). The propagation of genes that promote transfer of maternal resources to the benefit of the offspring at the expense of the mother's future reproductive potential is maximized by expression from the padumnal (of paternal origin) allele and silencing of the madumnal (of maternal origin) allele. The propagation of genes that counteract the former benefits from padumnal silencing and madumnal expression. The most common and striking symptoms of imprinting perturbation are related to fetal and postnatal growth. Parental imprinting has been found in mammals and flowering plants, but not in birds.

Given this knowledge about imprinted genes, it was reasonable to speculate that part of the genetic variation for growth-related traits in livestock—one of the prime targets of many breeding programs—might be due to polymorphisms in imprinted genes. In fact, it was found that the inheritance of the callipyge muscular hypertrophy in sheep is characterized by a parent-of-origin effect—it can only be transmitted by father to offspring, never from mother to offspring (16)—and later discovered that the *CLPG* mutation indeed perturbs the expression of a cluster of imprinted genes, including padumnal growth enhancers and madumnal growth repressors (32). Moreover, a QTL affecting muscle mass and fat deposition on porcine chromosome 2 was found to be imprinted with exclusive padumnal expression (49, 73) and later shown to correspond to a regulatory mutation enhancing postnatal expression of the imprinted *IGF2* gene in striated muscle (100). These initial findings suggested that an unexpected proportion of QTL influencing traits of importance in livestock production might be subject to parental imprinting. These initial findings were followed by a flurry of papers reporting imprinted QTL (23). Reporting imprinted QTL has become nearly routine in the animal genetics literature, including in poultry, in which the occurrence of the corresponding biological phenomenon has never been demonstrated at the molecular level.

One of the common features of the studies reporting imprinted QTL is their intercross or F2 design. This is intriguing as the imprinting hypothesis can normally not be tested in a typical F2 cross obtained from inbred lines. Testing imprinting indeed consists in testing whether the difference between the allele substitution effect of the padumnal chromosomes and the allele substitution effect of the madumnal chromosomes, Δ_{IMP}, deviates significantly from zero. Δ_{IMP} is computed as

$$\begin{aligned}\Delta_{IMP} &= {}^{1}\!/_{2}[(\overline{11}-\overline{21})+(\overline{12}-\overline{22})] \\ &\quad -[(\overline{11}-\overline{12})+(\overline{21}-\overline{22})] \\ &= (\overline{12}-\overline{21})\end{aligned}$$

In this, 1 and 2 differentiate the chromosomes originating from the two parental lines, and $\overline{xx}$ correspond to the average phenotype of individuals with genotype *xx*, in which the first allele is madumnal, the second padumnal. Because it is not possible to distinguish 12 from 21 individuals, Δ_{IMP} cannot be measured and the imprinting hypothesis cannot be tested.

The reason it has been possible to test imprinting when working with domestic animals is that the corresponding F2 pedigrees are not generated from inbred strains but from divergent outbred breeds or selection lines. Their outbred nature manifests itself clearly by genetic markers that are polymorphic within the parental breeds. Contrary to F2 pedigrees obtained from inbred strains of mice where all F1 parents have the same 12 marker genotype, F1 parents obtained by mating breeds exhibit distinct marker genotypes (e.g., 12, 13, 24, . . .). Consequently, one can distinguish the four possible genotypes (homozygotes for line 1's allele, homozygotes for line 2's allele, heterozygotes receiving line 1's allele from the father, heterozygotes receiving line 1's allele from the mother) and thus test the imprinting hypothesis. This procedure has led to frequent detection of significant deviations from zero for Δ_{IMP}.

But do these results truly indicate that the underlying genes are subject to parental imprinting, as usually surmised? The caveat lies in the fact that the analyses are performed under the assumption that the parental breeds are fixed for alternate QTL alleles (for instance, *Q* and *q*), and thus that all F1 parents have the same *Qq* QTL genotype. In fact the within-breed variation that is typically observed at the marker loci indicates that the breeds will likely segregate at some QTL as well. F1 parents can thus be of a different QTL genotype, both homozygous or heterozygous. As the number of utilized F1 sires and dams is limited, there is no reason to expect that the average padumnal and madumnal allele substitution effect will be identical (or that $\Delta_{IMP} = 0$), even in the absence of genuine parental imprinting (24). This "artifact" is exacerbated by LD between marker loci and QTL as the matings for which the imprinting hypothesis can be tested (F1 sire and dam with different heterozygous marker genotype) are more likely to have different QTL genotypes as well (C. Sandor and M. Georges, in preparation).

The Importance of Epistatic Interactions

In general, QTL mapping in domestic species has been remarkably successful. As previously mentioned, QTL with moderate to large effects have been detected for most of the examined traits. However, with few exceptions, the identified QTL only explain a fraction of the genetic variance, which is likely to be overestimated in most cases. Indeed, estimates of QTL effects exceeding stringent statistical thresholds are biased upward as a result of the Beavis effect or the winner's curse (3, 31). The fact that most of the genetic variance remains unexplained is usually attributed to the fact that most experiments are underpowered for QTL with smaller effects, which are assumed to explain most of the genetic variance. This explains some of the observations, but is it the only factor? For instance, does underpowered design explain the fact that virtually no significant QTL have been found in at least four independent F2s that might account for the spectacular prolificacy of Chinese (~14 pigs per litter) when compared with European (~10 pigs per litter) sows? Or that the regions that showed suggestive QTL evidence in these studies were virtually always associated with cryptic effects, i.e., the European allele being superior to the Chinese allele (83)? Does lack of power explain the fact that QTL with minor effects only (jointly explaining less than 50% of the different between the parental lines) were found in an F2 cross counting 843 birds obtained from parental chicken lines that were divergently selected for body weight over 41 generations and differed by 1240 g at 56 days (45)?

That insufficient power does not account for the shortcomings of additive QTL models is also supported by the fact that the sum, over all chromosomes, of the difference between the phenotypic means of F2 animals that are homozygous for alternate nonrecombinant chromosomes (as determined from marker data) does not necessarily amount to the phenotypic difference between the parental lines (M. Draye, D. Desmecht and M. Georges, in preparation).

For the above-mentioned poultry experiment, Carlborg et al. (12) recently demonstrated that a considerable amount of "missing" genetic variance could be uncovered by modeling first-order gene-by-gene interaction. They identified a locus that, when homozygous for the low-line allele, completely masks the effects of at least five other QTL. Homozygosity for the high-line allele at this locus capacitates this radial network, which accounts for 13% of the phenotypic variance. This number has to be compared with 3% of the phenotypic variance, explained by the same loci under an additive model. None of the capacitated QTL exceeded the genome-wide significance thresholds when tested on their own. Only the capacitator locus' marginal effect proved significant, although it alone does not affect the phenotype. Preliminary results strongly suggest that modeling epistasis uncovers cryptic genetic variation for sow prolificacy in Chinese × European F2 crosses (J.L. Noguera, personal communication). These findings are consistent with results reported previously in plants, *Drosophila*, and mice that support the importance of epistasis in explaining genetic variation for many complex traits (11).

Modeling epistasis seems crucial to gaining a better understanding of the molecular architecture of complex traits in farm animals. However, testing even first-order epistasis requires considerable enlarged data sets because of the many pairs of loci to test and the larger number of genotype classes to contrast for each pair of loci. For example, a standard F2 QTL analysis models three genotype classes and $\sim 10^3$ locations, whereas a two-locus epistatic analysis models nine genotype classes and $\sim 10^6$ pairs of loci. Carlborg et al. (12) recommend at least a fourfold increase in F2 population size. The issue of number of genotype classes is even more acute in outbred populations. However, VC methods may be able to tackle epistasis in outbred populations (70).

QUANTITATIVE TRAIT LOCI FINE MAPPING IN DOMESTIC ANIMALS

Factors Limiting the Achievable Mapping Resolution

As mentioned above, many QTL influencing a myriad of traits have been mapped in domestic animals. However, the CIs for the QTL locations (estimated, for instance, by bootstrapping) are typically of the order of 20–40 cM, corresponding in mammals to ~20–40 million base pairs or 200 to 400 genes. The situation is a little bit brighter in poultry, where the centiMorgan/Megabase ratio is of the order of 2.8–6.4 rather than ~1, as in mammals (98). In all cases the road to identifying the quantitative trait nucleotide(s) [QTN(s)] remains a formidable task. Thus, the initial mapping step typically needs to be followed by a fine-mapping step.

To select the optimal fine-mapping strategy one needs to have a good understanding of what factors limit the achievable mapping resolution, the primary four factors being:

1. Marker density. Mapping consists of placing a QTL in a given marker interval. The more markers one has, the smaller the average interval size and, thus, the higher the map resolution.
2. Crossover density. Indeed, recombinant chromosomes are the only ones that provide mapping information.
3. QTL detectance. This corresponds to the accuracy with which one can infer the QTL genotype of a given individual or chromosome (104). Positioning a QTL with respect to a crossover requires knowledge of the QTL allele carried by the corresponding chromosome.
4. Molecular architecture of the QTL. Many QTL probably reflect the combined effect of not one, but several, linked QTNs. Approaching such a "composite" QTL using a model that assumes a single location may result in fuzzy positioning.

Increasing the Marker Density is Becoming Trivial

Increasing the marker density in a chromosome segment of interest is conceptually the easiest limiting factor to resolve. Until recently, however, developing markers that target specific regions was an arduous and time-consuming "wet lab" task. Fortunately, this has recently changed with the availability of the nearly complete genome sequences of the major domestic species. Microsatellite markers can be directly identified from the genomic sequences and suitable primers can be designed. In addition to the SNPs for which the reference genome proved heterozygous, all sequencing efforts have included substantial amounts of shallow sequencing of additional individuals from different origins, resulting in collections of 2.8 million SNPs in poultry (110), 2.5 million in the dog (58), 2 million in cattle (**http://www.hgsc.bcm.tmc.edu/projects/bovine/**), and soon a similar number in the pig. Although the validity of each SNP needs to be confirmed, the conversion rate is generally very high and the transportability across populations remarkably good (58, 110).

Increasing the Crossover Density: Current Recombinants

The most straightforward approach to increasing crossover density is to breed animals de novo. When working with experimental pedigrees one could, for instance, generate additional F2 or BC animals. However, the number of F2 or BC offspring required to bring the mapping resolution of a typical QTL down to 5 cM or less is ~5000 (18), as the probability for an individual to be recombinant in the interval of interest decreases with the size of the interval and many such individuals are needed to compensate for the poor QTL detectance. This is a barely viable proposition when working with mice, and a nontenable proposition when working with domestic animals.

One alternative is to work with advanced intercross lines (AILs), i.e., F3, F4, ... Fn generations (18). Using AIL, the CI for the QTL is reduced by a factor of ~2/n when compared with an F2 population of the same size, where n is the number of generations of intercrossing that characterize the AIL. Alternatively, the number of AIL animals required to map a QTL to a CI of a desired size is reduced by the same ~n/2 factor when working with a Fn AIL. However, Fn AIL must first be produced, which requires 50–100 breeding animals at each intermediate generation. Given the generation interval and housing costs, AIL remains an expensive and time-consuming option when working with most domestic species. Nevertheless, F8 and F9 have been generated for fine-mapping purposes in poultry (48; L. Andersson, personal communication), and F3 animals have been generated in cattle (R. Spelman, personal communication).

The above-mentioned large paternal half-sister pedigrees that are commonly encountered in dairy cattle as a result of AI offer some interesting opportunities for QTL fine mapping using current recombinants. Having identified a QTL using selective DNA pooling as described, one can genotype the extreme daughters individually for a pair of markers flanking the region of interest and identify those that inherited a recombinant chromosome from their sire. Extreme recombinant daughters can then be genotyped for additional informative markers within the interval to refine the map position of the QTL. We applied this approach to fine map a QTL influencing milk protein yield. A QTL was mapped to a chromosome segment of ~30 cM by selective DNA pooling of ~1500 cows from the top and bottom 10% of a half-sister pedigree of ~20,000 daughters. Using a pair of flanking markers, we identified 380 daughters that inherited a recombinant chromosome from their sire. These were genotyped for 22 additional microsatellites, allowing us to refine the map position to a 2-cM interval (W. Coppieters, P. Fisher, M. Georges, unpublished observations).

Increasing the Crossover Density: Historical Recombinants

The alternative to using current recombinants is to exploit the nonrandom allelic association between QTL and closely linked markers that is predicted to persist at the population level, i.e., LD. This requires a marker density commensurate with the reach of LD in the population of interest. In human genetics, it is of the order of 10–20 Kb, justifying the use of ~300,000–500,000 SNPs in genome-wide association studies. Because of the structure of domestic animal populations, and the ensuing long-range LD, the number of markers required may be an order of magnitude less. The structure of LD in domestic animals is best understood for dogs, which are likely representative of other species (58, 95). When studying haplotypes within breeds, useful LD extends over hundreds of thousands to millions of base pairs. When comparing haplotypes between breeds, the reach of LD is lower than that in human. Haplotype blocks have a ~10-Kb size on average, with ~4–5 distinct haplotypes across the entire dog population. Based on these findings, a two-stage mapping approach was proposed comprising an initial genome-wide within-breed mapping step with a panel of ~30000–50000 SNP markers, followed by a between-breed mapping step targeting chromosome regions of interest with locus-specific high-density SNP panels (see below).

One can perform either single-marker or multimarker association studies. In single-marker studies, the effect on the phenotype is tested separately for each marker. One can either model additive allelic effects or genotypic effects, thus allowing for dominance. An example of single-marker analysis is given in Grisart et al. (36), who studied a previously defined 3.8-cM CI and demonstrated that the QTL effect was strongest in the immediate vicinity of *DGAT1*, with close to single-gene resolution.

In many cases, the association with QTL alleles is likely stronger for marker haplotypes than for individual markers, hence justifying multimarker analyses. As one does not know in advance which marker combination will yield the strongest association, a pragmatic solution is to test all possible combinations of closely linked (adjacent or not) markers and to test the effect of all observed haplotypes given the selected markers (7). Similar approaches have been proposed to analyze human data (57).

However, the procedure most widely used in animal genetics builds on the VC approach for full-pedigree analysis described above. When performing linkage analyses, chromosomes are traced from parents to offspring by means of markers, allowing computation of pair-wise IBD probabilities at a given position conditional on marker genotypes. In these the founder chromosomes (i.e., the chromosomes of the pedigree members without known ancestors) are considered unrelated; their pair-wise IBD probabilities are set at zero. However, it is possible to exploit LD to compute IBD probabilities between founder chromosomes conditional on marker genotypes and a suitable demographic model (65). These IBD probabilities can then be combined with pedigree-based IBD values in a combined linkage plus LD analysis. Deterministic approximations that allow rapid computation of LD-based IBD probabilities have been proposed (66). Currently, the most commonly used algorithms account for recombination only. As marker density increases, mutations will have to be modeled, and work along those lines is progressing. Note that multimarker approaches require the determination of the linkage phase for all individuals. Given the typical pedigree structure used to map QTL in domestic species, this is usually not a major concern.

As in human genetics, a major issue when performing association studies with animals is population stratification. Animals that share IBD chromosomes at a tested map position are more likely to be IBD across the rest of the genome. This can cause significant marker trait associations in the absence of linkage between marker and QTN. An efficient way to control for this correlated coancestry is to include an individual animal effect accounting for the known genealogy in the analysis model (61). This is usually effective because pedigree records are accurately kept in many animal populations. In the absence of accurate pedigree data, one could estimate the level of coancestry from genome-wide marker data, as proposed by Yu et al. (112). An additional protection against population stratification that is commonly used in human genetics is to simultaneously test for association and linkage using the transmission disequilibrium test (TDT) (27). This is effectively mimicked by performing combined linkage and LD analysis within a VC setting, as described above.

Combined linkage and LD analysis in outbred cattle populations has proven effective in improving QTL mapping resolution in at least a few instances (7, 68, 76). It is intuitive that LD information can be exploited when analyzing "outbred" commercial pedigrees. What about experimental F2 and BC pedigrees? When such pedigrees are generated from inbred strains, as with mice, there is no LD signal to be exploited in addition to the linkage signal. However, breeds of domestic animals are not the same as inbred strains. As previously noted, they typically segregate both at the marker loci and QTL. If distinct QTL alleles are segregating within the parental breeds they can be associated with specific marker haplotypes, and this provides fine-mapping information that can be extracted using the VC methods described above. It is possible to extract LD information even for QTL that have been fixed in the parental breeds. QTL alleles fixed within strains as a result of selection will be embedded in regions of reduced haplotype diversity associated with correlated QTL effects and seen as such by the VC approach.

Increasing Quantitative Trait Locus Detectance: Locus-Specific Progeny Testing

The ability to position a QTL with respect to crossover breakpoints depends on the accuracy in inferring the QTL allele carried by the corresponding recombinant chromosomes or the QTL detectance (104). For complex traits, QTL detectance is typically poor, as each QTL accounts for only a minor part of the trait variance. We already mentioned two approaches used in animal genomics to increase QTL detectance: progeny testing (31) and selective genotyping (55).

The approach that has proven most valuable in fine mapping QTL in domestic animals is locus-specific progeny testing or marker-assisted segregation analysis (MASA). The QTL genotype of some chromosomes can be determined with high accuracy by examining their phenotypic effect in large progeny groups. For instance, if offspring that inherit the grandpaternal chromosome from their sire have a significantly higher phenotypic value than those inheriting the grandmaternal chromosome, then the former carries allele *Q* and the latter allele *q* (under a bi-allelic QTL model). The above-mentioned fine-mapping principles can then be applied on such a set of progeny-tested chromosomes.

For example, MASA has been used in combination with current generation recombinants to fine map a QTL underlying the muscular hypertrophy of Texel sheep (15). Progeny were generated from a Romanov x Texel F2 ram that inherited (*a*) a nonrecombining Texel chromosome from one parent and (*b*) a hybrid Texel-Romanov chromosome recombining in the 10-cM CI for the QTL. As the recombinant chromosome clearly increased muscle mass in the offspring when compared with the nonrecombining chromosome, the ram had to be of Qq genotype, hence positioning the QTL in the 2.5-cM subsegment of the CI, for which the recombinant chromosome was of Romanov descent. A similar approach has been used with BC3 to BC7 boars carrying Wild Boar–Large White hybrid chromosomes recombining in the CI for the *FAT1* QTL, resulting in its fine mapping to a 3.3-cM interval (5).

MASA has also been used in combination with historical recombinants. For example, five chromosomes shown by MASA to be of "Q" genotype for an imprinted QTL located on proximal SSC2 (Sus scrofa, chr2) shared a haplotype spanning 250 Kb that was predicted to contain the QTL, which was an important step towards the identification of the corresponding QTN (74). Haplotype sharing combined with locus-specific progeny testing has also been instrumental in the positional identification of QTN influencing milk composition on BTA14 (Bos taurus, chr14) (29) and BTA6 (Bos taurus, chr 6) (17).

Dissecting Composite Quantitative Trait Loci: Multi-Quantitative Trait Loci Models

Although in some cases increasing marker and crossover density improves the definition of the QTL location, this is not always the case. A more common occurrence is a rather uniform increase of the location scores across large chromosome segments. This could indicate that the observed QTL reflects the cumulative effect of a cluster of QTNs. QTL mappers working with experimental crosses derived from inbred lines have primarily addressed such composite QTLs using composite interval mapping (CIM) or multiple QTL mapping (MQM) (46). Although it has proven difficult to transfer CIM and MQM directly to domestic animal populations, multi-QTL models were recently proposed within a Bayesian-enhanced VC framework (60, 69). Although more work is needed to improve our understanding of the behavior of such methods when dealing with real data, they have already performed outstandingly in fine mapping a QTL influencing milk protein percentage to a 420-Kb region on BTA6 (76).

Detecting Signatures of Selection

One hallmark feature of domestic animals is that they have and continue to undergo stringent selection for a range of phenotypes. One would expect this to cause dramatic shifts in allele frequencies at the loci influencing the traits under selection, leaving detectable signatures. These might be searched for in order to map or fine map the corresponding genes (85).

Wong et al. (110) used whole-genome sequence data obtained on red jungle fowl as well as three domestic chicken breeds to search for selective sweeps for adaptive alleles that occurred before the divergence of modern domestic poultry breeds. Such sweeps would have generated extended regions of haplotype sharing among extant domestic breeds not found in red jungle fowl. No such regions could be identified on length scales of 100 Kb or greater. The authors speculated that this might be due to the historically large effective population size of domestic chickens combined with the high centiMorgan to Megabase ratio in chicken. Deeper sequence data on larger numbers of individuals might be needed to detect such sweeps at a <100 Kb scale.

To the best of my knowledge, no other genome scans for selective sweeps have been conducted in domestic animals. This is mainly due to the fact that the tools needed to perform such genome-wide scans are only just becoming available and a number of experiments are underway.

However, signatures of selection have been searched for in specific regions. For example, a local reduction in genetic variation resulting from a selective sweep was exploited to refine the map position of the mutation underlying the muscular hypertrophy of Texel sheep (15). We also demonstrated that the K allele of the *DGAT1* K232A polymorphism is characterized in the HF dairy cattle population by unusually long "extended haplotype homozygosity" when compared with the A allele. This is consistent with recent strong positive selection and thus the K allele being the causal mutation (35, 36). Retrospective analyses have shown changes in allele frequencies of the *GHR* F279Y (7; R. Spelman, personal communication) and *ABCG2* Y581S (17) mutations consistent with the evolution of the breeding objectives over the past 25 years.

Whether these signatures of selection stand sufficiently clearly above the genome-wide background to be detected ab initio remains to be determined. The bottlenecks, founder effects, expansions, and subdivisions that characterize the demographic history of many domestic populations could considerably obscure the signal. However, because a sufficient number of chromosome regions, having undergone selective sweeps, are now known in domestic animals, one should soon be able to systematically evaluate the potential of such ab initio approaches. Recently, Gunnarsson et al. (39) demonstrated retrospectively that all poultry breeds with the *SLC45A2* 347M allele causing the *Silver* phenotype shared a minimum haplotype in the 15–35 Kb range, whereas White Leghorn lines, having fixed the *SLC45A2* 277C mutation, shared a different >100-Kb haplotype. Additional examples of chromosome regions where signatures of selection are likely to exist and which could be used to evaluate the power of selective sweep detection are the *myostatin* locus in Belgian Blue Cattle (37, 38) and the *IGF2* locus in European pigs (100).

Note that selection is expected to cause LD between loci affecting the selected trait: the so-called "Bulmer effect." Directional and stabilizing selection, for instance, tend to generate negative gametic-phase disequilibrium (alleles increasing the character value at one locus preferentially associated with alleles decreasing the character value at other loci and vice versa), while disruptive selection and directional selection on characters displaying certain patterns of epistasis will generate positive gametic-phase disequilibrium (28). It will be interesting to examine whether the Bulmer effect has left detectable signatures in domestic animal populations.

Bypassing the Mapping Step: Whole-Genome Association Studies

Until recently, the habitual strategy of QTL hunters working with outbred populations was to first perform a genome scan using linkage and low-density microsatellite panels, followed by LD fine mapping of the most prominent QTL via high-density SNP maps that had to be developed de novo for each region of interest. As mentioned above, high-density genome-wide SNP maps are, or will soon become, available for all domestic animals. Tens to hundreds of thousands of SNPs can now be genotyped using either ordered or random microarrays for the same price as a few hundred microsatellites. This will drastically affect the way QTL mapping is performed. The previously separate mapping and fine-mapping steps will be merged into a single combined linkage and LD analysis, or replaced by direct whole-genome association mapping, as in human (42). The optimal density and choice of SNPs remain to be determined more rigorously for the different species and populations. However, as genotyping costs continue to diminish, it is likely that panels of hundreds of thousands of SNPs will soon be the preferred and standard reagent in all species. One advantage of dense marker panels enabling direct interrogation of LD is that it relaxes the requirements for specific pedigree relationships, which are indispensable for linkage studies. This considerably facilitates sample collection, even if population structure requires careful consideration when designing the experiment. Whole-genome association studies have already proven effective for the rapid positional identification of genes underlying monogenic traits in domestic animals. For example, we used a whole-genome association approach to identify the gene causing congenital muscular dystonia (CMD) in Belgian Blue Cattle. Direct fine mapping based on haplotype sharing followed by mutation detection allowed us to demonstrate that CMD encompasses two distinct genetic entities caused by missense mutations in the sarcoplasmic reticulum-specific *ATP2A1* Ca^{2+} pump and presynaptic *SLCA5* glycine transporter (C. Charlier and M. Georges, in preparation).

Based on an in-depth analysis of haplotype structure and sharing within and between dog breeds, Lindblad-Toh et al. (58) proposed a two-tiered approach to identify genes underlying phenotypes segregating in multiple breeds. Their premise is that such situations reflect the sharing of IBD causal mutations. These causal mutations are assumed to be embedded in shared short-range (<-10-Kb) haplotypes predating breed creation, and in breed-specific long-range (>100-Kb) haplotypes created at the time of breed formation. Panels of 30 to 50,000 SNPs should be dense enough to pick up the breed-specific haplotypes. This initial within-breed mapping step could then be followed up by the saturation of the chromosome segments lighting up in two or more related breeds to identify the ancestral short-range haplotypes and hence fine map to single-gene resolution.

The feasibility and effectiveness of this approach for monogenic traits has been demonstrated by whole-genome association mapping of several monogenic traits, including white coat color in Boxers, Bullterriers, and Dalmatians, the "ridge" in Rhodesian Ridgebacks, and primary hyperparathyroidism in Keeshonden (K. Lindblad-Toh and L. Andersson, personal communication). The white coat color mapping demonstrates how genome-wide association mapping using only 10 white and 10 solid Boxers can identify a ~1-Megabase (Mb) associated region, followed by the positioning of the causal mutation by across-breed analysis in both Boxers and Bullterriers to a ~100-Kb segment encompassing only two exons of a single gene, *MITF* (Karlsson K, Hillbertz N, Wade C, Andersson L, Lindblad-Toh K, et al. in preparation). It remains to be seen how effective the approach will be at tackling traits with complex inheritance.

QUANTITATIVE TRAIT NUCLEOTIDE IDENTIFICATION IN DOMESTIC ANIMALS

Success Stories: Positional Identification of Regulatory Mutations

Currently, four causal QTNs with strong support have been identified by positional cloning in domestic animals: the *DGAT1* K232A (35, 36, 109) and *ABCG2* Y581S (17) mutations influencing milk composition in cattle, and the *IGF2* intron 3-3072(G-A) (100) and *MSTN* 3′ untranslated region (UTR) g+6723(G-A) (15) mutations influencing muscle mass in pigs and sheep, respectively. A small number of additional putative QTNs have been reported but require additional scrutiny in independent samples (88, 94). This number may not seem like much, but it compares favorably with results achieved in other organisms, including man and mice, which receive considerably more attention and support (34).

What is most striking in domestic animal studies is the ability to identify regulatory mutations (**Figure 3**). It is much more difficult to predict the effect a putative regulatory mutation has on gene function than the effect of a mutation that alters the primary protein sequence. To the *IGF2* intron 3-3072(G-A) and *MSTN* 3′UTR g+6723(G-A) regulatory mutations, one may add the *CLPG* mutation, although the corresponding callipyge trait was approached more as a binary than as a quantitative trait (30, 90).

The intron 3-3072(G-A) mutation in the imprinted porcine *IGF2* gene invalidates a muscle-specific silencer located in intron 3 operating after birth. The mutation abrogates binding of an as-of-yet unidentified *trans*-acting factor. As a result, animals inheriting the intron 3-3072(G-A) mutation from their father have threefold higher levels of *IGF2* mRNA in muscle, which leads to increased muscle mass and decreased fat deposition (100).

The Texel *MSTN* g+6723(G-A) mutation creates an illegitimate target site for microRNAs *miR-1* and *miR-206* in the gene's 3′UTR. As a result, mutant *MSTN* mRNAs are translationally repressed, causing decreased levels of the MSTN protein and increased muscle mass. The g+6723(G-A) mutation is particularly interesting as it is the prototype of a new class of mutations susceptible to generate both hypo- [gain of a microRNA (miRNA) target site] and

Figure 3

Positional identification of three regulatory quantitative trait nucleotides (QTNs) influencing muscle mass in livestock. (*a*) The intron 3-3072(G-A) mutation in the porcine *IGF2* gene abrogates binding of a *trans*-acting factor to a muscle-specific silencer element, hence causing an increase in IGF2 levels with resulting muscular hypertrophy. The mutation appeared on a chromosome of Asian descent introduced in European pig populations (100). (*b*) The *CLPG* mutation is an A to G mutation (*shown as a star*) in the 90-Kb *DLK1-GTL2* intergenic region that causes ectopic expression of a set of imprinted genes in skeletal muscle. When inherited from the mother only, the *CLPG* mutation leads to overexpression of maternally expressed long noncoding RNA genes hosting tandem clusters of C/D small nucleolar RNAs (snoRNAs) and microRNAs (miRNAs) (*shown in red*). When inherited from the father only, the *CLPG* mutation causes overexpression of the paternally expressed protein-encoding genes *DLK1* and *PEG11*. Ectopic expression of the DLK1 protein in these animals causes the muscular hypertrophy of callipyge sheep. In homozygous *CLPG/CLPG* animals, both maternally and paternally expressed genes are ectopically expressed. Post-transcriptional *trans*-inhibition of the paternally expressed protein-encoding genes, supposedly by maternally expressed miRNA genes, explains the absence of phenotypic expression in these animals. (e.g., 32). (*c*) In wild-type sheep, *MSTN* transcripts (*shown in blue*) do not interact with the skeletal muscle-specific *miR-1* and *miR-206* miRNAs. The g+6723(G-A) mutation in the *MSTN* 3′ untranslated region (3′UTR) of Texel sheep (*shown in red*) creates an illegitimate target site for these miRNAs, leading to myostatin downregulation and, hence, muscular hypertrophy (15).

hypermorphic (loss of a miRNA target site) alleles that might make significant contributions to the heritability of a broad range of complex traits (15, 33).

The *CLPG* mutation also inactivates a muscle-specific silencer that controls the expression of the *DLK1-GTL2* cluster of imprinted genes after birth. The *CLPG* mutation is located in the middle of the 90-Kb intergenic region separating *DLK1* and *GTL2*. Heterozygous animals receiving the *CLPG* mutation from their father exhibit ectopic expression of the paternally expressed *DLK1* mRNA and DLK1 protein in muscle, which causes the callipyge muscular hypertrophy (13, 19). Intriguingly, homozygous *CLPG/CLPG* animals show increased levels of *DLK1* mRNA, but not of DLK1 protein in skeletal muscle and hence do not express the callipyge phenotype. There is increasing evidence that this is due to translational repression of the padumnal *DLK1* transcripts by madumnal miRNAs (20). It is remarkable that both the Texel and callipyge muscular

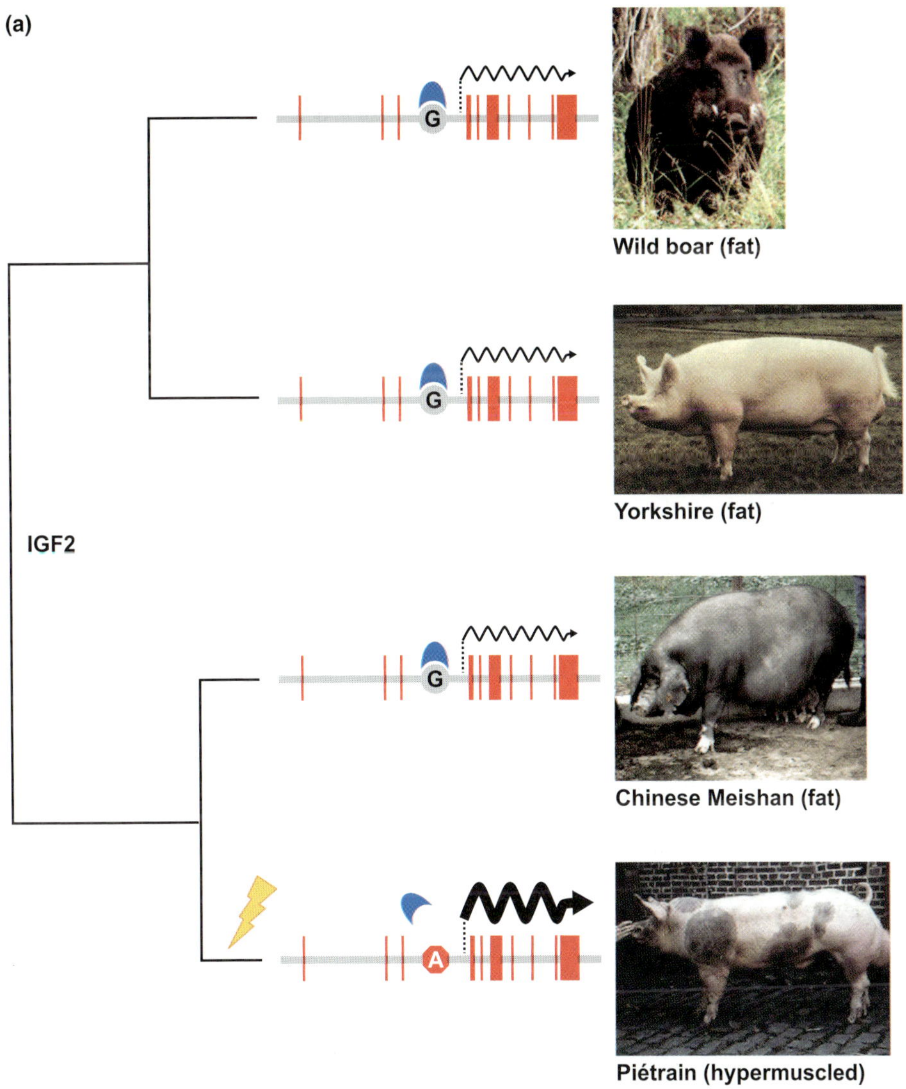

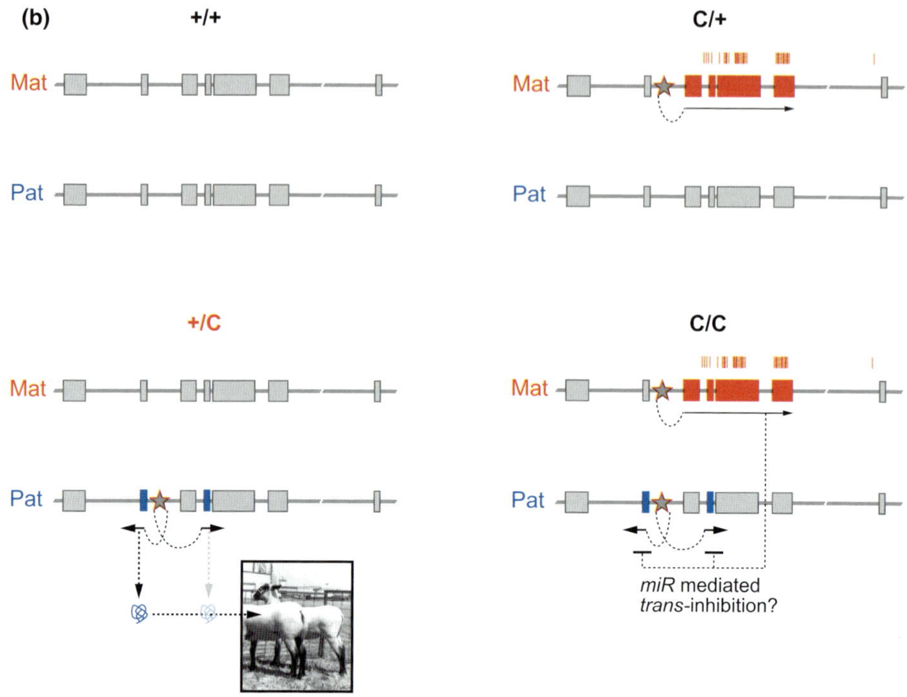

Figure 3

(Continued)

hypertrophies involve miRNA-mediated mechanisms, highlighting the importance of this newly discovered mode of gene regulation.

What enabled the demonstration that these three, at first glance anodyne single-base substitutions, were causative QTN? In addition to dealing with muscular development, the three scenarios have at least two features in common. The first is that the three QTL had a simple molecular architecture, which considerably facilitated their molecular dissection: a single mutational event underlying the QTL effect in each case. The second is the careful selection and analysis of unique haplotypes with known QTL genotype, deduced either by locus-specific progeny testing (*IGF2* and *MSTN*) or from individual phenotypes (*CLPG*).

In the case of the intron 3-3072 mutation, sequencing 30 Kb spanning the *IGF2* and *INS* genes from five progeny-tested Q chromosomes and nine progeny-tested q chromosomes identified 258 SNPs. Only one of these showed a perfect allelic segregation between the Q and q groups: the intron 3-3072 *IGF2* mutation. The unusually high density of SNPs is due to the fact that all Q chromosomes were of Asian descent, whereas all but one q chromosome was of European descent. Being phylogenetically closely related to the clade of Q chromosomes, the Asian q chromosome was particularly informative, enabling the exclusion of all but the causal SNPs.

For the Texel g+6723(G-A) *MSTN* mutation, the comparison of five progeny-tested Q and four progeny-tested q chromosomes allowed exclusion of 19 out of the 20 SNPs identified in an 10.5-Kb segment spanning the *MSTN* gene. The A allele of the remaining g+6723(G-A) SNP was Texel specific, thus strongly supporting its causality (15, 33).

To identify the *CLPG* mutation, we selected animals expressing the callipyge phenotype, which was bound to be heterozygous $+/CLPG^{Pat}$, but homozygous for closely linked markers. By doing so, we focused our analysis on phylogenetically related, yet functionally distinct, haplotypes, thereby minimizing the number of noncausal SNPs we had

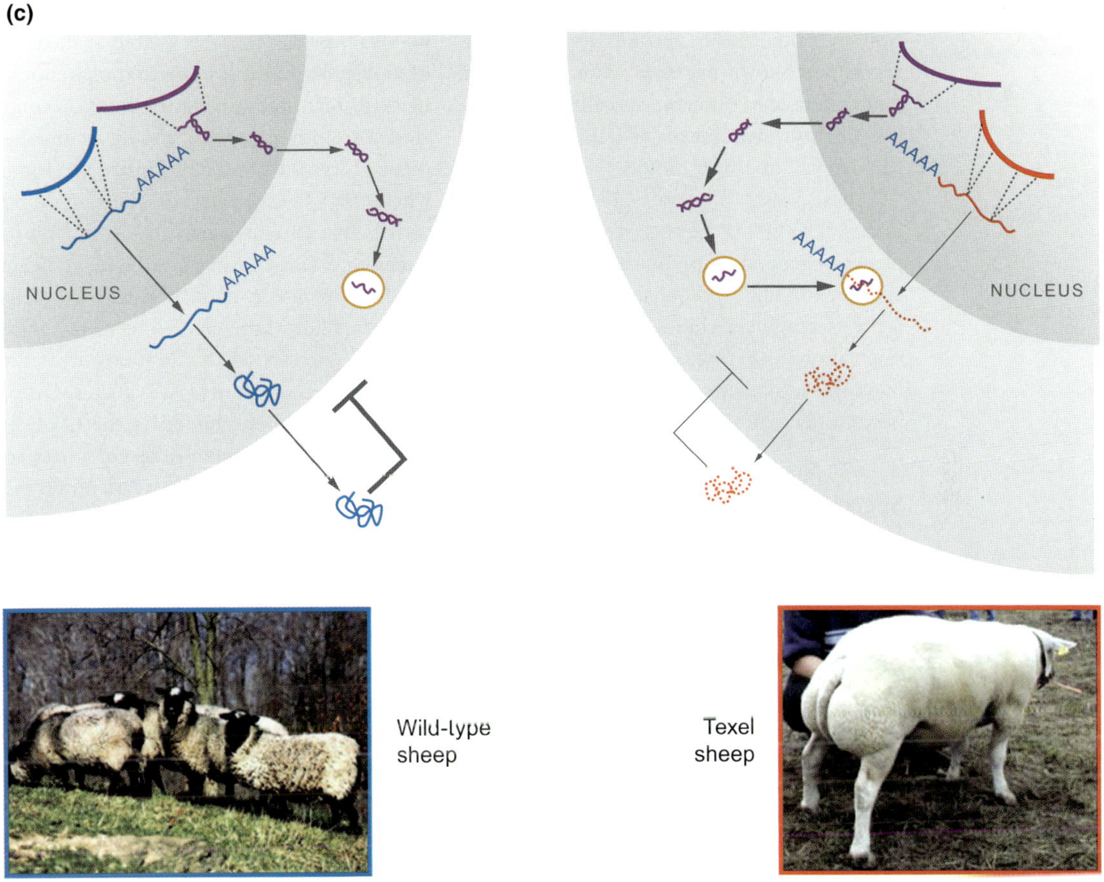

Figure 3

(Continued)

to sift through. Sequencing 200 Kb in the corresponding individuals identified a single mutation differentiating the mutant and wild-type chromosome: the *CLPG* A to G substitution (30, 90).

In all three cases, in vivo and in vitro functional tests supported the causality of the QTN. In addition to increasing skeletal muscle transcript levels on paternal transmission, the *IGF2* intron 3-3072 mutation was shown by electrophoretic mobility shift assay (EMSA) to abrogate binding of a nuclear factor, and by luciferase reporter assay to reduce the silencing activity conferred by the corresponding wild-type sequence (100).

Texel sheep were shown to have threefold lower circulating MSTN levels, steady-state levels of mutant *MSTN* transcripts were shown to be reduced 1.5-fold when compared with wild-type transcripts in heterozygous animals, and the g+6723(G-A) mutation was shown to promote *miR-1*- and *miR-122*-dependent downregulation of luciferase reporter constructs (15).

The *CLPG* mutation was shown to cause ectopic transcription of genes from the *DLK1-GTL2* domain in skeletal muscle (13), and to alter the muscular epigenotype of the *DLK1-GTL2* intergenic region in *cis*, (including hypomethylation, acquisition of novel DNase-I hypersensitive sites, and enhanced bidirectional long range) (72, 96). Ectopic expression of the DLK1 protein in transgenic mice was shown to cause a muscular hypertrophy

(19), whereas introducing the *CLPG* mutation by homologous recombination in the mouse was shown to recapitulate part of the phenomenology observed in callipyge sheep (L. Pirottin, H. Takeda, M. Goerges and C. Charlier, in preparation).

These cases illustrate two important points. The first is the remarkable discerning power of genetics in pinpointing QTNs irrespective of their mode of action. The second is the exceptional richness of domestic animal populations in diverse, and hence informative, haplotypes, allowing full exploitation of the power of genetics.

Success Stories: Trees Hiding the Forest

Despite the unique features that make domestic animal populations particularly valuable resources for the positional identification of QTN, it would be a fallacy to pretend that the process will become routine anytime soon. The few success stories described above hide a much larger number of QTL that are proving to be extremely difficult to crack. As mentioned above, the most commonly encountered issue is the lack of improved mapping resolution following augmentation of the marker and crossover density. The most plausible explanation of this common occurrence is the fact that the QTL effect reflects the combined action of multiple linked QTNs. It remains to be seen whether a further increase of marker density combined with multi-QTL analysis models will resolve at least some of these problematic cases.

Another major issue when identifying QTNs is the lack of a practical approach that provides formal proof of causality for a candidate SNP. Although the accumulated evidence supporting the causality of the five QTNs described above is fairly convincing, one cannot formally exclude that the causal mutation(s) are actually distinct and in strong LD with the ones identified. Functional tests are often considered the panacea, yet seldom if ever provide sufficient evidence on their own. Observing an effect of a mutation in an in vitro test by no means proves that it is related to the chain of molecular events leading to the phenotype in vivo. Quantitative complementation comes closest to formal proof of the direct involvement of a given gene (62) (**Figure 4**). To test whether a QTL of interest reflects modified activity of a candidate gene, one compares the phenotypes of Q/Δ and q/Δ hemizygous individuals, where Δ corresponds to a loss-of-function allele of the gene to be tested, while Q and q correspond to chromosomes carrying functionally distinct QTL alleles. Only if the inactivated gene underlies the QTL effect will the phenotypic difference between the two groups be larger than the difference between Q/+ and q/+ individuals. The test is not perfect as interactions induced by the Δ mutation may cause the contrasts to vary even in the absence of direct involvement of the interrogated gene. Nevertheless, it is a step in the right direction and demonstrates once more the pertinence of genetics. Quantitative complementation has been used in yeast (25, 92), Drosphila (62), and mice (111), organisms for which gene knockouts can be produced with relative ease. The method has provided remarkable insights in the molecular architecture of QTL underlying high temperature growth and sporulation efficiency in yeast. It has also unveiled how frighteningly complicated some QTL are likely to be. Unfortunately, it is difficult to see how this approach could be effectively adapted for domestic animals.

The Candidate Gene Approach

One attractive feature of the positional cloning approach is its generic nature, its potential to assign novel functions to genes or other components in any genome. For example, miRNAs were first discovered by positional cloning in *Caenorhabditis elegans* (56). For a minority of genes, prior knowledge about function suggests that they might contribute to the genetic variation observed for a given trait. It is therefore logical to readily look for correlations between phenotypic variation and DNA sequence variation in these candidate genes. This has led to well established and replicated associations between bovine milk protein variants (caseins, β-lactoglobulin, α-lactalbumin) and milk yield, composition, and processability (107), between major histo-compatibility (MHC) variants and disease susceptibility, or between the porcine melanocortin-4 receptor (MC4R) and fatness, growth, and feed intake (53).

Unfortunately, the candidate gene literature is reputedly polluted with too many nonreproducible results. This is unfortunate because it discredits the entire field of complex trait analysis and marker assisted selection. Too often candidate gene studies identify a very limited set of polymorphisms in a supposedly valid candidate gene, genotype cohorts of limited size (sometimes not appropriate at all for association studies, e.g., F2 pedigrees), test the effect on several traits, and report supposedly significant associations on the basis of nominal p-values, without addressing functional questions.

It seems that everybody would benefit from raising the standards used in candidate gene studies. A thorough evaluation of a candidate gene should (*a*) utilize significance thresholds that more thoroughly account for multiple testing and prior probability of success, (*b*) exhaustively characterize the haplotype diversity and structure that exists in the population of interest, (*c*) include replication of the positive associations in an independent cohort, and (*d*) include resequencing data of functionally distinct haplotypes to identify putative causal SNPs.

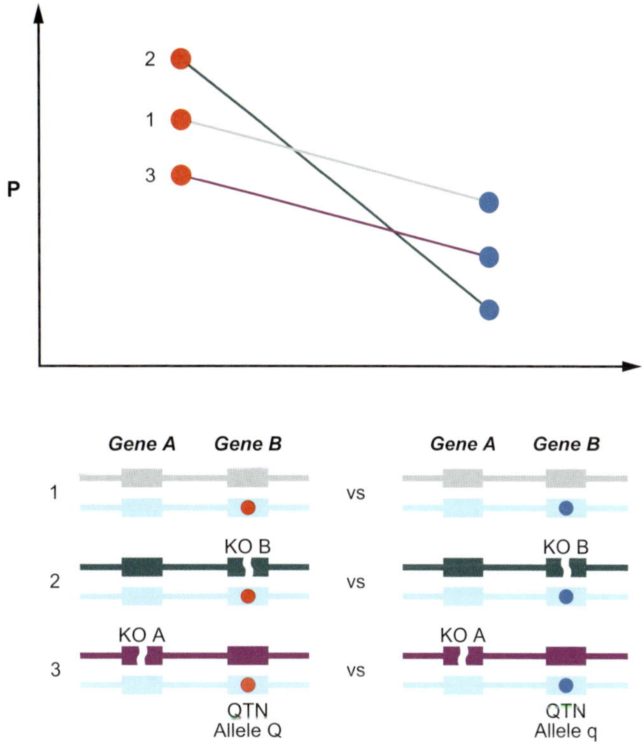

Figure 4

Principle of the quantitative complementation assay. Assume that a quantitative trait locus (QTL) has been mapped to a region harboring genes A and B, and that progeny-tested chromosomes carrying both Q [*red* quantitative trait nucleotide (QTN) allele] and q (*blue* QTN allele) alleles are available. To determine whether gene A or B is underlying the QTL effect, one generates two series of heterozygote individuals, with one homolog carrying the Q (series 1) and q (series 2) alleles, respectively, and the other homolog being (*a*) wild type for gene A and B (++, *gray*), (*b*) wild type for gene A and knockout for gene B (+0, *aqua*), or (*c*) knockout for gene A and wild type for gene B (0+, *purple*). The phenotypic difference Δ_1 between the Q/++ and q/++ animals shows the effect of a single-dose Q to q allele substitution effect. The phenotypic difference Δ_3 between Q/0+ and q/0+ animals is the same as Δ_1, indicating that gene A is not underlying the QTL effect. The Q/++ and Q/0+ means and q/++ and q/0+ means differ by the same amount and reflect the phenotypic effect of knocking out gene A, which is thus independent of the QTL genotype. By contrast, the phenotypic difference Δ2 between Q/+0 and q/+0 animals is much larger than Δ_1 and Δ_3. This interaction strongly suggests that gene B underlies the QTL effect.

The Promises of Genetical Genomics

Transcriptome analysis with microarrays is opening exciting possibilities for the genetic dissection of complex traits. In an approach dubbed genetical genomics, the expression levels of many, if not all, genes are measured in one or more tissues assumed to be relevant with regards to the organismal phenotype of interest. The transcript levels of individual genes are treated as quantitative traits and subjected to QTL mapping to identify expression QTL (eQTL). eQTL are typically sorted into local eQTL when the affected gene lies within the CI of the eQTL, as opposed to distant eQTL when not. Local and distant eQTL are also respectively referred to as *cis*- versus *trans*-acting eQTL. One possible explanation for a local eQTL is a *cis*-acting regulatory mutation that directly controls the transcript level of the corresponding gene, whereas a distant eQTL necessarily implies a *trans*-acting molecular mechanism (47, 82).

Local eQTL that colocalize with QTL affecting the phenotype of interest (hereafter referred to as pQTL) point toward possible causal genes. Genetical genomics thus provides a highly parallelized shortcut bypassing–at least in some instances—tedious QTL fine mapping. For instance, Hubner et al. (44) used this approach to establish a list of candidate genes that might influence blood pressure in the rat.

It is important to realize that finding local eQTL overlapping a pQTL—a common occurrence given the abundance of local eQTL in most experiments—provides interesting candidates, but does not establish causal connection. A correlation between the corresponding expression traits and phenotype in the studied population does not establish causal connection either. Assuming that the eQTN and pQTN are distinct, the expected LD between them will generate a fortuitous correlation between the traits. One possible strategy to distinguish fortuitous from causal correlation is to apply conditional correlation measures (**Figure 5**). If transcript levels of the candidate gene directly affect the phenotype, one will find a correlation between transcript levels and the phenotype both across and within genotypes; if transcript levels and

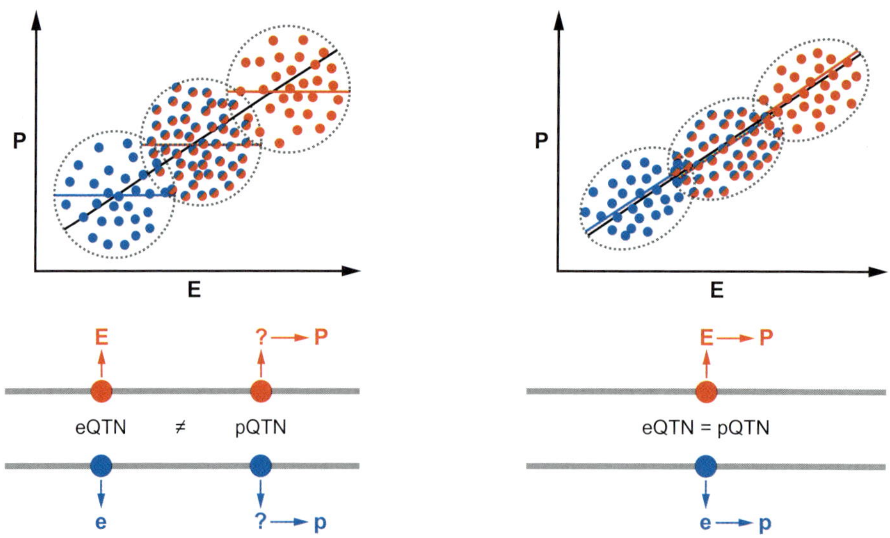

Figure 5

Principle of conditional correlation measures to distinguish between fortuitously colocalized phenotypic quanitative trait nucleotides (pQTNs) and local expression quantitative trait nucleotides (eQTNs) (*left panel*) versus eQTNs that cause pQTL (*right panel*). Imagine a situation with two alleles (*red and blue*) segregating in an F2 population. The red allele increases the expression level (E) of the culprit gene and the phenotypic value (P), whereas the blue allele decreases both gene expression level (e) and phenotypic value (p). In the left panel, the effects on gene expression and phenotype are independent; in the right panel, the effect on the phenotype results directly from the effect on gene expression. For independent pQTN and eQTN, one finds a correlation between P and E across the entire data set, but not within genotypes (*blue-blue, blue-red, and red-red*). When eQTL and pQTN are the same, one finds a correlation between P and E both across and within genotypes. In the left panel, the distance separating eQTN and pQTN is too small for recombination to be an issue.

the phenotype are not causally linked, the correlation will be observed across, but not within, genotypes (87). Alternatively, one can attempt to specifically perturb the candidate gene either genetically (for instance, knockout or knockdown strategies) or nongenetically (for example, via drugs) and measure the effects on the phenotype.

The contribution of genetical genomics to the molecular dissection of complex traits is not limited to facilitating the discovery of causal genes. Whereas local eQTL coinciding with pQTL point toward primary events, colocalized distant eQTL may help unravel the networks that connect primary events and phenotypes. The phenotype may be controlled by the products of the genes regulated in *trans*. Just as for overlapping local eQTL, one has to be wary of fortuitous QTL overlap and resulting trait correlations. The same conditional correlation measures and gene perturbations may be applied to probe into the nature of the observed correlations.

Despite its many attractive features, genetical genomics has its limitations. It can only detect effects that are mediated by alterations in transcript levels. Moreover, it can only detect effects that manifest themselves in the panel of examined tissues, which are usually limited.

Genetical genomics applied to domestic animals is only in its infancy. A number of experiments are underway and will hopefully yield exciting results in the near future (R. Spelman, personal communication). In hindsight, one can assume that a genetical genomics approach might have accelerated the positional identification of the ovine *CLPG* and porcine *IGF2* intron 3-3072(G-A) regulatory mutations leading to >20-fold and threefold upregulation of the *DLK1-GTL2* gene cluster (13, 103) and *IGF2* genes (100), respectively. The ovine *MSTN* 3′UTR g+6723(G-A) mutation would more than likely not have been noticed using conventional microarray analysis, although its 1.5-fold effect on transcript levels was detected using a test for allelic imbalance. Such a test might be converted to a high-throughput, microarray-based format and advantageously complement the genetical genomics tool box (78).

BYPASSING QUANTITATIVE TRAIT NUCLEOTIDE IDENTIFICATION: GENOMIC SELECTION

One of the main motivations of QTL mapping in domestic animals is MAS. The usual paradigm is to map QTL, fine map QTL, and, if possible, clone the corresponding QTN. This creates opportunities for increasingly effective versions of MAS, using markers linked to the causal mutation(s), followed by markers in LD with the causal mutation(s) and ultimately the causal mutation(s) per se. Rigorous statistical thresholds are manipulated along the way to control the level of false discovery.

One major issue with this conservative approach is that the QTL that exceed the chosen significance thresholds usually only account for a minor fraction of the trait variance, let alone the breeding objectives. This considerably limits the scope and potential impact of MAS. Nevertheless, the approach was easily justified on the basis that fine mapping and QTN identification required considerable investments (of the same order of magnitude as the initial genome scan for each QTL), thus demanding high levels of confidence in the QTL.

Meuwissen et al. (67) recently proposed a different approach toward MAS, called genomic selection (GS). The approach made the now-verified assumptions that genome-wide SNP panels with tens of thousands of genetic markers would become available in parallel with high-throughput cost-effective genotyping procedures. Meuwissen et al. also assumed that the marker density was sufficient for most QTNs to be in strong LD with flanking markers, and that QTL effects were additive. The key difference with previous proposals was to not limit to QTL effects that would exceed some stringent statistical significance threshold but instead consider all markers as potentially informative. To capture a higher proportion of the genetic variance, Meuwissen et al. allowed themselves to enter the "gray zone," where true and false positives cohabitate. However, across the entire genome, the cumulative effect of false positives should average zero, leaving only the true signal (67). In one version of GS, the effect assigned to a given haplotype is computed by MCMC from its posterior distribution conditional on the effects of all other markers. The prior distribution of the QTL effect has a gamma distribution that assumes many small effects and few large ones. In their initial simulations, four effects were estimated per marker interval, corresponding to each of the four possible haplotypes with two SNPs. More recently, a separate interval effect was estimated for each chromosome in the population constrained by a covariance structure reflecting pair-wise IBD probabilities conditional on marker genotypes, as described above for fine-mapping purposes (65). When simulating a trait with 50% heritability and applying GS, Meuwissen et al. (67) obtained correlations up to 0.85 between predicted and true breeding values.

The proposed approach has met with a lot of enthusiasm and is being implemented by several breeding companies around the world. Initial results are positive, albeit not as brilliant as anticipated from the simulations. It is currently unclear whether the reasons for this are primarily insufficient marker density or statistical models that do not fit the underlying biology closely enough. Because of its promises, one can expect considerable investments, and hence improvements, in GS applied to livestock breeding in the near future.

The principles underlying GS may also be interesting in human genetics. They may lead to more accurate genome-wide relative risks to develop a specific complex disease or to exhibit adverse reactions towards specific drugs, etc. As in animal breeding, estimates of relative risk that exploit information from markers that are not as clearly associated with the condition of interest might nevertheless be superior to those obtained by limiting oneself to the major gene effects. Work along those lines is in progress.

CONCLUSIONS

The truth probably is that, despite 20 years of intense efforts and as stated by Rutherford & Henikoff (84), "The nature of quantitative-trait variation" remains "one of the last unexplored frontiers in genetics, awaiting the future cloning and definitive identification of quantitative-trait determinants, whether they be genetic or epigenetic." Although many QTL have been mapped, the underlying molecular architecture remains much of a mystery for the vast majority.

During this period, "–omics" technologies have made remarkable strides. The most advanced methods and reagents are rapidly becoming available for studies on domestic animals; even whole-genome resequencing is being considered when planning future experiments in animal genomics (4). Although the "–omics" field may sometimes seem like an endless "fuite en avant," there is hope that the complex trait fortress will soon yield. There is no doubt that the rewards—both in terms of basic biological understanding and useful medical and agronomic applications—will be worth the invested efforts. The most recent results are revealing fascinating underlying biology, for example in the area of epigenetics (81). Key advances will not be driven by technology alone, but by the intelligence with which new methodologies are applied to address well-formulated and incisive questions.

Domestic animal resources have the potential to make the most worthwhile contribution in deciphering the molecular architecture of complex traits. Let us make sure that we put them to good use!

DISCLOSURE STATEMENT

The author is not aware of any biases that might be perceived as affecting the objectivity of this review.

ACKNOWLEDGMENTS

I am grateful to Leif Andersson and Kerstin Lindblad-Toh for sharing unpublished data, and to Carole Charlier, Haruko Takeda and Leif Andersson, for their constructive comments on the manuscript. Results described in this manuscript were obtained with support from the FW7 EU Callimir poject, from the Belgian SSTC (PAI), from the Communauté Française de Belgique (ARC), from the Walloon Ministry of Agriculture, and from the University of Liège.

LITERATURE CITED

1. Andersson L, Haley CS, Ellegren H, Knott SA, Johansson M, et al. 1994. Genetic mapping of quantitative trait loci for growth and fatness in pigs. *Science* 263:1771–74
2. Andersson L, Georges M. 2004. Domestic animal genomics: deciphering the genetics of complex traits. *Nat. Rev. Genet.* 5:202–12
3. Beavis WD. 1998. QTL analysis: power, precision and accuracy. In *Molecular Dissection of Complex Traits*, ed. AH Patterson, pp. 145–62, New York: CRC
4. Bentley DR. 2006. Whole-genome resequencing. *Curr. Opin. Genet. Dev.* 16:545–52
5. Berg F, Stern S, Andersson K, Andersson L, Moller M. 2006. Refined localization of the FAT1 quantitative trait locus on pig chromosome 4 by marker-assisted backcrossing. *BMC Genet.* 7:17
6. Blangero J, Williams JT, Almasy L. 2001. Variance component methods for detecting complex trait loci. *Adv. Genet.* 42:151–81
7. Blott S, Kim JJ, Moisio S, Schmidt-Kuntzel A, Cornet A, et al. 2003. Molecular dissection of a quantitative trait locus: a phenylalanine-to-tyrosine substitution in the transmembrane domain of the bovine growth hormone receptor is associated with a major effect on milk yield and composition. *Genetics* 163:253–66
8. Boichard D. 1996. Analyse généalogique des races bovines laitières françaises. *INRA Prod. Anim.* 9:323–35
9. Bovenhuis H, Spelman RJ. 2000. Selective genotyping to detect quantitative trait loci for multiple traits in outbred populations. *J. Dairy Sci.* 83:173–80
10. Brown SDM, Peters J. 1996. Combining mutagenesis and genomics in the mouse: closing the phenotype gap. *Trends Genet.* 12:433–35
11. Carlborg O, Haley CS. 2004. Epistasis: too often neglected in complex trait studies? *Nat. Rev. Genet.* 5:618–25
12. Carlborg O, Jacobsson L, Ahgren P, Siegel P, Andersson L. 2006. Epistasis and the release of genetic variation during long-term selection. *Nat. Genet.* 38:418–20
13. Charlier C, Segers K, Karim L, Shay T, Gyapay G, et al. 2001. The callipyge mutation enhances the expression of coregulated imprinted genes in cis without affecting their imprinting status. *Nat. Genet.* 27:367–69
14. Deleted in proof
15. Clop A, Marcq F, Takeda H, Pirottin D, Tordoir X, et al. 2006. A mutation creating a potential illegitimate microRNA target site in the myostatin gene affects muscularity in sheep. *Nat. Genet.* 38:813–18

16. Cockett NE, Jackson SP, Shay TL, Farnir F, Berghmans S, et al. 1996. Polar overdominance at the ovine callipyge locus. *Science* 273:236–38
17. Cohen-Zinder M, Seroussi E, Larkin DM, Loor JJ, Everts-van der Wind A, et al. 2005. Identification of a missense mutation in the bovine *ABCG2* gene with a major effect on the QTL on chromosome 6 affecting milk yield and composition in Holstein cattle. *Genome Res.* 15:936–44
18. Darvasi A. 1998. Experimental strategies for the genetic dissection of complex traits in animal models. *Nat. Genet.* 18:19–24
19. Davis E, Jensen CH, Schroder HD, Farnir F, Shay-Hadfield T, et al. 2004. Ectopic expression of DLK1 protein in skeletal muscle of padumnal heterozygotes causes the callipyge phenotype. *Curr. Biol.* 14:1858–62
20. Davis E, Caiment F, Tordoir X, Cavaille J, Ferguson-Smith A, et al. 2005. RNAi-mediated allelic *trans*-interaction at the imprinted Rtl1/Peg11 locus. *Curr Biol.* 15:743–49
21. Dekkers JCM, Dentine MR. 1991. Quantitative genetic variance associated with chromosomal markers in segregating populations. *Theor. Appl. Genet.* 81:212–20
22. Dekkers JCM, Hospital F. 2002. The use of molecular genetics in the improvement of agricultural populations. *Nat. Rev. Genet.* 3:23–32
23. de Koning DJ, Rattink AP, Harlizius B, van Arendonk JA, Brascamp EW, Groenen MA. 2000. Genome-wide scan for body composition in pigs reveals important role of imprinting. *Proc. Natl. Acad. Sci. USA* 97:7947–50
24. de Koning DJ, Bovenhuis H, van Arendonk JA. 2002. On the detection of imprinted quantitative trait loci in experimental crosses of outbred species. *Genetics* 161:931–38
25. Deutschbauer AM, Davis RW. 2005. Quantitative trait loci mapped to single-nucleotide resolution in yeast. *Nat. Genet.* 37:1333–40
26. Doerge RW, Churchill GA. 1996. Permutation tests for multiple loci affecting a quantitative character. *Genetics* 142:285–94
27. Ewens WJ, Spielman RS. 2003. The Transmission-Disequilibrium Test. In *Handbook of Statistical Genetics*, ed. DJ Balding, M Bishop, C Cannings, pp. 961–72. 2nd ed.
28. Farnir F, Coppieters W, Arranz JJ, Berzi P, Cambisano N, et al. 2000. Extensive genome-wide linkage disequilibrium in cattle. *Genome Res.* 10:220–27
29. Farnir F, Grisart B, Coppieters W, Riquet J, Berzi P, et al. 2002. Simultaneous mining of linkage and linkage disequilibrium to fine map quantitative trait loci in outbred half-sib pedigrees: revisiting the location of a quantitative trait locus with major effect on milk production on bovine chromosome 14. *Genetics* 161:275–87
30. Freking BA, Murphy SK, Wylie AA, Rhodes SJ, Keele JW, et al. 2002. Identification of the single base change causing the callipyge muscle hypertrophy phenotype, the only known example of polar overdominance in mammals. *Genome Res.* 12:1496–506
31. Georges M, Nielsen D, Mackinnon M, Mishra A, Okimoto R, et al. 1995. Mapping quantitative trait loci controlling milk production in dairy cattle by exploiting progeny testing. *Genetics* 139:907–20
32. Georges M, Charlier C, Smit M, Davis E, Shay T, et al. 2004. Toward molecular understanding of polar overdominance at the ovine callipyge locus. *Cold Spring Harb. Symp. Quant. Biol.* 69:477–83
33. Georges M, Clop A, Marcq F, Takeda H, Pirottin D, et al. 2006. Polymorphic miRNA-target interactions: a novel source of phenotypic variation. *Cold Spring Harb. Symp. Quant. Biol.* 71: 343–350
34. Glazier AM, Nadeau JH, Aitman TJ. 2002. Finding genes that underlie complex traits. *Science* 298:2345–49

35. Grisart B, Coppieters W, Farnir F, Karim L, Ford C, et al. 2002. Positional candidate cloning of a QTL in dairy cattle: identification of a missense mutation in the bovine *DGAT1* gene with major effect on milk yield and composition. *Genome Res.* 12:222–31
36. Grisart B, Farnir F, Karim L, Cambisano N, Kim JJ, et al. 2004. Genetic and functional confirmation of the causality of the DGAT1 K232A quantitative trait nucleotide in affecting milk yield and composition. *Proc. Natl. Acad. Sci. USA* 101:2398–403
37. Grobet L, Martin LJ, Poncelet D, Pirottin D, Brouwers B, et al. 1997. A deletion in the bovine myostatin gene causes the double-muscled phenotype in cattle. *Nat. Genet.* 17:71–74
38. Grobet L, Poncelet D, Royo LJ, Brouwers B, Pirottin D, et al. 1998. Molecular definition of an allelic series of mutations disrupting the myostatin function and causing double-muscling in cattle. *Mamm. Genome* 9:210–13
39. Gunnarsson. 2007. Mutations in SLC45A2 cause plumage color variation in chicken and Japanese Quail. *Genetics* 175:867–877
40. Haley CS, Knott SA. 1992. A simple regression method for mapping quantitative trait loci in line crosses using flanking markers. *Heredity* 69:315–24
41. Hayes B, Goddard ME. 2001. The distribution of the effects of genes affecting quantitative traits in livestock. *Genet. Sel. Evol.* 33:209–29
42. Hirschhorn JN, Daly MJ. 2005. Genome-wide association studies for common diseases and complex traits. *Nat. Rev. Genet.* 6:95–108
43. Hoeschele I. 2003. Mapping quantitative trait loci in outbred pedigrees. In *Handbook of Statistical Genetics*, ed. DJ Balding, M Bishop, C Cannings, pp. 477–525. 2nd ed.
44. Hubner N, Wallace CA, Zimdahl H, Petretto E, Schulz H, et al. 2005. Integrated transcriptional profiling and linkage analysis for identification of genes underlying disease. *Nat. Genet.* 37:243–53
45. Jacobsson L, Park HB, Wahlberg P, Fredriksson R, Perez-Enciso M, et al. 2005. Many QTLs with minor additive effects are associated with a large difference in growth between two selection lines in chickens. *Genet Res.* 86:115–25
46. Jansen RC. 2003. Quantitative trait loci in inbred lines. In *Handbook of Statistical Genetics*, ed. DJ Balding, M Bishop, C Cannings, pp. 443–76. 2nd ed.
47. Jansen RC. 2003. Studying complex biological systems using multifactorial perturbation. *Nat. Rev. Genet.* 4:145–51
48. Jennen DG, Vereijken AL, Bovenhuis H, Crooijmans RM, van der Poel JJ, Groenen MA. 2005. Confirmation of quantitative trait loci affecting fatness in chickens. *Genet. Sel. Evol.* 37:215–28
49. Jeon JT, Carlborg O, Tornsten A, Giuffra E, Amarger V, et al. 1999. A paternally expressed QTL affecting skeletal and cardiac muscle mass in pigs maps to the IGF2 locus. *Nat. Genet.* 21:157–58
50. Deleted in proof
51. Kerje S, Carlborg O, Jacobsson L, Schutz K, Hartmann C, et al. 2003. The twofold difference in adult size between the red junglefowl and White Leghorn chickens is largely explained by a limited number of QTLs. *Anim. Genet.* 34:264–74
52. Kile BT, Hilton DJ. 2005. The art and design of genetic screens: mouse. *Nat. Rev. Genet.* 6:557–67
53. Kim KS, Larsen N, Short T, Plastow G, Rothschild MF. 2000. A missense variant of the porcine melanocortin-4 receptor (*MC4R*) gene is associated with fatness, growth, and feed intake traits. *Mamm. Genome* 11:131–35
54. Knott SA, Elsen JM, Haley CS. 1996. Methods for multiple-marker mapping of quantitative trait loci in half-sib populations. *Theor. Appl. Genet.* 93:71–80

55. Lander ES, Botstein D. 1989. Mapping Mendelian factors underlying quantitative traits using RFLP linkage maps. *Genetics* 121:185–99
56. Lee RC, Feinbaum RL, Ambros V. 1993. The *C. elegans* heterochronic gene lin-4 encodes small RNAs with antisense complementarity to lin-14. *Cell* 75:843–54
57. Lin S, Chakravarti A, Cutler DJ. 2004. Exhaustive allelic transmission disequilibrium tests as a new approach to genome-wide association studies. *Nat. Genet.* 36:1181–88
58. Lindblad-Toh K, Wade CM, Mikkdelsen TS, Karlsson EK, Jaffe DB, et al. 2005. Genome sequence, comparative analysis and haplotype structure of the domestic dog. *Nature* 438:803–19
59. Lipkin E, Mosig MO, Darvasi A, Ezra E, Shalom A, et al. 1998. Quantitative trait locus mapping in dairy cattle by means of selective milk DNA pooling using dinucleotide microsatellite markers: analysis of milk protein percentage. *Genetics* 149:1557–67
60. Lund MS, Sorensen P, Guldbrandtsen B, Sorensen DA. 2003. Multitrait fine mapping of quantitative trait loci using combined linkage disequilibria and linkage analysis. *Genetics* 163:405–10
61. Lynch M, Walsh B. 1998. *Genetics and Analysis of Quantitative Traits.* Sunderland, MA: Sinauer Assoc.
62. Mackay TF. 2001. Quantitative trait loci in *Drosophila*. *Nat. Rev. Genet.* 2:11–20
63. Margulies EH, Blanchette M. 2003. NISC Identification and characterization of multi-species conserved sequences. *Genome Res.* 13:2507–18
64. McRae AF, McEwan JC, Dodds KG, Wilson T, Crawford AM, Slate J. 2002. Linkage disequilibrium in domestic sheep. *Genetics* 160:1113–22
65. Meuwissen TH, Goddard ME. 2000. Fine mapping of quantitative trait loci using linkage disequilibria with closely linked marker loci. *Genetics* 155:421–30
66. Meuwissen TH, Goddard ME. 2001. Prediction of identity by descent probabilities from marker-haplotypes. *Genet. Sel. Evol.* 33:605–34
67. Meuwissen TH, Hayes BJ, Goddard ME. 2001. Prediction of total genetic value using genome-wide dense marker maps. *Genetics* 157:1819–29
68. Meuwissen TH, Karlsen A, Lien S, Olsaker I, Goddard ME. 2002. Fine mapping of a quantitative trait locus for twinning rate using combined linkage and linkage disequilibrium mapping. *Genetics* 161:373–79
69. Meuwissen TH, Goddard ME. 2004. Mapping multiple QTL using linkage disequilibrium and linkage analysis information and multitrait data. *Genet. Sel. Evol.* 36:261–79
70. Mitchell BD, Ghosh S, Schneider JL, Birznieks G, Blangero J. 1997. Power of variance component linkage analysis to detect epistasis. *Genet. Epidemiol.* 14:1017–22
71. Mosig MO, Lipkin E, Khutoreskaya G, Tchourzyna E, Soller M, Friedmann A. 2001. A whole genome scan for quantitative trait loci affecting milk protein percentage in Israeli-Holstein cattle, by means of selective milk DNA pooling in a daughter design, using an adjusted false discovery rate criterion. *Genetics* 157:1683–98
72. Murphy SK, Nolan CM, Huang Z, Kucera KS, Freking BA, et al. 2006. Callipyge mutation affects gene expression in *cis*: a potential role for chromatin structure. *Genome Res.* 16:340–46
73. Nezer C, Moreau L, Brouwers B, Coppieters W, Detilleux J, et al. 1999. An imprinted QTL with major effect on muscle mass and fat deposition maps to the IGF2 locus in pigs. *Nat. Genet.* 21:155–56
74. Nezer C, Collette C, Moreau L, Brouwers B, Kim JJ, et al. 2003. Haplotype sharing refines the location of an imprinted quantitative trait locus with major effect on muscle mass to a 250-kb chromosome segment containing the porcine IGF2 gene. *Genetics* 165:277–85

75. Nsengimana J, Baret P, Haley CS, Visscher PM. 2004. Linkage disequilibrium in the domesticated pig. *Genetics* 166:1395–404
76. Olsen HG, Lien S, Gautier M, Nilsen H, Roseth A, et al. 2005. Mapping of a milk production quantitative trait locus to a 420-kb region on bovine chromosome 6. *Genetics* 169:275–83
77. Orr HA. 2005. The genetic theory of adaptation: a brief history. *Nat. Rev. Genet.* 6:119–27
78. Pant PV, Tao H, Beilharz EJ, Ballinger DG, Cox DR, Frazer KA. 2006. Analysis of allelic differential expression in human white blood cells. *Genome Res.* 16:331–39
79. Peltonen L, Palotie A, Lange K. 2000. Use of population isolates for mapping complex traits. *Nat. Rev. Genet.* 1:182–90
80. Reik W, Walter J. 2001. Genomic imprinting: parental influence on the genome. *Nat. Rev. Genet.* 2:21–32
81. Richards EJ. 2006. Inherited epigenetic variation–revisiting soft inheritance. *Nat. Rev. Genet.* 7:395–401
82. Rockman MV, Kruglyak L. 2006. Genetics of global gene expression. *Nat. Rev. Genet.* 7:862–72
83. Rohrer GA, Ford JJ, Wise TH, Vallet JL, Christenson RK. 1999. Identification of quantitative trait loci affecting female reproductive traits in a multigeneration Meishan-White composite swine population. *J. Anim. Sci.* 77:1385–91
84. Rutherford SL, Henikoff S. 2003. Quantitative epigenetics. *Nat. Genet.* 33:6–8
85. Sabeti PC, Schaffner SF, Fry B, Lohmueller J, Varilly P, et al. 2006. Positive natural selection in the human lineage. *Science* 312:1614–20
86. Sandor C, Farnir F, Hansoul S, Coppieters W, Meuwissen T, Georges M. 2006. Linkage disequilibrium on the bovine X chromosome: characterization and use in quantitative trait locus mapping. *Genetics* 173:1777–86
87. Schadt EE, Lamb J, Yang X, Zhu J, Edwards S, et al. 2005. An integrative genomics approach to infer causal associations between gene expression and disease. *Nat. Genet.* 37:710–17
88. Schnabel RD, Kim JJ, Ashwell MS, Sonstegard TS, Van Tassell CP, et al. 2005. Fine-mapping milk production quantitative trait loci on BTA6: analysis of the bovine osteopontin gene. *Proc. Natl. Acad. Sci. USA* 102:6896–901
89. Shifman S, Darvasi A. 2001. The value of isolated populations. *Nat. Genet.* 28:309–10
90. Smit M, Segers K, Carrascosa LG, Shay T, Baraldi F, et al. 2003. Mosaicism of solid gold supports the causality of a noncoding A-to-G transition in the determinism of the callipyge phenotype. *Genetics* 163:453–56
91. Shuster DE, Kehrli ME Jr, Ackermann MR, Gilbert RO. 1992. Identification and prevalence of a genetic defect that causes leukocyte adhesion deficiency in Holstein cattle. *Proc. Natl. Acad. Sci. USA* 89:9225–29
92. Steinmetz LM, Sinha H, Richards DR, Spiegelman JI, Oefner PJ, et al. 2002. Dissecting the architecture of a quantitative trait locus in yeast. *Nature* 416:326–30
93. Storey JD, Tibshirani R. 2003. Statistical significance for genomewide studies. *Proc. Natl. Acad. Sci. USA* 100:9440–45
94. Sugimoto M, Fujikawa A, Womack JE, Sugimoto Y. 2006. Evidence that bovine forebrain embryonic zinc finger-like gene influences immune response associated with mastitis resistance. *Proc. Natl. Acad. Sci. USA* 103:6454–59
95. Sutter NB, Eberle MA, Parker HG, Pullar BJ, Kirkness EF, et al. 2004. Extensive and breed-specific linkage disequilibrium in *Canis familiaris*. *Genome Res.* 14:2388–96

96. Takeda H, Caiment F, Smit M, Hiard S, Tordoir X, et al. 2006. The callipyge mutation enhances bidirectional long-range DLK1-GTL2 intergenic transcription in cis. *Proc. Natl. Acad. Sci. USA* 103:8119–24
97. Terwilliger JD, Weiss KM. 1998. Linkage disequilibrium mapping of complex disease: fantasy or reality? *Curr. Opin. Biotechnol.* 9:578–94
98. The International Genome Sequencing Consortium. 2004. Sequence and comparative analysis of the chicken genome provide unique perspectives on vertebrate evolution. *Nature* 432:695–716
99. Thomsen B, Horn P, Panitz F, Bendixen E, Petersen AH, et al. 2006. A missense mutation in the bovine SLC35A3 gene, encoding a UDP-N-acetylglucosamine transporter, causes complex vertebral malformation. *Genome Res.* 16:97–105
100. Van Laere AS, Nguyen M, Braunschweig M, Nezer C, Collette C, et al. 2003. A regulatory mutation in IGF2 causes a major QTL effect on muscle growth in the pig. *Nature* 425:832–36
101. Visscher PM, Haley CS. 1996. Detection of putative quantitative trait loci in line crosses under infinitesimal genetic models. *Theor. Appl. Genet.* 93:691–702
102. Visscher PM, Thompson R, Haley CS. 1996. Confidence intervals in QTL mapping by bootstrapping. *Genetics* 143:1013–20
103. Vuocolo T, Byrne K, White J, McWilliam S, Reverter A, et al. 2007. Identification of a gene network contributing to hypertrophy in callipyge skeletal muscle. *Physiol Genomics.* 28:253–72.
104. Weiss KM, Terwilliger JD. 2000. How many diseases does it take to map a gene with SNPs? *Nat. Genet.* 26:151–57
105. Weller JI, Kashi Y, Soller M. 1990. Power of daughter and granddaughter designs for determining linkage between marker loci and quantitative trait loci in dairy cattle. *J. Dairy Sci.* 73:2525–37
106. Weller JI, Song JZ, Heyen DW, Lewin HA, Ron M. 1998. A new approach to the problem of multiple comparisons in the genetic dissection of complex traits. *Genetics* 150:1699–706
107. Welch RAS, Burns DJW, Davis SR, Popay AI, Prosser CG. 1997. *Milk Composition, Production and Biotechnology. Biotechnology in Agriculture Series 18.* Wallingford, UK: Cab Intl.
108. Wilkins JF, Haig D. 2003. What good is genomic imprinting: the function of parent-specific gene expression. *Nat. Rev. Genet.* 4:359–68
109. Winter A, Kramer W, Werner FA, Kollers S, Kata S, et al. 2002. Association of a lysine-232/alanine polymorphism in a bovine gene encoding acyl-CoA:diacylglycerol acyltransferase (*DGAT1*) with variation at a quantitative trait locus for milk fat content. *Proc. Natl. Acad. Sci. USA* 99:9300–5
110. Wong GK, International Chicken Polymorphism Map Consortium. 2004. A genetic variation map for chicken with 2.8 million single-nucleotide polymorphisms. *Nature* 432:717–22
111. Yalcin B, Willis-Owen SA, Fullerton J, Meesaq A, Deacon RM, et al. 2004. Genetic dissection of a behavioral quantitative trait locus shows that Rgs2 modulates anxiety in mice. *Nat. Genet.* 36:1197–202
112. Yu J, Pressoir G, Briggs WH, Vroh Bi I, Yamasaki M, et al. 2006. A unified mixed-model method for association mapping that accounts for multiple levels of relatedness. *Nat. Genet.* 38:203–8

Host Genetics of Mycobacterial Diseases in Mice and Men: Forward Genetic Studies of BCG-osis and Tuberculosis

A. Fortin,[1] L. Abel,[2,3] J.L. Casanova,[2,3,4] and P. Gros[5]

[1]Emerillon Therapeutics, Montréal, Canada, H3A-1L2; [2]Laboratory of Human Genetics of Infectious Diseases, University of Paris René Descartes, Faculty of Medicine René Descartes, Necker, 75015 Paris, France, European Union; [3]INSERM U550, Paris, France, European Union; [4]Pediatric Hematology-Immunology Unit, Necker Hospital, 75015 Paris, France, European Union; [5]Department of Biochemistry, and Center for Host Resistance, McGill University, Montréal, Canada, H3G-1Y6; email: philippe.gros@mcgill.ca

Annu. Rev. Genomics Hum. Genet. 2007. 8:163–92

First published online as a Review in Advance on May 10, 2007.

The *Annual Review of Genomics and Human Genetics* is online at genom.annualreviews.org

This article's doi:
10.1146/annurev.genom.8.080706.092315

1527-8204/07/0922-0163$20.00

Key Words

mycobacteria, forward genetics, genes, loci, human, mouse strains, infectious diseases

Abstract

In humans, genetic factors have long been suspected to contribute to the onset and outcome of tuberculosis. Such effects are difficult to identify owing to their complex inheritance, and to the confounding impact of environmental factors, notably pathogen-associated virulence determinants. Recently, forward genetic approaches in mouse models and in human populations have been used to elucidate a molecular basis for predisposition to mycobacterial diseases. The genetic dissection of host predisposition to infection with *Mycobacterium bovis* BCG and *M. tuberculosis* will help to define the key molecules involved in host antituberculous immunity and should provide new insights into this important infectious disease.

INTRODUCTION

Tuberculosis (TB) remains one of the leading causes of morbidity and mortality by infectious diseases worldwide (65). In 2004, the World Health Organization (WHO) estimated the number of prevalent cases of TB worldwide at 14.6 million, including 8.9 million new cases (173). More than 80% of these new cases were concentrated in Africa, southeast Asia, and the western Pacific regions. The dramatic impact of TB on global health occurs despite vaccination programs using *Mycobacterium bovis* BCG (Bacillus Calmette-Guérin) that have reached ~85% of children worldwide, and despite efficacious and readily available chemotherapy with antimycobacterial antibiotics.

TB is primarily a pulmonary disease caused by infection with the bacterium *Mycobacterium tuberculosis* transmitted by the aerosol route from the cough and sputum of infected individuals. *M. tuberculosis* is a slow-growing facultative intracellular pathogen that can survive and multiply inside macrophages. Upon entry into the lung, *M. tuberculosis* is rapidly taken up by alveolar macrophages. The organism survives in the macrophage by inhibiting phago-lysosome fusion, a process particularly well studied in mouse macrophages (149). The infected macrophages migrate to peripheral lymph nodes, resulting in priming of T cells, which then initiate the formation of granulomas in the lung. Granulomas are protective structures composed of macrophages, lymphocytes, and epithelioid cells that appear to physically contain the infection. Subsequently, fibrosis develops within the granuloma and in the surrounding parenchyma, producing macroscopic nodules called tubercles. Histologically, the hallmarks of tuberculous lesions are cavitation and the formation of caseum, a liquid containing multiple, highly contagious bacilli. Progression of the infection leads to clinical symptoms such as fever, cough, and weight loss. As the lungs become more necrotic, the infection can spread to other organs, like the central nervous system and the bone.

The host immune system does not stay idle, although the molecular details of "protective" immunity against TB are still poorly understood. Host response to *M. tuberculosis* is initiated when parasitized macrophages secrete interleukin-12 (IL-12) to activate and recruit T lymphocytes to the site of infection. Subsequently, $CD4^+$ and $CD8^+$ T cells recognize bacterial antigens presented by phagocytes in association with Class II and Class I major histocompatibility complex (MHC) molecules, respectively, and secrete type I cytokines, most importantly interferon-γ (IFN-γ) and tissue necrosis factor-α (TNF-α). Activated macrophages produce various bacteriostatic and bactericidal molecules and proteins, such as reactive oxygen radicals, nitric oxide, and iron chelators (lipocalins), which enhance the antimycobacterial activity of these cells. T cells produce perforin and granulysin, which additionally contribute to microbicidal activity (reviewed in 92, 129). The robust immune response to *M. tuberculosis* serves to restrict bacterial replication and prevent development of active disease. However, *M. tuberculosis* can also stay dormant for many months or years, and may cause disease later in life through a poorly understood process called reactivation. The observation that immunosuppression may lead to reactivation of infection supports the contention that the immune response is also important in prevention of this resurgence.

Infection with *M. tuberculosis* may lead to different clinical outcomes (3), including primary disseminated TB, in which clinical symptoms develop within the first 1–2 years of infection. This type of infection represents the majority of pediatric cases. Alternatively, infection can lead to chronic, slowly progressive TB in which clinical symptoms develop after more than 2 years of infection. Finally, in 90% of cases, the infection remains latent and totally asymptomatic. The latter two groups of infected individuals constitute the reservoir of *M. tuberculosis*. Host and microbial factors influencing the rate of progression to clinical disease are largely unknown. Several environmental factors, such as alcoholism, nutritional deficiency, and, more importantly, immunosuppression, multiply the risks of contracting, developing, and succumbing to active TB. The incidence of clinical TB has markedly increased with the AIDS epidemic: In 2004, an estimated 1.7 million people died from TB, including about 250,000 people coinfected with HIV (173).

The large inter-individual variability in outcome of infection with *M. tuberculosis* together with the reported familial clustering of TB cases led to the belief in the eighteenth and the nineteenth century that TB was a hereditary disease (45). Even though Pasteur's microbial theory and Koch's subsequent identification of *M. tuberculosis* overturned such theories, it was not before the 1930s that rigorous genetic epidemiological studies provided strong evidence for the contribution of genetic factors to disease susceptibility. Twin studies showed higher concordance rates of TB in monozygotic than in dizygotic twin pairs (140). In addition, although it was long known that the incidence of TB was particularly high in newly exposed populations (45), a genetic interpretation was provided when the ancestors of susceptible individuals, unlike those of resistant individuals living in the same environment, were found to be more likely to have originated from TB-free areas (156, 157).

Although the role of genetic determinants in susceptibility of humans to mycobacterial diseases in general, and TB in particular, is now well established, identifying the genes involved has proven extremely difficult. This is not only due to the complexity of the host genetic component, but also the important role played by "environmental" factors, including the pathogenicity determinants of the microbe itself and the level of exposure. To find the susceptibility genes in humans, association studies have been performed using candidate genes. More recently, whole-genome scanning has been used to identify new genes contributing to disease susceptibility. Studies using inbred strains with differences in susceptibility to TB have supported the presence of a major genetic component to differences in disease severity. Recently some of the genes causing these differences have been identified by positional cloning. A large body of genetic data has also been obtained with the vaccinating *M. bovis* BCG strain. Thus, forward genetic approaches in inbred mouse strains have resulted in the identification of a number of genes, chromosomal regions, and linkages controlling susceptibility to either *M. tuberculosis* or BCG.

GENETIC FACTORS IN SUSCEPTIBILITY TO BCG INFECTION

Genetic Studies in Mouse Models

To identify genes underlying differential host response to mycobacterial diseases, two major types of studies were conducted in mouse: The first is a "reverse genetics" approach, where the role of individual genes is tested by infecting mice bearing loss-of-function mutations at the corresponding locus. This type of study encompasses a vast literature that was recently reviewed (5, 60, 91, 129) so it will not be included here. The focus of this review is on the genetic analysis of differential susceptibility of inbred mouse strains to mycobacterial infections. The advantage of this "forward genetics" approach is that the impact of the genetic effect on susceptibility is validated *in vivo*. An obvious disadvantage is that the genetic effect may be complex and difficult to decipher in standard mouse crosses. Nevertheless, in the study of other infectious diseases, this approach led to the discovery of key innate response pathways and molecules including the Toll-like receptor family (17), the microbial-sensing proteins of the NBS-LRR family (81), phagosomal metal chelating proteins (113), and several others (76, 170, 176).

At the beginning of the twentieth century, Albert Calmette and Camille Guérin cultured virulent *M. bovis* for 230 serial passages *in vitro* in a broth containing potato extracts, glycerol, and ox bile (20). The resulting *M. bovis* BCG strain progressively lost its virulence in animals. Substrains of BCG have been used worldwide as live vaccines against TB (143) with varying degrees of protective efficacy, but with a remarkable record of safety for live vaccines. Molecular biology studies have shown that the BCG genome harbors 16 major deletions compared with the genome of *M. tuberculosis* (14), including the critical RD1 region that encodes a novel secretory system in which mutations decrease virulence in vivo and reduce ability to lyse pneumocytes (162).

The mouse has been used to study BCG vaccination, and to define molecularly the factors contributing to protective immunity against virulent *M. tuberculosis* (131). As a less virulent microbe the attenuated *M. bovis* BCG was also used as a model organism to study general host response to mycobacterial infections. Infection models using high doses of *M. bovis* BCG ($>1 \times 10^7$ CFU) introduced by the intraperitoneal (i.p.) or subcutaneous (s.c.) route were developed in an attempt to mirror BCG vaccination in humans (123). However, most genetic studies in inbred mouse strains were conducted using infection with small BCG doses ($1–5 \times 10^4$ CFU) introduced by the intravenous (i.v.) route. In this protocol, transient bacillary replication in the spleen and liver occurs during the first three weeks, and is followed by a curative phase associated with T-cell-mediated immune response, granuloma formation, and elimination of the bacterial load (134).

Nramp1/Slc11a1. The phenotype used to map the first gene in mice that contributes to differences in response to *M. bovis* BCG was permissiveness to microbial replication in the spleen after i.v. injection. Inbred mouse strains are either permissive or nonpermissive to *M. bovis* BCG replication in the spleen in the early stages of infection (63). Susceptibility is inherited as a recessive trait that segregates as a single locus (*Bcg*) mapping to proximal chromosome 1 (73, 111, 153). *Bcg* is identical to two other loci that control resistance to *Leishmania donovani* (*Lsh*) and *Salmonella typhimurium* (*Ity*) (153). Experiments in vitro demonstrated that *Bcg* controls the extent of intracellular microbial replication in macrophages (11, 35, 101). A positional cloning approach was used to identify *Bcg* (168). The gene was designated *Nramp1* (Natural resistance-associated macrophage protein 1, now known as *Slc11a1*) (**Table 1**). *Nramp1* is expressed primarily in macrophages and codes for a membrane protein with 12 transmembrane domains. Mouse strains susceptible to *M. bovis* BCG infection carry a single G169D mutation in the predicted TM4 region, which affects protein folding, impairs maturation, and causes rapid degradation of the protein (112, 168, 171). Allelism between *Nramp1* and *Bcg/Ity/Lsh* was verified through the creation of loss (knockout) or gain-of-function alleles of *Nramp1*, either in vivo in transgenic mice or in vitro in transfected RAW macrophages (69, 70, 169). Polymorphic variants at or near the human *NRAMP1* gene were identified and shown to be associated with susceptibility to a number of infections and inflammatory conditions (see below and 137).

The Nramp1 protein is expressed in the membrane of lysosomes in macrophages and of tertiary granules in neutrophils (22, 74), and is rapidly recruited to the membrane of phagosomes formed in these cells (22, 74) (**Figure 1**). Nramp1 is also expressed in the membrane of gelatinase, and vacuolar H^+/ATPase positive tertiary granules of primary neutrophils. Microfluorescence-imaging experiments in macrophages with labeled latex beads, and direct transport studies in transfected cells, demonstrated that Nramp1 acts as an efflux pump for Fe^{2+} and Mn^{2+} ions at the phagosomal membrane (62, 83). Nramp1-mediated Fe^{2+}/Mn^{2+} efflux is pH dependent and occurs down a proton gradient generated by the vacuolar H^+/ATPase. Nramp1-mediated depletion of divalent cations would inhibit intracellular bacterial replication by: a) removing a metal essential for microbial metabolic activity, b) enhancing

Table 1 Mouse genes/loci affecting susceptibility to mycobacterial infections

Gene/loci	Infectious agent	Cross(es)	Gender effect	Linkage position	Peak marker	Corresponding human region*	References
Nramp1/ Slc11a1	*BCG/ S. typhimurium/ l. donovanni*	[C57L X C57Bl/6J] X C57Bl/6J	M+F	prox 1	*D1Mcg4*	2q35	(111, 168)
Modifiers of Nramp1	*BCG*	[SPRET/EiJ X C57BL/6J] X C57BL6/J	M	prox 4	*D4Mit150*	6q16	(166)
				central X	*DXMit249*	Xp22	
			F	central 9	*D9Mit77*	3q22	
				prox 17	*D17Mit81*	6p21	
Ity2	*S. typhimurium*	[MOLF/Ei X C57BL/6J]F2	M, F	central 11	*D11Mit5*	17p13	(152)
Ity3				central 1	*D1Mit100*	1q31	
Ity4	*S. typhimurium*	[AcB61 × 129S6]F2	M, F	central 3	*rs3672755*	1p12	(148)
Ity5				distal 2	*rs3687512*	20q13	
Ity6		[AcB64 X DBA2/J]F2	M, F	14	*rs3659450*	8p23	
Ity7				7	*rs3726275*	11p15	
Ity8				15	*rs3676400*	12q12	
Icsbp/IRF8	*BCG, S. typhimurium, P. chabaudi*	[Balb/c X BXH-2]F2/ [A/J X BXH-2]F2	M, F	distal 8	*D8Mit14/ D8Mit13*	16q24	(164, 165)
Trl-1	*M. tuberculosis*	[DBA/2J X C57Bl/6J]F2	M, F	dist 1	*D1Mit396*	1q25	(119)
Trl-2				central 3	*D3Mit51*	3q31-q32	
Trl-3				prox 7	*D7Mit270*	19q12	
Trl-4	*M. tuberculosis*	[DBA/2J X C57Bl/6J]F2	M, F	distal 19	*D19Mit54*	10q25	(120)
Tbs1	*M. tuberculosis*	[A/sn X I/st] X I/stSn	F	distal 3	*D3Mit215*	Xp22	(95, 151)
Tbs2				prox 9	*D9Mit89*	11q25	
sst1/Ipr1	*M. tuberculosis, L. monocytogenes*	[C3HeB/FeJ X C57Bl/6J]F2	M, F	1	*D1Mit10*	2q37	(93, 133)
Modifiers of sst1	*M. tuberculosis*	[C3H.B6sst1^{r} X B6]F2	M, F	prox 7	*D7Mit194*	15q25-q26	(175)
				distal 12	*D12Mit263*	14q32	
				distal 15	*D15Mit239*	22q13	
				prox 17	*D17Mit21*	6p21	

*Based on correspondence between mouse (February 2006) (mm8) and human (March 2006) (hg18) assemblies (**http://genome.ucsc.edu/**).

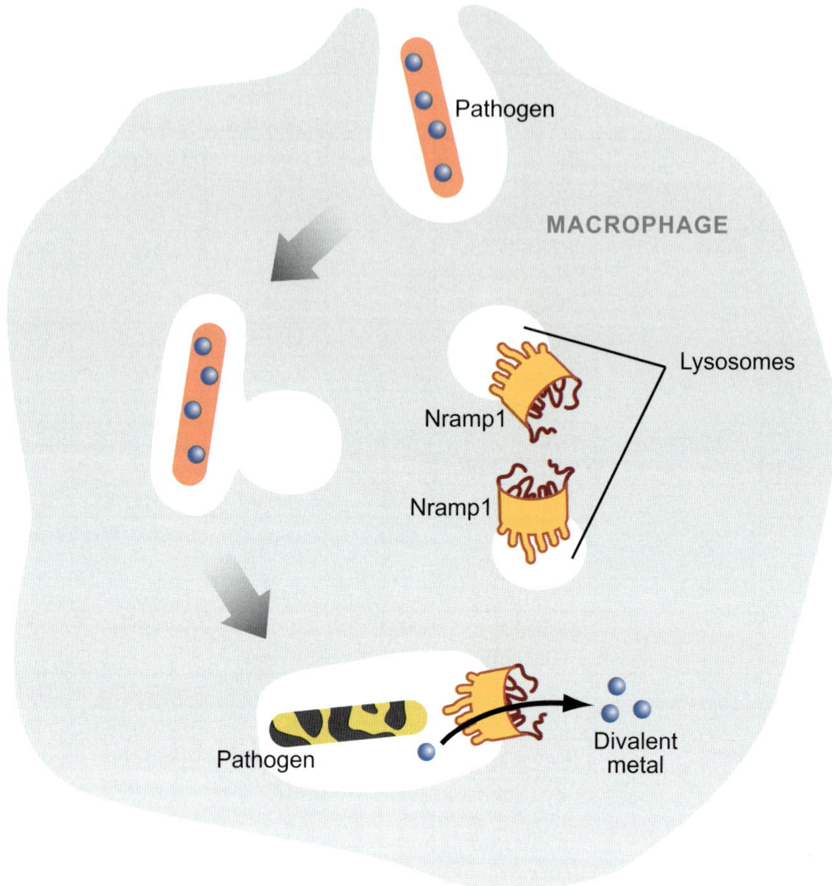

Figure 1

Proposed mechanism of action of Nramp1 (Slc11a1). In macrophages, Nramp1 is expressed in lysosomes, and is recruited to the membrane of maturing phagosomes containing either live bacteria or inert particles. Nramp1 functions as a phagosomal metal efflux pump that transports divalent cations (Mn^{2+}, Fe^{2+}, and possibly others) across the phagosomal membrane. Nramp1-mediated depletion of phagosomal Fe2^{+}/Mn^{2+} has profound consequences on the intracellular survival mechanisms normally deployed by selected pathogens. Live pathogens are shown in pink while dead pathogens are shown in yellow and grey.

macrophage bactericidal defenses or c) antagonizing microbial survival functions.

M. bovis BCG and *Mycobacterium avium* can survive intracellularly by inhibiting phagolysosome fusion and reside in a vacuole that acidifies poorly. Recruitment of Nramp1 to mycobacterial phagosomes causes increased acidification, increased delivery of lysosomal markers, and reduced intracellular survival of the organism (**Figure 1**) (64, 75). Likewise, Nramp1 antagonizes the ability of *Salmonella* to sequester itself in a vacuole that acquires neither manose-6-phosphate receptor (marker for delivery of lysosomal cargo) nor endosomal markers (33), an effect that can be recapitulated in *Nramp1*-defective macrophages exposed to membrane permeant iron chelators (82). Thus, recruitment of Nramp1 to the phagosomal membrane and chelation of essential metal ions antagonizes the ability of different pathogens to express their individual intracellular survival strategy

(**Figure 1**). Recent studies indicate that Nramp1-mediated depletion of phagosomal Fe^{2+}/Mn^{2+} is sensed by the intracellular microbe. In reponse to the cellular depletion of Fe^{2+}/Mn^{2+} *Salmonella* activate expression of several virulence genes (177), and *Leishmania* increase the expression of iron acquisition pathways of the parasite (80).

Genetic modifiers of Nramp1/Slc11a1. Genetic modifiers of *Nramp1*-mediated resistance to infections have been mapped in mice using either *M. bovis* BCG (166) or *S. typhimuirum* infection models (152) (**Table 1**). Despite the presence of a resistant *Nramp1*^{G169} allele, the wild-derived strain *Mus spretus* (SPRET/EiJ) shows an intermediate level of susceptibility to *M. bovis* BCG infection compared with resistant A/J and susceptible C57BL/6J controls. The presence of genetic modifiers of *Nramp1* in *Mus spretus* was analyzed by whole-genome scanning in a (C57BL/6J X SPRET/EiJ) X C57BL/6J cross. In addition to *Nramp1*, which showed a major effect on bacterial replication, additional single marker effects were identified on chromosomes 4 and X in males, and on chromosomes 9 and 17 in females. Using a two-locus epistasis quantitative trait locus (QTL) linkage mapping approach, a strong interaction between *Nramp1* and the MHC (chromosome 17) was noted in female mice (166). The wild-derived strain *Mus musculus molossinus* (MOLF/Ei) is extremely susceptible to *S. typhimurium* infection, despite the presence of resistance alleles at *Nramp1*(G169). Using informative (C57Bl/6J X MOLF/Ei)F2 mice, two loci controlling susceptibility of MOLF/Ei mice were mapped to chromosomes 1 (*Ity3*) and 11 (*Ity2*) (152). More recent studies in AcB/BcA recombinant congenic strains derived from resistant A/J and susceptible B6 parents suggest the presence of as many as five additional *Ity* loci (*Ity5* to *Ity8*) acting as potential modifiers of *Nramp1*, either alone or through epistatic interactions (148). The identity of the genes responsible for these effects has not been established.

Icsbp1/IRF8. BXH2 is a recombinant inbred strain derived from C3H/HeJ (*Nramp1*^{G169}) and C57Bl/6J (*Nramp1*^{D169}). Despite the presence of a resistant *Nramp1*^{G169} allele, BXH2 shows susceptibility to *M. bovis* infection. BXH-2 also displays a myeloproliferative syndrome with infiltration of spleen, bone marrow, and lymph nodes by $GR1^+/Mac1^+$ granulocyte precursors (164). Genetic studies in informative crosses localized the gene responsible for both effects to chromosome 8 (164), and subsequent positional cloning experiments identified the gene responsible as *Icsbp/IRF8* (165) (**Table 1**). Icsbp is a transcription factor and a member of the interferon response factor (IRF) family that plays a key role in myeloid cell commitment and maturation.

Icsbp associates with other transcription factors to positively or negatively regulate transcription of some IFN-inducible genes (**Figure 2**). BXH-2 mice carry a loss-of-function mutation (R294C) within the predicted IRF association domain of ICSBP, and BXH-2 splenocytes are unable to produce IL-12 and IFN-γ in response to activating stimuli (165). BXH-2 (*Icsbp1*^{C294}) shows impaired control of BCG multiplication both early and late in the infection, with uncontrolled replication linked to the inability to form granulomas in infected liver and spleen. Studies in informative [BXH-2 X BALB/cJ] F2 mice show that homozygosity for *Icsbp1*^{C294} causes susceptibility to *Salmonella typhimurium* to a level comparable to that seen for mice lacking functional *Nramp1* or *TLR4*. Finally, impaired *Icsbp1*^{C294} function is associated with (*a*) increased replication of the *Plasmodium chabaudi AS* malarial parasite during the first burst of blood parasitemia and (*b*) recurring waves of high blood parasitemia late during infection (P. Gros, unpublished work). These results show that Icsbp1 is required for orchestrating the early innate response and also long-term immune protection against unrelated intracellular pathogens, including *Mycobacterium* (**Figure 2**).

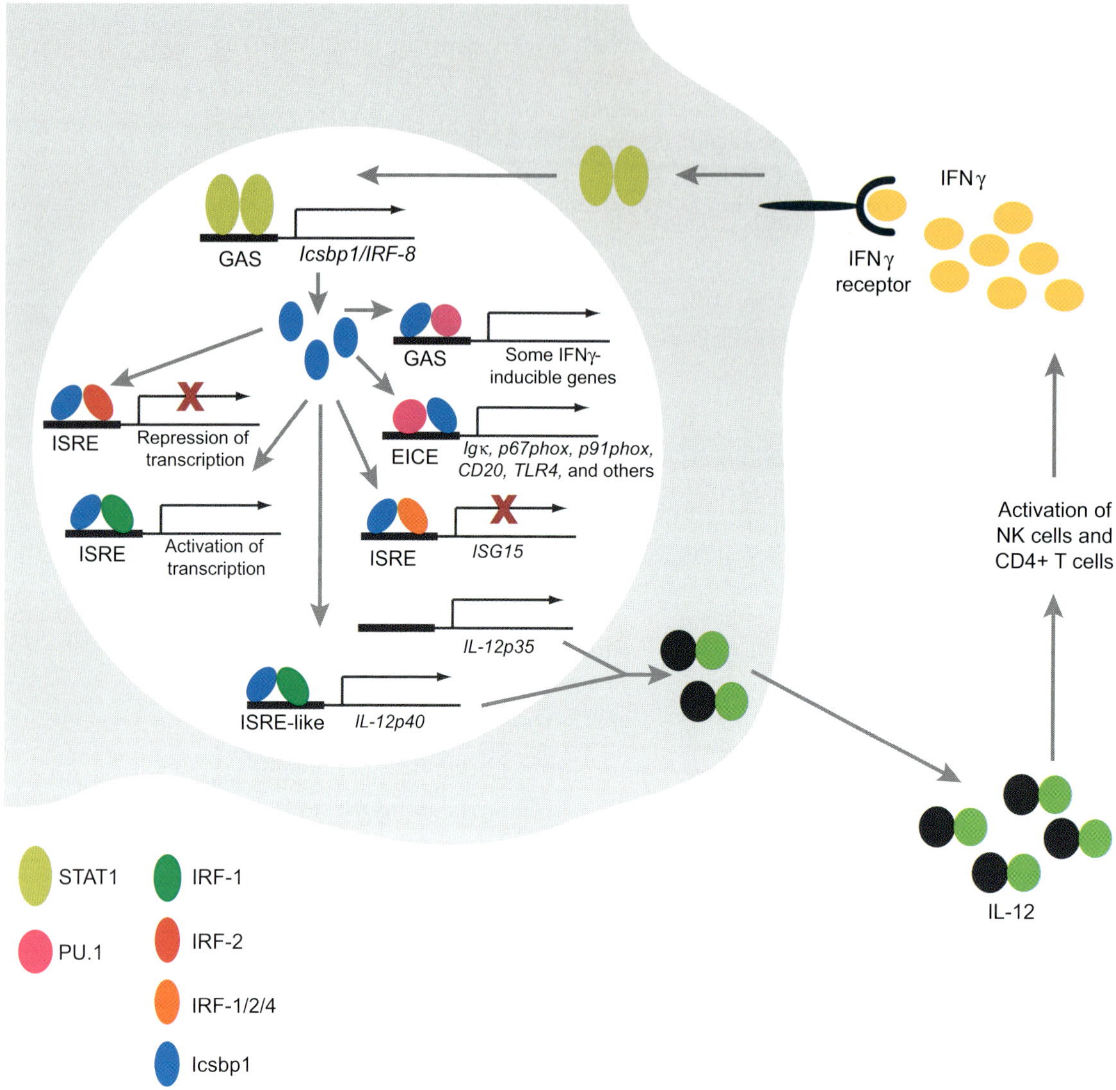

Figure 2

Regulatory role of Icsbp1/IRF-8 in macrophages. IFNγ induces expression of the *Icsbp1/IRF-8* gene through the activation of Jak1 and Jak2 kinase to stimulate formation of the Stat1 homodimer. Icsbp1/IRF-8 acts by associating with other transcription factors and possesses a dual activity that is dictated by an interacting partner, either an IRF or a non-IRF member. The *IL-12p40* gene, encoding the p40 subunit of the cytokine IL-12, is one target of Icsbp1/IRF-8, together with IRF-1. Secretion of IL-12 by macrophages activates NK and Th1 cells which secrete IFNγ, creating a positive regulatory loop.

BCG-osis in Humans

Live BCG vaccines can cause severe clinical disease in children with inborn errors of immunity often referred to as BCG-osis (24–26, 57). BCG-osis may affect otherwise healthy children whose infections were designated as "idiopathic" (24–26, 50, 57). Mendelian susceptibility to mycobacterial diseases (MSMD) (MIM 209950) (77) is described as a broader syndrome defined by the development of

Table 2 Known etiology of Mendelian susceptibility to mycobacterial diseases*

Gene	Inheritance	Defect	Protein	References
IFNGR1	AR	C	E+	(88)
	AR	C	E−	(86, 126)
	AD	P	E+	(88, 43)
	AR	P	E+	(57, 87)
IFNGR2	AR	C	E+	(172)
	AR	C	E−	(41, 172)
	AR	P	E+	(38)
STAT1	AD	P	E+ P−	(46)
	AD	P	E+ B−	(30)
IL12B	AR	C	E−	(9, 135)
IL12RB1	AR	C	E+	(53)
	AR	C	E	(7, 8, 23, 36, 121, 132, 150)
NEMO	XR	P	E+	(58)

*Modes of inheritance are either autosomal dominant (AD), autosomal recessive (AR), or X-linked recessive (XR). The functional defects are either complete (C) or partial (P). The mutant proteins are either expressed (E+) or not (E−), being not phosphorylated (P−) or not binding DNA (B−) upon IFN stimulation.

severe clinical conditions upon infection with weakly virulent mycobacterial species, such as BCG vaccines and nontuberculous, environmental mycobacteria (EM), in otherwise healthy individuals (24–26, 42, 96). As expected, patients with MSMD are also susceptible to the more virulent species *M. tuberculosis* (3, 7, 23, 87, 132). Severe disease caused by nontyphoidal and, to a lesser extent, typhoidal *Salmonella* serotypes is observed in nearly half of MSMD patients (24, 42, 107). Although associated with a narrow range of infections, "idiopathic" BCG-osis and EM diseases were often diagnosed in consanguineous and/or multiplex kindreds, strongly suggesting a Mendelian inheritance. In 1994, a candidate gene approach was undertaken in children with BCG-osis, with the goal of providing insights into the genetic basis for differences in susceptibility to TB (**Table 2**) (57).

IFN-γR1 deficiency. Interferons (IFNα, β, γ) are part of a large class of small glycoproteins known as cytokines, which are produced by cells of the immune system in response to challenges by foreign agents such as viruses, bacteria, and parasites. Intereferon γ is recognized by two cell surface receptors (IFN-γR1, IFN-γR2) expressed on macrophages and other immune cells. IFN-γR1 deficiency is the second most common etiology of MSMD: ~39% of MSMD patients with a genetic lesion (57, 86, 126).

Recessive complete (RC) IFN-γR1 deficiency was first identified in 1996 (86, 126). Subsequently other kindreds with this disorder have been reported (57). Even though most mutations in this receptor interfere with IFN-γR1 expression on the cell surface (126), a few mutant alleles encode proteins that reach the cell surface but fail to recognize IFN-γ (88) (**Figure 3**). In all cases, cells fail to respond to IFN-γ stimulation (43, 47). Complete IFN-γR1 deficiency is a life-threatening, early-onset condition (43). Children are mostly infected by BCG and rapidly growing EM. Salmonellosis has been documented in only three patients (43, 86, 126), and a few other types of infections have also been described (21, 44). Most affected individuals die in childhood and less than 20% of patients reach the age of 12 (43). Antibiotic treatment does not induce sustained clinical remission in these patients However, hematopoietic stem cell transplantation

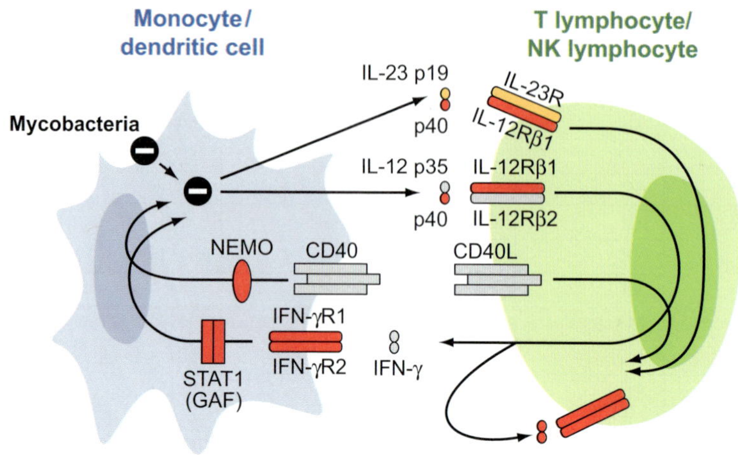

Figure 3

Mendelian susceptibility to mycobacterial disease (MSMD)-causing gene products in the IL-12/IFN-γ circuit. Schematic representation of cytokine production and cooperation between monocytes/dendritic cells and NK/T cells. The IL-12/IFNγ loop and the CD40L-activated CD40 pathway corresponding to cooperation between T cells and monocyte/dendritic cells are crucial for providing protective immunity to mycobacterial infection in humans. IL-12 production is controlled by both IFN-γ and CD40-NEMO (IkappaB kinase-gamma) signaling. Mutant molecules in patients with MSMD are indicated in red. Allelic heterogeneity of the five autosomal disease-causing genes results in the definition of 12 genetic disorders, and specific alleles of NEMO leucine zipper (LZ) domain cause the X-linked form of MSMD as they impair the CD40-dependent induction of IL-12. IL-23 is a much less potent inducer than IL-12.

(HSCT) was curative in three children (29, 56, 144).

One mutation in the *IFNGR1* gene (I87T) was associated with a recessive form of partial (RP) IFN-γR1 deficiency (57, 87). In these patients, cells have a residual response to IFN-γ (47, 52, 87). A dominant form of partial (DP) IFN-γR1 deficiency due to a truncation in the cytoplasmic domain of the receptor causes the accumulation of dominant-negative IFN-γR1 proteins at the cell surface (88). The 818del4 mutation is the most frequent dominant *IFNGR1* mutation (43, 88) and represented the first hot spot for small deletions identified in the human genome (88). Other dominant mutations in *IFNGR1* have been identified in isolated patients (43, 88).

RP- and DP-IFN-γR1 deficiencies are less severe than complete IFN-γR1 deficiency and show later onset. (43). Patients are susceptible to BCG and EM, but infections with rapidly growing bacteria are rarely seen. Salmonellosis has been documented in only 5% of patients, and other infections were documented in a single patient. A diagnosis of mycobacterial osteomyelitis is often associated with DP-IFN-γR1 deficiency (6, 10, 43).

IFN-γR2 deficiency. IFN-γR2 deficiency is a rare cause of MSMD (38, 41, 146, 172). The first patient with recessive complete (RC)-IFN-γR2 deficiency was reported in 1998 (41, 57). Some patients with RC-IFN-γR2 deficiency have no detectable cell surface expression of the protein (41, 172), while others express nonfunctional receptors (**Figure 3**) (172). Recessive partial (RP)-IFN-γR2 deficiency due to a homozygous R114C mutation impairs, but does not completely abolish, cellular responses to IFN-γ (38). A T168N null recessive mutation in IFN-γR2, which creates a new N-glycosylation site, results in the synthesis of an additional polysaccharide branched to the IFN-γR2 chain (on Asn 168) (172). This additional N-carbohydrate accounts for the pathogenic effect of the mutation since mutant IFN-γR2-expressing cells are functionally complemented upon treatment with PNG-ase F or tunicamycin (172).

Clinically, complete IFN-γR2 deficiency is as severe as complete IFN-γR1 deficiency (41, 146, 172). However, so far, the only child reported to have a RP form of IFN-γR2 had a modest clinical phenotype similar to that of children with RP-IFN-γR1 deficiency (38). Overall, the level of IFN-γ responsiveness seems strongly correlated to clinical phenotype in all disorders of the IFN-γR1 and IFN-γR2 chains (47).

Stat-1 deficiency. The first reported germline mutation (L706S) in *STAT1* (30, 46) was shown to impair the response to IFN-γ, but not to IFN-α/β, in heterozygous cells from the patients (**Figure 3**) (46). This mutation impairs the phosphorylation of Tyr 701, and behaves as a loss-of-function mutation for activation of both the IFN-γ and IFN-α/β pathways. In fact, the L706S allele is deleterious for the two interferon pathways but is dominant for one and recessive for the other (46). Two other mutations, E320Q and Q463H, both located in the DNA-binding domain of STAT-1, were recently found in other patients (30). These mutations defined a novel form of partial STAT-1 deficiency since in these cases Tyr 701 phosphorylation proceeds normally. These mutations result in a failure of the nuclear-translocated STAT-1-containing complexes to bind correctly to DNA regulatory elements. Like L706S, the E320Q and Q463H *STAT1* alleles are dominant for IFNγ-mediated antimycobacterial immunity, but recessive for IFNα/β-mediated antiviral immunity, accounting for the patients' clinical phenotype of MSMD without susceptibility to viral diseases (30). Children with DP-STAT-1 deficiency have relatively mild clinical disease, resembling that of children with RP-IFN-γR1 and RP-IFN-γR2 deficiency.

Other mutations in *STAT1* have been implicated in a related syndrome of susceptibility to mycobacteria and viruses, due to impaired IFN-γ- and IFN-α/β-mediated immunity, respectively (31, 48). Three homozygous mutations which cause loss of expression and loss of function were reported as a RC-STAT-1 deficiency. These children are not only susceptible to mycobacterial infections but are also prone to lethal viral infections (30, 31, 46, 48).

IL-12p40 deficiency. IL-12 is a cytokine that plays a critical role in innate immune responses. IL-12p40 deficiency remains the only known immunodeficiency resulting from a cytokine gene defect (9, 49, 135). All known *IL12B* mutations are recessive, loss of function defects that result in a RC-IL-12p40 deficiency (9, 135). IL-12 deficiency causes low IFN-γ production, probably accounting for mycobacterial susceptibility (9, 49, 135). Founder effects were documented for two of the four known *IL12B* mutations (135). IL-12p40-deficient patients suffer from BCG and/or EM disease (9, 49, 135). Half of the cases were infected with *Salmonella*, often together with mycobacterial disease (9, 54, 107, 135). One child, who was not vaccinated with BCG, developed recurrent and disseminated infection caused by nontyphoidal *Salmonella* (135). Similar clinical manifestations were reported in patients with IL-12Rβ1 deficiency (see below). In contrast, very few cases (~6%) of *Salmonella* infection were observed among MSMD patients bearing mutations affecting the IFN-γ-signaling pathway (24, 43, 107), and isolated *Salmonella* infections have not yet been reported in patients with IFN-γ-signaling defects. These observations suggest that IL-12 or IL-23 or both play a key role in protective immunity against *Salmonella*, probably in part via IFN-γ-independent mechanisms (**Figure 3**).

IL-12Rβ1 deficiency. Nearly 100 IL-12Rβ1-deficient patients have been described in the literature (7, 8, 23, 36, 49, 53, 55, 57, 121, 132, 150), making IL-12Rβ1 deficiency the most frequent known genetic etiology of MSMD. All mutant alleles are loss of function and cause RC-IL-12Rβ1 deficiency. None of the patients tested respond to IL-12 or IL-23 (**Figure 3**) (52, 53). Mycobacterial infection and salmonellosis are the most frequent infectious diseases found in these patients (7, 8, 23, 36, 49, 53, 55, 107, 121, 150, 155), but other infectious diseases have also been observed, although rarely.

The penetrance of IL-12Rβ1 deficiency for the case-definition phenotypes of disseminated BCG/EM and/or nontyphoidal systemic salmonellosis is low, at about 40% (55). Interindividual variability of IL-12Rβ1-deficient patients may be due to modifier genes, but environmental factors are critical, as BCG vaccination confers resistance to EM disease (55). These observations strongly suggest that IL-12/23 is critical for primary, but not secondary, immunity to mycobacteria. In contrast, given the long duration of salmonellosis in patients, IL-12/23 is equally important for primary and secondary immunity to *Salmonella*. Finally, it is intriguing that no p35-deficient and IL-12Rβ2-deficient patient has yet been reported.

Mutations in the NEMO (IκB kinase-gamma) leucine zipper domain. Hypomorphic *NEMO* mutations in male patients are associated with anhidrotic ectodermal dysplasia with immunodeficiency (EDA-ID), resulting in various infections including mycobacterial disease (39, 84, 94, 139). In 2006, two specific mutations (E315A and R319Q) in the leuzine zipper (NEMO LZ) domain were shown to be associated with X-linked recessive MSMD (58). The XR form of MSMD was first clinically described in 1991 in a multiplex kindred (66, 67, 78, 79, 124). A profound defect in IL-12 (and secondary IFN-γ) production was observed in in vitro cultures of monocytes from these patients (58, 66, 67). Specifically, the LZ-NEMO mutations were responsible for impairing IL-12 production in response to CD40 signaling, with delayed nuclear accumulation of the signaling molecule c-Rel, but not RelA (58). E315 and R319 amino acids form a highly conserved salt bridge within the LZ domain of NEMO, suggesting that mutations in these two amino acids may disturb the local plasticity of the LZ helix of NEMO, interfering with the CD40-NEMO-NF-κB-signaling pathway (58). However, CD40 signaling is not completely abolished in X-linked MSMD, which accounts for the differences observed between these patients and those with complete CD40 and CD40L deficiency. The selective impairment of CD40 signaling in monocytes and dendritic cells, and the subsequent defect in the production of IL-12 and IFN-γ, account for mycobacterial diseases in patients with NEMO-LZ mutations (58).

TUBERCULOSIS IN MICE AND HUMAN

Susceptibility to Tuberculosis: Genetic Studies in Mice

Studying the complex genetic component of susceptibility to TB is inherently difficult in humans. Alternatively, animal models of disease can be used to analyze the genetic control of susceptibility under defined experimental conditions, in which route of introduction, dose, strain, and virulence of *M. tuberculosis*, as well as environmental factors such as diet and stress, can be tightly controlled. Because the wide spectrum of disease outcomes seen in human TB is difficult to mimic in any single animal model, experimental infections in mouse, hamster, rabbit, primates, and fish have been used to study different aspects of *M. tuberculosis* pathogenesis (60, 131).

Guinea pigs are extremely susceptible to *M. tuberculosis* infection, developing rapidly fatal disease highlighted by the presence of necrotic granulomas similar to those seen in human TB. Early genetic studies in inbred strains of guinea pigs and in crosses derived from them identified pronounced differences in the type of ulcerative lesions (174) and survival time following *M. tuberculosis* infection (97). Rabbits are also very susceptible to infection with *M. tuberculosis* and *M. bovis.* Classical studies by Lurie in inbred strains of rabbits selected for resistance and susceptibility to infection showed that the type and number of pulmonary lesions observed postinfection were genetically controlled (103, 104). Although these early studies clearly established a role for genetic factors in innate susceptibility to TB, the paucity of both immunological and molecular reagents precluded the identification of the genes involved. In zebrafish, *M. marinum* can induce the formation of granulomas, and although the use of genetic tools in fish holds great potential to study granuloma formation and host response, this system is just beginning to be explored (32, 34).

The mouse has been most extensively used to study the genetics and pathogenesis of susceptibility to TB for a number of reasons. First, mice infected with *M. tuberculosis* develop a fatal pulmonary disease that pathologically resembles that seen in humans. Moreover, the innate and acquired host immune responses to *M. tuberculosis* in mice resemble those seen in humans, including the protective roles played by $CD4^+$ T cells, IFN-γ, TNF-α, and IL-12 (reviewed in 129). In addition, sufficient numbers of mice can be used to monitor multiple phenotypic aspects of disease susceptibility and severity, providing enough statistical power for genetic studies, and yet can be housed at a reasonable cost in a controlled environment. Importantly, a large number of well-defined inbred, recombinant, and mutant strains of mice with varying degrees of TB susceptibility are available for sampling genetic diversity. Mice are uniquely suited for genetic analysis, including the production of large informative crosses for genome scans, due to a fully sequenced genome, single nucleotide polymorphism and haplotype maps, and the relative ease with which gain- and loss-of-function mutants can be generated by germline modification to evaluate positional candidates. Finally, critical genes and proteins that affect TB susceptibility in mouse models can be tested as potential risk factors in parallel association and linkage studies in humans from areas of endemic disease.

Various strains of *M. tuberculosis* (H37Rv, Erdman, Vallee, Ravenel) have been used to infect mice via different infection routes, including intraperitoneal, intravenous, intranasal, and aerosol (60). Although high doses of bacilli ($>10^5$ CFUs) introduced by the i.v. route induce a rapidly fatal acute form of TB, aerosol infection with small doses ($\sim10^2$ CFUs) is more physiologic, and the pulmonary disease it induces most closely resembles human TB. Infection by either the i.v. or aerosol route results in transient replication of *M. tuberculosis* in the spleen, liver, and lungs, reaching a plateau at around four weeks. Subsequently, bacterial loads tend to remain stable or decrease in spleen/liver, while there may be continuing replication in the lungs leading to severe, and ultimately fatal, pulmonary disease. The extent of pulmonary replication, the type and severity of lung pathology, and the efficacy of the immune response are genetically controlled in mice (reviewed in 129).

It is important to note that although experimental infections in mouse can mimic several aspects of the pathophysiology and host response to *M. tuberculosis* seen in humans, other features are species-specific. For example, the long latency period typically observed in humans does not occur in the mouse model. In humans, granulomas are well organized with acid-fast bacilli-positive macrophages surrounded by lymphocytes; human granulomas are heterogeneous and can become necrotic, calcified, and caseous, a hallmark of human TB. In mice, granulomas are less well organized and necrosis is rarely observed (60).

Studies in inbred mouse strains. Strain differences in susceptibility to TB were first noted more than 50 years ago in ancestor inbred strains, following infection with various preparations of *M. tuberculosis* (40, 71, 136). In these studies, bacilli were introduced at fairly high doses (10^4–10^6 CFU) by the intravenous, aerosol, or intranasal routes, and survival time was generally used as a phenotypic measure of susceptibility. Strain C57 was susceptible, whereas strains such as C3H and H.I were ranked as intermediate, and Swiss albino and M.U.A strains were highly resistant (40, 71, 136). A first genetic analysis of susceptibility to *M. tuberculosis* infection (106) was carried out using C57BL and Swiss as susceptible [median survival time (MST) 28 days] and resistant (MST 53 days) strains following i.v. infection with 3×10^6 CFUs. Lynch et al. noted that the resistance/susceptibility trait was influenced by age, and was also completely dependent on the infectious dose, with C57BL mice becoming more resistant than Swiss at very low infectious doses. Informative [F1 X C57BL] backcross mice segregated into two nonoverlapping groups showing MST of 29 days and 78 days and a 0.6:1.0 segregation ratio, respectively, suggesting a major gene effect segregating in this cross.

Modern inbred strains were recently surveyed for susceptibility to *M. tuberculosis* (H37Rv) (114, 115, 122, 130). Using survival as a phenotypic readout, strains were classified as susceptible (S) (CBA, DBA/2, C3H, 129/Sv) and resistant (R) (C57BL/10, C57BL/6, BALB/c) to infection following i.v. infection with $\sim 10^5$ bacilli. Similar results were obtained following infection by the aerosol route (10^2 CFUs). Differential susceptibility of DBA/2 (S) and BALB/c (R) mice to i.v. infection was dose dependent and completely disappeared at high infectious doses (10^7 CFUs). In a number of F1 hybrids tested (B6D2, CD2, CB6, B6129), resistance was inherited in a dominant fashion. Studies in congenic lines identified a modest effect of haplotypes at the H-2 locus (115), but did not identify an effect of allelic combinations at the host resistance locus *Nramp1*, a finding verified in informative F2 mice (114) and in a 129sv *Nramp1*-targeted mutant (130). In this and other studies of modern inbred strains, susceptibility is generally associated with failure to contain bacterial growth in the lungs, inflammatory reaction causing consolidation of the lungs, and reduction of surface area available for oxygen exchange, leading to death. On the other hand, resistant animals limit bacterial growth and can prevent massive tissue injury. At the cellular level, lesions in resistant animals show large aggregates of lymphocytes penetrating granulomas, whereas lesions in susceptible animals display few scattered lymphocytes, but contain large foamy macrophages, are infiltrated with neutrophils, and are often necrotic (114, 115).

Genetic linkage studies and positional cloning of quantitative trait loci. To investigate the genes responsible for the large differences in host response noted between DBA/2J (S) and C57BL/6J (S) strains, a whole-genome scan was carried out in informative [C57BL/6J X DBA/2J] F2 mice infected with *M. tuberculosis* H37Rv (10^5 CFU, i.v.) (119). Using survival time as a phenotypic marker, two significant linkages on distal chromosome 1 (*Trl-1*; LOD 4.8) and proximal chromosome 7 (*Trl-3*; LOD 4.7), and a third suggestive linkage on the proximal portion of chromosome 3 (*Trl-2*; LOD 3.9) were detected. At each locus, homozygosity for B6 alleles was associated with increased resistance to infection. Individually, each locus accounted for ~20% of the phenotypic variance of the F2s, whereas the combined effect of the three loci was additive (no epistasis detected) and explained 45% of the variance in the cross.

A second genome scan was conducted in (C57BL/6J X DBA/2) F2 mice infected with 10^2 *M. tuberculosis* H37Rv by the aerosol route, and the extent of bacterial replication in the lungs (CFU) at 90 days was used as a phenotypic measure of susceptibility (120). Multipoint linkage analysis identified a strong linkage on the central portion of chromosome 19 (*Trl-4*; LOD 5.6), with B6 alleles at this locus associated with increased resistance in a codominant fashion. Weaker linkages were also noted with the *Trl-3* region (LOD 3.1), but also with chromosome 5 (LOD 3.7) and chromosome 10 (LOD 3.4). A strong additive effect was detected between *Trl-3* and *Trl-4*, with two-loci linkage analysis yielding a LOD 10.1, thus explaining 38% of the variation in pulmonary load. In this study, mice homozygous for B6 alleles at *Trl-3* and *Trl-4* were completely resistant and showed CFU counts in the range of B6 controls, whereas mice homozygous for D2 alleles were completely susceptible and showed CFU counts similar to those seen in the susceptible D2 controls.

Following infection with *M. tuberculosis* (10^5, i.v.), I/St mice are highly susceptible compared with resistant A/Sn, as measured by MST (~20 days versus ~45 days), increased bacterial loads in spleen and lungs, and body weight loss at day 20 (20% in I/St, versus none in A/Sn) (128). Compared with A/Sn mice, I/St animals are also more permissive to replication of attenuated *M. tuberculosis* H37Ra and *M. bovis* Ravenel indicative of hypersensitivity of the I/St strains to mycobacterial infections. An initial genome-wide scan was conducted in an [A/Sn X I/St] X I/St backcross population, using loss in body weight following infection with *M. tuberculosis* (5×10^5, i.v.) as a phenotypic marker (95). In females, significant linkages were identified on proximal chromosome 9 (LOD 6.7) and distal chromosome 3 (LOD 3.9), with additional suggestive linkages detected on proximal chromosome 8 (LOD 3.0) and chromosome 17 (LOD 2.9). In males, only suggestive linkages were detected on proximal chromosome 5 (LOD 3) and distal chromosome 10 (LOD 2.3) (95). Additional genetic studies in an [A/Sn X I/St]F2 cross (n = 406) confirmed the presence of QTLs on distal chromosome 3 (designated *Tbs1*; LOD 1.8), chromosome 9 (designated *Tbs2*; LOD 2.6 in males, LOD 1.9 in females), and chromosome 17, the site of the H2 locus (LOD 2.5) (151).

A new coding mutation in the TNFα leader sequence in TB-sensitive I/St mice that causes higher secretion levels of soluble TNFα may represent the chromosome 17 contribution to pathophsyiology in I/St mice (90). The differential pathophysiology and host response to infection controlled by these loci have been studied. Susceptibility in I/St was associated with increased Th2 polarization of the T-cell response, including increased production of IL-10 in the lungs compared with increased production of IFN-γ and increased accumulation of $CD8^+$ T cells in A/Sn mice (105). On the other hand, explanted I/St lung macrophages were more permissive to mycobacterial replication than A/Sn cells, suggesting that these cells play an important role in the expression of susceptibility (109). In addition, adaptive transfer of bone marrow cells from [I/St X A/Sn]F1 into I/St recipients increased survival, augmented antimycobacterial functions of macrophages and was concomitant to enhanced production of IFN-γ in the lungs and decreased infiltration of $CD4^+$ and $Ly6G^+$ cells in response to *M. tubertculosis* (108). I/St neutrophils do not seem to express CD95 (Fas), are resistant to apoptosis, and have a longer lifespan than those of A/Sn mice. Studies in segregating [A/Sn X I/St]F2 mice show that homozygosity for the I/St allele at *tbs1* is associated with increased influx of $Ly6G^+$ neutrophils in the lungs of infected mice at three weeks postinfection (51).

An independent study of inbred strains infected with *M. tuberculosis* Erdman (10^6, i.v.), using median survival as a discriminating trait, classified CBA/N, DBA/2, and, in particular, C3HeB/FcJ as susceptible, BALB/c and 129Sv as intermediate, and C57BL6/J as highly resistant to infection. Studies in informative F1 and F2 mice located a major gene effect on central chromosome 1 (LOD 10.4), designated *sst1* (93). Studies in congenic C3HeB/FeJ mice bearing B6 $sst1^r$ alleles (C3H.B6-*sst1*) showed that *sst1* has pleiotropic effects on other intracellular pathogens conferring increased protection in vivo against *Listeria monocytogenes*, with explanted *sst1* congenic macrophages showing increased IFN-γ dependent listericidal activity in vitro (19). The gene underlying the *sst1* effect was recently identified by positional cloning (133). Using a pair of congenic lines, the authors first showed that resistance was associated with increased capacity of macrophages to restrict replication of *M. tuberculosis* in vitro linked to induction of apoptosis in *M. tuberculosis*-infected resistant macrophages. Scrutiny of the *sst1* genetic interval for positional candidates expressed in macrophages identified a transcript, designated *Ipr1* (intracellular pathogen resistance 1), which is present in resistant macrophages and is induced upon *M. tuberculosis* infection, but is absent from susceptible cells. The *sst1* genetic interval includes part of a homogeneously staining region (HSR), a large unstable repeat of amplified DNA that contains several rearranged copies of *Ipr1*-related sequences. In C3HeB/FeJ, this amplification may have caused rearrangement(s) of the *Ipr1* gene, leading to an absence of mRNA expression. The *Ipr1* candidacy for *sst1* was validated by showing that expression of the full-length *Ipr1* transcript in transgenic C3HeB/FeJ mice can partially suppress *M. tuberculosis* replication in the lungs in vivo and in macrophages in vitro. *Ipr1* codes for Ifi75, a protein that harbors several sequence motifs, indicating a nuclear localization and transcriptional regulatory activity. Ifi75 is a relative of the human protein SP110 (89), a transcription factor regulated by interferon that binds to the retinoic acid receptor. In addition, SP110 interacts with certain viral proteins, including proteins from the Hepatitis C virus and the Epstein-Barr virus (127). These findings suggest that Ifi75 participates in transcriptional activation in macrophages in response to intracellular pathogens.

In congenic lines, transfer of the $sst1^r$ or $sst1^s$ alleles on C3HeB/FeJ (C3H.B6$sst1^r$) and B6 (B6.C3H$sst1^s$) genetic backgrounds, respectively, only partly confers resistance or susceptibility, suggesting that other genetic loci are responsible for the phenotype of parental strains. Such loci were investigated by a whole-genome scan in [C3H.B6$sst1^r$ X B6]F2 mice (175). Four additional loci mapping to chromosomes 7 (proximal; LOD 4.8), 12 (distal; LOD 6.6), 15 (distal; LOD 4.6), and 17 (proximal; LOD 5.5) influenced survival to infection. Additional studies of positional candidates for the chromosome 12 locus showed that deleting the immunoglobulin heavy chain (Igh) reduced survival to *M. tuberculosis* infection, causing a reduction of the number and size of B-cell clusters in TB lesions. The chromosome 17 QTL encompasses the H2 complex, and studies in B10 congenic lines harboring different H2 haplotypes showed reduced survival to infection in B10 mice bearing the H2k haplotype (derived from C3H) (175). The H2 complex includes a large number of positional candidates, including the TNFα gene, in which specific promotor polymorphism possibly modulating host response to TB has been described (90).

Human Genetics of Tuberculosis

Studies in human populations exposed to TB have attempted to identify host genetic factors determining disease susceptibility. Because host response to *M. tuberculosis* infection is a complex process influenced by multiple factors, several complementary strategies have been developed to identify the genes involved. Association-based candidate gene studies, in which gene variants believed to be potentially important are typed in case and control subjects, were generally used to verify the contribution of specific candidate genes in disease susceptibility. This method can detect a disease-susceptibility locus when the polymorphisms that are tested include either the causal allele or variant in strong linkage disequilibrium with the causal allele. Alternatively, genome-wide linkage analyses have been used to identify regions in the genome containing major disease susceptibility loci without any a priori assumption on the nature of the genes involved. Although this approach is more systematic, it can only detect loci that have a substantial effect on the disease phenotypes that are commonly denoted as major loci. Sibling-pair families containing two or more affected siblings are needed in whole-genome linkage studies whereas unrelated single-case and control individuals are typically used in association studies.

Complex predisposition to tuberculosis: association studies with candidate genes. Thus far, the vast majority of work conducted to identify the molecular basis of TB susceptibility has been association studies investigating the role of specific candidate genes. Some of these genes were selected as candidates because they were orthologs of murine genes found to influence the response to experimental mycobacterial infections. The most emblematic gene in that category is *NRAMP1* (alias *SLC11A1*), the human ortholog of murine *Nramp1* (27). Association between TB and *NRAMP1* variants has been reported in many studies following the initial finding in a Gambian population (16). A meta-analysis of these studies showed that several *NRAMP1* polymorphisms were significantly associated with pulmonary TB in African and Asian populations, but not in populations of European descent (98). As an example, this meta-analysis estimated the pooled odds ratio for developing TB associated to the D543N *NRAMP1* polymorphism at 1.69 (95% confidence interval 1.14–2.50) and 1.86 (1.33–2.62) in African and Asian subjects, respectively.

Two studies also provided evidence for the role of *NRAMP1* in early-onset, i.e., primary, TB. The first one showed strong linkage to the *NRAMP1* gene (LOD 3.5) in a large Canadian Aboriginal family that experienced a TB outbreak (72), and the second found association of *NRAMP1* polymorphisms with pediatric TB among children, mostly of Hispanic and African origin, living in Texas (110). *NRAMP1* was also involved in the two other common mycobacterial diseases, leprosy (1, 4) and Buruli ulcer (158). Overall, these studies strongly support the influence of *NRAMP1* in susceptibility to mycobacteria with an effect displaying heterogeneity across populations, epidemiological settings, and clinical phenotypes.

The *SP110* gene is the nearest homolog of the *Ipr1* gene discussed previously and has been shown to affect the host response to tuberculosis in mice (133). The corresponding protein, SP110, is a nuclear body protein with a role in transcription (18), showing 41% identity to the murine protein encoded by *Ipr1*. Two recent association studies between TB and *SP110* variants conducted in pulmonary TB patients from West Africa reported conflicting results. In a first, family-based study from Gambia, three *SP110* polymorphisms were associated with TB (161). The association of two of these polymorphisms, including a nonsynonymous change in exon 11, was also replicated in two other familial samples from Guinea-Bissau and the Republic of Guinea. The observation failed to be replicated in a large case-control study of more than 1000 HIV-negative TB patients and 1000 controls from Ghana. No significant associations were found between TB and 21 *SP110* variants, including the same nonsynonymous change in exon 11 as in the first study (160). Additional ongoing association studies will clarify these results. Note that loss-of-function mutations of *SP110* were recently shown to be responsible for a recessive primary immunodeficiency with hepatic vascular occlusion and fibrosis, denoted as VODI (OMIM235550), and none of the six reported patients presented with any mycobacterial infections (145).

Most of the candidate genes investigated in human TB were selected on the basis of their supposed role in the immune response against mycobacteria. As in many other infectious diseases, a very large number of studies were conducted with the MHC genes (24). The most consistent results have been obtained with alleles of the HLA class II DR2 serotype and HLA-DQB1 variants (37, 68). In particular, the association between pulmonary TB and homozygosity for alleles encoding an aspartic acid at codon 57 of the HLA-DQ β-chain was reported to be related, with a reduced ability for HLA-DQ β57-Asp to bind a peptide from the central region of the mycobacterial ESAT-6 protein (37). Associations with numerous other candidate genes, such as those encoding the vitamin D receptor and mannose-binding lectin (*MBL2*), have been reported, but not clearly reproduced or confirmed (reviewed in 141). A recent population genetic study showed that *MBL2*-deficient/low-producing haplotypes have not been targeted by natural selection, which strongly suggests a largely redundant immunological role for MBL and argues against a substantial role of this gene in susceptibility to TB (167).

With respect to the results found in MSMD, the influence of genes in the IL12/23-IFN-γ pathway has also been explored. So far, the most convincing result was obtained with a +874 A/T polymorphism located in a NKκB-binding site of *IFNG*. The A allele was found to predispose to TB in several populations (99, 102, 147, 163), an observation consistent with a report showing that the A allele is associated with lower expression of IFN-γ (138). The odds ratio of developing TB for subjects homozygous for this A allele varies from 1.46 (1.12–1.91) in South African subjects (147) to 3.79 (1.93–7.45) in Hong Kong Chinese subjects (163).

Finally, two recent studies provide evidence for the association of adult pulmonary TB with two other candidate genes. Two promoter variants in *CD209*, the gene encoding DC-SIGN (dendritic cell specific ICAM-3 grabbing non-integrin) which is a *M. tuberculosis* receptor on human dendritic cells (159), were found associated with TB in a South African population (13). Another study found a strong association with a 5′UTR A/G polymorphism of the gene encoding the chemokine monocyte chemoattractant protein-1 (MCP-1) in a Mexican and in a South Korean population (59), although the same variant was previously reported not to be associated with TB in a Brazilian cohort (85). The G allele was very common (0.49 and 0.36 in Mexican and Korean controls, respectively), and the odds ratios of TB among individuals with AG and GG versus AA genotypes were estimated over 2 and 5, respectively, in both populations. Overall, this study reports the most substantial impact ever described of a human allele on adult TB at the population level and needs further replications in other populations.

These types of association studies have a number of limitations (2). Multiple testing of numerous polymorphisms can lead to false positive results, which may be difficult to appreciate because negative (nonsignificant) results are often not reported. Independent replication studies are fundamental to confirm reported associations and should account for population-specific linkage disequilibrium patterns and relevant covariates to avoid false negative results. Finally, two critical points are the phenotype definition for patients, and, in case-control studies, the choice of controls. Misspecification of these points can markedly affect both robustness and power of the analysis. One way to avoid the control selection bias is to perform family-based association studies such as the Transmission Disequilibrium Test (154) and related approaches.

Mendelian predisposition to tuberculosis. Common polymorphisms providing moderate individual risks for disease susceptibility usually represent complex genetic predisposition. A fascinating recent observation is that TB can also reflect a Mendelian predisposition (3). Following the findings obtained in MSMD, several children with genetic defects of the IL-12/23-IFNγ pathway, in particular those with complete IL-12Rβ1 deficiency, had culture-proven severe TB as the sole infectious phenotype. In one family, an IL-12Rβ1-deficient patient developed abdominal TB (8). She had been vaccinated three times with BCG with no adverse effects, whereas her brother, who was also IL-12Rβ1 deficient, developed BCG disease following immunization. In another family, an IL-12Rβ1-deficient girl developed disseminated TB (23). Her IL-12Rβ1-deficient sister had a history of nontyphoidal extraintestinal salmonellosis and, despite prophylactic treatment with isoniazid, also developed TB. Finally, a girl with IL-12Rβ1 deficiency, but no relevant personal or familial history of mycobacteriosis or salmonellosis, developed disseminated TB (132).

These cases of severe TB in children with IL-12Rβ1 deficiency provided proof of principle that TB can be a Mendelian disease. Interestingly, more common *IL12RB1* polymorphisms can also be associated with adult pulmonary TB, as indicated by the first study in Morocco (142). However, Mendelian predisposition to TB is not limited to IL-12Rβ1 deficiency, as one child with a partial IFN-γR1 deficiency also suffered from TB (87). The proportion of children with disseminated TB due to Mendelian predisposition remains to be experimentally determined, but has been estimated at 3–30% by Bayesian statistics (3). This prediction is experimentally testable by investigating groups of children with TB living in endemic areas.

Major gene predisposition to tuberculosis: whole-genome scans. Although these findings suggest both a simple and complex mode of inheritance for genetic susceptibility to TB, no major genes were reported to be involved until 2006. Some chromosomal regions provide suggestive (LOD scores around 2), but not significant, evidence for linkage to TB. These chromosomal regions include regions of chromosomes 15 and X identified in African families from Gambia and South Africa (15), and two regions on chromosomes 11 and 20 previously identified in Brazilian families (116). Only recently was the first major locus identified by a genome-wide screen mapped to chromosome 8q12-q13 in a Moroccan population (12). This locus predisposed to adult pulmonary TB with a dominant mode of inheritance (LOD 3.6), providing novel insight into evolutionary population genetics.

It has been suggested that prolonged exposure to TB results in the genetic adaptation of exposed populations accounting for the rapid decline in TB mortality rates in Europe during the nineteenth century before any specific measures against the disease were taken (45). Not all genetic models of TB susceptibility are consistent with this theory. For example, if resistant alleles were rare at the beginning of the European epidemic (early seventeenth century), 300 years would not have been long enough for significant genetic selection (100). In contrast, if rare dominant susceptibility alleles, such as the 8q12-q13 locus, accounted for a substantial proportion of cases, then natural selection would have had a strong impact on the genetic composition of an exposed host population even over a short period of time. The precise identification of this gene is ongoing following a linkage disequilibrium strategy similar to that which led to the successful positional cloning of the first major locus in leprosy (117, 118). Overall, these recent findings in TB provide the proof of concept that human genetics of common infectious diseases involves a continuous spectrum from Mendelian to complex predisposition with intermediate major gene situations.

CONCLUSION

Population studies and experiments in animal models have clearly established a role of genetic factors in predisposition to TB. Furthermore, the relatively low percentage of exposed individuals progressing to active TB suggests genetic effects on disease progression. Understanding the molecular mechanisms responsible for successful protection against TB is critical for the development of effective vaccines against this disease, but may also provide novel targets for immunomodulation and for antimicrobial drug discovery. Forward genetic approaches with clinically validated human populations and carefully selected mouse strains have yielded complementary results and are starting to reveal some of the major gene effects involved in onset and progression of disease. These point to a critical role for macrophages, lymphocytes, and secreted mediators in innate and immune response to TB infection. Independently, the genetic dissection of the molecular basis of "idiopathic" BCG-osis and MSMD has had important clinical, genetic, and immunological implications. In immunological terms, the most surprising observation is that patients with lesions in the IL-12-IFN-γ loop are apparently resistant to most infectious agents, possibly suggesting unsuspected compartmentalization of the host immune response to this and other groups of intracellular pathogens. In addition, patients with MSMD can now be offered molecular diagnosis, improving the prediction of individual clinical outcome and facilitating treatment based on a rational understanding of the pathogenesis of infections. Finally, the more complex problem of disease reactivation has yet to be successfully characterized by a genetic approach.

DISCLOSURE STATEMENT

The authors are not aware of any biases that might be perceived as affecting the objectivity of this review.

ACKNOWLEDGMENTS

Authors apologize to colleagues who could not be cited owing to space limitations. A.F. worked with J.L.C. and L.A. under a fellowship from the Novartis Foundation. Work in P.G.'s laboratory is supported by grants from the Canadian Institutes of Health Research and the National Institutes of Health. P.G. is a distinguished scientist of the CIHR. Work in L.A. and J.L.C.'s laboratories is supported in part by grants from the European Community (QLK2-CT-2002-00846), INSERM, and the ANR. J.L.C. is an International Scholar of the Howard Hughes Medical Institute. The authors also wish to express their appreciation to Karine Turcotte for preparing **Figure 2**.

LITERATURE CITED

1. Abel L, Sanchez FO, Oberti J, Thuc NV, Hoa LV, et al. 1998. Susceptibility to leprosy is linked to the human *NRAMP1* gene. *J. Infect. Dis.* 177:133–45
2. Alcais A, Abel L. 2004. Application of genetic epidemiology to dissecting host susceptibility/resistance to infection illustrated with the study of common mycobacterial infections. In *Susceptibility to Infectious Diseases: The Importance of Host Genetics*, ed. R Bellamy, pp. 7–44. Cambridge, UK; New York: Cambridge Univ. Press
3. Alcais A, Fieschi C, Abel L, Casanova JL. 2005. Tuberculosis in children and adults: two distinct genetic diseases. *J. Exp. Med.* 202:1617–21
4. Alcais A, Sanchez FO, Thuc NV, Lap VD, Oberti J, et al. 2000. Granulomatous reaction to intradermal injection of lepromin (Mitsuda reaction) is linked to the human *NRAMP1* gene in Vietnamese leprosy sibships. *J. Infect. Dis.* 181:302–8
5. Algood HM, Chan J, Flynn JL. 2003. Chemokines and tuberculosis. *Cytokine Growth Factor Rev.* 14:467–77
6. Allende LM, Lopez-Goyanes A, Paz-Artal E, Corell A, Garcia-Perez MA, et al. 2001. A point mutation in a domain of gamma interferon receptor 1 provokes severe immunodeficiency. *Clin. Diagn. Lab. Immunol.* 8:133–37
7. Altare F, Durandy A, Lammas D, Emile JF, Lamhamedi S, et al. 1998. Impairment of mycobacterial immunity in human interleukin-12 receptor deficiency. *Science* 280:1432–35
8. Altare F, Ensser A, Breiman A, Reichenbach J, Baghdadi JE, et al. 2001. Interleukin-12 receptor beta1 deficiency in a patient with abdominal tuberculosis. *J. Infect. Dis.* 184:231–36

9. Altare F, Lammas D, Revy P, Jouanguy E, Doffinger R, et al. 1998. Inherited interleukin 12 deficiency in a child with bacille Calmette-Guerin and *Salmonella enteritidis* disseminated infection. *J. Clin. Invest.* 102:2035–40
10. Arend SM, Janssen R, Gosen JJ, Waanders H, de Boer T, et al. 2001. Multifocal osteomyelitis caused by nontuberculous mycobacteria in patients with a genetic defect of the interferon-gamma receptor. *Neth. J. Med.* 59:140–51
11. Arias M, Rojas M, Zabaleta J, Rodriguez JI, Paris SC, et al. 1997. Inhibition of virulent *Mycobacterium tuberculosis* by Bcg(r) and Bcg(s) macrophages correlates with nitric oxide production. *J. Infect. Dis.* 176:1552–58
12. Baghdadi JE, Orlova M, Alter A, Ranque B, Chentoufi M, et al. 2006. An autosomal dominant major gene confers predisposition to pulmonary tuberculosis in adults. *J. Exp. Med.* 203:1679–84
13. Barreiro LB, Neyrolles O, Babb CL, Tailleux L, Quach H, et al. 2006. Promoter variation in the DC-SIGN-encoding gene CD209 is associated with tuberculosis. *PLoS Med.* 3:e20
14. Behr MA, Wilson MA, Gill WP, Salamon H, Schoolnik GK, et al. 1999. Comparative genomics of BCG vaccines by whole-genome DNA microarray. *Science* 284:1520–23
15. Bellamy R, Beyers N, McAdam KP, Ruwende C, Gie R, et al. 2000. Genetic susceptibility to tuberculosis in Africans: a genome-wide scan. *Proc. Natl. Acad. Sci. USA* 97:8005–9
16. Bellamy R, Ruwende C, Corrah T, McAdam KPWJ, Whittle HC, Hill AVS. 1998. Variations in the *NRAMP1* gene and susceptibility to tuberculosis in West Africans. *New Engl. J. Med.* 338:640–44
17. Beutler B, Jiang Z, Georgel P, Crozat K, Croker B, et al. 2006. Genetic analysis of host resistance: Toll-like receptor signaling and immunity at large. *Annu. Rev. Immunol.* 24:353–89
18. Bloch DB, Nakajima A, Gulick T, Chiche JD, Orth D, et al. 2000. Sp110 localizes to the PML-Sp100 nuclear body and may function as a nuclear hormone receptor transcriptional coactivator. *Mol. Cell Biol.* 20:6138–46
19. Boyartchuk V, Rojas M, Yan BS, Jobe O, Hurt N, et al. 2004. The host resistance locus sst1 controls innate immunity to *Listeria* monocytogenes infection in immunodeficient mice. *J. Immunol.* 173:5112–20
20. Calmette A, Guerin C, Negre L, Boquet A. 1927. Sur la vaccination preventive des enfants nouveaux-nes contre la tuberculose par le BCG. *Annu. Inst. Pasteur.* 3:201
21. Camcioglu Y, Picard C, Lacoste V, Dupuis S, Akcakaya N, et al. 2004. HHV-8-associated Kaposi sarcoma in a child with IFNgammaR1 deficiency. *J. Pediatr.* 144:519–23
22. Canonne-Hergaux F, Calafat J, Richer E, Cellier M, Grinstein S, et al. 2002. Expression and subcellular localization of NRAMP1 in human neutrophil granules. *Blood* 100:268–75
23. Caragol I, Raspall M, Fieschi C, Feinberg J, Larrosa MN, et al. 2003. Clinical tuberculosis in 2 of 3 siblings with interleukin-12 receptor beta1 deficiency. *Clin. Infect. Dis.* 37:302–6
24. Casanova JL, Abel L. 2002. Genetic dissection of immunity to mycobacteria: the human model. *Annu. Rev. Immunol.* 20:581–620
25. Casanova JL, Blanche S, Emile JF, Jouanguy E, Lamhamedi S, et al. 1996. Idiopathic disseminated bacillus Calmette-Guerin infection: a French national retrospective study. *Pediatrics* 98:774–78
26. Casanova JL, Jouanguy E, Lamhamedi S, Blanche S, Fischer A. 1995. Immunological conditions of children with BCG disseminated infection. *Lancet* 346:581
27. Cellier M, Govoni G, Vidal S, Kwan T, Groulx N, et al. 1994. Human natural resistance-associated macrophage protein: cDNA cloning, chromosomal mapping, genomic organization, and tissue-specific expression. *J. Exp. Med.* 180:1741–52

28. Cellier M, Gros P, eds. 2004. *The Nramp Family*. New York: Kluwer Acad./Plenum. **http://www.Eurekah.com**
29. Chantrain CF, Bruwier A, Brichard B, Largent V, Chapgier A, et al. 2006. Successful hematopoietic stem cell transplantation in a child with active disseminated *Mycobacterium* fortuitum infection and interferon-gamma receptor 1 deficiency. *Bone Marrow Transplant.* 38:75–76
30. Chapgier A, Boisson-Dupuis S, Jouanguy E, Vogt G, Feinberg J, et al. 2006. Novel STAT1 alleles in otherwise healthy patients with Mycobacterial disease. *PLoS Genet.* 2(8):e131
31. Chapgier A, Wynn RF, Jouanguy E, Filipe-Santos O, Zhang S, et al. 2006. Human complete Stat-1 deficiency is associated with defective type I and II IFN responses in vitro but immunity to some low virulence viruses in vivo. *J. Immunol.* 176:5078–83
32. Cosma CL, Humbert O, Ramakrishnan L. 2004. Superinfecting mycobacteria home to established tuberculous granulomas. *Nat. Immunol.* 5:828–35
33. Cuellar-Mata P, Jabado N, Liu J, Furuya W, Finlay BB, et al. 2002. Nramp1 modifies the fusion of *Salmonella typhimurium*-containing vacuoles with cellular endomembranes in macrophages. *J. Biol. Chem.* 277:2258–65
34. Davis JM, Clay H, Lewis JL, Ghori N, Herbomel P, Ramakrishnan L. 2002. Real-time visualization of mycobacterium-macrophage interactions leading to initiation of granuloma formation in zebrafish embryos. *Immunity* 17:693–702
35. de Chastellier C, Frehel C, Offredo C, Skamene E. 1993. Implication of phagosome-lysosome fusion in restriction of *Mycobacterium avium* growth in bone marrow macrophages from genetically resistant mice. *Infect. Immun.* 61:3775–84
36. de Jong R, Altare F, Haagen IA, Elferink DG, Boer T, et al. 1998. Severe mycobacterial and *Salmonella* infections in interleukin-12 receptor-deficient patients. *Science* 280:1435–38
37. Delgado JC, Baena A, Thim S, Goldfeld AE. 2006. Aspartic acid homozygosity at codon 57 of HLA-DQ beta is associated with susceptibility to pulmonary tuberculosis in Cambodia. *J. Immunol.* 176:1090–97
38. Döffinger R, Jouanguy E, Dupuis S, Fondanèche MC, Stéphan JL, et al. 2000. Partial interferon gamma receptor signaling chain deficiency in a patient with bacille Calmette-Guérin and *Mycobacterium* abscessus infection. *J. Infect. Dis.* 181:379–84
39. Doffinger R, Smahi A, Bessia C, Geissmann F, Feinberg J, et al. 2001. X-linked anhidrotic ectodermal dysplasia with immunodeficiency is caused by impaired NF-kappaB signaling. *Nat. Genet.* 27:277–85
40. Donovick R, McKee CM, Jambor WP, Rake G. 1949. The use of mouse in a standardized test for antituberculous activity of compounds of natural or synthetic origin. II. Choice of mouse strain. *Am. Rev. Tuber.* 60:109–20
41. Dorman SE, Holland SM. 1998. Mutation in the signal-transducing chain of the interferon-gamma receptor and susceptibility to mycobacterial infection. *J. Clin. Invest.* 101:2364–69
42. Dorman SE, Holland SM. 2000. Interferon-gamma and interleukin-12 pathway defects and human disease. *Cytokine Growth Factor Rev.* 11:321–33
43. Dorman SE, Picard C, Lammas D, Heyne K, van Dissel JT, et al. 2004. Clinical features of dominant and recessive interferon gamma receptor 1 deficiencies. *Lancet* 364:2113–21
44. Dorman SE, Uzel G, Roesler J, Bradley JS, Bastian J, et al. 1999. Viral infections in interferon-γ receptor deficiency. *J. Pediatr.* 135:640–43
45. Dubos RJ, Dubos J. 1952. *The White Plague: Tuberculosis, Man, and Society*. Boston: Little Brown. 277 pp.

46. Dupuis S, Dargemont C, Fieschi C, Thomassin N, Rosenzweig S, et al. 2001. Impairment of mycobacterial but not viral immunity by a germline human STAT1 mutation. *Science* 293:300–3
47. Dupuis S, Doffinger R, Picard C, Fieschi C, Altare F, et al. 2000. Human interferon-gamma-mediated immunity is a genetically controlled continuous trait that determines the outcome of mycobacterial invasion. *Immunol. Rev.* 178:129–37
48. Dupuis S, Jouanguy E, Al-Hajjar S, Fieschi C, Al-Mohsen IZ, et al. 2003. Impaired response to interferon-alpha/beta and lethal viral disease in human STAT1 deficiency. *Nat. Genet.* 33:388–91
49. Elloumi-Zghal H, Barbouche MR, Chemli J, Bejaoui M, Harbi A, et al. 2002. Clinical and genetic heterogeneity of inherited autosomal recessive susceptibility to disseminated *Mycobacterium bovis* bacille calmette-guerin infection. *J. Infect. Dis.* 185:1468–75
50. Emile JF, Patey N, Altare F, Lamhamedi S, Jouanguy E, et al. 1997. Correlation of granuloma structure with clinical outcome defines two types of idiopathic disseminated BCG infection. *J. Pathol.* 181:25–30
51. Eruslanov EB, Lyadova IV, Kondratieva TK, Majorov KB, Scheglov IV, et al. 2005. Neutrophil responses to *Mycobacterium tuberculosis* infection in genetically susceptible and resistant mice. *Infect. Immun.* 73:1744–53
52. Feinberg J, Fieschi C, Doffinger R, Feinberg M, Leclerc T, et al. 2004. *Bacillus* Calmette Guerin triggers the IL-12/IFN-gamma axis by an IRAK-4- and NEMO-dependent, noncognate interaction between monocytes, NK, and T lymphocytes. *Eur. J. Immunol.* 34:3276–84
53. Fieschi C, Bosticardo M, de Beaucoudrey L, Boisson-Dupuis S, Feinberg J, et al. 2004. A novel form of complete IL-12/IL-23 receptor beta1 deficiency with cell surface-expressed nonfunctional receptors. *Blood* 104:2095–101
54. Fieschi C, Casanova JL. 2003. The role of interleukin-12 in human infectious diseases: only a faint signature. *Eur. J. Immunol.* 33:1461–64
55. Fieschi C, Dupuis S, Catherinot E, Feinberg J, Bustamante J, et al. 2003. Low penetrance, broad resistance, and favorable outcome of interleukin 12 receptor beta1 deficiency: medical and immunological implications. *J. Exp. Med.* 197:527–35
56. Fieschi C, Dupuis S, Picard C, Smith CI, Holland SM, Casanova JL. 2001. High levels of interferon gamma in the plasma of children with complete interferon gamma receptor deficiency. *Pediatrics* 107:E48
57. Filipe-Santos O, Bustamante J, Chapgier A, Vogt G, de Beaucoudrey L, Feinberg J, Jouanguy E, Boisson-Dupuis S, Fieschi C, Picard C, Casanova JL. 2006. Inborn errors of IL-12/23- and IFN-gamma-mediated immunity: molecular, cellular, and clinical features. *Semin. Immunol.* 18:347–61
58. Filipe-Santos O, Bustamante J, Haverkamp MH, Vinolo E, Ku CL, et al. 2006. X-linked susceptibility to mycobacteria is caused by mutations in NEMO impairing CD40-dependent IL-12 production. *J Exp. Med.* 203:1745–59
59. Flores-Villanueva PO, Ruiz-Morales JA, Song CH, Flores LM, Jo EK, et al. 2005. A functional promoter polymorphism in monocyte chemoattractant protein-1 is associated with increased susceptibility to pulmonary tuberculosis. *J. Exp. Med.* 202:1649–58
60. Flynn JL. 2006. Lessons from experimental *Mycobacterium tuberculosis* infections. *Microbes Infect.* 8:1179–88
61. Forbes JR, Gros P. 2001. Divalent-metal transport by NRAMP proteins at the interface of host-pathogen interactions. *Trends Microbiol.* 9:397–403
62. Forbes JR, Gros P. 2003. Iron, manganese, and cobalt transport by Nramp1 (Slc11a1) and Nramp2 (Slc11a2) expressed at the plasma membrane. *Blood* 102:1884–92

63. Forget A, Skamene E, Gros P, Miailhe AC, Turcotte R. 1981. Differences in response among inbred mouse strains to infection with small doses of *Mycobacterium bovis* BCG. *Infect. Immun.* 32:42–47
64. Frehel C, Canonne-Hergaux F, Gros P, De Chastellier C. 2002. Effect of Nramp1 on bacterial replication and on maturation of *Mycobacterium avium*-containing phagosomes in bone marrow-derived mouse macrophages. *Cell Microbiol.* 4:541–56
65. Frieden TR, Sterling TR, Munsiff SS, Watt CJ, Dye C. 2003. Tuberculosis. *Lancet* 362:887–99
66. Frucht DM, Holland SM. 1996. Defective monocyte costimulation for IFN-gamma production in familial disseminated *Mycobacterium avium* complex infection: abnormal IL-12 regulation. *J. Immunol.* 157:411–16
67. Frucht DM, Sandberg DI, Brown MR, Gerstberger SM, Holland SM. 1999. IL-12-independent costimulation pathways for interferon-γ production in familial disseminated *Mycobacterium avium* complex infection. *Clin. Immunol.* 91:234–41
68. Goldfeld AE, Delgado JC, Thim S, Bozon MV, Uglialoro AM, et al. 1998. Association of an HLA-DQ allele with clinical tuberculosis. *JAMA* 279:226–28
69. Govoni G, Canonne-Hergaux F, Pfeifer CG, Marcus SL, Mills SD, et al. 1999. Functional expression of Nramp1 in vitro in the murine macrophage line RAW264.7. *Infect. Immun.* 67:2225–32
70. Govoni G, Vidal S, Gauthier S, Skamene E, Malo D, Gros P. 1996. The Bcg/Ity/Lsh locus: genetic transfer of resistance to infections in C57BL/6J mice transgenic for the Nramp1 Gly169 allele. *Infect. Immun.* 64:2923–29
71. Gray DF, Graham-Smith H, Noble JL. 1960. Variations in natural resistance to tuberculosis. *J. Hyg. (Lond.)* 58:215–27
72. Greenwood CM, Fujiwara TM, Boothroyd LJ, Miller MA, Frappier D, et al. 2000. Linkage of tuberculosis to chromosome 2q35 loci, including NRAMP1, in a large aboriginal Canadian family. *Am. J. Hum. Genet.* 67:405–16
73. Gros P, Skamene E, Forget A. 1981. Genetic control of natural resistance to *Mycobacterium bovis* (BCG) in mice. *J. Immunol.* 127:2417–21
74. Gruenheid S, Pinner E, Desjardins M, Gros P. 1997. Natural resistance to infection with intracellular pathogens: the Nramp1 protein is recruited to the membrane of the phagosome. *J. Exp. Med.* 185:717–30
75. Hackam DJ, Rotstein OD, Zhang W, Gruenheid S, Gros P, Grinstein S. 1998. Host resistance to intracellular infection: mutation of natural resistance-associated macrophage protein 1 (Nramp1) impairs phagosomal acidification. *J. Exp. Med.* 188:351–64
76. Haller O, Frese M, Kochs G. 1998. Mx proteins: mediators of innate resistance to RNA viruses. *Rev. Sci. Tech.* 17:220–30
77. Hamosh A, Scott AF, Amberger JS, Bocchini CA, McKusick VA. 2005. Online Mendelian Inheritance in Man (OMIM), a knowledgebase of human genes and genetic disorders. *Nucleic Acids Res.* 33:D514–17
78. Holland SA, Dorman SE, Kwon A, Pitha-Rowe IF, Frucht DM, et al. 1998. Abnormal regulation of interferon gamma, interleukin 12, and tumor necrosis factor alpha in interferon gamma receptor 1 deficiency. *J. Infect. Dis.* 178:1095–104
79. Holland SM, Eisenstein EM, Kuhns DB, Turner ML, Fleisher TA, et al. 1994. Treatment of refractory disseminated nontuberculous mycobacterial infection with interferon gamma. A preliminary report. *N. Engl. J. Med.* 330:1348–55
80. Huynh C, Sacks DL, Andrews NW. 2006. A *Leishmania amazonensis* ZIP family iron transporter is essential for parasite replication within macrophage phagolysosomes. *J. Exp. Med.* 203:2363–75

81. Inohara N, Chamaillard M, McDonald C, Nunez G. 2005. NOD-LRR proteins: role in host-microbial interactions and inflammatory disease. *Annu. Rev. Biochem.* 74:355–83
82. Jabado N, Cuellar-Mata P, Grinstein S, Gros P. 2003. Iron chelators modulate the fusogenic properties of *Salmonella*-containing phagosomes. *Proc. Natl. Acad. Sci. USA* 100:6127–32
83. Jabado N, Jankowski A, Dougaparsad S, Picard V, Grinstein S, Gros P. 2000. Natural resistance to intracellular infections: natural resistance-associated macrophage protein 1 (Nramp1) functions as a pH-dependent manganese transporter at the phagosomal membrane. *J. Exp. Med.* 192:1237–48
84. Jain A, Ma CA, Liu S, Brown M, Cohen J, Strober W. 2001. Specific missense mutations in NEMO result in hyper-IgM syndrome with hypohydrotic ectodermal dysplasia. *Nat. Immunol.* 2:223–28
85. Jamieson SE, Miller EN, Black GF, Peacock CS, Cordell HJ, et al. 2004. Evidence for a cluster of genes on chromosome 17q11-q21 controlling susceptibility to tuberculosis and leprosy in Brazilians. *Genes Immun.* 5:46–57
86. Jouanguy E, Altare F, Lamhamedi S, Revy P, Emile JF, et al. 1996. Interferon-gamma-receptor deficiency in an infant with fatal bacille Calmette-Guerin infection. *N. Engl. J. Med.* 335:1956–61
87. Jouanguy E, Lamhamedi-Cherradi S, Altare F, Fondaneche MC, Tuerlinckx D, et al. 1997. Partial interferon-gamma receptor 1 deficiency in a child with tuberculoid bacillus Calmette-Guerin infection and a sibling with clinical tuberculosis. *J. Clin. Invest.* 100:2658–64
88. Jouanguy E, Lamhamedi-Cherradi S, Lammas D, Dorman SE, Fondaneche MC, et al. 1999. A human IFNGR1 small deletion hotspot associated with dominant susceptibility to mycobacterial infection. *Nat. Genet.* 21:370–78
89. Kadereit S, Gewert DR, Galabru J, Hovanessian AG, Meurs EF. 1993. Molecular cloning of two new interferon-induced, highly related nuclear phosphoproteins. *J. Biol. Chem.* 268:24432–41
90. Kahler AK, Persson AS, Sanchez F, Kallstrom H, Apt AS, et al. 2005. A new coding mutation in the Tnf-alpha leader sequence in tuberculosis-sensitive I/St mice causes higher secretion levels of soluble TNF-alpha. *Genes Immun.* 6:620–27
91. Kaufmann SH. 2001. How can immunology contribute to the control of tuberculosis? *Nat. Rev. Immunol.* 1:20–30
92. Kaufmann SH. 2003. Immune response to tuberculosis: experimental animal models. *Tuberculosis (Edinb.)* 83:107–11
93. Kramnik I, Dietrich WF, Demant P, Bloom BR. 2000. Genetic control of resistance to experimental infection with virulent *Mycobacterium tuberculosis*. *Proc. Natl. Acad. Sci. USA* 97:8560–65
94. Ku CL, Dupuis-Girod S, Dittrich AM, Schulze I, Chang HH, et al. 2005. NEMO mutations in two unrelated boys with severe infections and conical teeth. *Pediatrics* 115:615–19
95. Lavebratt C, Apt AS, Nikonenko BV, Schalling M, Schurr E. 1999. Severity of tuberculosis in mice is linked to distal chromosome 3 and proximal chromosome 9. *J. Infect. Dis.* 180:150–55
96. Levin M, Newport MJ, D'Souza S, Kalabalikis P, Brown IN, et al. 1995. Familial disseminated atypical mycobacterial infection in childhood: a human mycobacterial susceptibility gene? *Lancet* 345:79–83
97. Lewis PA, Loomis D. 1927. Ulcerative types as determined by inheritance and as related to natural resistance against tuberculosis: an experimental study on inbred guinea pigs. *J. Exp. Med.* 47:449–68

98. Li HT, Zhang TT, Zhou YQ, Huang QH, Huang J. 2006. SLC11A1 (formerly NRAMP1) gene polymorphisms and tuberculosis susceptibility: a meta-analysis. *Int. J. Tuberc. Lung. Dis.* 10:3–12
99. Lio D, Marino V, Serauto A, Gioia V, Scola L, et al. 2002. Genotype frequencies of the +874T → A single nucleotide polymorphism in the first intron of the interferon-gamma gene in a sample of Sicilian patients affected by tuberculosis. *Eur. J. Immunogenet.* 29:371–74
100. Lipsitch M, Sousa AO. 2002. Historical intensity of natural selection for resistance to tuberculosis. *Genetics* 161:1599–607
101. Lissner CR, Swanson RN, O'Brien AD. 1983. Genetic control of the innate resistance of mice to *Salmonella typhimurium*: expression of the Ity gene in peritoneal and splenic macrophages isolated in vitro. *J. Immunol.* 131:3006–13
102. Lopez-Maderuelo D, Arnalich F, Serantes R, Gonzalez A, Codoceo R, et al. 2003. Interferon-gamma and interleukin-10 gene polymorphisms in pulmonary tuberculosis. *Am. J. Respir. Crit. Care Med.* 167:970–75
103. Lurie MB. 1950. Native and acquired resistance to tuberculosis. *Am. J. Med.* 9:591–610
104. Lurie MB, Zappasodi P, Dannenberg AM, Veiss GH. 1952. On the mechanism of genetic resistance and its mode of inheritance. *Am. J. Med.* 4:302–14
105. Lyadova IV, Eruslanov EB, Khaidukov SV, Yeremeev VV, Majorov KB, et al. 2000. Comparative analysis of T lymphocytes recovered from the lungs of mice genetically susceptible, resistant, and hyperresistant to *Mycobacterium tuberculosis*-triggered disease. *J. Immunol.* 165:5921–31
106. Lynch CJ, Pierce-Chase CH, Dubos R. 1965. A genetic study of susceptibility to experimental tuberculosis in mice infected with mammalian tubercle bacilli. *J. Exp. Med.* 121:1051–70
107. MacLennan C, Fieschi C, Lammas DA, Picard C, Dorman SE, et al. 2004. Interleukin (IL)-12 and IL-23 are key cytokines for immunity against *Salmonella* in humans. *J. Infect. Dis.* 190:1755–57
108. Majorov KB, Eruslanov EB, Rubakova EI, Kondratieva TK, Apt AS. 2005. Analysis of cellular phenotypes that mediate genetic resistance to tuberculosis using a radiation bone marrow chimera approach. *Infect. Immun.* 73:6174–78
109. Majorov KB, Lyadova IV, Kondratieva TK, Eruslanov EB, Rubakova EI, et al. 2003. Different innate ability of I/St and A/Sn mice to combat virulent *Mycobacterium tuberculosis*: phenotypes expressed in lung and extrapulmonary macrophages. *Infect. Immun.* 71:697–707
110. Malik S, Abel L, Tooker H, Poon A, Simkin L, et al. 2005. Alleles of the *NRAMP1* gene are risk factors for pediatric tuberculosis disease. *Proc. Natl. Acad. Sci. USA* 102:12183–88
111. Malo D, Vidal SM, Hu J, Skamene E, Gros P. 1993. High-resolution linkage map in the vicinity of the host resistance locus Bcg. *Genomics* 16:655–63
112. Malo D, Vogan K, Vidal S, Hu J, Cellier M, et al. 1994. Haplotype mapping and sequence analysis of the mouse Nramp gene predict susceptibility to infection with intracellular parasites. *Genomics* 23:51–61
113. Marquis JF, Forbes JR, Canonne-Hergaux F, Horth C, Gros P. 2006. Metal transport at the interface of host: pathogen interactions. In *Genetic Susceptibility to Infectious Diseases*, ed. RA Kaslow, JM McNicholl, AVS Hill. Oxford University Press, Oxford, UK
114. Medina E, North RJ. 1996. Evidence inconsistent with a role for the Bcg gene (*Nramp1*) in resistance of mice to infection with virulent *Mycobacterium tuberculosis*. *J. Exp. Med.* 183:1045–51

115. Medina E, North RJ. 1998. Resistance ranking of some common inbred mouse strains to *Mycobacterium tuberculosis* and relationship to major histocompatibility complex haplotype and Nramp1 genotype. *Immunology* 93:270–74
116. Miller EN, Jamieson SE, Joberty C, Fakiola M, Hudson D, et al. 2004. Genome-wide scans for leprosy and tuberculosis susceptibility genes in Brazilians. *Genes Immun.* 5:63–67
117. Mira MT, Alcais A, di Pietrantonio T, Thuc NV, Phuong MC, et al. 2003. Segregation of HLA/TNF region is linked to leprosy clinical spectrum in families displaying mixed leprosy subtypes. *Genes Immun.* 4:67–73
118. Mira MT, Alcais A, Nguyen VT, Moraes MO, Di Flumeri C, et al. 2004. Susceptibility to leprosy is associated with PARK2 and PACRG. *Nature* 427:636–40
119. Mitsos LM, Cardon LR, Fortin A, Ryan L, LaCourse R, et al. 2000. Genetic control of susceptibility to infection with *Mycobacterium tuberculosis* in mice. *Genes Immun.* 1:467–77
120. Mitsos LM, Cardon LR, Ryan L, LaCourse R, North RJ, Gros P. 2003. Susceptibility to tuberculosis: a locus on mouse chromosome 19 (Trl-4) regulates *Mycobacterium tuberculosis* replication in the lungs. *Proc. Natl. Acad. Sci. USA* 100:6610–15
121. Moraes-Vasconcelos D, Grumach AS, Yamaguti A, Andrade ME, Fieschi C, et al. 2005. *Paracoccidioides brasiliensis* disseminated disease in a patient with inherited deficiency in the beta1 subunit of the interleukin (IL)-12/IL-23 receptor. *Clin. Infect. Dis.* 41:e31–37
122. Musa SA, Kim Y, Hashim R, Wang GZ, Dimmer C, Smith DW. 1987. Response of inbred mice to aerosol challenge with *Mycobacterium tuberculosis*. *Infect. Immun.* 55:1862–66
123. Nakamura RM, Tokunaga T. 1978. Strain difference of delayed-type hypersensitivity to BCG and its genetic control in mice. *Infect. Immun.* 22:657–64
124. Nedorost ST, Elewski B, Tomford JW, Camisa C. 1991. Rosacea-like lesions due to familial *Mycobacterium avium*-intracellulare infection. *Int. J. Dermatol.* 30:491–97
125. Nevo Y, Nelson N. 2006. The NRAMP family of metal-ion transporters. *Biochim. Biophys. Acta* 1763:609–20
126. Newport MJ, Huxley CM, Huston S, Hawrylowicz CM, Oostra BA, et al. 1996. A mutation in the interferon-gamma receptor gene and susceptibility to mycobacterial infection. *N. Engl. J. Med.* 335:1941–49
127. Nicewonger J, Suck G, Bloch D, Swaminathan S. 2004. Epstein-Barr virus (EBV) SM protein induces and recruits cellular Sp110b to stabilize mRNAs and enhance EBV lytic gene expression. *J. Virol.* 78:9412–22
128. Nikonenko BV, Averbakh MM Jr, Lavebratt C, Schurr E, Apt AS. 2000. Comparative analysis of mycobacterial infections in susceptible I/St and resistant A/Sn inbred mice. *Tuberc. Lung. Dis.* 80:15–25
129. North RJ, Jung YJ. 2004. Immunity to tuberculosis. *Annu. Rev. Immunol.* 22:599–623
130. North RJ, LaCourse R, Ryan L, Gros P. 1999. Consequence of Nramp1 deletion to *Mycobacterium tuberculosis* infection in mice. *Infect. Immun.* 67:5811–14
131. Orme IM. 2005. The use of animal models to guide rational vaccine design. *Microbes Infect.* 7:905–10
132. Özbek N, Fieschi C, Yilmaz BT, De Beaucoudrey L, Bikmaz YE, Casanova JL. 2005. Interleukin-12 receptor beta 1 chain deficiency in a child with disseminated tuberculosis. *Clin. Infect. Dis.* 40:e55–e58.
133. Pan H, Yan BS, Rojas M, Shebzukhov YV, Zhou H, et al. 2005. Ipr1 gene mediates innate immunity to tuberculosis. *Nature* 434:767–72
134. Pelletier M, Forget A, Bourassa D, Gros P, Skamene E. 1982. Immunopathology of BCG infection in genetically resistant and susceptible mouse strains. *J. Immunol.* 129:2179–85

135. Picard C, Fieschi C, Altare F, Al-Jumaah S, Al-Hajjar S, et al. 2002. Inherited interleukin-12 deficiency: IL12B genotype and clinical phenotype of 13 patients from six kindreds. *Am. J. Hum. Genet.* 70:336–48

136. Pierce C, Dubos R, Middlebrook G. 1947. Infection of mice with mammalian tubercle bacilli grown in tween-albumin liquid medium. *J. Exp. Med.* 86:159–65

137. Poon A, Schurr E. 2004. The *NRAMP* genes and human susceptibility to common diseases. In *The NRAMP Family*, ed. M Cellier, P Gros, pp. 29–43. New York: Kluwer Acad./Plenum. **http://www.Eurekah.com**

138. Pravica V, Perrey C, Stevens A, Lee JH, Hutchinson IV. 2000. A single nucleotide polymorphism in the first intron of the human IFN-gamma gene: absolute correlation with a polymorphic CA microsatellite marker of high IFN-gamma production. *Hum. Immunol.* 61:863–66

139. Puel A, Reichenbach J, Bustamante J, Ku CL, Feinberg J, et al. 2006. The NEMO mutation creating the most-upstream premature stop codon is hypomorphic because of a reinitiation of translation. *Am. J. Hum. Genet.* 78:691–701

140. Puffer R. 1944. *Familial Susceptibility to Tuberculosis: Its Importance as a Public Health Problem*. Cambridge, MA: Harvard Univ. Press. 106 pp.

141. Remus N, Alcais A, Abel L. 2003. Human genetics of common mycobacterial infections. *Immunol. Res.* 28:109–29

142. Remus N, El Baghdadi J, Fieschi C, Feinberg J, Quintin T, et al. 2004. Association of IL12RB1 polymorphisms with pulmonary tuberculosis in adults in Morocco. *J. Infect. Dis.* 190:580–87

143. Roche PW, Triccas JA, Winter N. 1995. BCG vaccination against tuberculosis: past disappointments and future hopes. *Trends Microbiol.* 3:397–401

144. Roesler J, Horwitz ME, Picard C, Bordigoni P, Davies G, et al. 2004. Hematopoietic stem cell transplantation for complete IFN-gamma receptor 1 deficiency: a multi-institutional survey. *J. Pediatr.* 145:806–12

145. Roscioli T, Cliffe ST, Bloch DB, Bell CG, Mullan G, et al. 2006. Mutations in the gene encoding the PML nuclear body protein Sp110 are associated with immunodeficiency and hepatic veno-occlusive disease. *Nat. Genet.* 38:620–22

146. Rosenzweig SD, Dorman SE, Uzel G, Shaw S, Scurlock A, et al. 2004. A novel mutation in IFN-gamma receptor 2 with dominant negative activity: biological consequences of homozygous and heterozygous states. *J. Immunol.* 173:4000–8

147. Rossouw M, Nel HJ, Cooke GS, van Helden PD, Hoal EG. 2003. Association between tuberculosis and a polymorphic NFkappaB binding site in the interferon gamma gene. *Lancet* 361:1871–72

148. Roy MF, Riendeau N, Loredo-Osti JC, Malo D. 2006. Complexity in the host response to *Salmonella* Typhimurium infection in AcB and BcA recombinant congenic strains. *Genes Immun.* 7(8):655–666

149. Russell DG. 2001. *Mycobacterium tuberculosis*: here today, and here tomorrow. *Nat. Rev. Mol. Cell Biol.* 2:569–77

150. Sanal O, Turul T, De Boer T, Van de Vosse E, Yalcin I, et al. 2006. Presentation of interleukin-12/-23 receptor beta1 deficiency with various clinical symptoms of *Salmonella* infections. *J. Clin. Immunol.* 26:1–6

151. Sanchez F, Radaeva TV, Nikonenko BV, Persson AS, Sengul S, et al. 2003. Multigenic control of disease severity after virulent *Mycobacterium tuberculosis* infection in mice. *Infect. Immun.* 71:126–31

152. Sebastiani G, Olien L, Gauthier S, Skamene E, Morgan K, et al. 1998. Mapping of genetic modulators of natural resistance to infection with *Salmonella typhimurium* in wild-derived mice. *Genomics* 47:180–86
153. Skamene E, Gros P, Forget A, Kongshavn PA, St Charles C, Taylor BA. 1982. Genetic regulation of resistance to intracellular pathogens. *Nature* 297:506–9
154. Spielman RS, McGinnis RE, Ewens WJ. 1993. Transmission test for linkage disequilibrium: the insulin gene region and insulin-dependent diabetes mellitus (IDDM). *Am. J. Hum. Genet.* 52:506–16
155. Staretz-Haham O, Melamed R, Lifshitz M, Porat N, Fieschi C, et al. 2003. Interleukin-12 receptor beta1 deficiency presenting as recurrent *Salmonella* infection. *Clin. Infect. Dis.* 37:137–40
156. Stead WW. 1992. Genetics and resistance to tuberculosis. Could resistance be enhanced by genetic engineering? *Ann. Intern. Med.* 116:937–41
157. Stead WW, Senner JW, Reddick WT, Lofgren JP. 1990. Racial differences in susceptibility to infection by *Mycobacterium tuberculosis*. *N. Engl. J. Med.* 322:422–27
158. Stienstra Y, van der Werf TS, Oosterom E, Nolte IM, van der Graaf WT, et al. 2006. Susceptibility to Buruli ulcer is associated with the SLC11A1 (NRAMP1) D543N polymorphism. *Genes Immun.* 7:185–89
159. Tailleux L, Schwartz O, Herrmann JL, Pivert E, Jackson M, et al. 2003. DC-SIGN is the major *Mycobacterium tuberculosis* receptor on human dendritic cells. *J. Exp. Med.* 197:121–27
160. Thye T, Browne EN, Chinbuah MA, Gyapong J, Osei I, et al. 2006. No associations of human pulmonary tuberculosis with Sp110 variants. *J. Med. Genet.* 43:e32
161. Tosh K, Campbell SJ, Fielding K, Sillah J, Bah B, et al. 2006. Variants in the SP110 gene are associated with genetic susceptibility to tuberculosis in West Africa. *Proc. Natl. Acad. Sci. USA* 103:10364–68
162. Trajkovic V, Natarajan K, Sharma P. 2004. Immunomodulatory action of mycobacterial secretory proteins. *Microbes Infect.* 6:513–19
163. Tso HW, Ip WK, Chong WP, Tam CM, Chiang AK, Lau YL. 2005. Association of interferon gamma and interleukin 10 genes with tuberculosis in Hong Kong Chinese. *Genes Immun.* 6:358–63
164. Turcotte K, Gauthier S, Mitsos LM, Shustik C, Copeland NG, et al. 2004. Genetic control of myeloproliferation in BXH-2 mice. *Blood* 103:2343–50
165. Turcotte K, Gauthier S, Tuite A, Mullick A, Malo D, Gros P. 2005. A mutation in the *Icsbp1* gene causes susceptibility to infection and a chronic myeloid leukemia-like syndrome in BXH-2 mice. *J. Exp. Med.* 201:881–90
166. Turcotte K, Loredo-Osti JC, Fortin P, Schurr E, Morgan K, Gros P. 2006. Complex genetic control of susceptibility to *Mycobacterium bovis* (Bacille Calmette-Guerin) infection in wild-derived Mus spretus mice. *Genes Immun.* 7(8):684–87
167. Verdu P, Barreiro LB, Patin E, Gessain A, Cassar O, et al. 2006. Evolutionary insights into the high worldwide prevalence of MBL2 deficiency alleles. *Hum. Mol. Genet.* 15:2650–58
168. Vidal S, Malo D, Vogan K, Skamene E, Gros P. 1993. Natural resistance to infection with intracellular parasites: isolation of a candidate for Bcg. *Cell* 73:469–85
169. Vidal S, Tremblay ML, Govoni G, Gauthier S, Sebastiani G, et al. 1995. The Ity/Lsh/Bcg locus: natural resistance to infection with intracellular parasites is abrogated by disruption of the Nramp1 gene. *J. Exp. Med.* 182:655–66
170. Vidal SM, Lanier LL. 2006. NK cell recognition of mouse cytomegalovirus-infected cells. *Curr. Top. Microbiol. Immunol.* 298:183–206

171. Vidal SM, Pinner E, Lepage P, Gauthier S, Gros P. 1996. Natural resistance to intracellular infections: Nramp1 encodes a membrane phosphoglycoprotein absent in macrophages from susceptible (Nramp1 D169) mouse strains. *J. Immunol.* 157:3559–68
172. Vogt G, Chapgier A, Yang K, Chuzhanova N, Feinberg J, et al. 2005. Gains of glycosylation comprise an unexpectedly large group of pathogenic mutations. *Nat. Genet.* 37:692–700
173. WHO. 2004. Tuberculosis. *Nat. Rev. Microbiol.* 2:930–32
174. Wright S, Lewis PA. 1921. Factors in the resistance of guinea pigs to tuberculosis with special regard to inbreeding and heredity. *Am. Natural.* 55:20–50
175. Yan BS, Kirby A, Shebzukhov YV, Daly MJ, Kramnik I. 2006. Genetic architecture of tuberculosis resistance in a mouse model of infection. *Genes Immun.* 7:201–10
176. Yokoyama WM, Plougastel BF. 2003. Immune functions encoded by the natural killer gene complex. *Nat. Rev. Immunol.* 3:304–16
177. Zaharik ML, Vallance BA, Puente JL, Gros P, Finlay BB. 2002. Host-pathogen interactions: host resistance factor Nramp1 up-regulates the expression of *Salmonella* pathogenicity island-2 virulence genes. *Proc. Natl. Acad. Sci. USA* 99:15705–10

Computation and Analysis of Genomic Multi-Sequence Alignments

Mathieu Blanchette

McGill Centre for Bioinformatics, McGill University, Montreal, Quebec, Canada, H3A 2B4; email: blanchem@mcb.mcgill.ca

Annu. Rev. Genomics Hum. Genet. 2007. 8:193–213

First published online as a Review in Advance on May 9, 2007.

The *Annual Review of Genomics and Human Genetics* is online at genom.annualreviews.org

This article's doi: 10.1146/annurev.genom.8.080706.092300

1527-8204/07/0922-0193$20.00

Key Words

multiple sequence alignment, comparative genomics, whole-genome alignment, computational genome annotation, genome evolution

Abstract

Multi-sequence alignments of large genomic regions are at the core of many computational genome-annotation approaches aimed at identifying coding regions, RNA genes, regulatory regions, and other functional features. Such alignments also underlie many genome-evolution studies. Here we review recent computational advances in the area of multi-sequence alignment, focusing on methods suitable for aligning whole vertebrate genomes. We introduce the key algorithmic ideas in use today, and identify publicly available resources for computing, accessing, and visualizing genomic alignments. Finally, we describe the latest alignment-based approaches to identify and characterize various types of functional sequences. Key areas of research are identified and directions for future improvements are suggested.

Multiple sequence alignment (MSA): predicting homology of DNA bases from different genomes or from different regions of the same genome

Evolutionarily correct alignment: alignment where all and only homologous bases are aligned

INTRODUCTION

Although the first complete draft of the human genome was publicly released more than five years ago (75), its functional interpretation remains in its infancy. Much of the human DNA appears to be nonfunctional, and the role of most putatively functional regions remains poorly understood. One of the best ways to accurately annotate our genome (i.e., to delineate and characterize each type of functional element it contains) is through comparative genomics. By comparing our genome to that of other species and studying how different regions have evolved, one can often make educated guesses about the functions of these regions. This is one of the key motivations behind the sequencing of other vertebrate genomes. As of today, "complete" vertebrate genome sequences are available for human (75), chimpanzee (23), mouse (129), rat (57), dog (77), chicken (61), macaque (1), and Takifugu (3). Several other genome sequences are about to be released (including cow, African frog, opossum, zebrafish, Tetraodon) and others are in the process of being sequenced at a low coverage (2X) (see **http://www.genome.gov/10002154**). Large amounts of genomic sequence are also available for several species of insects (in particular fruit flies), worms, yeasts, plants, and a plethora of eubacteria and archebacteria.

To take advantage of this data for the purposes of doing genome annotation and evolutionary studies, these sequences must be meaningfully compared to each other, homologous regions identified, and their evolutionary history inferred. This is the problem of whole-genome sequence alignment, or, when more than two genomes are compared (which is when comparative genomics approaches gain most of their power), whole-genome multiple sequence alignment (MSA).

In this review, we give an overview of the latest algorithm advances for genomic multi-sequence alignment, discuss the limitations of these approaches, their accuracy evaluation, and their use for genome annotation purposes. We identify resources available for computing, accessing, and analyzing these alignments. Throughout, we outline key issues that remain to be addressed.

THE MULTIPLE SEQUENCE ALIGNMENT PROBLEM

MSA was one of the earliest topics studied by computational biologists (78, 109). Although the basic alignment principles apply to DNA, RNA, and amino acid sequences, specialized algorithms have been developed to address the particular challenges posed by each type of biomolecule. Here we focus on alignment of multiple, large (e.g., >1 Mb) genomic DNA sequences. Reviews of related topics can be found in Batzoglou (5), Dewey & Pachter (39), and Miller (88). For reviews of MSA of protein sequences, we refer the reader to Edgar & Batzoglou (46) and Wallace et al. (125).

Evolutionarily Correct Alignments

The input to the MSA problem usually consists of a set of genomic sequences, either completely assembled genomes, partially assembled draft genomes, or large finished genomic regions such as the ENCODE regions (49). Typically, the sequences being compared come from different species, but they can also be from the same species if they arose from duplication events. In many cases, a phylogenetic tree specifying the evolutionary relationships among the sequences being considered is adjoined to the input. This species tree might not accurately represent the history of every sequence region because of incomplete lineage sorting, gene duplications, gene conversion, or lateral gene transfer. If these confounding issues are temporarily ignored (as they mostly have been until recently), an evolutionarily correct MSA is an alignment where nucleotide i of species X is aligned to nucleotide j of species Y only if i and j are homologous, i.e., if, and only if, they were derived from a common ancestral nucleotide, either through faithful inheritance or through one or more substitutions. If the two nucleotides diverged because of a speciation event, they are called orthologous; they are called paralogous if their divergence results from a duplication [see Dewey & Pachter (39) for a more detailed classification]. Thus, one can see an MSA as a collection of sets of homologous nucleotides. This representation indirectly specifies the gaps in the alignment as positions where one sequence contains no nucleotide homologous to that of another sequence,[1] as a result of an insertion or a deletion (or incomplete data). An evolutionarily correct alignment thus provides much information about the evolutionary processes that led to the sequences considered. However, it does not unambiguously specify the history of all alignment bases, in particular because it does not differentiate between gaps due to insertions and those due to deletions (9, 24).

When duplications events are considered, defining the correct alignment becomes more difficult. The simple model of "one alignment row per species" fails to capture the possibility that several nucleotides from the same genome may be homologous. One solution is to allow each alignment column to contain more than one nucleotide per species, the goal being to identify any set of nucleotides (within species and across species) that share a common ancestry. Because no whole-genome MSA program published to date handles duplications in a completely satisfactory manner, we will postpone our discussion of this important aspect of evolution.

For several reasons, obtaining an evolutionarily correct MSA is not always possible. When sequences have diverged beyond a certain point, homologous bases cannot be reliably distinguished from nonhomologous ones. Information about homology is slowly but irremediably lost during evolution, and there is a limit to the accuracy of MSA algorithms. For example, very little noncoding DNA can be reliably aligned between human and fish because most of the vertebrate ancestral DNA that has not been deleted along one of the two lineages has usually diverged beyond recognition. Thus, the fact that a very small fraction of the human genome can be aligned to that of other nonmammalian species reflects a combination of the actual loss of homologous DNA and the increasing difficulty in reliably identifying it.

TFBS: transcription factor binding site

Function-Based Alignments

An alternative to searching for an evolutionarily correct alignment is to search for one that aligns regions that are functionally, but not necessarily evolutionarily, related. In species that are not too diverged (e.g., different mammals), functional homology often approximately equates to evolutionary homology, in particular for large, complex regions such as protein-coding genes. However, smaller functional regions such as transcription factor binding sites (TFBSs) can be functionally homologous without sharing a common ancestry. These functional regions are sufficiently short that they can arise from neutral drift and become fixed in a population because of selective pressure on the newly acquired function (37, 93). Many local multiple alignment programs (usually described as motif discovery algorithms) have been developed to identify such short functional regions. However, because the questions arising from this research area are quite distinct from those related to larger-scale evolutionary alignments, we refer the reader to Pavesi et al. (97) for more details. For the rest of this review, we focus on the identification, evaluation, and analysis of alignments in an evolutionary context.

Mathematical Objective Functions

Although the definition of an evolutionarily correct alignment is relatively simple, it

[1] It should be noted that this representation as sets of homologous nucleotides can often be depicted visually by several equivalent multiple sequence alignment (MSA) representations.

is not practical, as it is generally impossible to test. Instead, mathematical objective functions have been defined to approximate this notion. The most theoretically sound approach relies on stochastic models of sequence evolution specifying the probability of events such as substitutions, insertions, and deletions. Once such a model is defined, one evaluates a given MSA with its likelihood under the model by implicitly summing the probabilities of all evolutionary scenarios that are compatible with the alignment. The MSA sought is the one that maximizes this likelihood function. Although this idea is attractive from a theoretical standpoint, it quickly runs into practical difficulties. Evaluating the likelihood of a given MSA, even assuming a fixed tree topology and no duplications, is a NP-hard problem (24), i.e., one for which no efficient algorithm is believed to exist. For this reason, objective functions have historically been based on simpler schemes such as the sum-of-pairs score or a weighted version thereof (44). Today, most algorithms do not explicitly try to optimize any kind of global objective function, but simply follow a reasonable sequence of operations that usually produce relatively accurate alignments.

Global, Local, and "Glocal" Multiple Sequence Alignment

Alignment approaches have historically been divided into two groups: global and local. In a global alignment, the entirety of the sequences given as input are aligned in a single alignment, thus imposing the constraint that orthologous regions have to be colinear (i.e., no genome rearrangements and no duplications are allowed). If the sequences considered are closely related and are expected to have been completely derived from a common ancestor through a set of insertions, deletions, and substitutions, a global alignment is appropriate. In contrast, a local alignment predicts the homology of a set of fragments of the input sequences, leaving the rest of the sequences unaligned. In general, alignment programs identify a set of such local alignments, resulting in a mosaic of homology predictions, in which the segments aligned are not necessarily colinear in the input sequences. This enables one to find homologous segments that have undergone rearrangements. When a sizeable fraction of the sequences considered are expected to be highly diverged, a local alignment approach may also be preferable because by ignoring unalignable regions it reduces the risk of these regions causing misalignments of the whole sequence (26). However, the flexibility afforded by local alignment comes at a cost: By dropping ordering constraints, we increase the risk of aligning regions that are not true homologs (i.e., random hits). To maintain an acceptable level of false positives, one must accept a lower sensitivity.

The strategy adopted by most state-of-the-art alignment programs is a hybrid between local and global alignment [sometimes called a "glocal" alignment (16)] that takes advantage of the strengths of both methods. Most current MSA programs first identify sets of local alignments between pairs of sequences and then assemble the local alignments into chains of colinear sets of local alignments. Small local alignments that break the linearity of the surrounding regions are treated as likely false positives, thus improving specificity.

ALGORITHMS FOR MULTIPLE SEQUENCE ALIGNMENT

Algorithms for genomic MSA face a number of challenges, including:

1. **Computational complexity.** Optimal MSA is a NP-hard problem under most reasonable scoring schemes (47, 127), which means that the running time of all known exact algorithms for the problem is exponential in the number or the length of the sequences to be aligned.
2. **Large data sets.** Whole-genome sequence alignment programs have to be able to run on very large sequences. Despite the increasing computing power

available, this remains a very significant challenge and many compromises have to be made between accuracy and speed.

3. **Sensitivity.** Homologous nucleotides need to be identified for distantly related species.
4. **Specificity.** Nonhomologous nucleotides must not be aligned together.
5. **Model violations.** The evolutionary models and scoring schemes used by MSA programs are only approximations to reality. A good alignment program has to be robust with respect to these violations.

Different MSA programs make different trade-offs between these requirements, perhaps most obviously at the level of pair-wise alignment subroutines, upon which most MSA algorithms are based.

Seeded Pair-Wise Alignment

Most modern alignment methods rely on the ability to quickly and accurately align pairs of sequences or sequence "profiles" (see below). The accuracy of the multiple alignments produced depends in large part on that of the pair-wise alignment procedure used. Optimal local pair-wise alignments can be identified using the Smith-Waterman algorithm (117), but this approach is too slow for multimegabase sequences. For whole-genome alignments, nearly all existing approaches use a heuristic called seeded alignments, also at the base of the Blast algorithms (2), whereby local pair-wise alignments are explored only if they contain some type of highly conserved short match. Details vary from program to program: Blastz (110) explores exact matches over a set of nearby but nonconsecutive positions (called spaced seeds), LAGAN (15) relies on the CHAOS program (14) to identify short inexact (but consecutive) matches, AVID (12) searches for maximal exact matches (but allows mismatches at third codon positions), and MUMmer (34) uses suffix trees to identify maximal unique matches. The design of fast and sensitive seeded alignment algorithms remains a very active research area (79, 83, 121). When the aligned sequences are expected to largely consist of protein-coding regions (e.g., for prokaryote genomes), accuracy gains can be obtained by computing seeded pair-wise alignments at the translated level (35). For noncoding regulatory regions, Berezikov et al. (7) showed that sensitivity gains could be obtained by using seeds conserved for predicted TFBSs.

Seeded pair-wise alignment algorithm: heuristic used to align pairs of very large sequences, where local alignments are considered only if they contain a short, highly (and often perfectly) conserved match, called the seed

Once a good seed alignment is found, it is extended to the left and the right using various variants on the Smith-Waterman dynamic programming algorithm, this time allowing for gaps to be inserted. Note that at this stage the arrangement and orientation of pair-wise local alignments are unrestricted and one region can align to more than one other region, thus enabling one to find homology between sequences that have undergone genomic rearrangements and duplications.

Sets of pair-wise local alignments can be chained together into sets of colinear regions (63, 119, 134), using programs like Chaining/Netting (71), GRIMM-synteny (99), MAUVE (33), and MERCATOR (Colin Dewey, **http://www.biostat.wisc.edu/~cdewey/mercator/**). Within a chain, more sensitive settings can be used to discover weaker pair-wise alignments located between anchors (31), and the process can be repeated in a recursive manner (15). However, such approaches are inapplicable for unassembled genomes [e.g., those sequenced at a 2X coverage (87)], thus posing significant new challenges to alignment algorithms.

Multiple Pair-Wise Alignments versus True Multiple Alignments

MSAs are often computed for the sole purpose of annotating one of the sequences (called the reference sequence—for example, the human sequence) using the others. The main interest is to find orthology relationships between regions of the reference sequence and those of the other species, but not to find the relationships between two nonreference species. In this case, why not simply compute pair-wise alignments between the human sequence and each other species and assemble these alignments into a human-centric multiple alignment? This was the strategy used by Multi-Pipmaker (111) and Vista (43), and it is adequate for closely related sequences. However, when sequences are more diverged, the MSA obtained in this way is less accurate than one obtained from an algorithm that attempts to identify orthology relationships among all species (21, 85). This is because correctly aligning two nonhuman sequences provides homology information that can help align them correctly to human. For this reason, most modern MSA programs instead use a phylogenetically guided progressive multiple alignment procedure.

Progressive Multiple Sequence Alignment

Many modern MSA programs are based on a simple scheme called "progressive alignment" (44), popularized by the Clustal package (22, 60). In a progressive alignment algorithm, a phylogenetic tree is first inferred on the basis of pair-wise alignments, and then used as a guide for the multiple alignment procedure (the tree can also be provided by the user). The two species that are the most closely related are aligned first, and the process is repeated until all sequences are aligned. Because the phylogenetic tree is binary, each step consists of the pair-wise alignment of one group of aligned sequences (corresponding to one clade) against another (corresponding to a sister clade), a process referred to as profile-profile alignment. Variants lie in the choice of algorithm and scoring function for profile-profile alignment. TBA (10) bases its profile-profile on pair-wise alignments of extant sequences. MLAGAN (15) uses a weighted sum-of-pairs approach, whereas MAVID (13) represents each aligned clade with its most likely common ancestral sequence. This last approach is the closest to the evolutionary view of multiple alignment. Whereas all these MSA algorithms use seed-based pair-wise sequence alignments at every step of the progressive alignment procedure, Flannick & Batzoglou (52) have proposed an algorithm that performs seed-based profile-profile alignment. By dynamically selecting more sensitive seeds in regions of each profile that are the most likely to yield matches (e.g., those corresponding to highly conserved regions), they report an improved sensitivity over standard seed-based approaches. Although the idea of progressive alignment was proposed more than 20 years ago, the devil is in the details, and the optimal trade-off between efficiency and accuracy has not yet been reached.

Other Multiple Alignment Strategies

Although progressive multiple alignment is intuitively appealing in an evolutionary context, nonphylogenetic alignment procedures have also been successfully developed. One of them, called consistency-based MSA, was pioneered by Morgenstern's group with the Dialign family of MSA programs (91, 101) and followed by others (122, 131). Like most progressive alignment approaches, the algorithm relies on the identification of pair-wise anchors (seed alignment). However, it weighs these pair-wise alignments based not only on the similarity of the two sequences involved, but also on the set of all other pair-wise anchors they are involved in with other species. Thus, if region R_x of species X has a weak alignment with region R_y of species Y, but both R_x and R_y have a strong alignment to region R_z of species Z, the reliability of the anchor alignment (R_x, R_y) is increased. Dialign follows this reweighting of anchors by a greedy procedure that attempts to identify the maximum weight set of collinear anchors. By using the entire set of anchors to decide which to retain in the final alignment, Dialign can afford a higher false-positive rate at the pair-wise level, and consequently a higher sensitivity. Recently, Do et al. (40) integrated a probabilistic version of the consistency-based approach into a progressive MSA algorithm for protein sequences called ProbCons. Paten & Birney (96) extended this promising idea to genomic MSA in the program PECAN.

In a different direction, Zhang & Waterman (133) formulated the local MSA problem as one identifying Eulerian paths in a graph, a problem for which efficient algorithms exist, and used this approach to identify new families of functional elements and retrotransposons. Raphael et al. (105) improved the approach to make it suitable for the alignment of highly diverged, rearranged, repetitive genomic regions, and the results are competitive with more traditional progressive alignment methods.

Refining Alignments

To compute alignments reasonably quickly for large genomes, genomic MSA programs have to cut corners, which sometimes results in suboptimal alignments. For example, with a progressive MSA approach, the subalignment produced for a certain clade is not revised later in the light of other sequences, which sometimes leads to nonoptimal alignments. Although the overall structure of the alignment produced may often be correct, details may not be. Many authors have proposed postprocessing algorithms to refine multiple alignments—a process also known as realignment (20). Most approaches assume that the initial alignment is approximately correct and attempt to fine-tune the position of gaps. By assuming global correctness, computational complexity is reduced so more sensitive approaches can be used. A simple procedure proposed by several authors and implemented in CLUSTALW and MLAGAN is to repeatedly remove individual sequences from the MSA and realign them to the remaining n-1 aligned sequences. When the data set considered contains closely related sequences, a more effective approach might be to remove a complete subtree (e.g., the primate clade) and realign it to the remaining sequences. Other approaches include making local changes to the alignment by moving gaps to optimize some objective function (for example, using a genetic algorithm) (126).

Uncertainty in Alignments

One drawback of the fast alignment algorithms described above is that they do not provide an assessment of the uncertainty related to particular regions of the alignment. Some global alignment methods (e.g., MLAGAN) attempt to align every nucleotide of every sequence, and others leave some regions unaligned, but none quantitate confidence in the alignment correctness. A small step in that direction is a procedure called "alignment cleaning" (81), which identifies alignment regions that do not meet certain statistical significance thresholds (104), and then either dissolves these blocks or breaks them into highly supported subalignments.

Meaningful confidence measures of alignment accuracy require a probabilistic model of sequence evolution. One can then ask questions like: Given a set of sequences $X_1, \ldots, X_n$, what is the probability that nucleotide i of sequence X_a is orthologous to nucleotide j of sequence X_b? This type of question is addressed by a new type of alignment problem called statistical alignment. Although the existing algorithms for statistical MSA (59) remain too slow to be executed on a large scale, they bear the promise of a richer representation of MSA, with important applications in phylogeny, functional sequence annotation, and ancestral genome reconstruction.

BENCHMARKING MULTIPLE SEQUENCE ALIGNMENT PROGRAMS

Objective and quantitative benchmarking procedures are necessary to evaluate existing MSA algorithms, point out weaknesses, and suggest improvements. MSA benchmarking is challenging because the evolutionary correctness of a MSA cannot be tested for actual biological sequences. Furthermore, the mathematical scoring schemes described above are only approximations to the biological quality of an alignment, so using them as the basis of MSA evaluation might be misleading. For protein MSA, additional sources of information such as 3D structure and functional assays can be used to help determine the correct alignment; databases such as BaliBase (124) and Prefab (45) contain large numbers of such alignments against which protein MSA programs can be evaluated. Because of the lack of external data indicating the correct alignment, no such database exists for genomic sequences.

Two types of benchmarking approaches have been proposed for genomic MSA. In the first, a set of well-annotated orthologous genomic sequences are considered, and the quality of the MSA is assessed based on the fraction of the annotated functional regions that are correctly aligned. For example, Bray & Pachter (13) have used known coding exons as landmarks. A drawback of this approach is that well-annotated functional regions (e.g., coding exons) are often the easiest to align properly because of their relatively large size, high degree of conservation, and preserved linearity of the gene structure. Thus, estimates based on the success of aligning such regions may be optimistic. Evaluations based on other types of functional regions (e.g., regulatory regions) have also been considered (55) but remain problematic because of the lack of well-annotated noncoding functional regions. However, large-scale, unbiased experimental data from projects such as ENCODE (49) are now providing an excellent basis for this type of benchmarking (86).

The second type of benchmarking involves synthetic sequences (102). Here the evolution of a synthetic DNA sequence is computationally simulated along the branches of a given phylogenetic tree. By keeping track of the mutations performed along each branch, the evolutionarily correct alignment can be constructed. The leaf sequences (synthetic extant sequences) are then fed to the alignment program being evaluated, and the alignment obtained is compared to the correct one. Various approaches have been proposed to compare the two alignments (10, 89), but most revolve around the evaluation of the sensitivity (fraction of orthologous sites that are correctly aligned) and specificity (fraction of aligned sites that are orthologous) of the predicted alignment. The usefulness of such benchmarking depends on the realism of the sequence evolution simulation and of the synthetic sequence chosen for the root. For example, Blanchette et al. (10) developed a fairly realistic simulation of neutrally evolving sequences, including variable mutation rates along different lineages as well as retrotransposon insertions, and used it to compare their MSA program with others. A shortcoming of their approach is that actual DNA sequences contain a mixture of slow- and fast-evolving sites, something that is not modeled in their simulation. More sophisticated evolution simulators allowing both neutrally evolving and constrained elements, as well as tandem and segmental duplications and DNA polymerase slippage, are thus necessary for an improved assessment of alignment accuracy.

Whether benchmarking is done on real or synthetic sequences, the evaluation of alignment accuracy has to be carried out keeping in mind the context in which the alignment will be used. If the alignment is computed to seek out highly conserved functional regions such as coding regions one may care less about the accuracy of the alignment in neutrally evolving regions and more about the sensitivity of local alignments. However, if the alignment is used to identify weaker signals such as TFBSs, the accuracy of the alignment on noncoding regions is key (32). Finally, for genome evolution studies and ancestral genome reconstruction efforts (9), the alignment accuracy for neutrally evolving regions is paramount, and modeling the effects of alignment errors on the analysis is important (82, 106).

USING MULTIPLE SEQUENCE ALIGNMENT FOR GENOME ANNOTATION

One of the most powerful approaches to separate functional from nonfunctional genomic regions and identify the putative role of the former is through the analysis of the "evolutionary signatures" of functional regions. Nonfunctional regions mostly evolve neutrally and tend to accumulate mutations at rates that are relatively well understood (27). Functional regions are under selective pressure to preserve their function.[2] Here we briefly review the types of functional sequence analysis approaches enabled by MSA. This is only a partial list and virtually every aspect of computational genome annotation has benefited from the availability of multiple alignments [see (42) for a related treatment of the topic].

Identification of Regions Under Selective Pressure

A large fraction of many metazoan genomes is believed to be nonfunctional in the sense that it does not encode a beneficial function for the host other than perhaps maintaining an appropriate spacing between functional elements. For example, by comparing the human and mouse genomes, Waterston et al. (129) estimated that only about 5% of the DNA of each genome is under selection. One of the original incentives to sequence more mammalian genomes was to help identify and characterize these functional regions.

Many approaches have been developed to scan a genomic MSA and identify regions under selection. Most of these methods are phylogenetically aware, i.e., they use an evolutionary model and a phylogenetic tree to evaluate sequence conservation. Some use a sliding window approach (54) and perform a hypothesis test to try to reject the null model of neutral evolution [e.g., binCons (84)]. Others estimate position-specific substitution rates and identify regions under selection using a hidden Markov model (HMM) approach [PhastCons (115), eShadow (94)] or some other flexible window approach [GERP (30)]. These have the advantage of being able to detect long, weakly conserved regions as well as short, highly conserved ones. Various genome browsers offer precomputed conservation scores for whole-genome alignments (e.g., PhastCons on the UCSC Genome Browser and GERP on the Ensembl Genome Browser). Although the methodological differences between the various approaches are important, the agreement among their predictions tends to be rather high, and the main source of uncertainty in the identification of conserved regions remains the difference between alignment programs (86). A number of genomic regions have been dissected and analyzed based on MSA (e.g., 6, 21, 65, 86), generally resulting in the identification of many more constrained regions than can be accounted for by protein-coding selective pressure. Although the precise identification of regions under selective pressure, especially shorter ones, remains a challenge, we believe that the more promising research direction now lies in the interpretation of evolutionary signatures to predict the putative function of each region.

Evolutionary signature: pattern of interspecies sequence conservation/variation typical of a particular type of function (protein-coding gene, RNA gene, regulatory region, etc.)

Identifying Protein-Coding Regions

Although the constitutive coding exons of most human genes have now been accurately identified, much uncertainty remains regarding the number and location of alternatively spliced exons, especially those rarely expressed and thus poorly represented in transcript databases. Alignment-based gene prediction has proven to be a useful complement to gene annotation based on experimental data because the evolutionary signature of coding exons is clear, strong, and well understood. High-synonymous to nonsynonymous

[2]This, however, does not mean that their sequence is perfectly or even highly conserved as a whole, but rather that its functional features tend to be.

substitution rate ratios, absence of in-frame stop codons, and conserved splice sites are signatures sought by a number of gene-finding algorithms, using either pair-wise (90, 132) or MSA (18, 38, 58). These algorithms generally far outperform single-organism gene finders. Because coding exons tend to be well conserved throughout vertebrates but most other types of functional and nonfunctional regions are not, vertebrate multiple alignment is extremely informative for human gene finding, although species- or clade-specific exons will systematically be missed by such approaches. All of these approaches depend on an accurate MSA, and Dewey et al. (38) have proposed to circumvent this problem by performing joint alignment (using a translated seed-based alignment) and gene prediction.

Identifying RNA Genes

Structural RNA genes (tRNAs, rRNAs, microRNA precursors, etc.) also have a specific evolutionary signature that can be detected based on an accurate MSA. A number of RNA gene predictors attempt to detect compensatory mutations in RNA molecules (those compensating one substitution by another, to preserve Watson-Crick complementarity), the most recent being Evofold (98, 128) and RNAz (41, 128). However, RNA gene prediction presents very specific challenges as the primary sequence often lacks detectable conservation, resulting in the difficulty of obtaining accurate MSA. It was suggested very early on (108) that a better way to proceed might be to jointly predict multiple alignment and RNA secondary structure. However, the algorithms proposed thus far (114) remain too slow to be used on a large scale.

Identifying Regulatory Regions

One of the most promising uses of MSA is in identifying transcriptional regulatory regions, an approach coined "phylogenetic footprinting." Generally located in noncoding regions, TFBSs often stand out by their interspecies conservation. Thus, the simplest approach is to use sequence conservation as a preliminary filter before searching for TFBSs using position weight matrix scans [e.g., ConSite (107)]. Most binding sites are flexible and do not require perfect sequence conservation to remain functional, and many recent approaches based on position weight matrix scanning [rVISTA (80), PreMod (51), and others] use conservation of estimated binding affinity rather than raw sequence conservation, or TFBS-specific evolutionary models [Monkey (92)] to detect putative aligned binding sites. However, it is now recognized that binding site turnover, in which random mutations create and destroy binding sites, is quite common (37), and that lack of conservation does imply lack of function (32). There is clearly a need for more accurate models of binding-site turnover and for algorithms using them for binding-site prediction [see (93) for a good step in this direction].

De novo motif discovery algorithms have also started taking advantage of genomic multiple alignments, either to favor the identification of motifs whose occurrences tend to be conserved across species (70, 103, 113, 116), or to use reliable alignment anchors as constraints for the identification of shorter conserved motifs (11, 50). Finally, we note that mammalian regulatory regions appear to have an evolutionary signature that can be detected even without explicitly referring to TFBSs, as was demonstrated by the Hardison group (48, 72, 73, 123).

Genome Evolution Studies

Because they attempt to recapitulate the evolutionary relationships between genomes, genomic multi-sequence alignments are at the core of many phylogenomics studies (36, 70, 82, 100, 118) and genome evolution studies (71, 130). These alignments also underly the interesting prospect of accurately reconstructing the sequence of ancestral genomes from the genome of extant species (9, 24, 64).

Although phylogenetic studies are outside the scope of this review, we warn the reader against the pitfalls of phylogenetic inference and evolutionary studies based on dubious alignments. When dealing with sets of sequences whose correct alignment is not obvious, alignments are often sensitive to particular parameters values of the alignment program (e.g., gap open and extend penalties), and downstream analyses may be biased (27, 28, 106). One elegant solution to this problem is statistical alignment, which allows sampling alignments from their posterior probability distribution to estimate evolutionary distance and trees (62, 82).

RESOURCES

Despite the high quality of alignment programs available today, constructing large multiple alignments remains a challenging project that requires expertise and substantial computing power. Although the tools available are becoming increasingly accurate, they are generally not trivial to use and often involve the execution of a number of intermediate steps that are not always well described. For this reason, most whole-genome MSA users rely on publicly available precomputed alignments or on alignment Web servers for analysis. Here we briefly review those that are most commonly used (see also 53).

Precomputed Genomic Multi-Sequence Alignments

Because of the computational power and infrastructure required, whole-genome multiple alignment remains out of reach for the typical biology laboratory. However, for the most studied genomes, whole-genome multiple alignments have been computed using one or more of the programs discussed above and are being made available to the community. **Table 1** lists a set of whole-genome MSAs currently available for various groups of species. In many cases, alignments are integrated into genome browsers such as the UCSC Genome Browser (74) and the Ensembl Genome Browser (8), which enable visualization of the alignment itself, as well as of many other types of sequence annotations.

Although computation of large MSAs requires an impressive feat of computational engineering, handling and analyzing them are also not simple tasks. Whole-genome alignments files often consist of several gigabytes

Table 1 Publicly available precomputed whole-genome alignments

Species	Resource	MSA tool(s) used	URL
- 17 vertebrate genomes - ENCODE regions - 9 insect genomes - 2 nematode genomes	UCSC Genome Browser	TBA/MULTIZ	**http://genome.ucsc.edu/**
- 9 vertebrate genomes	Ensembl Genome Browser	PECAN	**http://www.ensembl.org/index.html**
- 5 vertebrate genomes - 3 plant genomes - 2 Ciona genomes - 2 Phytophtora genomes	Vista Browser	Shuffle-LAGAN	**http://pipeline.lbl.gov/cgi-bin/gateway2**
- 3 nematode genomes	Wormbase	MLAGAN	**http://wormbase.org/**
- 9 vertebrate genomes - 6 *Drosophila* genomes	ECR Browser	Blastz	**http://ecrbrowser.dcode.org/**
- 4 yeast genomes	Broad Institute	Kellis et al. (70)	**http://www.broad.mit.edu/annotation/**
- 12 *Drosophila* genomes	Eisen Lab	Mercator/MAVID, MAUVE, MLAGAN, MULTIZ	**http://rana.lbl.gov/drosophila/**

Table 2 Genomic multiple sequence alignment (MSA) Web servers

MSA tool	URL
CLUSTALW (22, 60)	**http://www.ebi.ac.uk/clustalw**
MAVID (13)	**http://baboon.math.berkeley.edu/mavid**
CHAOS/DIALIGN (14)	**http://dialign.gobics.de**
MAFFT (69)	**http://timpani.genome.ad.jp/~mafft/server**
MLAGAN	**http://lagan.stanford.edu/lagan_web/index.shtml**
Mulan (95)	**http://mulan.dcode.org**
aliWABA (66)	**http://aba.nbcr.net**

of data, and extracting meaningful information from them requires efficient tools. Several groups provide computer source code or Web servers to handle and analyze large alignments. Web-based tools include the UCSC Table Browser (67) and Galaxy (56), whereas popular open-source software includes Jim Kent's source tree (**http://hgwdev.cse.ucsc.edu/~kent**), a set of C programs at the core of the UCSC Genome Browser, and the EnsMart/BioJava/BioPerl (68) code underlying some of the Ensembl Genome Browser. The latter has the advantage of being able to remotely query Ensembl data sets without having to install them on a local machine, although this approach runs into efficiency problems for complex or numerous queries.

Multiple Sequence Alignment Web Servers

All multiple alignment programs mentioned in this review can be freely downloaded (for academic use) and can be run locally. However, in many cases (see **Table 2**), there are Web servers that compute alignments for a set of user-provided sequences, thus avoiding the hassle of software installation while taking advantage of dedicated machines optimized for that purpose. Note, however, that input sequences are usually limited to a few megabases. The SinicView program (112) allows the simultaneous computation and comparison of MSAs computed using several of the tools listed in **Table 2**.

Alignment Visualization

Visualization of multiple alignments is crucial for developing an intuitive understanding of the evolution of a particular genomic region, for integrating information from different sources, and for detecting possible alignment errors. A number of tools have been developed to address different aspects of visualization. Genome browsers such as the UCSC Genome Browser (74) and the Ensembl Genome Browser (8) are designed to visually integrate information from different sources, including multiple alignment. These browsers are based on the visualization of data related to a given reference sequence and typically exclude or indirectly represent bases that do not align to bases in the reference sequence.

One drawback of these tools is that it is currently not possible to visualize user-provided alignments or alignments computed on the fly from user-provided sequences. Tools allowing visualization and manipulation of user-provided MSA are listed in **Table 3** (programs analyzing only pair-wise alignments are omitted). Most visualization tools perform basic sequence analysis tasks such as the detection of conserved regions, and most allow the display of user-provided annotation. Many allow variable zoom-in resolution.

FUTURE CHALLENGES

The Holy Grail of the genomic MSA community is to design efficient algorithms to

Table 3 Visualization tools for multiple sequence alignments (MASs)

Visualization Tool	Functionality	URL
ABC (29)	Multiresolution views; user-provided annotations; conserved region detection	**http://mendel.stanford.edu/sidowlab/**
VISTA family (17, 43)	Visualization of alignment of user-provided sequence against precomputed MSA; multiresolution views; user-provided annotation; conserved region detection	**http://genome.lbl.gov/vista/index.shtml**
K-Browser (19)	True MSA view (i.e., not projected on a reference sequence); visualization of genome annotation on multiple genomes simultaneously	**http://hanuman.math.berkeley.edu/kbrowser**
COMPAM (76)	Visualization of multiple pair-wise alignments; automatic COG-based gene annotation; conserved regions detection and clustering; synteny analysis	**http://bio.informatics.indiana.edu/projects/compam/**
MaM (1)	Automated functional annotation of MSA; motif detection; extraction of subalignments based on annotation	**http://compbio.cs.sfu.ca/MAM.htm**
JalView (25)	MSA editing; conserved region detection; basic phylogenetic inference. Mostly designed for protein sequences	**http://www.jalview.org**

identify maximum likelihood MSA based on a realistic multiscale probabilistic model of sequence evolution that allows for substitutions, insertions, deletions, retrotransposon activity, duplications, and genome rearrangements. Although various groups have been developing increasingly accurate sequence evolution models for each type of mutation, these models have not yet been integrated into genomic MSA tools. For example, there appears to be much to gain from using a context-dependent insertion and deletion model of DNA polymerase slippage (4). However, at some point, remote homology detection becomes impossible with a generic model of evolution and requires multiple alignment algorithms tailored to specific types of functional regions [e.g., SLAM (38) for coding regions, or CONREAL (7) for regulatory regions]. We expect that improvement in alignment accuracy will come from a combination of fast, existing approaches and slower, domain-specific methods (e.g., 120).

We believe that one of the keys to progress in the MSA area is the development of algorithms that explicitly quantify the uncertainty in the alignment produced. This would allow the identification of more speculative alignments, which could then be included or excluded at will by the user. However, representing uncertainty is not simply a matter of providing a number describing the probability of correctness of a given region of the alignment, but also requires producing and representing the many possible alternate solutions. Genome annotation tools based on MSA should be developed that take this uncertainty into consideration.

Finally, we believe that MSA-based genome annotation approaches (in particular, RNA gene finding, discovery of regulatory regions, etc.) are only starting to take full advantage of the evolutionary signatures contained in these alignments. The large quantity of high-quality data coming out of projects like the ENCODE project (49) will allow the development of improved evolutionary models on which the next generation of computational genome annotation algorithms will be based.

SUMMARY POINTS

1. Genomic MSAs aim to recapitulate evolutionary history by identifying the regions that share a common ancestry in genomes of different species.
2. Because of the computational complexity of the problem and the very large size of the genomes to be aligned, computer programs have to trade off accuracy for efficiency. However, for most recent algorithms, the loss in accuracy is minimal.
3. Accurate genomic multiple alignments greatly facilitate the computational prediction of protein-coding genes, RNA genes, and regulatory regions. This is achieved by detecting patterns of sequence conservation that are characteristic of a particular type of function.
4. Many of the best MSA programs are run as Web servers and their results can be analyzed using interactive visualization tools. Pre-computed whole-genome multiple alignments can be downloaded for most groups of sequenced genomes.

FUTURE ISSUES TO BE RESOLVED

1. Genomic multi-sequence alignment remains an open problem. Even the best programs available are fairly slow, have a relatively poor sensitivity for diverged sequences, and often erroneously align nonhomologous nucleotides. Algorithms using more realistic evolutionary models are likely to improve the situation.
2. Benchmarking alignment of programs remains challenging. Functional and comparative genomics projects such as ENCODE are now providing data to evaluate and improve alignment programs.
3. Each type of functional region evolves under a different type of selective pressure that induces a different type of evolutionary signature. Characterizing and detecting such signatures for noncoding functional regions are keys to improving computational genome annotation.
4. Alignment tools representing the uncertainty in the alignment being reported need to be developed, and this uncertainty needs to be considered by alignment-based computational annotation approaches.

DISCLOSURE STATEMENT

The authors are not aware of any biases that might be perceived as affecting the objectivity of this review.

LITERATURE CITED

1. Alkan C, Tuzun E, Buard J, Lethiec F, Eichler EE, et al. 2005. Manipulating multiple sequence alignments via MaM and WebMaM. *Nucleic Acids Res.* 33:W295–98
2. Altschul SF, Gish W, Miller W, Myers EW, Lipman DJ. 1990. Basic local alignment search tool. *J. Mol. Biol.* 215:403–10

3. Aparicio S, Chapman J, Stupka E, Putnam N, Chia JM, et al. 2002. Whole-genome shotgun assembly and analysis of the genome of Fugu rubripes. *Science* 297:1301–10
4. Ball EV, Stenson PD, Abeysinghe SS, Krawczak M, Cooper DN, Chuzhanova NA. 2005. Microdeletions and microinsertions causing human genetic disease: common mechanisms of mutagenesis and the role of local DNA sequence complexity. *Hum. Mutat.* 26:205–13
5. Batzoglou S. 2005. The many faces of sequence alignment. *Brief Bioinform.* 6:6–22
6. Bejerano G, Pheasant M, Makunin I, Stephen S, Kent WJ, et al. 2004. Ultraconserved elements in the human genome. *Science* 304:1321–25
7. Berezikov E, Guryev V, Plasterk RH, Cuppen E. 2004. CONREAL: conserved regulatory elements anchored alignment algorithm for identification of transcription factor binding sites by phylogenetic footprinting. *Genome Res.* 14:170–78
8. Birney E, Andrews D, Caccamo M, Chen Y, Clarke L, et al. 2006. Ensembl 2006. *Nucleic Acids Res.* 34:D556–61
9. Blanchette M, Green ED, Miller W, Haussler D. 2004. Reconstructing large regions of an ancestral mammalian genome in silico. *Genome Res.* 14:2412–23
10. Blanchette M, Kent WJ, Riemer C, Elnitski L, Smit AF, et al. 2004. Aligning multiple genomic sequences with the threaded blockset aligner. *Genome Res.* 14:708–15
11. Blanchette M, Tompa M. 2002. Discovery of regulatory elements by a computational method for phylogenetic footprinting. *Genome Res.* 12:739–48
12. Bray N, Dubchak I, Pachter L. 2003. AVID: a global alignment program. *Genome Res.* 13:97–102
13. Bray N, Pachter L. 2004. MAVID: constrained ancestral alignment of multiple sequences. *Genome Res.* 14:693–99
14. Brudno M, Chapman M, Gottgens B, Batzoglou S, Morgenstern B. 2003. Fast and sensitive multiple alignment of large genomic sequences. *BMC Bioinform.* 4:66
15. Brudno M, Do CB, Cooper GM, Kim MF, Davydov E, et al. 2003. LAGAN and Multi-LAGAN: efficient tools for large-scale multiple alignment of genomic DNA. *Genome Res.* 13:721–31
16. Brudno M, Malde S, Poliakov A, Do CB, Couronne O, et al. 2003. Glocal alignment: finding rearrangements during alignment. *Bioinformatics* 19(Suppl. 1):i54–62
17. Brudno M, Poliakov A, Salamov A, Cooper GM, Sidow A, et al. 2004. Automated whole-genome multiple alignment of rat, mouse, and human. *Genome Res.* 14:685–92
18. Carter D, Durbin R. 2006. Vertebrate gene finding from multiple-species alignments using a two-level strategy. *Genome Biol.* 7(Suppl. 1):S6 1–12
19. Chakrabarti K, Pachter L. 2004. Visualization of multiple genome annotations and alignments with the K-BROWSER. *Genome Res.* 14:716–20
20. Chakrabarti S, Lanczycki CJ, Panchenko AR, Przytycka TM, Thiessen PA, Bryant SH. 2006. State of the art: refinement of multiple sequence alignments. *BMC Bioinform.* 7:499
21. Chapman MA, Donaldson IJ, Gilbert J, Grafham D, Rogers J, et al. 2004. Analysis of multiple genomic sequence alignments: a web resource, online tools, and lessons learned from analysis of mammalian SCL loci. *Genome Res.* 14:313–18
22. Chenna R, Sugawara H, Koike T, Lopez R, Gibson TJ, et al. 2003. Multiple sequence alignment with the Clustal series of programs. *Nucleic Acids Res.* 31:3497–500
23. Chimpanzee Sequencing and Analysis Consortium. 2005. Initial sequence of the chimpanzee genome and comparison with the human genome. *Nature* 437:69–87

24. Chindelevitch L, Li Z, Blais E, Blanchette M. 2006. On the inference of parsimonious indel evolutionary scenarios. *J. Bioinform. Comput. Biol.* 4:721–44
25. Clamp M, Cuff J, Searle SM, Barton GJ. 2004. The Jalview Java alignment editor. *Bioinformatics* 20:426–27
26. Cliften P, Sudarsanam P, Desikan A, Fulton L, Fulton B, et al. 2003. Finding functional features in Saccharomyces genomes by phylogenetic footprinting. *Science* 301:71–76
27. Cooper GM, Brudno M, Green ED, Batzoglou S, Sidow A. 2003. Quantitative estimates of sequence divergence for comparative analyses of mammalian genomes. *Genome Res.* 13:813–20
28. Cooper GM, Brudno M, Stone EA, Dubchak I, Batzoglou S, Sidow A. 2004. Characterization of evolutionary rates and constraints in three mammalian genomes. *Genome Res.* 14:539–48
29. Cooper GM, Singaravelu SA, Sidow A. 2004. ABC: software for interactive browsing of genomic multiple sequence alignment data. *BMC Bioinform.* 5:192
30. Cooper GM, Stone EA, Asimenos G, Green ED, Batzoglou S, Sidow A. 2005. Distribution and intensity of constraint in mammalian genomic sequence. *Genome Res.* 15:901–13
31. Couronne O, Poliakov A, Bray N, Ishkhanov T, Ryaboy D, et al. 2003. Strategies and tools for whole-genome alignments. *Genome Res.* 13:73–80
32. Deleted in proof
33. Darling AC, Mau B, Blattner FR, Perna NT. 2004. Mauve: multiple alignment of conserved genomic sequence with rearrangements. *Genome Res.* 14:1394–403
34. Delcher AL, Kasif S, Fleischmann RD, Peterson J, White O, Salzberg SL. 1999. Alignment of whole genomes. *Nucleic Acids Res.* 27:2369–76
35. Delcher AL, Phillippy A, Carlton J, Salzberg SL. 2002. Fast algorithms for large-scale genome alignment and comparison. *Nucleic Acids Res.* 30:2478–83
36. Delsuc F, Brinkmann H, Philippe H. 2005. Phylogenomics and the reconstruction of the tree of life. *Nat. Rev. Genet.* 6:361–75
37. Dermitzakis ET, Clark AG. 2002. Evolution of transcription factor binding sites in mammalian gene regulatory regions: conservation and turnover. *Mol. Biol. Evol.* 19:1114–21
38. Dewey C, Wu JQ, Cawley S, Alexandersson M, Gibbs R, Pachter L. 2004. Accurate identification of novel human genes through simultaneous gene prediction in human, mouse, and rat. *Genome Res.* 14:661–64
39. Dewey CN, Pachter L. 2006. Evolution at the nucleotide level: the problem of multiple whole-genome alignment. *Hum. Mol. Genet.* 15(Spec. No. 1):R51–56
40. Do CB, Mahabhashyam MS, Brudno M, Batzoglou S. 2005. ProbCons: probabilistic consistency-based multiple sequence alignment. *Genome Res.* 15:330–40
41. Dowell RD, Eddy SR. 2006. Efficient pairwise RNA structure prediction and alignment using sequence alignment constraints. *BMC Bioinform.* 7:400
42. Down TA, Hubbard TJ. 2004. What can we learn from noncoding regions of similarity between genomes? *BMC Bioinform.* 5:131
43. Dubchak I, Ryaboy DV. 2006. VISTA family of computational tools for comparative analysis of DNA sequences and whole genomes. *Methods Mol. Biol.* 338:69–89
44. Durbin R, Eddy S, Krogh A, Mitchison G. 1998. *Biological Sequence Analysis: Probabilistic Models of Proteins and Nucleic Acids*. Cambridge, UK: Cambridge Univ. Press
45. Edgar RC. 2004. MUSCLE: a multiple sequence alignment method with reduced time and space complexity. *BMC Bioinform.* 5:113

46. Edgar RC, Batzoglou S. 2006. Multiple sequence alignment. *Curr. Opin. Struct. Biol.* 16:368–73
47. Elias I. 2006. Settling the intractability of multiple alignment. *J. Comput. Biol.* 13:1323–39
48. Elnitski L, Hardison RC, Li J, Yang S, Kolbe D, et al. 2003. Distinguishing regulatory DNA from neutral sites. *Genome Res.* 13:64–72
49. ENCODE Consortium. 2004. The ENCODE (ENCyclopedia Of DNA Elements) Project. *Science* 306:636–40
50. Fang F, Blanchette M. 2006. FootPrinter3: phylogenetic footprinting in partially alignable sequences. *Nucleic Acids Res.* 34:W617–20
51. Ferretti V, Poitras C, Bergeron D, Coulombe B, Robert F, Blanchette M. 2006. PReMod: a database of genome-wide mammalian cis-regulatory module predictions. *Nucleic Acids Res.* 35(Database issue): D122–26
52. Flannick J, Batzoglou S. 2005. Using multiple alignments to improve seeded local alignment algorithms. *Nucleic Acids Res.* 33:4563–77
53. Frazer KA, Elnitski L, Church DM, Dubchak I, Hardison RC. 2003. Cross-species sequence comparisons: a review of methods and available resources. *Genome Res.* 13:1–12
54. Frazer KA, Pachter L, Poliakov A, Rubin EM, Dubchak I. 2004. VISTA: computational tools for comparative genomics. *Nucleic Acids Res.* 32:W273–79
55. Giardine B, Elnitski L, Riemer C, Makalowska I, Schwartz S, et al. 2003. GALA, a database for genomic sequence alignments and annotations. *Genome Res.* 13:732–41
56. Giardine B, Riemer C, Hardison RC, Burhans R, Elnitski L, et al. 2005. Galaxy: a platform for interactive large-scale genome analysis. *Genome Res.* 15:1451–55
57. Gibbs RA, Weinstock GM, Metzker ML, Muzny DM, Sodergren EJ, et al. 2004. Genome sequence of the Brown Norway rat yields insights into mammalian evolution. *Nature* 428:493–521
58. Gross SS, Brent MR. 2006. Using multiple alignments to improve gene prediction. *J. Comput. Biol.* 13:379–93
59. Hein J, Wiuf C, Knudsen B, Moller MB, Wibling G. 2000. Statistical alignment: computational properties, homology testing and goodness-of-fit. *J. Mol. Biol.* 302:265–79
60. Higgins DG, Sharp PM. 1988. CLUSTAL: a package for performing multiple sequence alignment on a microcomputer. *Gene* 73:237–44
61. Hillier LW, Miller W, Birney E, Warren W, Hardison RC, et al. 2004. Sequence and comparative analysis of the chicken genome provide unique perspectives on vertebrate evolution. *Nature* 432:695–716
62. Holmes I. 2005. Using evolutionary Expectation Maximization to estimate indel rates. *Bioinformatics* 21:2294–300
63. Huang X, Chao KM. 2003. A generalized global alignment algorithm. *Bioinformatics* 19:228–33
64. Hudek AK, Brown DG. 2005. Ancestral sequence alignment under optimal conditions. *BMC Bioinform.* 6:273
65. Hughes JR, Cheng JF, Ventress N, Prabhakar S, Clark K, et al. 2005. Annotation of *cis*-regulatory elements by identification, subclassification, and functional assessment of multispecies conserved sequences. *Proc. Natl. Acad. Sci. USA* 102:9830–35
66. Jones NC, Zhi D, Raphael BJ. 2006. AliWABA: alignment on the web through an A-Bruijn approach. *Nucleic Acids Res.* 34:W613–16

67. Karolchik D, Hinrichs AS, Furey TS, Roskin KM, Sugnet CW, et al. 2004. The UCSC Table Browser data retrieval tool. *Nucleic Acids Res.* 32:D493–96
68. Kasprzyk A, Keefe D, Smedley D, London D, Spooner W, et al. 2004. EnsMart: a generic system for fast and flexible access to biological data. *Genome Res.* 14:160–69
69. Katoh K, Kuma K, Toh H, Miyata T. 2005. MAFFT version 5: improvement in accuracy of multiple sequence alignment. *Nucleic Acids Res.* 33:511–18
70. Kellis M, Patterson N, Endrizzi M, Birren B, Lander ES. 2003. Sequencing and comparison of yeast species to identify genes and regulatory elements. *Nature* 423:241–54
71. Kent WJ, Baertsch R, Hinrichs A, Miller W, Haussler D. 2003. Evolution's cauldron: duplication, deletion, and rearrangement in the mouse and human genomes. *Proc. Natl. Acad. Sci. USA* 100:11484–89
72. King DC, Taylor J, Elnitski L, Chiaromonte F, Miller W, Hardison RC. 2005. Evaluation of regulatory potential and conservation scores for detecting *cis*-regulatory modules in aligned mammalian genome sequences. *Genome Res.* 15:1051–60
73. Kolbe D, Taylor J, Elnitski L, Eswara P, Li J, et al. 2004. Regulatory potential scores from genome-wide three-way alignments of human, mouse, and rat. *Genome Res.* 14:700–7
74. Kuhn RM, Karolchik D, Zweig AS, Trumbower H, Thomas DJ, et al. 2006. The UCSC genome browser database: update 2007. *Nucleic Acids Res.* 35(Database issue):D668–73
75. Lander ES, Linton LM, Birren B, Nusbaum C, Zody MC, et al. 2001. Initial sequencing and analysis of the human genome. *Nature* 409:860–921
76. Lee D, Choi JH, Dalkilic MM, Kim S. 2006. COMPAM: visualization of combining pairwise alignments for multiple genomes. *Bioinformatics* 22:242–44
77. Lindblad-Toh K, Wade CM, Mikkelsen TS, Karlsson EK, Jaffe DB, et al. 2005. Genome sequence, comparative analysis and haplotype structure of the domestic dog. *Nature* 438:803–19
78. Lipman DJ, Altschul SF, Kececioglu JD. 1989. A tool for multiple sequence alignment. *Proc. Natl. Acad. Sci. USA* 86:4412–15
79. Lippert RA, Zhao X, Florea L, Mobarry C, Istrail S. 2005. Finding anchors for genomic sequence comparison. *J. Comput. Biol.* 12:762–76
80. Loots GG, Ovcharenko I. 2004. rVISTA 2.0: evolutionary analysis of transcription factor binding sites. *Nucleic Acids Res.* 32:W217–21
81. Loytynoja A, Milinkovitch MC. 2001. SOAP, cleaning multiple alignments from unstable blocks. *Bioinformatics* 17:573–74
82. Lunter G, Miklos I, Drummond A, Jensen JL, Hein J. 2005. Bayesian coestimation of phylogeny and sequence alignment. *BMC Bioinform.* 6:83
83. Ma B, Tromp J, Li M. 2002. PatternHunter: faster and more sensitive homology search. *Bioinformatics* 18:440–45
84. Margulies EH, Blanchette M, Haussler D, Green ED. 2003. Identification and characterization of multi-species conserved sequences. *Genome Res.* 13:2507–18
85. Margulies EH, Chen CW, Green ED. 2006. Differences between pair-wise and multi-sequence alignment methods affect vertebrate genome comparisons. *Trends Genet.* 22:187–93
86. Margulies EH, Cooper G, Asimenos G, Thomas DJ, Dewey CN, et al. 2007. Analyses of deep mammalian sequence alignments and constraint predictions for 1% of the human genome. *Genome Res.* In press
87. Margulies EH, Vinson JP, Miller W, Jaffe DB, Lindblad-Toh K, et al. 2005. An initial strategy for the systematic identification of functional elements in the human genome by low-redundancy comparative sequencing. *Proc. Natl. Acad. Sci. USA* 102:4795–800

88. Miller W. 2001. Comparison of genomic DNA sequences: solved and unsolved problems. *Bioinformatics* 17:391–97
89. Morgenstern B, Goel S, Sczyrba A, Dress A. 2003. AltAVisT: comparing alternative multiple sequence alignments. *Bioinformatics* 19:425–26
90. Morgenstern B, Rinner O, Abdeddaim S, Haase D, Mayer KF, et al. 2002. Exon discovery by genomic sequence alignment. *Bioinformatics* 18:777–87
91. Morgenstern B, Werner N, Prohaska SJ, Steinkamp R, Schneider I, et al. 2005. Multiple sequence alignment with user-defined constraints at GOBICS. *Bioinformatics* 21:1271–73
92. Moses AM, Chiang DY, Pollard DA, Iyer VN, Eisen MB. 2004. MONKEY: identifying conserved transcription-factor binding sites in multiple alignments using a binding site-specific evolutionary model. *Genome Biol.* 5:R98
93. Moses AM, Pollard DA, Nix DA, Iyer VN, Li XY, et al. 2006. Large-scale turnover of functional transcription factor binding sites in *Drosophila*. *PLOS Comput. Biol.* 2:e130
94. Ovcharenko I, Boffelli D, Loots GG. 2004. eShadow: a tool for comparing closely related sequences. *Genome Res.* 14:1191–98
95. Ovcharenko I, Loots GG, Giardine BM, Hou M, Ma J, et al. 2005. Mulan: multiple-sequence local alignment and visualization for studying function and evolution. *Genome Res.* 15:184–94
96. Paten B, Birney E. 2006. Pecan. **http://www.ebi.ac.uk/~bjp/pecan/**
97. Pavesi G, Mauri G, Pesole G. 2004. In silico representation and discovery of transcription factor binding sites. *Brief Bioinform.* 5:217–36
98. Pedersen JS, Bejerano G, Siepel A, Rosenbloom K, Lindblad-Toh K, et al. 2006. Identification and classification of conserved RNA secondary structures in the human genome. *PLOS Comput. Biol.* 2:e33
99. Pevzner P, Tesler G. 2003. Genome rearrangements in mammalian evolution: lessons from human and mouse genomes. *Genome Res.* 13:37–45
100. Philippe H, Blanchette M. 2007. Overview of the First International Conference on Phylogenomics. *BMC Evol. Biol.* 7(Suppl 1):S1
101. Pohler D, Werner N, Steinkamp R, Morgenstern B. 2005. Multiple alignment of genomic sequences using CHAOS, DIALIGN and ABC. *Nucleic Acids Res.* 33:W532–34
102. Pollard DA, Bergman CM, Stoye J, Celniker SE, Eisen MB. 2004. Benchmarking tools for the alignment of functional noncoding DNA. *BMC Bioinform.* 5:6
102a. Pollard DA, Moses AM, Iyer VN, Eisen MB. 2006. Detecting the limits of regulatory element conservation and divergence estimation using pairwise and multiple alignments. *BMC Bioinform.* 7:1–18
103. Prakash A, Blanchette M, Sinha S, Tompa M. 2004. Motif discovery in heterogeneous sequence data. *Pac. Symp. Biocomput.* 348–59
104. Prakash A, Tompa M. 2005. Statistics of local multiple alignments. *Bioinformatics* 21(Suppl. 1):i344–50
105. Raphael B, Zhi D, Tang H, Pevzner P. 2004. A novel method for multiple alignment of sequences with repeated and shuffled elements. *Genome Res.* 14:2336–46
106. Rosenberg MS. 2005. Multiple sequence alignment accuracy and evolutionary distance estimation. *BMC Bioinform.* 6:278
107. Sandelin A, Wasserman WW, Lenhard B. 2004. ConSite: web-based prediction of regulatory elements using cross-species comparison. *Nucleic Acids Res.* 32:W249–52
108. Sankoff D. 1985. Simultaneous solution of the RNA folding, alignment and protosequence problems. *SIAM J. Appl. Math.* 45:810–25

109. Sankoff D, Cedergren RJ. 1983. Simultaneous comparison of three or more sequences related by a tree. In *Time Warps, String Edits and Macromolecules: The Theory and Practice of Sequence Comparison*, ed. Sankoff D and Kruskal JB, pp. 253–63. Reading, MA: Addison-Wesley
110. Schwartz S, Kent WJ, Smit A, Zhang Z, Baertsch R, et al. 2003. Human-mouse alignments with BLASTZ. *Genome Res.* 13:103–7
111. Schwartz S, Zhang Z, Frazer KA, Smit A, Riemer C, et al. 2000. PipMaker–a web server for aligning two genomic DNA sequences. *Genome Res.* 10:577–86
112. Shih AC, Lee DT, Lin L, Peng CL, Chen SH, et al. 2006. SinicView: a visualization environment for comparisons of multiple nucleotide sequence alignment tools. *BMC Bioinform.* 7:103
113. Siddharthan R, Siggia ED, van Nimwegen E. 2005. PhyloGibbs: a Gibbs sampling motif finder that incorporates phylogeny. *PLOS Comput. Biol.* 1:e67
114. Siebert S, Backofen R. 2005. MARNA: multiple alignment and consensus structure prediction of RNAs based on sequence structure comparisons. *Bioinformatics* 21:3352–59
115. Siepel A, Bejerano G, Pedersen JS, Hinrichs AS, Hou M, et al. 2005. Evolutionarily conserved elements in vertebrate, insect, worm, and yeast genomes. *Genome Res.* 15:1034–50
116. Sinha S, Blanchette M, Tompa M. 2004. PhyME: a probabilistic algorithm for finding motifs in sets of orthologous sequences. *BMC Bioinform.* 5:170
117. Smith TF, Waterman MS. 1981. Identification of common molecular subsequences. *J. Mol. Biol.* 147:195–97
118. Snel B, Huynen MA, Dutilh BE. 2005. Genome trees and the nature of genome evolution. *Annu. Rev. Microbiol.* 59:191–209
119. Sobel E, Martinez HM. 1986. A multiple sequence alignment program. *Nucleic Acids Res.* 14:363–74
120. Stocsits RR, Hofacker IL, Fried C, Stadler PF. 2005. Multiple sequence alignments of partially coding nucleic acid sequences. *BMC Bioinform.* 6:160
121. Sun Y, Buhler J. 2006. Choosing the best heuristic for seeded alignment of DNA sequences. *BMC Bioinform.* 7:133
122. Szklarczyk R, Heringa J. 2006. AuberGene–a sensitive genome alignment tool. *Bioinformatics* 22:1431–36
123. Taylor JTS, King DC, Hardison RC, Miller W, Chiaromonte F. 2006. ESPERR: learning strong and weak signals in genomic sequence alignments to identify functional elements. *Genome Res.* 16:1596–604
124. Thompson JD, Koehl P, Ripp R, Poch O. 2005. BAliBASE 3.0: latest developments of the multiple sequence alignment benchmark. *Proteins* 61:127–36
125. Wallace IM, Blackshields G, Higgins DG. 2005. Multiple sequence alignments. *Curr. Opin. Struct. Biol.* 15:261–66
126. Wang C, Lefkowitz EJ. 2005. Genomic multiple sequence alignments: refinement using a genetic algorithm. *BMC Bioinform.* 6:200
127. Wang L, Jiang T. 1994. On the complexity of multiple sequence alignment. *J. Comput. Biol.* 1:337–48
128. Washietl S, Hofacker IL, Stadler PF. 2005. Fast and reliable prediction of noncoding RNAs. *Proc. Natl. Acad. Sci. USA* 102:2454–59
129. Waterston RH, Lindblad-Toh K, Birney E, Rogers J, Abril JF, et al. 2002. Initial sequencing and comparative analysis of the mouse genome. *Nature* 420:520–62

130. Yang S, Smit AF, Schwartz S, Chiaromonte F, Roskin KM, et al. 2004. Patterns of insertions and their covariation with substitutions in the rat, mouse, and human genomes. *Genome Res.* 14:517–27
131. Ye L, Huang X. 2005. MAP2: multiple alignment of syntenic genomic sequences. *Nucleic Acids Res.* 33:162–70
132. Zhang L, Pavlovic V, Cantor CR, Kasif S. 2003. Human-mouse gene identification by comparative evidence integration and evolutionary analysis. *Genome Res.* 13:1190–202
133. Zhang Y, Waterman MS. 2005. An Eulerian path approach to local multiple alignment for DNA sequences. *Proc. Natl. Acad. Sci. USA* 102:1285–90
134. Zhang Z, Raghavachari B, Hardison RC, Miller W. 1994. Chaining multiple-alignment blocks. *J. Comput. Biol.* 1:217–26

microRNAs in Vertebrate Physiology and Human Disease

Tsung-Cheng Chang and Joshua T. Mendell

The McKusick-Nathans Institute of Genetic Medicine, Johns Hopkins University School of Medicine, Baltimore, Maryland, 21205; email: tchang12@jhmi.edu, jmendell@jhmi.edu

Annu. Rev. Genomics Hum. Genet. 2007. 8:215–39

First published online as a Review in Advance on May 21, 2007.

The *Annual Review of Genomics and Human Genetics* is online at genom.annualreviews.org

This article's doi:
10.1146/annurev.genom.8.080706.092351

1527-8204/07/0922-0215$20.00

Key Words

noncoding RNAs, RNA interference, post-transcriptional regulation, cancer

Abstract

Over the past five years, the importance of a diverse class of 18–24 nucleotide RNA molecules, known as microRNAs (miRNAs) has increasingly been recognized. These highly conserved RNAs regulate the stability and translational efficiency of complementary target messenger RNAs. The human genome is now predicted to encode nearly 1,000 miRNAs that likely regulate at least one third of all human transcripts. Despite rapid progress in miRNA discovery, the physiologic functions of only a small number have been definitively established. In this review, we discuss the principles of miRNA function that have emerged from the studies performed thus far in vertebrates. We also discuss known and potential roles for miRNAs in human disease states and discuss the influence of human genetic variation on miRNA-mediated regulation.

THE DISCOVERY OF microRNAs

MicroRNA (miRNA)-mediated regulation of gene expression was not discovered as a result of directed research aimed at better understanding regulatory RNAs. Rather, like many other fundamental cellular processes, our first glimpse at the importance of this mode of regulation came from forward genetic analyses of particular phenotypes in a model organism. Specifically, work from the laboratories of Ambros and Ruvkun in the early 1990s revealed that a miRNA controlled a specific step in developmental timing in *Caenorhabditis elegans* by downregulating a conventional protein-coding gene (89, 140). Only later was it appreciated that this regulatory mechanism is widely utilized to control diverse pathways in plants and animals.

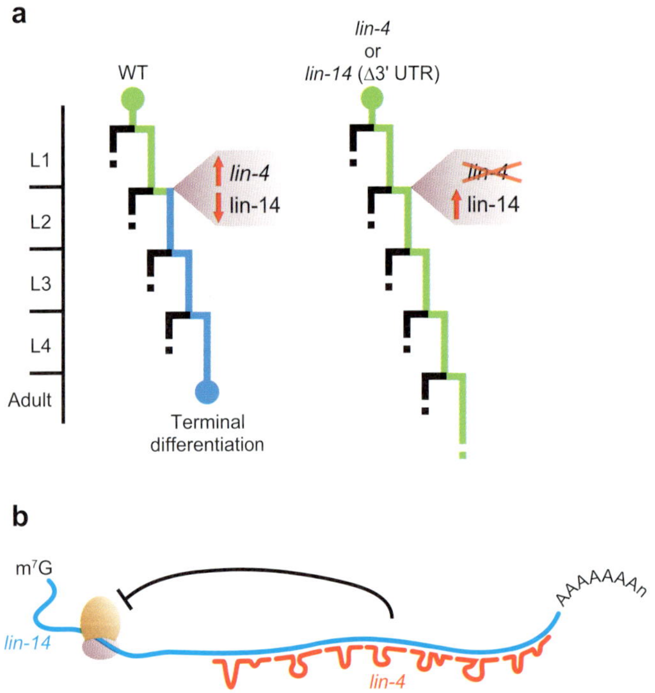

Figure 1

Discovery of the first microRNA (miRNA). (*a*) Schematic representation of the function of the *lin-4* miRNA in the *C. elegans* seam-cell lineage. In wild-type worms, *lin-4* is induced at the transition between the L1 and L2 larval stages. This correlates with a decrease in lin-14 protein levels. In worms with mutations in *lin-4* or deletions in the *lin-14* 3′ untranslated region (UTR), lin-14 protein is not downregulated at this stage. As a consequence, cell divisions characteristic of L1 worms reiterate throughout development. (*b*) Cloning of the *lin-4* gene revealed that it encodes a small RNA that can base pair with imperfect complementarity to multiple sites in the 3′ UTR of the *lin-14* transcript. This results in translational repression of *lin-14*.

Two key *C. elegans* mutants were central to the discovery of the first miRNA: *lin-4*, initially described by Horvitz & Sulston (60), and *lin-14*, also discovered in the Horvitz laboratory (48). *lin-4* loss of function results in failure of diverse cell lineages to differentiate properly as worms mature through larval development. Instead, early cell divisions are reiterated repeatedly. This is well illustrated by the seam-cell lineage that normally undergoes a defined series of divisions until adulthood (**Figure 1*a***). At this time, seam cells normally stop dividing, terminally differentiate, and form part of the hypodermis. In *lin-4* mutants, the earliest division is repeated and proper seam-cell differentiation does not occur. A deletion in the 3′ untranslated region (UTR) of the *lin-14* gene (a protein-coding transcript) results in a very similar phenotype. Both *lin-4* (loss of function) and the *lin-14* 3′ UTR deletion mutants show inappropriately high expression of the lin-14 protein after the first larval division. It was therefore proposed that *lin-4* downregulates *lin-14* by interacting with its 3′ UTR. The real surprise came when causative mutations were identified in *lin-4* worms and found to result in deletion or mutation of a 22-nucleotide RNA molecule. Immediately thereafter, Ambros and Ruvkun recognized that the *lin-14* 3′ UTR harbors multiple sites of imperfect complementarity to the *lin-4* small RNA. They proposed that *lin-4* binds to these sites and blocks *lin-14* translation, precisely foreshadowing our current understanding of animal miRNA function (**Figure 1*b***).

Forward genetics also identified a second miRNA in *C. elegans*, known as *let-7*, that regulates developmental timing (118). This miRNA is highly conserved both in sequence and in its pattern of expression during development in various animal species (112). This observation prompted several laboratories to hypothesize that *lin-4* and *let-7* may be founding members of a much more diverse class of small regulatory RNAs. This idea was also undoubtedly catalyzed by the elucidation of the RNA interference (RNAi) mechanism. RNAi refers to the potent inhibition of gene expression that occurs in most eukaryotic organisms when double-stranded RNA is introduced into a cell (49). A critical insight into the RNAi pathway was provided by studies demonstrating that long double-stranded RNA is first processed into small ~21-nucleotide RNA molecules known as short-interfering RNAs (siRNAs), which guide the RNAi machinery to target messages (44, 145). This mechanism in general, and siRNAs in particular, seemed reminiscent of *lin-4* and *let-7*, fostering the idea that these regulatory RNAs may be natural RNAi triggers. This idea was ultimately proven correct (64). Shortly after these ideas crystallized, several groups constructed and sequenced libraries of cloned small RNAs from diverse sources including *C. elegans*, *Drosophila*, and mammals, confirming that *lin-4* and *let-7* are members of an abundant class of small regulatory RNAs (81, 86, 88).

Prior to the discovery of the first miRNAs and RNAi, it was generally believed that the most important components of the transcriptome were predominantly large RNA molecules. This way of thinking, rather than any technical limitation, was primarily responsible for the long delay in the appreciation that most eukaryotic genomes encode numerous small RNA molecules that function in diverse pathways. Once this mindset was overcome, identifying the first wave of new miRNAs through cloning was relatively rapid and straightforward. Small RNA cloning strategies generally involve purification of low-molecular-weight RNA, ligation of linkers to the 5′ and 3′ ends of these molecules, and amplification by polymerase chain reaction (PCR) (81, 86, 88). Newly identified miRNA candidates are then mapped to their positions in the genome and their sequence contexts are examined. miRNAs are encoded by transcripts that fold back on themselves to form double-stranded hairpin structures (see microRNA Biogenesis and Mechanism, below). Thus, newly identified miRNA candidates are first screened by determining if they are surrounded by genomic sequence which can form this structure. Final validation that a candidate miRNA exists usually involves direct demonstration that the RNA is expressed using northern blotting, reverse transcriptase polymerase chain reaction (RT-PCR), or microarray approaches. These methods have now been applied to diverse tissues and cell types from several species, leading to the identification of several hundred miRNAs (6, 33, 61, 75, 103).

After the discovery of the first few hundred miRNAs by cloning, sufficient miRNA sequences were available to begin applying bioinformatic analyses to scan genomes for additional miRNA genes. Although numerous methods exist (11, 12, 84, 95, 137), they generally rely on three main principles for miRNA recognition. First, miRNAs are contained within regions that fold into characteristic hairpin structures. Second, miRNAs tend to be highly conserved in related species. Third, the specific pattern of conservation displayed by miRNAs is very distinctive, with extreme conservation of the hairpin stems and less conservation of the loop of the hairpin and flanking sequences. Applying these methods to scans of the human genome has revealed ~800–1000 potential human miRNAs, including those previously identified by cloning. Both direct sequencing and bioinformatics efforts to comprehensively identify miRNAs are ongoing at a rapid pace and, undoubtedly, the existence of many more miRNAs will be confirmed by the time this review is published.

The most carefully curated set of miRNA sequences is maintained by the Sanger Institute in a publicly available database termed miRBase (**http://microrna.sanger.ac.uk/**) (53). This resource provides a searchable interface for all known miRNAs supported by experimental evidence, supplying sequence

pri-miRNAs: primary microRNA transcripts; generally several thousand nucleotides in length or longer

miRNA cluster: one or more miRNAs that are located near one another in the genome and are co-transcribed as part of a single pri-miRNA

information, genomic coordinates, potential targets, and literature citations. miRBase currently has entries for 474 human miRNAs and 373 mouse miRNAs (release 9.0). Although miRNAs seem to be missing from lower eukaryotes such as *Saccharomyces cerevisiae*, all studied multicellular eukaryotes appear to possess miRNAs.

Despite these great advances in miRNA discovery, our understanding of the role of miRNAs in physiologic and pathophysiologic processes is at a very early stage. Here we provide examples that illuminate principles of miRNA function in normal developmental and cellular pathways. We also describe our current understanding of the links between miRNAs and human disease and the impact of natural genetic variation on miRNA function. It is important to keep in mind that we are just beginning to elucidate these connections and many basic principles of miRNA function remain to be described. Before delving into these areas, we first provide a description of miRNA biogenesis and molecular function.

miRNA BIOGENESIS AND MECHANISM

Despite the obvious differences between the biology of miRNAs and mRNAs, all available evidence suggests that these transcripts share common mechanisms of transcriptional regulation. In fact, at least 100 known miRNAs are located within introns of protein-coding genes and are believed to be processed from intron lariats liberated by splicing (120). miRNAs located in intergenic regions are also transcribed by RNA polymerase II (pol II). These noncoding primary miRNA transcripts, referred to as pri-miRNAs, have a 5′ 7-methylguanosine cap, a poly(A) tail, and may be a few kilobases to several hundred kilobases in length (17, 91). The transcription factors that regulate pri-miRNA expression appear to overlap substantially with those that control protein-coding genes. For example, the highly studied mammalian transcription factors c-Myc, cAMP-response element binding protein (CREB), and MyoD are now known to regulate specific miRNAs (108, 117, 134, 149). These findings illustrate the importance of considering miRNAs as potential targets to fully elucidate the functions of a given transcription factor.

Most noncoding pri-miRNAs are efficiently processed and therefore not easily detectable. Although this has hindered their characterization, a few common features have emerged from those that have been studied thus far (76, 120). First, a single pri-miRNA may contain one or more individual miRNAs. A group of co-transcribed miRNAs is usually referred to as a miRNA cluster. Surveys of tissue-specific miRNA expression patterns suggest that human miRNAs located within 50 kb of one another are likely co-transcribed (10). Second, noncoding pri-miRNAs are frequently spliced with the miRNA located in either introns or exons. Third, although most mature miRNA sequences are highly conserved among vertebrates, pri-miRNA sequences are not generally well conserved outside the vicinity of the mature miRNA. Moreover, the precise genomic organization of pri-miRNAs (e.g., the number of introns and exons) is also not a highly conserved feature. Nevertheless, if a miRNA is contained within an intron or exon, orthologous miRNAs in other vertebrates also tend to be intronic or exonic, respectively. This implies that the presence of a miRNA in an intron or exon, regardless of precise sequence context, may subject it to as-yet-undescribed conserved mechanisms of post-transcriptional regulation. Further characterization of pri-miRNA structure and regulation remains an important area for future research.

One other aspect of miRNA organization within animal genomes merits discussion. miRNAs frequently exist in multiple highly related or identical copies distributed throughout the genome of a given species. The precise expression patterns of these independently transcribed loci have not been systematically evaluated, so some of these related miRNAs may function in distinct cell types. Nonetheless, this organization undoubtedly produces significant redundancy, as has been selectively demonstrated (2). This redundancy has likely contributed to the relative paucity of miRNAs identified through forward genetics, although their small size also makes them infrequent targets for mutation. Disease-causing mutations may also rarely occur in miRNAs for similar reasons. miRNA-binding sites in mRNAs, which are nonredundant, may still be important sites of mutation, as discussed below (see The Influence of Genetic Variation on miRNA Function).

miRNA processing has been reviewed extensively and we present only the basic scheme here (for further detail, see 76, 123). Mature miRNA sequences are contained within regions of pri-miRNAs that fold back on themselves to form approximately 60–80 nucleotide stem-loop structures (**Figure 2**). These hairpin structures are excised from pri-miRNAs in the nucleus by the microprocessor complex that includes the RNA endonuclease Drosha and its binding partner DGCR8 as core components (37, 52, 85, 90). The excised hairpins, referred to as pre-miRNAs, feature a short stem of ~22 base pairs and a two-nucleotide 3′ overhang, which serve as structural requirements for recognition by the nuclear export factor exportin 5 (14, 99, 144, 146). Following export to the cytoplasm, pre-miRNAs undergo further processing by the RNA endonuclease Dicer, which removes the terminal loop, yielding a double-stranded 18–24 nucleotide RNA duplex (54, 62, 73). This fully processed miRNA duplex is subsequently incorporated into a multicomponent protein complex known as the RNA-induced silencing complex (RISC). During this process, one strand of the miRNA duplex is selected as the mature miRNA (referred to as the guide strand) and remains stably associated with RISC. The other strand, known as the passenger strand, or miRNA*, is rapidly removed and degraded. Selection of the appropriate strand is primarily determined by the strength of base pairing at the ends of the miRNA/miRNA* duplex. The strand with less-stable base pairing at its 5′ end (i.e., more A:U base pairs or mismatches) is usually destined to become the mature miRNA and is chosen for incorporation into RISC (74, 122). It is worth noting that fully processed miRNA duplexes are functionally equivalent to experimentally introduced siRNAs, which are commonly used to silence gene expression (43). siRNAs are loaded into RISC exactly as described for miRNAs and function in an identical manner. Consequently, current algorithms for choosing effective siRNAs usually incorporate asymmetric thermodynamic stability into the RNA duplexes to promote loading of the desired strand into RISC (74, 122).

pre-miRNAs: approximately 60–80 nucleotide stem-loop RNA structures containing the mature miRNA sequence; excised from pri-miRNAs

RISC: RNA-induced silencing complex

Once loaded with the guide strand miRNA, RISC accomplishes gene silencing through two major mechanisms: mRNA cleavage or translational repression. The choice of mechanism is governed by the degree of complementarity between a miRNA and its target (39, 61, 147). In plants, most miRNAs exhibit perfect or near-perfect complementarity to their target mRNAs (97, 119). This perfect base pairing leads to RISC-mediated endonucleolytic cleavage of the mRNA. In contrast, miRNAs in *C. elegans*, *Drosophila melanogaster*, and vertebrates usually recognize one or more imperfectly complementary binding sites in the target mRNA 3′ UTR. This imperfect base pairing prevents RISC-mediated cleavage and instead results primarily in translational repression of the mRNA. To a lesser extent, this imperfect base pairing can also lead to decreased target mRNA abundance (8, 94), although this is probably a result of accelerated turnover of the mRNA through general cellular RNA decay pathways rather than RISC-mediated cleavage.

The core protein components of RISC that interact with miRNAs and carry out miRNA-mediated functions are the Argonaute proteins (22). Mammals possess four Argonaute family members (Ago1-4), which are widely coexpressed. Each Argonaute protein appears to be capable of binding any

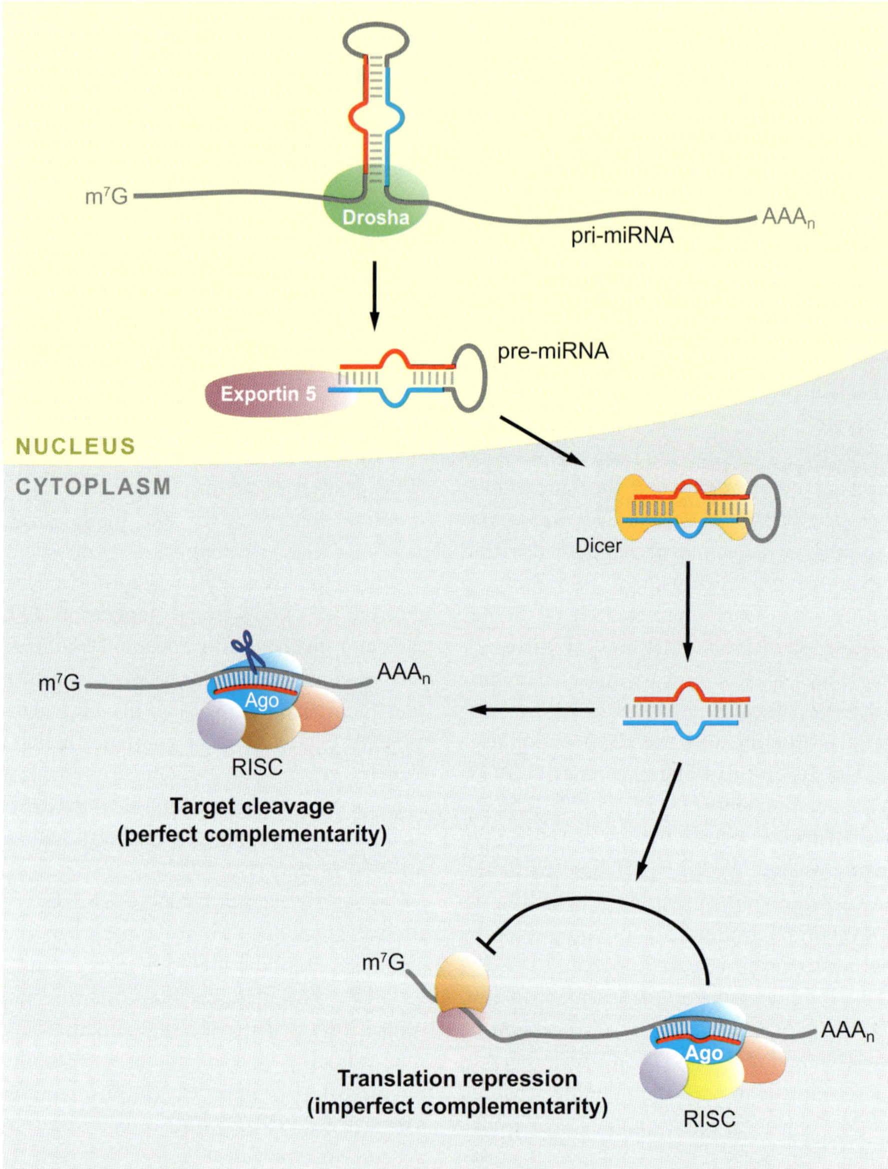

Figure 2

miRNA biogenesis and function. miRNAs are initially transcribed as long primary transcripts (pri-miRNAs) that are capped and polyadenylated. The miRNA sequence is contained within a 60–80-nucleotide hairpin structure that is excised from the pri-miRNA by the endonuclease Drosha. This liberated stem-loop, or pre-miRNA, is transported to the cytoplasm by exportin 5, where it undergoes further processing by the Dicer endonuclease. One strand of the resulting duplex is selectively loaded into the RNA-induced silencing complex (RISC) and guides target transcript silencing through mRNA cleavage or translational repression.

miRNA, although only Ago2 can direct miRNA-mediated target mRNA cleavage (96, 102). A defining feature of Argonaute proteins is the PIWI domain, which structurally resembles RNase H. It is this domain of Ago2 that provides the RISC endonuclease activity (111, 128). Although the other mammalian Argonautes are catalytically inactive and cannot cleave mRNA targets, all four Argonaute proteins can likely mediate translational repression through a mechanism that is still unresolved (96, 114). Evidence currently exists for miRNA-mediated inhibition of translation both at the level of translation initiation as well as at steps after initiation (109, 113, 115).

miRNA TARGET PREDICTION AND VALIDATION

Elucidation of the function of a miRNA requires identification of the mRNA targets that it regulates. As such, significant effort has been devoted to predicting and validating miRNA targets. In plants, these efforts have yielded rapid progress, due primarily to the perfect base pairing that typifies miRNA:target interactions in these species. This has dramatically simplified genome-wide searches for miRNA-binding sites (70, 148). Identifying potential miRNA targets is considerably more challenging in animal species because miRNAs are usually imperfectly complementary to their targets. In animals, the most consistent feature of miRNA:target interaction is base pairing at the 5′ end of the miRNA. In particular, nucleotides 2–7 or 2–8 of the miRNA have been termed the seed sequence and complementarity to this region is considered most important, although not always essential, for target recognition (16, 83) (**Figure 3**).

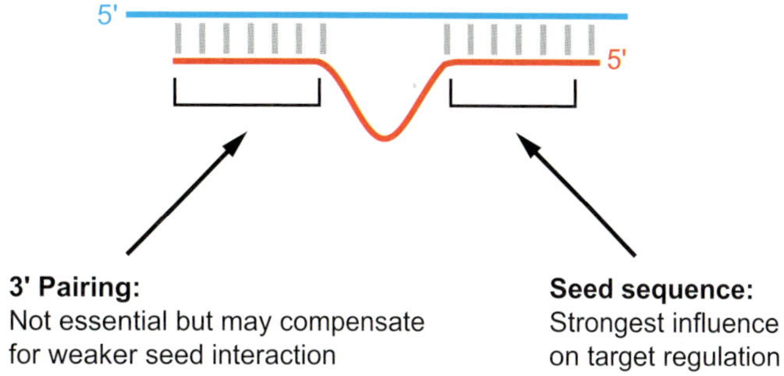

Figure 3

Anatomy of an animal miRNA:target mRNA interaction. Animal miRNAs base pair with their targets with imperfect complementarity. The miRNA is shown in red and the target is shown in blue. The seed sequence represents nucleotides 2–7 or 2–8 of the miRNA.

In spite of the challenges, a diverse array of bioinformatic approaches have now been applied to identify the mRNAs regulated by animal miRNAs (125). Most of these prediction tools are available online for public use. In general, the different algorithms utilize at least some of the following criteria to identify and prioritize putative targets: (*a*) complementarity between the miRNA seed sequence and the 3′ UTR of the target mRNA; (*b*) the overall stability of putative miRNA:target duplexes; (*c*) miRNA target site conservation between closely related species; (*d*) multiple binding sites for a single miRNA within a given target 3′ UTR; and (*e*) weak or no secondary structure in the target at the miRNA-binding site. These programs predict hundreds of targets for each miRNA and estimate that between 30–90% of human mRNAs are subjected to miRNA-mediated regulatory control. The output from these diverse algorithms can be overwhelming to individual investigators interested in identifying likely targets for miRNAs of interest. Moreover, the true sensitivity and specificity of these methods is not clear. To address these issues, Sethupathy et al. recently evaluated the performance of several widely utilized algorithms in predicting a set of experimentally verified targets (125). In general, they found that no single program is sufficient to accurately predict all targets, yet combining predictions from multiple algorithms produces an exhaustive list with probable high false positive rates. One approach is to first focus on a high-yield set derived from the intersection of multiple predictions (assembled by these authors at **http://www.diana.pcbi.upenn.edu/cgi-bin/TargetCombo.cgi**). This, of course, will miss many bona fide targets. To expand this list, we suggest careful examination of all predictions to identify those targets that may fit with the known biology of the miRNA under study.

Experimental identification of novel miRNA targets is also effective. These strategies generally involve expressing or inhibiting a miRNA of interest in a cell line or animal and examining changes in global mRNA expression profiles using microarrays. This approach relies on the detection of the relatively small-magnitude changes in mRNA abundance that result from miRNA-mediated regulation (8, 94). Once a set of mRNAs with altered expression is identified, the 3′ UTRs can be directly searched for sites

miRNA seed sequence: nucleotides 2–7 or 2–8 of a mature miRNA. This region has the strongest influence on target selection

complementary to the seed sequence of the miRNA of interest. Although this method has been quite successful (79, 136), it is not comprehensive because some targets may change dramatically at the protein level but fail to show detectable changes in mRNA abundance.

Despite successes in computational or microarray-based identification of putative targets, subsequent experimental verification of these targets has advanced much more slowly. A recent compilation of experimentally supported human miRNA targets reports fewer than 100, despite thousands of published predictions (124). Further progress in this area is critical to refine and improve the accuracy of computational methods. Two major approaches are utilized to verify targets. The most common method involves placing a putative miRNA-binding site in the 3′ UTR of a reporter transcript such as luciferase or green fluorescent protein (GFP) (16, 93). The behavior of the reporter in cell lines or animals is examined with or without mutations in the binding site or with ectopic expression of the miRNA. Although this approach may demonstrate the potential for direct interaction between a miRNA and target of interest, it is important to consider that these assays rarely recapitulate the expression levels of the endogenous RNAs. Thus, the magnitude of effects observed may be dramatically different than that which occurs in vivo. Ideally, reporter studies should be coupled with loss-of-function studies to determine whether inhibition of the miRNA leads to upregulation of the target mRNA or protein. Inhibition of miRNAs in cell lines may be accomplished by introducing 2′-*O*-methyl oligoribonucleotides complementary to the miRNA (63, 101). Recently, similar antisense oligos were developed for systemic delivery to whole animals (79). Although somewhat less reliable than loss of function, overexpression of a miRNA in cell lines or animals may also be informative for target validation. Combined approaches incorporating reporter assays and in vivo expression studies are recommended.

PHYSIOLOGIC FUNCTIONS OF VERTEBRATE microRNAs

The importance of miRNAs in select developmental pathways in model organisms is clear from the *lin-4* and *let-7* examples described above. But how generalizable are these observations? Do miRNAs play widespread, important roles in normal physiology in vertebrates? One way to address this question is to mutate or delete *Dicer*, thus removing all miRNAs. Although this experimental approach will likely reveal the consequences of a block in miRNA biogenesis, an important caveat that must always be considered when interpreting these data is that *Dicer* may be performing additional functions unrelated to miRNA biology. Thus, these results must be interpreted with caution.

Dicer loss of function results in profound developmental defects in both zebrafish and mice (13, 139). Zebrafish embryos lacking *Dicer* die approximately two weeks postfertilization due to what appears to be a general growth arrest rather than a failure in the development of one or more specific organ systems. The survival of *Dicer*$^{-/-}$ embryos to this stage of development likely reflects the activity of maternal *Dicer*. To interrogate the functions of *Dicer* at earlier stages of development, when the major steps of morphogenesis and organogenesis occur, Schier and colleagues used a germ-line replacement technique to produce wild-type zebrafish with a *Dicer*$^{-/-}$ germ line (51). Intercrosses of such fish produced offspring lacking both maternal and zygotic *Dicer*. These embryos exhibited severe defects most prominantly in gastrulation, brain morphogenesis, and cardiac development. Remarkably, restoring expression of a single miRNA (miR-430) by injecting a fully processed synthetic miRNA duplex into early *Dicer*$^{-/-}$ embryos was sufficient to largely rescue the defects in brain morphogenesis. These results definitely establish an important role for miRNAs in neural development in zebrafish and likely indicate that these molecules provide critical functions in other pathways during embryogenesis.

Reminiscent of the zebrafish *Dicer*-null phenotype, mouse embryos lacking this protein die at embryonic day 8.5 (13). One prominent defect in these embryos is a lack of pluripotent stem cells. To further explore the consequences of *Dicer* deletion, several laboratories have generated mice harboring conditional *Dicer* alleles (30, 55, 71, 105). Such mice have been used to demonstrate essential functions for *Dicer* in T-cell differentiation (30, 104) as well as for morphogenesis of the lung (56), limb (55), skin, and hair follicles (5, 143). Taken together with the data from zebrafish, it appears that the miRNA pathway is not generally required for cellular viability but plays a prominent role in the differentiation of various tissue-specific cell types and the morphogenesis of embryonic structures.

Although these studies emphasize the importance of the miRNA pathway in vertebrate development, most do not provide information regarding the functions of specific miRNAs. Studies aimed at elucidating the role of individual miRNAs in various biologic processes are increasingly being performed and a few examples that illustrate principles of miRNA function are presented here. The diverse functions of vertebrate miRNAs are reflected in their expression patterns. The Plasterk laboratory recently used in situ hybridization to study the expression of more than 100 zebrafish miRNAs during embryogenesis (138). Despite their small size, miRNAs can be detected with high sensitivity and specificity using modified oligonucleotides known as locked-nucleic-acid (LNA) probes. These studies revealed that most miRNAs are expressed with precise tissue specificity late in development, again pointing to an important role for these RNAs in differentiation or maintenance of tissue-specific cell types.

There are now several known vertebrate miRNAs that participate in such tissue-specific functions. For example, one of the first mammalian miRNAs to be carefully studied was miR-181. Expression of this miRNA is highly enriched in B lymphocytes and ectopic expression of miR-181 in hematopoietic progenitor cells skews lymphocyte differentiation toward the B-cell lineage (25). The critical targets underlying the functions of miR-181 in lymphocyte development have yet to be identified. miR-181 was also identified in a screen for miRNAs that are upregulated during muscle cell differentiation and regeneration (106). This miRNA promotes muscle differentiation in part by downregulating Hox-A11 (an inhibitor of differentiation). These studies emphasize the potential for a single miRNA to participate in distinct pathways in different tissues. Given that the function of a miRNA is dictated by the milieu of targets that are coexpressed, distinct tissue-specific roles for individual miRNAs may emerge as a common theme.

Another miRNA that is critical for normal muscle development is miR-1, one of the most highly conserved miRNAs studied to date. This miRNA is nearly identical in sequence in *C. elegans*, *Drosophila*, and vertebrates and shows skeletal muscle and heart-specific expression in these species (82, 88, 138). Two groups have independently studied the consequences of *Drosophila* miR-1 loss of function. Sokol & Ambros (127) observed the death of second instar *Drosophila* larvae lacking this miRNA, due predominantly to a failure in postmitotic muscle growth. Kwon et al. (80) described a somewhat more severe phenotype resulting from miR-1 deletion with a higher prevalence of embryonic lethality and a more pronounced muscle differentiation phenotype. miR-1 also plays an important role in muscle physiology in mammals. miR-1 is directly transcriptionally upregulated by the mammalian MyoD and myogenin transcription factors, which are important positive regulators of myogenic differentiation (117). Overexpression or inhibition of miR-1 promotes or inhibits mammalian in vitro muscle cell differentiation, respectively (26, 107). These effects appear to be due in part to miR-1-mediated downregulation of the inhibitor of myogenic differentiation HDAC4. Thus, miR-1 provides highly conserved functions essential for normal muscle development in diverse animal species. Other miRNAs with roles in tissue-specific differentiation pathways have been studied and are reviewed elsewhere (4, 9, 78).

These examples illustrate the ability of miRNAs to dramatically influence the generation and behavior of tissue-specific cell types. As described above, studies to date have identified a few key targets that seem to be important in mediating these functions. However, most miRNAs are predicted to regulate hundreds of targets each (68, 77, 92), so the target transcript networks that underlie the phenotypes under miRNA control are likely much more complex. A fascinating study by Lim and colleagues (94) sheds light on the potential roles for miRNAs in regulating large sets of transcripts. In this series of experiments, the human muscle-specific miRNA miR-1 and the brain-specific miRNA miR-124 were expressed ectopically in HeLa cells. This resulted in the downregulation of ~100–200 transcripts for each miRNA, as revealed by microarray-based expression profiling. These downregulated transcripts, many of which appear to be direct targets of these miRNAs, were greatly enriched for those that are expressed poorly in muscle (for miR-1) and brain (for miR-124). Thus, expression of these miRNAs skewed the expression profiles of HeLa cells toward that of the tissue in which the miRNA is normally expressed. These data point to an important role for miRNAs in establishing and/or maintaining gene expression patterns characteristic of specific tissues. This activity of miR-1 and miR-124 does not appear to be unique as examination of large-scale gene expression data sets across various mammalian tissues reveals that many miRNAs and their predicted targets are reciprocally expressed (47, 129).

Further insight into the mechanisms through which miRNAs establish tissue-specific expression patterns was provided by studies focused on the RE1 silencing transcription factor (REST). This protein inhibits expression of miR-124 in neuronal precursor cells and non-neuronal cells (31). Upon commitment of precursor cells to a neuronal differentiation pathway, repression of miR-124 by REST is relieved. These observations suggest a model whereby neuronal differentiation is coupled to the induction of miR-124, which globally tunes the transcriptome toward that of the terminally differentiated cell type.

Before concluding this section, it is important to emphasize that although a prominent role for miRNAs in tissue development and identity is clear, there are several known examples of vertebrate miRNAs that participate in cell-autonomous functions not related to development. For example, human miR-375 is expressed specifically in pancreatic islet β-cells and regulates insulin secretion (116). The liver-specific miRNA miR-122 regulates cholesterol homeostasis (45, 79). miRNAs have also been implicated in the regulation of neuronal synaptic function and neurite outgrowth (121, 134). Finally, specific miRNAs control cellular proliferation and apoptosis and accordingly play important roles in cancer. miRNAs with these latter properties are discussed in detail below. In sum, the relatively few miRNAs that have been studied in detail thus far suggest that these molecules participate broadly in normal physiologic processes. The logical extension of this conclusion is that abnormalities in miRNA function should influence human disease phenotypes. Accumulating evidence suggests that this is the case.

miRNAs IN CANCER

Although studies linking miRNA dysfunction to human disease are in their infancy, a great deal of data already exists establishing an important role for miRNAs in the pathogenesis of cancer. In retrospect, a role for miRNAs in this group of disorders may not have been surprising given the phenotype of the founding miRNA, *lin-4*. In the absence of this miRNA, the *C. elegans* seam-cell lineage fails to differentiate properly and instead these cells reiterate early larval divisions throughout development and into adulthood (89). This phenotype is reminiscent of many human malignancies, such as several derived from hematopoietic lineages, which result from the expansion of a population of precursor cells that have failed to execute a program of differentiation. Several miRNAs that regulate proliferation and apoptosis in *Drosophila* have also been described (15, 87, 141), suggesting widespread regulation of cancer-relevant pathways by miRNAs in model organisms.

Genome-wide miRNA expression profiling using specialized microarrays and other high-throughput technologies has now been broadly applied to the study of diverse cancer subtypes. It is clear from these surveys that abnormal patterns of miRNA expression are a typical, if not ubiquitous, feature of cancer cells (98, 135). Although these studies have illustrated the potential for miRNAs to act as novel diagnostic and prognostic markers for several types of cancer, a major question is whether these expression changes directly contribute to tumorigenesis. Indeed, given that miRNA expression is tightly regulated during development and cellular differentiation, abnormal miRNA expression in tumors may simply reflect the loss of normal cellular identity that accompanies malignant transformation. Nevertheless, several observations suggest that the dysfunction of select miRNAs may be a causative event in cancer pathogenesis. For example, miRNAs are greatly enriched at genomic loci known to undergo amplification, deletion, or rearrangement in human cancers (20). Additionally, a number of studies using human cancer cells and animal models have convincingly established that specific miRNAs possess tumor-suppressor or oncogenic activity. Select examples of likely tumor-promoting and tumor-suppressing miRNAs and the evidence supporting these activities are presented below. For further details and additional examples, we refer the reader to other reviews on this topic (46, 72).

miRNAs WITH TUMOR-SUPPRESSOR ACTIVITY

miR-15a and miR-16–1

The first miRNAs causatively linked to cancer development were miR-15a and miR-16–1, which are clustered together on human chromosome 13q14. This locus is deleted in over half of B-cell chronic lymphocytic leukemia (CLL) cases as well as in other malignancies and there has been significant effort to identify the putative tumor-suppressor gene located within this interval (40, 41). After refining the critical region of the deletion to an approximately 30-kb interval, Croce and colleagues (18) recognized that this region contained only miR-15a and miR-16–1 and no other identifiable genes. They subsequently demonstrated reduced expression of these miRNAs in over two thirds of CLL cases. Sequencing these miRNAs in 75 CLL patients later identified two individuals with the identical germ-line mutation near miR-16–1 that was not present in 160 control subjects (19). This mutation is associated with reduced expression of this miRNA, although the underlying mechanism of this effect is unclear. Insight into the mechanism through which miR-15a and miR-16–1 participate in tumorigenesis was provided by the demonstration that at least one target of these miRNAs is the antiapoptotic gene *BCL2* (28). This suggests that loss of function of miR-15a and miR-16–1 promotes high expression of Bcl2 and abnormal survival of CLL cells. Accordingly, overexpression of Bcl2 occurs frequently in CLL. Further studies are necessary to identify additional targets of these miRNAs that may be important in the pathogenesis of CLL and other cancers.

The let-7 Family

A variety of studies have now implicated human homologs of the *C. elegans let-7* miRNA as potential tumor suppressors. During mammalian evolution, the family of miRNAs related in sequence to *let-7* has dramatically expanded such that humans possess twelve *let-7* homologs organized in eight distinct clusters. At least four of these clusters are located in genomic regions known to be deleted in cancer (20). Reduced expression of human let-7 family members has also been observed in lung cancer and is associated with poor prognosis (131, 142). Consistent with these observations, expression of let-7 in lung adenocarcinoma cell lines reduces clonogenicity (131). Slack and colleagues (69) convincingly demonstrated that the highly studied oncogene *RAS* is a target of human let-7 family members. Low expression of let-7 was shown to correlate with high expression of Ras in a small set of lung tumors, providing a potential mechanism through which loss of function of these miRNAs promotes tumorigenesis. Of note, the high degree of nucleotide similarity between the members of the expanded human let-7 family has imposed unique challenges on studying these miRNAs. Microarrays, northern blotting, PCR approaches, and other techniques currently used to analyze miRNA expression do not readily discriminate between these related miRNAs. Consequently, the extent of redundancy between the let-7 clusters, their precise expression patterns, and the mechanisms that regulate their individual expression are not yet clear. Unraveling these issues will be critical to fully understand the role of let-7 in human cancers.

ONCOGENIC miRNAS

BIC/miR-155

Before the discovery of miRNAs, the B-cell integration cluster (BIC) noncoding RNA was shown to accelerate Myc-mediated lymphomagenesis in chicken (132, 133). It was later recognized that BIC is the pri-miRNA for miR-155 (42). Overexpression of BIC/miR-155 occurs frequently in diverse cancers including B-cell lymphoma and tumors of the breast, lung, colon, and thyroid (42, 58, 135). Strikingly, transgenic mice expressing miR-155 in B cells rapidly develop polyclonal B-cell malignancies, suggesting that the expression of this miRNA alone or in combination with minimal additional genetic "hits" is sufficient to induce lymphomagenesis (32). The available evidence therefore strongly implicates miR-155 as an oncogene.

The mir-17 Cluster

One of most extensively studied groups of miRNAs is the *mir-17* cluster, which includes miR-17-5p, miR-18a, miR-19a, miR-20a, miR-19b-1, and miR-92. The genomic locus encompassing these miRNAs (13q31-32) is frequently amplified in several types of lymphoma and solid tumors (110) and widespread overexpression of these miRNAs has been observed in diverse types of cancer (57, 59, 110, 135). Direct evidence for in vivo oncogenic activity of the *mir-17* cluster was obtained using *Eμ-myc* transgenic mice, a model of B-cell lymphoma (59). Introduction of the *mir-17* cluster into hematopoietic stem cells from *Eμ-myc* animals significantly accelerated lymphomagenesis and death of recipient animals. Moreover, it was recently demonstrated that expression of these miRNAs in a mouse model of colon cancer dramatically promotes tumor angiogenesis (38). Studies from O'Donnell et al. (108) provide further insight into the molecular mechanisms underlying overexpression of these oncogenic miRNAs in cancer cells. Using both model cell lines and primary cells, these authors demonstrated that the *mir-17* cluster is directly upregulated by the oncogenic transcription factor c-Myc. Together with the data obtained in animal models, these observations suggest that the *mir-17* cluster is an important downstream effector of the Myc oncogene, a protein that is pathologically activated in a large fraction of human malignancies (34).

miR-21

A potential role for miR-21 as an oncogene was first uncovered in a screen for altered miRNA expression in glioblastoma (24). These studies showed that miR-21 expression was highly elevated in human glioblastoma tumor tissues, primary tumor cultures, and established glioblastoma cell lines relative to normal fetal and adult brain tissue and primary cultured neurons and astrocytes. More recently, miR-21 was demonstrated to be widely overexpressed in an array of tumors including those derived from breast, colon, lung, pancreas, stomach, and prostate (135). Inhibition of miR-21 in glioblastoma or breast cancer cell lines using antisense oligonucleotides results in caspase activation, increased apoptotic cell death, and decreased tumor growth in a xenograft model (24, 126). These data provide strong evidence that overexpression of miR-21 promotes tumorigenesis by suppressing apoptosis.

THE ROLE OF miRNAs IN FRAGILE X SYNDROME AND SYNAPTIC FUNCTION

Although most attention has been focused on cancer, miRNA dysfunction may contribute to other human diseases. For example, recent findings have uncovered an unexpected link between Fragile X Syndrome and the miRNA pathway. Fragile X Syndrome represents one of the most common inherited forms of mental retardation and is caused by loss of function of the Fragile X Mental Retardation 1 (*FMR1*) gene. In the vast majority of cases, the causative mutation is a triplet repeat expansion in the *FMR1* 5′ UTR that leads to hypermethylation of the *FMR1* promoter and subsequent transcriptional silencing (50). Significant evidence points to an important role for the product of this gene, known as FMRP, as a negative regulator of local translation within dendrites (66). One critical aspect of FMRP function that has been extensively investigated is the mechanism through which this protein selects mRNA targets for translational repression. FMRP appears to recognize at least two RNA structures within its targets: a G-quartet, which is a special stem-loop structure formed by hydrogen bonding between four guanosine residues, and a more complex tertiary structure referred to as a FMRP-kissing complex (35, 36). New data suggest that miRNAs are intimately involved in FMRP-mediated translational repression, perhaps in the selection of target mRNAs.

The first indication of mechanistic linkage between Fragile X and miRNAs came from biochemical studies of the RISC protein complex, the molecular machinery that accomplishes miRNA- and siRNA-mediated regulation. Purification of this complex from *Drosophila* cells revealed that an abundant component is the fly ortholog of FMRP, dFMR1 (23, 65). These findings were extended to human cells where FMRP was also demonstrated to interact with RISC components and miRNAs (67). Compelling genetic evidence from fly also links the miRNA pathway to functions mediated by dFMR1. Overexpression of dFMR1 in the *Drosophila* eye leads to apoptosis, which can be largely suppressed by heterozygous loss of function of AGO1, a RISC component that is essential for miRNA function (67). Perhaps more relevant to the human phenotype, loss of dFMR1 leads to synaptic overgrowth at neuromuscular junctions, which is exacerbated by heterozygous loss of AGO1. The results suggest that functions mediated by dFMR1 or FMRP require RISC and possibly miRNAs. Warren and colleagues have therefore proposed a model whereby FMRP initially selects targets through low-affinity interactions with G-quartets and/or kissing complexes. FMRP-associated miRNAs may then provide additional specificity during translational silencing. These interactions may also indicate that FMRP and miRNAs share a common mechanism of translational repression. While this is an elegant hypothesis, a definitive link between human miRNAs, FMRP-mediated translational silencing, and the Fragile X phenotype remains to be established.

Although not explicitly related to a disease phenotype, new data from *Drosophila*

SNP: single nucleotide polymorphism

provide additional evidence that the miRNA pathway may be important in the control of local protein synthesis at synapses. To study these connections, Ashraf and colleagues (7) used a classic olfactory conditioning model whereby flies are trained to avoid certain odorants by exposing them to electric shock. Specific regions of the brain in flies trained in this manner show increased transport of *Calcium-calmodulin-dependent kinase II* (*CamKII*) mRNA to dendrites and increased local production of CamKII protein. Interestingly, this localization and regulated translation of *CamKII* is mediated by its 3′ UTR, which harbors predicted binding sites for multiple miRNAs. These sites are likely functional because flies with mutations in multiple components of RISC show dramatically increased CamKII protein expression. Loss of RISC activity also leads to increased transport of *CamKII* message to synapses. The authors further demonstrate that synaptic activity leads to proteosome-mediated destruction of RISC. These results support a compelling model in which miRNA-programmed RISC maintains *CamKII* and other localized messages in a translationally silenced state. Synaptic activity provides a signal to destroy RISC and these mRNAs are subsequently used as templates for local protein synthesis. A similar mechanism has been demonstrated to be operating in mammalian neurons, at least for some select miRNAs and their targets. For example, miR-134 has been demonstrated to maintain the *Lim-domain-containing protein kinase 1* (*LimK1*) mRNA, which encodes a positive regulator of dendritic growth, in a translationally silenced state prior to synaptic activity (121). Reminiscent of *Drosophila CamKII*, incoming synaptic stimuli somehow inactivate miR-134, releasing *LimKI* for local protein synthesis and leading to dendritic growth. An extremely abundant and diverse population of miRNAs are expressed in the mammalian brain (21), suggesting that these mechanisms are likely widespread.

THE INFLUENCE OF GENETIC VARIATION ON miRNA FUNCTION

The relationship between genetic variation in miRNAs or miRNA-binding sites and human phenotypes is only just beginning to be investigated. In theory, a single nucleotide polymorphism (SNP) in a miRNA could influence its function by affecting the efficiency of processing or by altering target specificity. Alternatively, SNPs in mRNA 3′ UTRs could strengthen or weaken miRNA:target interactions. These potential mechanisms are summarized in **Figure 4**. Although genetic variation in miRNAs appears to be very rare, SNPs in potential target sites have now been shown to influence human and animal phenotypes. For example, the neuropsychiatric disorder Tourette's syndrome (TS) was one of the first human diseases linked to alterations in a miRNA-binding site (3). Abelson and colleagues focused on *Slit and Trk-like 1 (SLITRK1)* as a candidate gene for this disease because it is located near a chromosomal breakpoint found in a TS patient harboring an inversion on chromosome 13q31. Subsequent sequencing of the *SLITRK1* gene in 174 unrelated TS patients identified a frameshift mutation and two independent occurrences of the identical mutation in the 3′ UTR. These variants were absent in thousands of screened control chromosomes. Interestingly, the 3′UTR variants strengthen a predicted binding site for miR-189, a miRNA that is expressed in a pattern that overlaps with *SLITRK1*. Luciferase reporter assays provided evidence that this is indeed a functional miR-189 site and, although the magnitude of effects were quite small, the rare variant seems to strengthen this miRNA:target interaction. Although further investigation of this hypothesis is necessary, these data are consistent with the TS phenotype being influenced by inappropriate downregulation of *SLITRK1* by miR-189.

A similar mechanism was revealed by studies aimed at mapping the quantitative trait loci

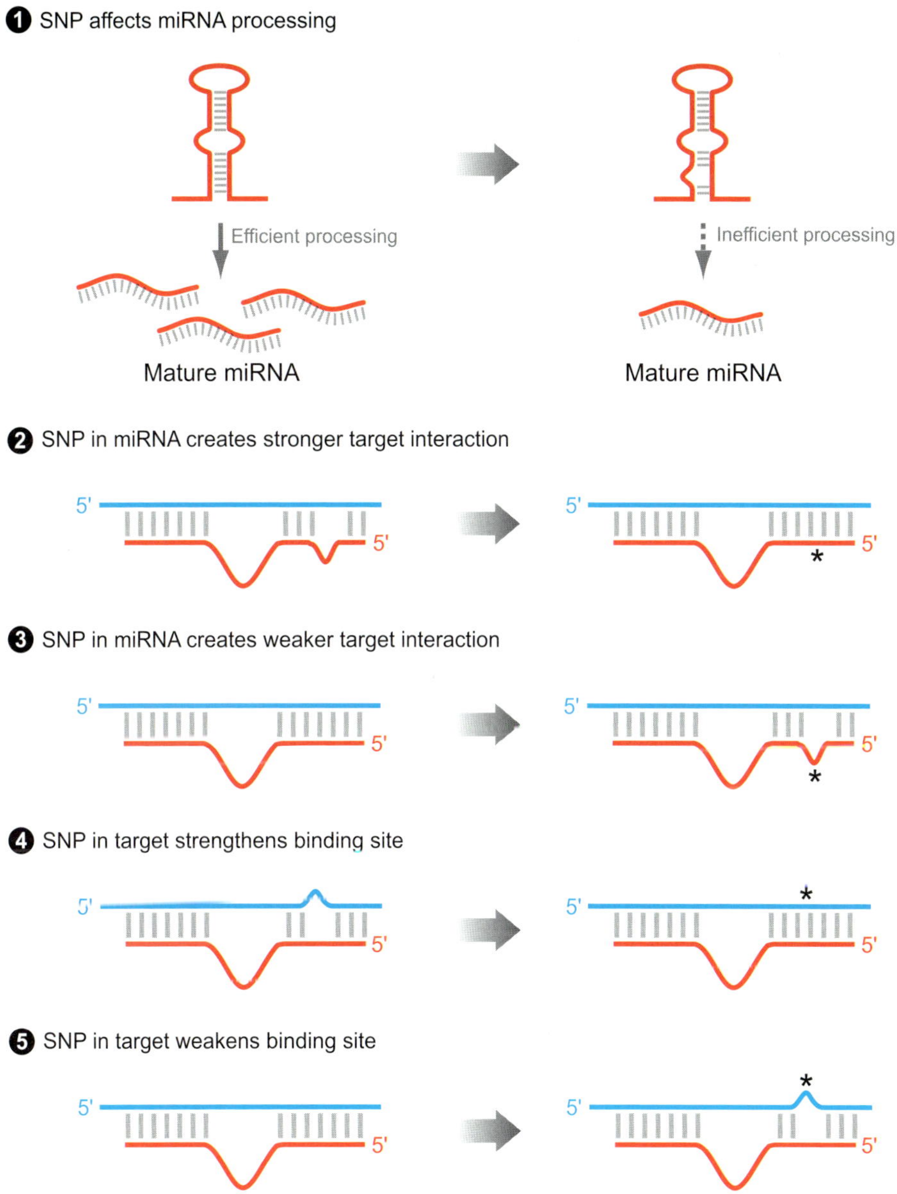

Figure 4

Potential effects of single nucleotide polymorphisms (SNPs) on microRNA (miRNA) biogenesis and function. Target sequences are shown in blue and miRNAs are shown in red. Asterisks indicate the position of sequence variants.

(QTLs) that regulate muscle mass in sheep (29). One locus that has a pronounced effect on musculature contains the well-known negative regulator of muscle growth myostatin (also known as *growth and differentiation factor 8*) (100). A naturally occurring variant in the 3′ UTR of the myostatin transcript is highly enriched in a muscular breed of sheep and appears to result in lower myostatin protein levels. This variant creates a potential binding site for two muscle-specific miRNAs, miR-1 and miR-206. Luciferase assays were again used to demonstrate that these miRNAs can regulate reporter transcripts harboring this newly created site. Based on these data, the authors formulated a model very similar to that proposed for *SLITRK1*, whereby fortuitous creation of a miRNA-binding site leads to downregulation of the myostatin protein and greater muscle mass.

These examples have led to speculation that polymorphic sequence variants in human populations that create or destroy miRNA-binding sites may have significant effects on phenotypic variation. Consistent with this possibility, ~2500 human SNPs that create potential miRNA-binding sites and ~2500 human SNPs that destroy potential miRNA-binding sites have been identified (29). These SNPs were deposited in a publicly available database that can be accessed at **http://www.patrocles.org**. Using SNP genotyping data provided by the International HapMap Project (1), Chen & Rajewsky (27) provided strong evidence for natural selection acting on a significant fraction of both conserved and nonconserved predicted miRNA-binding sites. Negative selection operating on these sites during human evolution has resulted in a relative decrease in SNP density. Nevertheless, these findings underscore the potential for variants in miRNA-binding sites to influence human phenotypes in a manner that clearly affects Darwinian fitness and likely human disease. The prevalence of miRNA-binding sites in human transcripts requires that these motifs be considered potential reservoirs of causative genetic variation when performing association studies or mutation screening.

CONCLUSIONS AND FUTURE DIRECTIONS

Over the past six years, the identification of miRNAs and our understanding of their biogenesis pathway and mechanisms of action have advanced at a remarkable pace. However, the phase of rapid miRNA discovery is likely reaching its pinnacle. Progress in elucidating the physiologic roles for these newly defined RNAs has advanced much more slowly. Undoubtedly, closing this knowledge gap between miRNA discovery and miRNA functional annotation represents a major priority and challenge for future research. These efforts must include improved methods for high-throughput miRNA target validation and loss-of-function studies in knockout mice. In parallel, continued exploration of the roles of miRNAs in human disease states and the influence of genetic variation in miRNAs on human phenotypes is necessary. This is particularly important given the recent development of technologies that allow systemic delivery of miRNAs or miRNA inhibitors to whole animals (79, 130). Application of these methods, in the context of improved understanding of physiologic and pathophysiologic roles for miRNAs, may enable the development of an entirely new class of miRNA-based therapeutics.

SUMMARY POINTS

1. miRNAs were originally discovered through genetic screens in *C. elegans* in the early 1990s. Since then, both experimental and bioinformatic approaches have led to the identification of thousands of miRNAs in diverse multicellular eukaryotes. The human genome probably encodes as many as 1000 miRNAs, if not more.
2. miRNAs are transcriptionally regulated in a manner identical to typical messenger RNAs. Mature miRNAs, which are 18–24-nucleotide single-stranded RNAs, are processed from long primary transcripts in a stepwise process involving a series of endonucleolytic cleavages. The mature miRNAs are loaded in a large protein complex known as the RNA-induced silencing complex (RISC). miRNAs guide RISC to complementary target messenger RNAs, which are cleaved or translationally repressed.

3. Animal miRNAs are usually imperfectly complementary to their targets, which has made bioinformatic identification of targets challenging. Many publicly available target prediction algorithms are available and should be used with caution.
4. miRNAs are involved in diverse physiologic processes. For example, they provide essential functions during animal development, regulate cellular differentiation, and help establish tissue-specific gene expression programs.
5. Aberrant miRNA function influences human disease. In particular, miRNAs are now known to participate in human malignancies by acting as novel oncogenes and tumor suppressors. miRNAs may also be involved in neuronal phenotypes, including Fragile X Syndrome.
6. Genetic variation in miRNA-binding sites in target mRNAs has been linked to dramatic phenotypes in humans and animals. A large number of SNPs that create or destroy potential miRNA-binding sites have been identified.

FUTURE ISSUES

1. Further characterization of primary miRNA transcripts is necessary to improve our understanding of the transcriptional and post-transcriptional mechanisms that govern miRNA expression.
2. More emphasis on loss-of-function studies of miRNAs in animal models is necessary to better establish the physiologic functions of these transcripts. These studies may be hindered by the redundancy that typifies miRNA organization in mammalian genomes.
3. High-throughput methods for miRNA target validation would not only expand our understanding of miRNA functions but also would allow refinement of target prediction algorithms to improve their sensitivity and specificity.
4. Efforts to investigate noncanonical roles for miRNAs in pathways distinct from post-transcriptional regulation of gene expression should be increased.

DISCLOSURE STATEMENT

The authors are not aware of any biases that might be perceived as affecting the objectivity of this review.

ACKNOWLEDGMENTS

The authors wish to thank Kathryn O'Donnell, Hun-Way Hwang, Judit Ribas-Fortuny, and Oliver Kent for critical reading of the manuscript. J.T.M. is a Rita Allen Foundation Scholar. The Mendell laboratory also receives research support from the Lustgarten Foundation for Pancreatic Cancer Research and the National Institutes of Health (R01CA120185). We apologize to those whose work we were unable to cite due to space limitations.

LITERATURE CITED

1. 2003. The International HapMap Project. *Nature* 426:789–96
2. Abbott AL, Alvarez-Saavedra E, Miska EA, Lau NC, Bartel DP, et al. 2005. The let-7 microRNA family members mir-48, mir-84, and mir-241 function together to regulate developmental timing in *Caenorhabditis elegans*. *Dev. Cell* 9:403–14
3. Abelson JF, Kwan KY, O'Roak BJ, Baek DY, Stillman AA, et al. 2005. Sequence variants in SLITRK1 are associated with Tourette's syndrome. *Science* 310:317–20
4. Ambros V. 2004. The functions of animal microRNAs. *Nature* 431:350–55
5. Andl T, Murchison EP, Liu F, Zhang Y, Yunta-Gonzalez M, et al. 2006. The miRNA-processing enzyme dicer is essential for the morphogenesis and maintenance of hair follicles. *Curr. Biol.* 16:1041–49
6. Aravin AA, Lagos-Quintana M, Yalcin A, Zavolan M, Marks D, et al. 2003. The small RNA profile during *Drosophila melanogaster* development. *Dev. Cell* 5:337–50
7. Ashraf SI, McLoon AL, Sclarsic SM, Kunes S. 2006. Synaptic protein synthesis associated with memory is regulated by the RISC pathway in *Drosophila*. *Cell* 124:191–205
8. Bagga S, Bracht J, Hunter S, Massirer K, Holtz J, et al. 2005. Regulation by let-7 and lin-4 miRNAs results in target mRNA degradation. *Cell* 122:553–63
9. Bartel DP. 2004. MicroRNAs: genomics, biogenesis, mechanism, and function. *Cell* 116:281–97
10. Baskerville S, Bartel DP. 2005. Microarray profiling of microRNAs reveals frequent coexpression with neighboring miRNAs and host genes. *RNA* 11:241–47
11. Bentwich I, Avniel A, Karov Y, Aharonov R, Gilad S, et al. 2005. Identification of hundreds of conserved and nonconserved human microRNAs. *Nat. Genet.* 37:766–70
12. Berezikov E, Guryev V, van de Belt J, Wienholds E, Plasterk RH, Cuppen E. 2005. Phylogenetic shadowing and computational identification of human microRNA genes. *Cell* 120:21–24
13. Bernstein E, Kim SY, Carmell MA, Murchison EP, Alcorn H, et al. 2003. Dicer is essential for mouse development. *Nat. Genet.* 35:215–17
14. Bohnsack MT, Czaplinski K, Gorlich D. 2004. Exportin 5 is a RanGTP-dependent dsRNA-binding protein that mediates nuclear export of premiRNAs. *RNA* 10:185–91
15. Brennecke J, Hipfner DR, Stark A, Russell RB, Cohen SM. 2003. bantam encodes a developmentally regulated microRNA that controls cell proliferation and regulates the proapoptotic gene hid in *Drosophila*. *Cell* 113:25–36
16. Brennecke J, Stark A, Russell RB, Cohen SM. 2005. Principles of microRNA-target recognition. *PLoS Biol.* 3:e85
17. Cai X, Hagedorn CH, Cullen BR. 2004. Human microRNAs are processed from capped, polyadenylated transcripts that can also function as mRNAs. *RNA* 10:1957–66
18. **Calin GA, Dumitru CD, Shimizu M, Bichi R, Zupo S, et al. 2002. Frequent deletions and down-regulation of micro-RNA genes miR15 and miR16 at 13q14 in chronic lymphocytic leukemia. *Proc. Natl. Acad. Sci. USA* 99:15524–29**

18. Demonstrates for the first time that deletion of select miRNAs in cancer cells likely promotes tumorigenesis.

19. Calin GA, Ferracin M, Cimmino A, Di Leva G, Shimizu M, et al. 2005. A microRNA signature associated with prognosis and progression in chronic lymphocytic leukemia. *N. Engl. J. Med.* 353:1793–801
20. Calin GA, Sevignani C, Dumitru CD, Hyslop T, Noch E, et al. 2004. Human microRNA genes are frequently located at fragile sites and genomic regions involved in cancers. *Proc. Natl. Acad. Sci. USA* 101:2999–3004
21. Cao X, Yeo G, Muotri AR, Kuwabara T, Gage FH. 2006. Noncoding RNAs in the mammalian central nervous system. *Annu. Rev. Neurosci.* 29:77–103

22. Carmell MA, Xuan Z, Zhang MQ, Hannon GJ. 2002. The Argonaute family: tentacles that reach into RNAi, developmental control, stem cell maintenance, and tumorigenesis. *Genes Dev.* 16:2733–42
23. Caudy AA, Myers M, Hannon GJ, Hammond SM. 2002. Fragile X-related protein and VIG associate with the RNA interference machinery. *Genes Dev.* 16:2491–96
24. Chan JA, Krichevsky AM, Kosik KS. 2005. MicroRNA-21 is an antiapoptotic factor in human glioblastoma cells. *Cancer Res.* 65:6029–33
25. Chen CZ, Li L, Lodish HF, Bartel DP. 2004. MicroRNAs modulate hematopoietic lineage differentiation. *Science* 303:83–86
26. Chen JF, Mandel EM, Thomson JM, Wu Q, Callis TE, et al. 2006. The role of microRNA-1 and microRNA-133 in skeletal muscle proliferation and differentiation. *Nat. Genet.* 38:228–33
27. Chen K, Rajewsky N. 2006. Natural selection on human microRNA binding sites inferred from SNP data. *Nat. Genet.* 38:1452–6
28. Cimmino A, Calin GA, Fabbri M, Iorio MV, Ferracin M, et al. 2005. miR-15 and miR-16 induce apoptosis by targeting BCL2. *Proc. Natl. Acad. Sci. USA* 102:13944–49
29. Clop A, Marcq F, Takeda H, Pirottin D, Tordoir X, et al. 2006. A mutation creating a potential illegitimate microRNA target site in the myostatin gene affects muscularity in sheep. *Nat. Genet.* 38:813–18
30. Cobb BS, Nesterova TB, Thompson E, Hertweck A, O'Connor E, et al. 2005. T cell lineage choice and differentiation in the absence of the RNase III enzyme Dicer. *J. Exp. Med.* 201:1367–73
31. Conaco C, Otto S, Han JJ, Mandel G. 2006. Reciprocal actions of REST and a microRNA promote neuronal identity. *Proc. Natl. Acad. Sci. USA* 103:2422–27
32. Costinean S, Zanesi N, Pekarsky Y, Tili E, Volinia S, et al. 2006. Pre-B cell proliferation and lymphoblastic leukemia/high-grade lymphoma in E(mu)-miR155 transgenic mice. *Proc. Natl. Acad. Sci. USA* 103:7024–29
33. Cummins JM, He Y, Leary RJ, Pagliarini R, Diaz LAJ, et al. 2006. The colorectal microRNAome. *Proc. Natl. Acad. Sci. USA* 103:3687–92
34. Dang CV. 1999. c-Myc target genes involved in cell growth, apoptosis, and metabolism. *Mol. Cell. Biol.* 19:1–11
35. Darnell JC, Fraser CE, Mostovetsky O, Stefani G, Jones TA, et al. 2005. Kissing complex RNAs mediate interaction between the Fragile-X mental retardation protein KH2 domain and brain polyribosomes. *Genes Dev.* 19:903–18
36. Darnell JC, Jensen KB, Jin P, Brown V, Warren ST, Darnell RB. 2001. Fragile X mental retardation protein targets G quartet mRNAs important for neuronal function. *Cell* 107:489–99
37. Denli AM, Tops BB, Plasterk RH, Ketting RF, Hannon GJ. 2004. Processing of primary microRNAs by the microprocessor complex. *Nature* 432:231–35
38. Dews M, Homayouni A, Yu D, Murphy D, Sevignani C, et al. 2006. Augmentation of tumor angiogenesis by a Myc-activated microRNA cluster. *Nat. Genet.* 38:1060–65
39. Doench JG, Petersen CP, Sharp PA. 2003. siRNAs can function as miRNAs. *Genes Dev.* 17:438–42
40. Dohner H, Stilgenbauer S, Benner A, Leupolt E, Krober A, et al. 2000. Genomic aberrations and survival in chronic lymphocytic leukemia. *N. Engl. J. Med.* 343:1910–16
41. Dong JT, Boyd JC, Frierson HFJ. 2001. Loss of heterozygosity at 13q14 and 13q21 in high grade, high stage prostate cancer. *Prostate* 49:166–71
42. Eis PS, Tam W, Sun L, Chadburn A, Li Z, et al. 2005. Accumulation of miR-155 and BIC RNA in human B cell lymphomas. *Proc. Natl. Acad. Sci. USA* 102:3627–32

43. Elbashir SM, Harborth J, Lendeckel W, Yalcin A, Weber K, Tuschl T. 2001. Duplexes of 21-nucleotide RNAs mediate RNA interference in cultured mammalian cells. *Nature* 411:494–98
44. Elbashir SM, Lendeckel W, Tuschl T. 2001. RNA interference is mediated by 21- and 22-nucleotide RNAs. *Genes Dev.* 15:188–200
45. Esau C, Davis S, Murray SF, Yu XX, Pandey SK, et al. 2006. miR-122 regulation of lipid metabolism revealed by in vivo antisense targeting. *Cell Metab.* 3:87–98
46. Esquela-Kerscher A, Slack FJ. 2006. Oncomirs—microRNAs with a role in cancer. *Nat. Rev. Cancer* 6:259–69
47. Farh KK, Grimson A, Jan C, Lewis BP, Johnston WK, et al. 2005. The widespread impact of mammalian microRNAs on mRNA repression and evolution. *Science* 310:1817–21
48. Ferguson EL, Sternberg PW, Horvitz HR. 1987. A genetic pathway for the specification of the vulval cell lineages of *Caenorhabditis elegans*. *Nature* 326:259–67
49. Fire A, Xu S, Montgomery MK, Kostas SA, Driver SE, Mello CC. 1998. Potent and specific genetic interference by double-stranded RNA in *Caenorhabditis elegans*. *Nature* 391:806–11
50. Garber K, Smith KT, Reines D, Warren ST. 2006. Transcription, translation and fragile X syndrome. *Curr. Opin. Genet. Dev.* 16:270–75
51. Giraldez AJ, Cinalli RM, Glasner ME, Enright AJ, Thomson JM, et al. 2005. MicroRNAs regulate brain morphogenesis in zebrafish. *Science* 308:833–38
52. Gregory RI, Yan KP, Amuthan G, Chendrimada T, Doratotaj B, et al. 2004. The microprocessor complex mediates the genesis of microRNAs. *Nature* 432:235–40
53. Griffiths-Jones S. 2006. miRBase: the microRNA sequence database. *Methods Mol. Biol.* 342:129–38
54. Grishok A, Pasquinelli AE, Conte D, Li N, Parrish S, et al. 2001. Genes and mechanisms related to RNA interference regulate expression of the small temporal RNAs that control *C. elegans* developmental timing. *Cell* 106:23–34
55. Harfe BD, McManus MT, Mansfield JH, Hornstein E, Tabin CJ. 2005. The RNaseIII enzyme Dicer is required for morphogenesis but not patterning of the vertebrate limb. *Proc. Natl. Acad. Sci. USA* 102:10898–903
56. Harris KS, Zhang Z, McManus MT, Harfe BD, Sun X. 2006. Dicer function is essential for lung epithelium morphogenesis. *Proc. Natl. Acad. Sci. USA* 103:2208–13
57. Hayashita Y, Osada H, Tatematsu Y, Yamada H, Yanagisawa K, et al. 2005. A polycistronic microRNA cluster, miR-17–92, is overexpressed in human lung cancers and enhances cell proliferation. *Cancer Res.* 65:9628–32
58. He H, Jazdzewski K, Li W, Liyanarachchi S, Nagy R, et al. 2005. The role of microRNA genes in papillary thyroid carcinoma. *Proc. Natl. Acad. Sci. USA* 102:19075–80
59. **He L, Thomson JM, Hemann MT, Hernando-Monge E, Mu D, et al. 2005. A microRNA polycistron as a potential human oncogene. *Nature* 435:828–33**

59. Provides one of the first demonstrations that a miRNA cluster can act as an oncogene in a mammalian animal model.

60. Horvitz HR, Sulston JE. 1980. Isolation and genetic characterization of cell-lineage mutants of the nematode *Caenorhabditis elegans*. *Genetics* 96:435–54
61. Houbaviy HB, Murray MF, Sharp PA. 2003. Embryonic stem cell-specific microRNAs. *Dev. Cell* 5:351–58
62. Hutvagner G, McLachlan J, Pasquinelli AE, Balint E, Tuschl T, Zamore PD. 2001. A cellular function for the RNA-interference enzyme Dicer in the maturation of the let-7 small temporal RNA. *Science* 293:834–38
63. Hutvagner G, Simard MJ, Mello CC, Zamore PD. 2004. Sequence-specific inhibition of small RNA function. *PLoS Biol.* 2:E98

64. Hutvagner G, Zamore PD. 2002. A microRNA in a multiple-turnover RNAi enzyme complex. *Science* 297:2056–60
65. Ishizuka A, Siomi MC, Siomi H. 2002. A *Drosophila* fragile X protein interacts with components of RNAi and ribosomal proteins. *Genes Dev.* 16:2497–508
66. Jin P, Alisch RS, Warren ST. 2004. RNA and microRNAs in fragile X mental retardation. *Nat. Cell Biol.* 6:1048–53
67. Jin P, Zarnescu DC, Ceman S, Nakamoto M, Mowrey J, et al. 2004. Biochemical and genetic interaction between the fragile X mental retardation protein and the microRNA pathway. *Nat. Neurosci.* 7:113–17
68. John B, Enright AJ, Aravin A, Tuschl T, Sander C, Marks DS. 2004. Human microRNA targets. *PLoS Biol.* 2:e363
69. Johnson SM, Grosshans H, Shingara J, Byrom M, Jarvis R, et al. 2005. RAS is regulated by the let-7 microRNA family. *Cell* 120:635–47
70. Jones-Rhoades MW, Bartel DP. 2004. Computational identification of plant microRNAs and their targets, including a stress-induced miRNA. *Mol. Cell* 14:787–99
71. Kanellopoulou C, Muljo SA, Kung AL, Ganesan S, Drapkin R, et al. 2005. Dicer-deficient mouse embryonic stem cells are defective in differentiation and centromeric silencing. *Genes Dev.* 19:489–501
72. Kent OA, Mendell JT. 2006. A small piece in the cancer puzzle: microRNAs as tumor suppressors and oncogenes. *Oncogene* 25:6188–96
73. Ketting RF, Fischer SE, Bernstein E, Sijen T, Hannon GJ, Plasterk RH. 2001. Dicer functions in RNA interference and in synthesis of small RNA involved in developmental timing in *C. elegans*. *Genes Dev.* 15:2654–59
74. Khvorova A, Reynolds A, Jayasena SD. 2003. Functional siRNAs and miRNAs exhibit strand bias. *Cell* 115:209–16
75. Kim J, Krichevsky A, Grad Y, Hayes GD, Kosik KS, et al. 2004. Identification of many microRNAs that copurify with polyribosomes in mammalian neurons. *Proc. Natl. Acad. Sci. USA* 101:360–65
76. Kim VN. 2005. MicroRNA biogenesis: coordinated cropping and dicing. *Nat. Rev. Mol. Cell. Biol.* 6:376–85
77. Kiriakidou M, Nelson PT, Kouranov A, Fitziev P, Bouyioukos C, et al. 2004. A combined computational-experimental approach predicts human microRNA targets. *Genes Dev.* 18:1165–78
78. Kloosterman WP, Plasterk RH. 2006. The diverse functions of microRNAs in animal development and disease. *Dev. Cell* 11:441–50
79. Krutzfeldt J, Rajewsky N, Braich R, Rajeev KG, Tuschl T, et al. 2005. Silencing of microRNAs in vivo with 'antagomirs'. *Nature* 438:685–89

79. Describes a new method to inhibit miRNA function in whole animals by systemic delivery of modified antisense oligonucleotides.

80. Kwon C, Han Z, Olson EN, Srivastava D. 2005. MicroRNA1 influences cardiac differentiation in *Drosophila* and regulates Notch signaling. *Proc. Natl. Acad. Sci. USA* 102:18986–91
81. Lagos-Quintana M, Rauhut R, Lendeckel W, Tuschl T. 2001. Identification of novel genes coding for small expressed RNAs. *Science* 294:853–58
82. Lagos-Quintana M, Rauhut R, Yalcin A, Meyer J, Lendeckel W, Tuschl T. 2002. Identification of tissue-specific microRNAs from mouse. *Curr. Biol.* 12:735–39
83. Lai EC. 2004. Predicting and validating microRNA targets. *Genome Biol.* 5:115
84. Lai EC, Tomancak P, Williams RW, Rubin GM. 2003. Computational identification of *Drosophila* microRNA genes. *Genome Biol.* 4:R42

85. Landthaler M, Yalcin A, Tuschl T. 2004. The human DiGeorge syndrome critical region gene 8 and its *D. melanogaster* homolog are required for miRNA biogenesis. *Curr. Biol.* 14:2162–67
86. Lau NC, Lim LP, Weinstein EG, Bartel DP. 2001. An abundant class of tiny RNAs with probable regulatory roles in *Caenorhabditis elegans*. *Science* 294:858–62
87. Leaman D, Chen PY, Fak J, Yalcin A, Pearce M, et al. 2005. Antisense-mediated depletion reveals essential and specific functions of microRNAs in *Drosophila* development. *Cell* 121:1097–08
88. Lee RC, Ambros V. 2001. An extensive class of small RNAs in *Caenorhabditis elegans*. *Science* 294:862–64
89. **Lee RC, Feinbaum RL, Ambros V. 1993. The *C. elegans* heterochronic gene lin-4 encodes small RNAs with antisense complementarity to lin-14. *Cell* 75:843–54**

89. Describes the identification of the first miRNA that was deleted or mutated in *C. elegans*, exhibiting a striking developmental phenotype.

90. Lee Y, Ahn C, Han J, Choi H, Kim J, et al. 2003. The nuclear RNase III Drosha initiates microRNA processing. *Nature* 425:415–19
91. Lee Y, Kim M, Han J, Yeom KH, Lee S, et al. 2004. MicroRNA genes are transcribed by RNA polymerase II. *EMBO J.* 23:4051–60
92. Lewis BP, Burge CB, Bartel DP. 2005. Conserved seed pairing, often flanked by adenosines, indicates that thousands of human genes are microRNA targets. *Cell* 120:15–20
93. Lewis BP, Shih IH, Jones-Rhoades MW, Bartel DP, Burge CB. 2003. Prediction of mammalian microRNA targets. *Cell* 115:787–98
94. Lim LP, Lau NC, Garrett-Engele P, Grimson A, Schelter JM, et al. 2005. Microarray analysis shows that some microRNAs downregulate large numbers of target mRNAs. *Nature* 433:769–73
95. Lim LP, Lau NC, Weinstein EG, Abdelhakim A, Yekta S, et al. 2003. The microRNAs of *Caenorhabditis elegans*. *Genes Dev.* 17:991–1008
96. Liu J, Carmell MA, Rivas FV, Marsden CG, Thomson JM, et al. 2004. Argonaute2 is the catalytic engine of mammalian RNAi. *Science* 305:1437–41
97. Llave C, Xie Z, Kasschau KD, Carrington JC. 2002. Cleavage of Scarecrow-like mRNA targets directed by a class of *Arabidopsis* miRNA. *Science* 297:2053–56
98. **Lu J, Getz G, Miska EA, Alvarez-Saavedra E, Lamb J, et al. 2005. MicroRNA expression profiles classify human cancers. *Nature* 435:834–38**

98. Demonstrates that the expression patterns of miRNAs allow accurate classification of diverse tumor types.

99. Lund E, Guttinger S, Calado A, Dahlberg JE, Kutay U. 2004. Nuclear export of microRNA precursors. *Science* 303:95–98
100. McPherron AC, Lawler AM, Lee SJ. 1997. Regulation of skeletal muscle mass in mice by a new TGF-beta superfamily member. *Nature* 387:83–90
101. Meister G, Landthaler M, Dorsett Y, Tuschl T. 2004. Sequence-specific inhibition of microRNA- and siRNA-induced RNA silencing. *RNA* 10:544–50
102. Meister G, Landthaler M, Patkaniowska A, Dorsett Y, Teng G, Tuschl T. 2004. Human Argonaute2 mediates RNA cleavage targeted by miRNAs and siRNAs. *Mol. Cell* 15:185–97
103. Mourelatos Z, Dostie J, Paushkin S, Sharma A, Charroux B, et al. 2002. miRNPs: a novel class of ribonucleoproteins containing numerous microRNAs. *Genes Dev.* 16:720–28
104. Muljo SA, Ansel KM, Kanellopoulou C, Livingston DM, Rao A, Rajewsky K. 2005. Aberrant T cell differentiation in the absence of Dicer. *J. Exp. Med.* 202:261–69
105. Murchison EP, Partridge JF, Tam OH, Cheloufi S, Hannon GJ. 2005. Characterization of Dicer-deficient murine embryonic stem cells. *Proc. Natl. Acad. Sci. USA* 102:12135–40

106. Naguibneva I, Ameyar-Zazoua M, Polesskaya A, Ait-Si-Ali S, Groisman R, et al. 2006. The microRNA miR-181 targets the homeobox protein Hox-A11 during mammalian myoblast differentiation. *Nat. Cell Biol.* 8:278–84
107. Nakajima N, Takahashi T, Kitamura R, Isodono K, Asada S, et al. 2006. MicroRNA-1 facilitates skeletal myogenic differentiation without affecting osteoblastic and adipogenic differentiation. *Biochem. Biophys. Res. Commun.* 350:1006–12
108. O'Donnell KA, Wentzel EA, Zeller KI, Dang CV, Mendell JT. 2005. c-Myc-regulated microRNAs modulate E2F1 expression. *Nature* 435:839–43
109. Olsen PH, Ambros V. 1999. The lin-4 regulatory RNA controls developmental timing in *Caenorhabditis elegans* by blocking LIN-14 protein synthesis after the initiation of translation. *Dev. Biol.* 216:671–80
110. Ota A, Tagawa H, Karnan S, Tsuzuki S, Karpas A, et al. 2004. Identification and characterization of a novel gene, *C13orf25*, as a target for 13q31-q32 amplification in malignant lymphoma. *Cancer Res.* 64:3087–95
111. Parker JS, Roe SM, Barford D. 2004. Crystal structure of a PIWI protein suggests mechanisms for siRNA recognition and slicer activity. *EMBO J.* 23:4727–37
112. Pasquinelli AE, Reinhart BJ, Slack F, Martindale MQ, Kuroda MI, et al. 2000. Conservation of the sequence and temporal expression of let-7 heterochronic regulatory RNA. *Nature* 408:86–89
113. Petersen CP, Bordeleau ME, Pelletier J, Sharp PA. 2006. Short RNAs repress translation after initiation in mammalian cells. *Mol. Cell* 21:533–42
114. Pillai RS, Artus CG, Filipowicz W. 2004. Tethering of human Ago proteins to mRNA mimics the miRNA-mediated repression of protein synthesis. *RNA* 10:1518–25
115. Pillai RS, Bhattacharyya SN, Artus CG, Zoller T, Cougot N, et al. 2005. Inhibition of translational initiation by Let-7 MicroRNA in human cells. *Science* 309:1573–76
116. Poy MN, Eliasson L, Krutzfeldt J, Kuwajima S, Ma X, et al. 2004. A pancreatic islet-specific microRNA regulates insulin secretion. *Nature* 432:226–30
117. Rao PK, Kumar RM, Farkhondeh M, Baskerville S, Lodish HF. 2006. Myogenic factors that regulate expression of muscle-specific microRNAs. *Proc. Natl. Acad. Sci. USA* 103:8721–26
118. Reinhart BJ, Slack FJ, Basson M, Pasquinelli AE, Bettinger JC, et al. 2000. The 21-nucleotide let-7 RNA regulates developmental timing in *Caenorhabditis elegans*. *Nature* 403:901–6
119. Rhoades MW, Reinhart BJ, Lim LP, Burge CB, Bartel B, Bartel DP. 2002. Prediction of plant microRNA targets. *Cell* 110:513–20
120. Rodriguez A, Griffiths-Jones S, Ashurst JL, Bradley A. 2004. Identification of mammalian microRNA host genes and transcription units. *Genome Res.* 14:1902–10
121. Schratt GM, Tuebing F, Nigh EA, Kane CG, Sabatini ME, et al. 2006. A brain-specific microRNA regulates dendritic spine development. *Nature* 439:283–89
122. Schwarz DS, Hutvagner G, Du T, Xu Z, Aronin N, Zamore PD. 2003. Asymmetry in the assembly of the RNAi enzyme complex. *Cell* 115:199–208
123. Seitz H, Zamore PD. 2006. Rethinking the microprocessor. *Cell* 125:827–29
124. Sethupathy P, Corda B, Hatzigeorgiou AG. 2006. TarBase: a comprehensive database of experimentally supported animal microRNA targets. *RNA* 12:192–97
125. Sethupathy P, Megraw M, Hatzigeorgiou AG. 2006. A guide through present computational approaches for the identification of mammalian microRNA targets. *Nat. Methods* 3:881–86
126. Si ML, Zhu S, Wu H, Lu Z, Wu F, Mo YY. 2006. miR-21-mediated tumor growth. *Oncogene* doi:10.1038/sj.onc.1210083

108. Provides the first demonstration that oncogenes can directly regulate miRNA expression, illuminating one mechanism through which miRNA expression becomes abnormal in cancer cells.

120. The available annotation of the human transcriptome is used to characterize a large number of primary miRNA transcripts, revealing how these transcription units are commonly organized.

125. An excellent overview of current miRNA target prediction methods and a step-by-step guide to applying the large number of bioinformatics tools that are publicly available for this purpose.

127. Sokol NS, Ambros V. 2005. Mesodermally expressed *Drosophila* microRNA-1 is regulated by Twist and is required in muscles during larval growth. *Genes Dev.* 19:2343–54
128. Song JJ, Smith SK, Hannon GJ, Joshua-Tor L. 2004. Crystal structure of Argonaute and its implications for RISC slicer activity. *Science* 305:1434–37
129. Sood P, Krek A, Zavolan M, Macino G, Rajewsky N. 2006. Cell-type-specific signatures of microRNAs on target mRNA expression. *Proc. Natl. Acad. Sci. USA* 103:2746–51
130. Soutschek J, Akinc A, Bramlage B, Charisse K, Constien R, et al. 2004. Therapeutic silencing of an endogenous gene by systemic administration of modified siRNAs. *Nature* 432:173–78
131. Takamizawa J, Konishi H, Yanagisawa K, Tomida S, Osada H, et al. 2004. Reduced expression of the let-7 microRNAs in human lung cancers in association with shortened postoperative survival. *Cancer Res.* 64:3753–56
132. Tam W, Ben-Yehuda D, Hayward WS. 1997. bic, a novel gene activated by proviral insertions in avian leukosis virus-induced lymphomas, is likely to function through its noncoding RNA. *Mol. Cell. Biol.* 17:1490–502
133. Tam W, Hughes SH, Hayward WS, Besmer P. 2002. Avian bic, a gene isolated from a common retroviral site in avian leukosis virus-induced lymphomas that encodes a noncoding RNA, cooperates with c-myc in lymphomagenesis and erythroleukemogenesis. *J. Virol.* 76:4275–86
134. Vo N, Klein ME, Varlamova O, Keller DM, Yamamoto T, et al. 2005. A cAMP-response element binding protein-induced microRNA regulates neuronal morphogenesis. *Proc. Natl. Acad. Sci. USA* 102:16426–31
135. Volinia S, Calin GA, Liu CG, Ambs S, Cimmino A, et al. 2006. A microRNA expression signature of human solid tumors defines cancer gene targets. *Proc. Natl. Acad. Sci. USA* 103:2257–61
136. Voorhoeve PM, le Sage C, Schrier M, Gillis AJ, Stoop H, et al. 2006. A genetic screen implicates miRNA-372 and miRNA-373 as oncogenes in testicular germ cell tumors. *Cell* 124:1169–81
137. Wang XJ, Reyes JL, Chua NH, Gaasterland T. 2004. Prediction and identification of *Arabidopsis thaliana* microRNAs and their mRNA targets. *Genome Biol.* 5:R65
138. Wienholds E, Kloosterman WP, Miska E, Alvarez-Saavedra E, Berezikov E, et al. 2005. MicroRNA expression in zebrafish embryonic development. *Science* 309:310–11
139. Wienholds E, Koudijs MJ, van Eeden FJ, Cuppen E, Plasterk RH. 2003. The microRNA-producing enzyme Dicer1 is essential for zebrafish development. *Nat. Genet.* 35:217–18
140. Wightman B, Ha I, Ruvkun G. 1993. Posttranscriptional regulation of the heterochronic gene lin-14 by lin-4 mediates temporal pattern formation in *C. elegans*. *Cell* 75:855–62

140. Describes the identification of the first miRNA target, *lin-14*, which is regulated by the *lin-4* miRNA during *C. elegans* development.

141. Xu P, Vernooy SY, Guo M, Hay BA. 2003. The *Drosophila* microRNA miR-14 suppresses cell death and is required for normal fat metabolism. *Curr. Biol.* 13:790–95
142. Yanaihara N, Caplen N, Bowman E, Seike M, Kumamoto K, et al. 2006. Unique microRNA molecular profiles in lung cancer diagnosis and prognosis. *Cancer Cell* 9:189–98
143. Yi R, O'Carroll D, Pasolli HA, Zhang Z, Dietrich FS, et al. 2006. Morphogenesis in skin is governed by discrete sets of differentially expressed microRNAs. *Nat. Genet.* 38:356–62
144. Yi R, Qin Y, Macara IG, Cullen BR. 2003. Exportin-5 mediates the nuclear export of premicroRNAs and short hairpin RNAs. *Genes Dev.* 17:3011–16
145. Zamore PD, Tuschl T, Sharp PA, Bartel DP. 2000. RNAi: double-stranded RNA directs the ATP-dependent cleavage of mRNA at 21 to 23 nucleotide intervals. *Cell* 101:25–33

146. Zeng Y, Cullen BR. 2004. Structural requirements for premicroRNA binding and nuclear export by Exportin 5. *Nucleic Acids Res.* 32:4776–85
147. Zeng Y, Yi R, Cullen BR. 2003. MicroRNAs and small interfering RNAs can inhibit mRNA expression by similar mechanisms. *Proc. Natl. Acad. Sci. USA* 100:9779–84
148. Zhang Y. 2005. miRU: an automated plant miRNA target prediction server. *Nucleic Acids Res.* 33:W701–4
149. Zhao Y, Samal E, Srivastava D. 2005. Serum response factor regulates a muscle-specific microRNA that targets Hand2 during cardiogenesis. *Nature* 436:214–20

Repetitive Sequences in Complex Genomes: Structure and Evolution

Jerzy Jurka, Vladimir V. Kapitonov, Oleksiy Kohany, and Michael V. Jurka

Genetic Information Research Institute, Mountain View, California 94043; email: jurka@girinst.org, vladimir@girinst.org, kohany@girinst.org, michael@girinst.org

Annu. Rev. Genomics Hum. Genet. 2007. 8:241–59

First published online as a Review in Advance on May 21, 2007.

The *Annual Review of Genomics and Human Genetics* is online at genom.annualreviews.org

This article's doi:
10.1146/annurev.genom.8.080706.092416

1527-8204/07/0922-0241$20.00

Key Words

transposable elements, repetitive DNA, regulation, speciation

Abstract

Eukaryotic genomes contain vast amounts of repetitive DNA derived from transposable elements (TEs). Large-scale sequencing of these genomes has produced an unprecedented wealth of information about the origin, diversity, and genomic impact of what was once thought to be "junk DNA." This has also led to the identification of two new classes of DNA transposons, *Helitrons* and *Polintons*, as well as several new superfamilies and thousands of new families. TEs are evolutionary precursors of many genes, including RAG1, which plays a role in the vertebrate immune system. They are also the driving force in the evolution of epigenetic regulation and have a long-term impact on genomic stability and evolution. Remnants of TEs appear to be overrepresented in transcription regulatory modules and other regions conserved among distantly related species, which may have implications for our understanding of their impact on speciation.

INTRODUCTION

The term "repetitive sequences" (repeats, DNA repeats, repetitive DNA) refers to homologous DNA fragments that are present in multiple copies in the genome. Repetitive DNA was originally discovered based on reassociation kinetics and classified into "highly" and "middle" repetitive sequences (14), roughly corresponding to tandem and interspersed repeats discussed below. This review is centered primarily on repeat research based on DNA sequence analysis and does not cover the so-called low copy repeats (LCRs), also known as segmental duplications, which represent a separate category of duplicated diverse chromosomal segments (105).

Repeats can be clustered into distinct families each traceable to a single ancestral sequence or a closely related group of ancestral sequences. In contrast to multigene families, which are defined based on their biological role, repetitive families are usually defined based on their active ancestors, called master or source genes, and on their generation mechanisms. Over time, individual elements from repetitive families may acquire diverse biological roles.

There are two basic types of repetitive sequences: interspersed repeats and tandem repeats. Interspersed repeats are DNA fragments with an upper size limit of 20–30 kb, inserted more or less at random into host DNA. In contrast, tandem repeats represent arrays of DNA fragments immediately adjacent to each other in head-to-tail orientation. This review focuses on interspersed repetitive DNA from eukaryotic genomes. Interspersed repeats are mostly inactive and often incomplete copies of transposable elements (TEs) inserted into genomic DNA. TEs are segments of DNA or RNA capable of being reproduced and inserted in the host genome. At the same time, genomes are essentially conservative structures that have evolved mechanisms to counteract such insertions. Therefore, TEs and host genomes are locked in a permanent antagonistic relationship resembling an "arms race." Eukaryotic hosts continuously suppress activities of TEs, but TE proliferation persists in virtually all known eukaryotic species. Of all eukaryotic genomes sequenced to date, only the genome *Plasmodium falciparum* appears not to host any active TEs (35).

Why do complex, conservative genomes tolerate the activities of inherently antagonistic elements? TEs cannot be easily eliminated and their endurance in the host can be compared to that of parasites. Furthermore, if TEs can provide evolutionary advantages to the host, their chances of survival increase. The view that TEs are beneficial to the host is not new (16, 44, 68, 87) but recent progress in the field puts it squarely at the center of the ongoing debate on eukaryotic evolution.

STRUCTURE AND SYSTEMATICS OF TRANSPOSABLE ELEMENTS

General Characteristics

Figure 1 presents a schematic structure of TEs. All types of TEs are represented by autonomous and nonautonomous variants. Whereas an autonomous element encodes a complete set of enzymes characteristic of its family and is self-sufficient in terms of transposition, a nonautonomous element transposes by borrowing the protein machinery encoded by its autonomous relatives. Despite their dazzling diversity, all eukaryotic TEs fall into two basic types: retrotransposons and DNA transposons. Retrotransposons are transposed through an RNA intermediate. Their messenger RNA (mRNA) is expressed in the host cell, reverse transcribed, and the resulting complementary DNA (cDNA) copy is integrated back into the host genome. Reverse transcription and integration are catalyzed by reverse transcriptase (RT) and endonuclease/integrase (EN/INT), which are encoded by autonomous elements. Unlike retrotransposons, DNA transposons

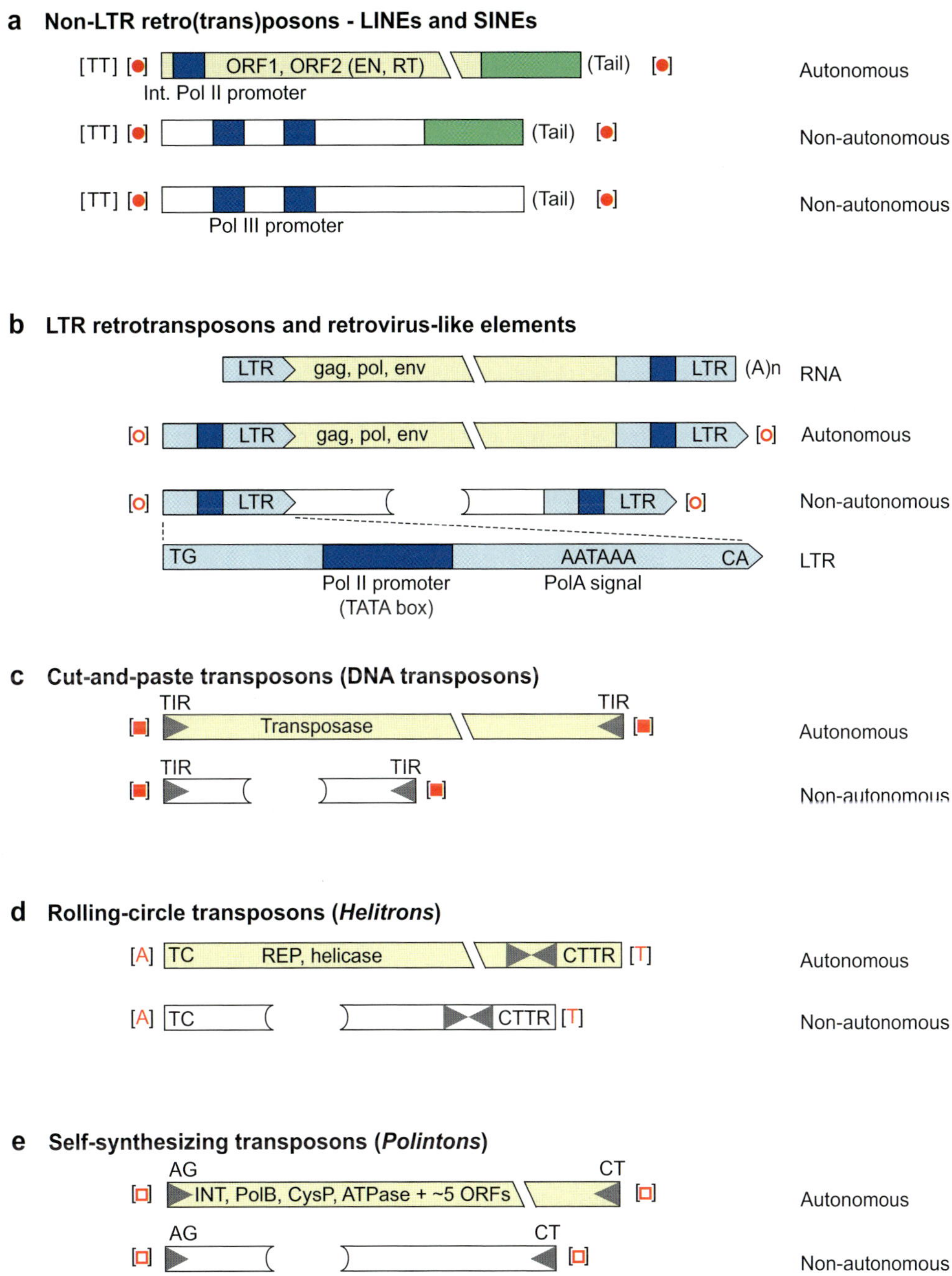

Figure 1

A schematic representation of major classes of transposable elements, including nonautonomous elements.

are transposed by moving their genomic DNA copies from one chromosomal location to another without any RNA intermediate. Most retrotransposons and DNA transposons are flanked by target site duplications (TSDs) resulting from fill-in repair of staggered nicks generated at the DNA target site upon insertion of TEs (42).

All currently known eukaryotic retrotransposons can be divided into four classes: non-long terminal repeat (LTR) retrotransposons, LTR retrotransposons, *Penelope*, and *DIRS* retrotransposons. Although the first two classes (**Figure 1*a*,*b***) are relatively well established and studied (29), the *Penelope* and *DIRS* classes were only recently introduced (2, 30, 81, 98). Members of all four classes of retrotransposons are present in the genomes of virtually all eukaryotic kingdoms: Protista, Plantae, Fungi, and Animalia. The only exception is *Penelope*, which, so far, has not been identified in plants.

Eukaryotic DNA transposons belong to three classes: "cut-and-paste" transposons, *Helitrons*, and *Polintons* (**Figure 1*c*,*d*,*e***). The corresponding mechanisms of transposition are cut-and-paste (23), rolling-circle replicative (60), and self-synthesizing (65), respectively. The cut-and-paste transposons and *Helitrons* cannot synthesize their own DNA; instead, they multiply using host replication machinery.

Non-Long Terminal Repeat and Long Terminal Repeat Retrotransposons

A typical autonomous non-LTR retrotransposon, commonly referred to as a long interspersed element (LINE), contains one or two open reading frames (ORFs). It includes an internal promoter in the 5′ terminal region that governs transcription of the retrotransposon inserted in the host genome. The mechanism of LINE retrotransposition and integration into the genome is well studied and is viewed as a coupled process called target-primed reverse transcription (TPRT). According to the TPRT model, reverse transcription is primed by the free 3′ hydroxyl group at the target DNA nick introduced by EN (29). The model was recently enhanced by the finding that initiation of the *L1* reverse transcription does not require base pairing between the primer and template (72). Moreover, as expected from the model, EN is not necessary for *L1* retrotransposition when free 3′-hydroxyl groups become available in disfunctional telomeres (91). Both RT and EN domains in *L1* are encoded by the same ORF. An mRNA expressed during transcription of a genomic copy of LINE retrotransposon serves as a template for reverse transcription, and the resulting cDNA is inserted in the genome.

Based on structural features of non-LTR retrotransposons and phylogeny of RTs, LINEs can be assigned to five groups, called *R2*, *L1*, *RTE*, *I*, and *Jockey*, which can be subdivided into 15 clades (29, 70). It is believed that the *R2* group is composed of the most ancient non-LTR retrotransposons, the *CRE*, *NeSL*, *R2*, and *R4* clades, which are characterized by a single ORF coding for RT and an EN C terminal to the RT domain. The *R2* EN is similar to different restriction enzymes, and all TEs from the *R2* group retrotranspose into highly specific target sites. Members of the remaining four groups encode the apurinic-apyrimidinic endonuclease (APE), which is always N terminal to the RT domain. In addition to RT and EN, members of the first group code for RNase H (29), including the *Ingi*, *I*, *LOA*, *R1*, and *Tad1* clades. Nonautonomous non-LTR retrotransposons are usually referred to as short interspersed elements (SINE) retrotransposons. Typically, they are mosaic structures derived from transfer RNA (tRNA) or 7SL or 5S ribosomal RNA, and contain 5′ internal pol III promoters involved in transcription. The 3′ ends of SINEs are either derived from LINE elements or contain poly(A) tails recognizable by *L1* elements. They may share common structural constraints (111). Their retrotransposition is catalyzed by RT/EN encoded by the autonomous non-LTR retrotransposons.

Non-LTR retrotransposons are transmitted vertically (i.e., from parents to offspring), with some notable exceptions (71).

An LTR retrotransposon (**Figure 1*b***) may carry three ORFs coding for the *gag*, *env*, and *pol* proteins. The *pol* protein is composed of the RT, EN, and aspartyl protease domains. The EN domain in LTR retrotransposons is usually called INT and is distantly related to the DDE transposase (named after two aspartate and one glutamate residues forming a catalytic triad), encoded by *Mariner* DNA transposons (20, 29). LTR retrotransposons can be transferred horizontally (49), although the extent of the process is not clear.

Penelope. The *Penelope* retrotransposons encode a single ORF composed of the RT and EN domains. The latter is similar to GIY-YIG intron-encoded ENs (2, 30, 84, 121), named after the conserved amino acid motif Gly-Ile-Tyr-Xn-Tyr-Ile-Gly. It appears that the *Penelope* RT is closer to telomerases and bacterial RTs than RTs encoded by non-LTR retrotransposons (2). Like many families of non-LTR retrotransposons, *Penelope* elements generate 10–15 base pair (bp) TSDs and probably follow the TPRT model of retrotransposition (29, 30). However, *Penelope* elements are characterized by unusual LTR-like or inverted terminal repeats not typical for standard non-LTR retrotransposons. Also, some *Penelope* elements from different species retain introns after their retrotransposition (2). Based on their structural and phylogenetic features, *Penelopes* are viewed as a separate class of retrotransposons. However, given the low resolution of phylogenetic trees built for extremely divergent and ancient RTs, an alternative view of *Penelopes* as the most ancient/basal group of non-LTR retrotransposons cannot yet be ruled out.

DIRS. The *Dictyostelium* intermediate repeat sequence (*DIRS*) retrotransposons encode a RT that is phylogenetically closer to that encoded by LTR retrotransposons than to the RT in non-LTR retrotransposons (29). Until recently, *DIRS* elements were viewed as an enigmatic class of INT-free retrotransposons characterized by an unusual structure of terminal repeats (19, 29). However, it turns out that *DIRS* elements encode a protein belonging to the INT family of tyrosine recombinases (tyrosine INT) (40). This observation and the unusual structure of termini led to the classification of *DIRS* elements as a separate class of LTR retrotransposons (28, 41). Given a wide distribution of highly diverse *DIRS* retrotransposons in different eukaryotic kingdoms, it appears that they are as ancient as LTR retrotransposons. It also appears that the *DIRS* RT is grouped phylogenetically with the *Gypsy* RT (29, 98) and separately from the *BEL* and *Copia* LTR retrotransposons, which are viewed as the most ancient LTR retrotransposons (29). Therefore, the most parsimonious scenario of *DIRS* origin is that they evolved from a *Gypsy*-like ancestral LTR retrotransposon after recruiting the tyrosine INT, which replaced the standard DDE INT. This scenario is consistent with the observation of tyrosine recombinase/INT-encoding DNA transposon-like elements in some fungi (*Crypton* transposons) (39) and in ciliates (*Tec* transposons) (27, 48). The suggested recruitment might have occurred following insertion of an ancient tyrosine recombinase-encoding DNA transposon into the *Gypsy*-like predecessor of *DIRS* elements. Analogously to *Penelope* retrotransposons, some *DIRS* elements retain introns in ORFs (41). Such intron retention could be important for the retrotranspositon of *Penelope* and *DIRS*. For instance, non-spliced *DIRS/Penelope* mRNA retained in the nucleus can be a better substrate for retrotransposition than the spliced one (30, 41).

Cut-and-Paste DNA Transposons

During its transposition, a cut-and-paste DNA transposon is excised (cut) from its original genomic location and inserted (pasted) into a new site (23). Both reactions are catalyzed by a transposase that binds the termini of a transposon and its target site and

Table 1 Superfamilies of "cut-and-paste" DNA transposons

Superfamily name	Related bacterial transposases	Size of target site duplications
En/Spm		3
hAT		8
Harbinger	IS5	3
IS4EU	IS4	2
Mariner	IS630	2
Merlin	IS1016	8–9
Mirage		2
MuDR	IS256	9–10
Novosib		8
P		7–8
piggyBac		4
Rehavkus		9
Transib		5

introduces DNA nicks. Most DNA transposons contain 10–400 bp long terminal inverted repeats (TIRs) at both ends. However, in some active transposons, TIRs are imperfect or absent (e.g., some *MuDR* transposons in *Arabidopsis thaliana*) (59). Eukaryotic "cut-and-paste" DNA transposons can be assigned to 13 superfamilies (**Table 1**). Each superfamily includes diverse families composed of autonomous and nonautonomous elements, whose transposition is catalyzed by superfamily-specific transposases. Transposases from different superfamilies are not similar to each other [i.e., position specific iterative basic local alignment and search tool (PSI-BLAST) expected values are higher than 0.05] (1). In addition to the superfamily-specific transposases, each superfamily is characterized by a specific length of TSDs (**Table 1**). However, some superfamilies, such as *En/Spm* and *Harbinger*, have the same length of TSDs. Autonomous DNA transposons from most superfamilies encode only one protein (transposase). They include *Mariner* (97), *hAT* (73), *P* (4), *piggyBac* (85), *Transib* (61, 64), *Merlin* (31), *Mirage*, *IS4EU*, *Novosib* and *Rehavkus* (36). Transposons from the *En/Spm* (73), *Harbinger* (59), and *MuDR* (122) superfamilies code for DNA-binding proteins in addition to transposases.

Harbinger was the first superfamily of DNA transposons discovered based on computational studies (59). The autonomous *Harbingers* encode two proteins: a ~400-amino acid (aa) *Harbinger* transposase and a ~200-aa DNA-binding protein that includes the conserved SANT/myb/trihelix motif. The *Harbinger* transposase is distantly related to transposases encoded by the IS5 group of bacterial transposons, including IS5, IS112, and ISL2. *Harbingers* are typically flanked by 3-bp TSDs, frequently TAA or TTA trinucleotides, but some *Harbingers* from the zebrafish genome show a striking preference for a 17-bp target site (AAAACACCWGGTCTTTT), longer than the target for any other DNA transposon family (62).

Helitrons. *Helitron* DNA transposons transpose via replicative rolling-circle transposition (60). *Helitrons* are present in the genomes of plants, fungi, insects, nematodes, and vertebrates. In some species, including *A. thaliana* and *Caenorhabditis elegans*, they constitute ~2% of the genome. Autonomous *Helitrons* encode the ~1500-aa, so-called Rep/Hel protein, composed of the replication initiator (Rep), and helicase (Hel) conserved domains. The Rep domain spans a ~160-aa region composed of the "two-His" (E-FYW-Q-K-R-G-LAV-PVH-X-H) and "KYK" (Yg-LVW-FAT-Kq-Y-X-X-K) motifs separated by ~130 aa. These motifs are conserved in Reps, which are encoded by plasmids and single-stranded DNA viruses replicating by rolling-circle mechanism. The Rep proteins perform both cleavage and ligation of DNA during rolling-circle replication, the same as transposases. The ~500-aa Hel domain is a helicase that belongs to the SF1 superfamily of DNA helicases. *Helitron* is the only known class of transposons in eukaryotes that integrates into the genome without introducing TSDs. Usually, the *Helitron* integration occurs precisely between A and T nucleotides in the host. *Helitrons* do not have TIRs, which are typically present in other DNA transposons. Instead, *Helitrons* have conserved 5′-TC and CTRR-3′

termini. They also contain a ~18-bp hairpin separated by 10–12 nucleotides from the 3′ end. Presumably, the hairpin serves as the terminator of rolling-circle replication, which is believed to be the mechanism for *Helitron*'s transposition. So far, only *Helitrons* found in the *Aspergillus nidulans* genome do not contain the 3′ hairpin (33).

Although no active *Helitrons* have been isolated and studied experimentally so far, the main features of *Helitron* transposition can be predicted a priori based on the structural invariants detected in different *Helitrons* and known properties of bacterial rolling-circle replicons (60). *Helitron* transposition starts from a site-specific Rep-encoded nicking of the transposon-plus strand. Next, the free 3′-OH end of the nicked-plus strand serves as a primer for leading-strand DNA synthesis facilitated by the *Helitron* helicase and some host replication proteins, including DNA polymerase and replication protein. A RPA-like single-stranded DNA-binding proteins. The newly synthesized leading-plus strand remains covalently linked to the 3′-OH end of the parent-plus strand during the continuous displacement of its 5′-OH end. When the leading strand makes a complete turn, Rep catalyzes a strand transfer reaction followed by the release of a single-stranded DNA intermediate, the parent-minus strand, and a double-stranded DNA *Helitron* composed of both the parental-plus and a newly synthesized strand (60).

Another interesting feature of *Helitron* is its ability to intercept host genes. For example, plant *Helitrons* encode RPA-like proteins, clearly derived from RPA encoded originally by the host genome (60). Given the conservation of RPA in *Helitrons*, this protein is almost certainly involved in *Helitron* transposition, presumably as a single-stranded DNA-binding protein. *Helitrons* present in sea anemone, sea urchin, fish, and frog carry EN derived from CR1-like non-LTR retrotransposons (63, 99). Again, the conservation of the EN in different *Helitrons* from different species shows that it must be necessary for the life cycle of *Helitrons*. Finally, numerous nonautonomous *Helitrons* in the corn genome harbor exon-/intron-coding portions from many different host genes (75, 90). Therefore, *Helitrons* may function as a powerful tool of evolution, by mediating duplication, shuffling, and recruitment of host genes.

Polintons. Like *Helitrons*, the third class of DNA transposons, *Polintons*, was discovered and characterized based on computational studies (65). *Polintons* are 15–20 kb long, with 6-bp TSDs and 100–1000 bp TIRs at both ends. They are the most complex eukaryotic transposons known to date. *Polintons* code for up to 10 proteins, including a family B DNA polymerase (POLB), a retroviral-like INT, an A transposase, and an adenoviral-like cysteine protease. The first three are universal for all autonomous *Polintons* identified in protists, fungi, and animals (65).

Polinton POLB belongs to a group of protein-primed DNA polymerases encoded by genomes of bacteriophages, adenoviruses, and linear plasmids from fungi and plants. POLB and its functional motifs are well defined (10, 24, 115), and their conservation in all extremely diverged *Polinton* POLBs indicates that the DNA-DNA polymerase and proofreading activities are necessary for *Polinton* transposition. The termini of *Polintons* are composed of short 1–3-bp tandem repeats, which are necessary for the slide-back mechanism in protein-primed DNA synthesis studied in bacteriophages (88). Based on these observations, it was proposed that *Polintons* propagate through protein-primed self-synthesis by POLB (65). First, during host genome replication, the INT-catalyzed excision of *Polinton* from the host DNA leads to an extrachromosomal single-stranded *Polinton* that forms a racket-like structure. Second, the *Polinton* POLB replicates the extrachromosomal *Polinton*. Finally, after the double-stranded *Polinton* is synthesized, the INT molecules bind to its termini and catalyze its integration into the host genome.

FROM TRANSPOSABLE ELEMENTS TO GENES

The first clear example of a functional protein-coding gene that evolved from a former TE was centromere protein B (CENP-B) (110, 116). CENP-B is conserved in mammals and binds a specific 17-bp site in the human α-satellite, but its exact function is still not clear. Based on gene knockout studies, it appears that CENP-B is involved in reproduction rather than in centromere-related activities, as was originally predicted (32, 112). Three yeast CENP-B homologs were reported (47), but given their low (<30%) identity to mammalian CENP-B and a similar range of identities between different transposases in mammals, plants, and fungi, they probably evolved in the yeast genome from the *Mariner/Pogo* transposase independently of the mammalian CENP-B.

Approximately 50–100 protein-coding genes in the mammalian genome evolved from coding sequences of DNA transposons and retrotransposons (12, 18, 62, 64, 66, 76, 120; V.V. Kapitonov & J. Jurka, unpublished work). Most of these genes ascended from transposases (*Mariner/Pogo*, *hAT*, *piggyBac*, *P*, *Harbinger*, and *Transib*). The RAG1 protein, which is a key player in V(D)J recombination (103, 113), is probably the most ancient known host protein derived from a transposable element (62, 64). RAG1 evolved some 500 million years ago (mya) in a common ancestor of jawed vertebrates from a *Transib* DNA transposase (64). It is also the only transposase-derived host gene with demonstrated nuclease-/transposase-like activities. Biological properties of the remaining transposase-derived genes are either not known or linked to DNA/RNA binding (52, 78, 114). The RAG1-based immune system is also the only example of a complex host machinery that evolved from transposase and TIRs from the same family of transposons (64). There are other genes that may be involved in DNA rearrangements, as they encode transposase-derived proteins sharing conserved catalytic amino acids with corresponding transposases. An example of such a conserved gene is *HARBI1*, which evolved from the *Harbinger* transposase in a common ancestor of fish, birds, frogs, and mammals (62).

Other potential sources of novel protein-coding genes are LTR retrotransposons. For instance, >50 protein-encoding genes syntenic between the human and mouse genomes evolved from the *gag* protein encoded by *Gypsy* LTR retrotransposons, which were active in ancestral genomes (12, 18, 66, 79, 94, 119). One of the *Gypsy*-derived genes, called PEG10 or KIAA1051, includes *Gypsy gag* and protease domains, which are fused together through the −1 ribosomal frame-shift mechanism typical for *Gypsy* elements (82, 94, 119). Although its exact function is still unknown, PEG10 is important for mouse parthenogenetic development based on observed embryonic lethality due to placental defects in PEG10 knockout mice (95). Although there are >30 examples of host genes evolved from DNA transposases, there is only one example of a recruited RT: the mammalian Rtl1 or PEG11 gene, which evolved from the *Gypsy gag* and RT (104). Interestingly, both PEG10 and PEG11 are paternally expressed genes, and more than 50% of all *gag*-derived genes reside on the X chromosome.

Finally, many microRNA genes appear to have evolved from TEs, and their involvement in gene regulation appears to be an outcome of the antagonistic relationship between TEs and the host genome. Expression of TEs and generation of repetitive DNA, including tandem repeats, are countered by RNA degradation and DNA methylation (17, 22, 100, 118) mediated by small RNAs (sRNAs) (∼20–26 bp) generated from the targeted repetitive DNA. Analogous processes are involved in modulating chromatin structure and regulating gene expression (7, 8, 15, 22, 101). Many of such processes are mediated by sRNAs derived from evolutionarily conserved precursors (21).

Most of the epigenetic regulation of endogenous genes in *A. thaliana* appears to have evolved from mechanisms to silence TEs (128). Furthermore, some mammalian precursors of microRNAs (miRNAs) appear to be derived from ancient MIR (SINE) and *L2* (LINE) elements (108), or even younger *Alu* (SINE) elements and processed pseudogenes (25, 109). Recent evidence that 5′ *Alus* can function as RNA polymerase promoters for miRNAs (11) further supports the contributions of TEs to the origin and expression of miRNAs involved in mammalian gene regulation.

OTHER HIGHLY CONSERVED TRANSPOSABLE ELEMENTS

Recent systematic comparisons of complete genomic sequences revealed the existence of noncoding sequences that are highly conserved across multiple species (6). They include LF-SINE (5), MER121 (58), AmnSINE1, and SINE3-1 (92, 127), which are SINE, or SINE-like, elements preserved in highly diverse vertebrates from *Latimeria* and reptiles to mammals. An additional 83 families of low and moderately repeated elements were reported recently and deposited in Repbase (38, 53). The list includes 20 Eulor families, 15 newly analyzed MER families, 31 UCON families, 14 LINE-like families (X*_LINEs), and 3 MARE families. Eulor families are relatively small, with self-complementary regions suggesting that they might have been derived from DNA transposons. Likewise, many MER elements also resemble nonautonomous DNA transposons. Furthermore, mammalian-specific MARE3 is a tRNA-derived SINE (38). X*_LINEs, where the asterisk stands for specification of one of the 14 families, were directly or indirectly derived from autonomous non-LTR retrotransposons, a fact supported by significant similarities between their translatable regions to diverse LINE elements (38).

Table 2 shows densities of the above-described families of repeats, including a moderately repetitive *L4* family (36), in five vertebrate genomes. Columns 2 and 3 show densities of the same families in human conserved sequences (106) and *cis*-regulatory modules (CRMs) (9). For some families, the densities of TEs in CRMs can be as much as a factor of magnitude higher than the average human genomic density. Similar

Table 2 Densities of selected repetitive families per 1 Mb of DNA sequence

	H.s.	Cons.	CRMs	E.t.	M.d.	G.g.	X.t.
AmnSINE1_GG	0.28	0.81	0.94	0.17	0.15	0.64	0.24
AmnSINE1_HS	0.17	1.72	1.65	0.13	0.27	1.10	0.02
Eulor*	0.42	5.00	6.16	0.35	0.44	1.66	0.15
L4	1.69	3.21	1.93	0.53	0.13	0.00	0.00
LF-SINE	0.18	1.96	2.19	0.13	0.29	1.48	0.07
MARE1-2	0.96	1.45	1.23	0.35	1.21	0.00	0.00
MARE3	0.19	0.91	0.93	0.07	0.45	0.00	0.00
MER121	0.30	3.39	0.33	0.20	0.30	0.00	0.00
MER122-136	1.22	8.69	9.52	0.63	1.53	1.85	0.07
SINE3-1*	0.03	0.07	0.08	0.04	0.11	0.02	0.17
UCONS1-31	0.60	6.63	6.96	0.51	0.65	2.55	0.15
X*_LINE	0.63	2.98	3.18	0.28	1.19	1.06	0.57

DNA origin: human (H.s., *Homo sapiens*), tenrec (E.t., *Echinops telfairi*), Brazilian gray short-tailed opposum (M.d., *Monodelphis domestica*), chicken (G.g., *Gallus gallus*), and the pipid frog (X.t., *Xenopus tropicalis*). Columns 2 and 3 represent densities in conserved human sequences (106) and *cis*-regulatory modules (CRMs) (9). Asterisks indicate additional subclassification.

overrepresentation can be found in conserved regions, which partially overlap with CRMs. Many CRMs are tissue specific (9), which may be significant for understanding developmental evolution, especially in light of recent evidence that TEs can affect developmental processes in mammalian oocytes and preimplantation embryos (96). The most overrepresented repeats are broadly conserved in mammals, chicken, and other vertebrates (e.g., LF-SINE). The overrepresentation may be due to the fact that nonconserved copies continue to decay over time, leading to lower overall genomic densities. However, the proportions of mammalian-specific repeat families reveal a more complex pattern, which can be seen in five mammalian families: MARE1-2, MARE3, *L4*, and MER121. These families are moderately repetitive as their copy numbers in the human genome range from ~600 (MARE3) to ~5000 (*L4*). The density of MARE1-2 repeats is about 50% higher in conserved regions than the overall human genomic density of these elements, whereas the density of MARE3 repeats is almost five times higher both in conserved regions and CRMs. In the conserved regions, the densities of *L4* and MER121 repeats are around 2 and 10 times higher, respectively, but in CRMs they are comparable with human genomic densities. In other words, MARE3 accumulated many times faster in both CRMs and conserved regions than MARE1-2. On the other hand, the accumulation of *L4* and MER121 in CRMs was marginal at best, whereas their overrepresentation in conserved regions was substantial, particularly MER121. A potential scenario of fixation of certain families in regulatory regions is further discussed in the last section in the context of speciation.

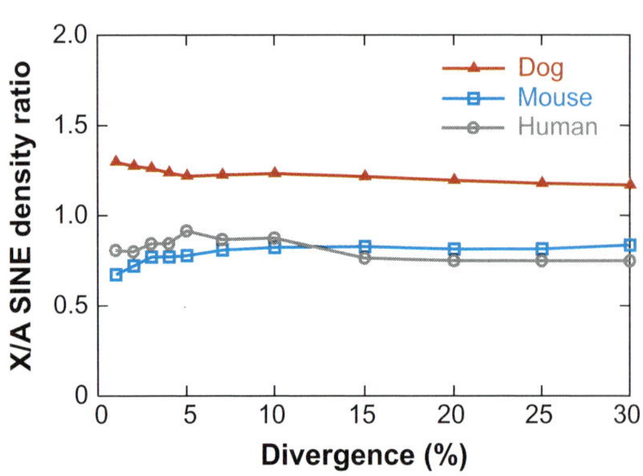

Figure 2

Ratios of short interspersed element (SINE) densities on chromosome X relative to autosomes from dog, mouse, and human.

CUMULATIVE IMPACT ON GENOMIC STRUCTURE

A systematic buildup of repetitive DNA in the genome is due to an excess of insertions over deletions of TEs. Over time it can lead to large-scale genomic changes, which were recently studied in some detail in mammalian genomes. The most prominent TEs in Euterian mammals are LINE1, (*L1*) non-LTR retroelements, and the associated nonautonoumous SINE elements. SINEs are particularly convenient for comparative studies of insertion and elimination of TEs in mammals due to their abundance, moderate lengths, and recent, or even ongoing, retrotransposition. Previous analyses of the human and mouse genomes indicate that the insertion and elimination of young SINE elements occur in male germlines (54–56). In the absence of selection, the systematic insertion of TEs in male germlines leads to their overrepresentation on chromosome Y and underrepresentation on chromosome X relative to autosomes. Analogous insertions in female germlines result in their overrepresentation on chromosome X relative to autosomes and their total absence on chromosome Y, except in regions undergoing X-Y recombination. **Figure 2** shows that proportions of SINE densities on chromosome X and autosomes (A) in human and mouse are close to 2/3, which is consistent with passing active SINEs through the male germline (54, 55). In the dog genome, the analogous X/A proportions are close to 4/3, the value predicted for transmission through female germ cells only (55, 124).

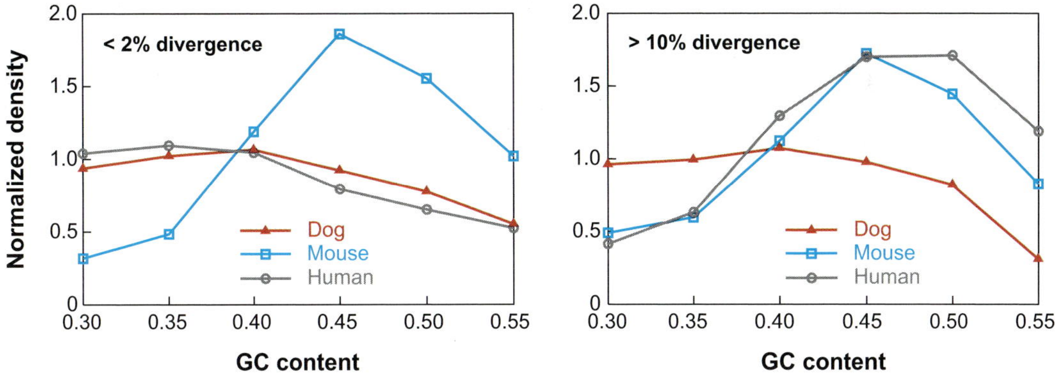

Figure 3

Normalized densities of younger (*left*) and older (*right*) short interspersed elements (SINEs) from dog (*red*), mouse (*blue*), and human (*gray*) genomes, plotted for different ranges of guanosine and cytosine (GC) content.

The biological impact of germline-specific transmission of TEs remains to be determined. Female-driven transmission could be among the initial mechanisms leading to an accumulation of *L1* elements on chromosome X, where they could be further accumulated by natural selection due to their potential involvement in spreading and amplifying the X-inactivation signal (83). Germline-specific expression is consistent with methylation patterns: Human *Alu*s are hypomethylated in sperm, whereas *L1*s are hypermethylated (45, 46, 123). Maternal and paternal methylation of repeats can provide genomic imprinting signatures guiding the epigenetic modification machineries to the imprinted regions (8, 123).

An emerging question is whether male and female modes of transposition can affect genetic stability of TEs due to the general premise that female germlines appear to be more genetically inert than male germlines (86, 89). The male-transmitted human *Alu* elements are originally prevalent in AT-rich DNA, but over time they tend to accumulate in GC-rich chromosomal regions (54, 76). In the mouse genome such an accumulation may be very fast (55). However, in the dog genome there is virtually no difference between the distribution of younger and older SINEs in DNA segments, indicating the lack of analogous postinsertion accumulation (**Figure 3**). Postinsertion changes in the distribution of SINE elements have been attributed to different SINE-SINE recombination patterns in gene-rich and gene-poor isochores (43, 54). This implies that recombinations between female-transmitted SINEs in the dog genome are much less frequent than those between male-transmitted SINEs in the human and mouse genomes. One of the outcomes of SINE-SINE recombination is the formation of hybrid SINEs that are no longer flanked by identical TSDs. The loss of TSDs over time has been observed in the human and mouse genomes, but not in the dog genome (37), which is consistent with the prediction of a low SINE-SINE recombination rate in dog. Finally, SINE-SINE recombinations can trigger the formation of long segmental duplications (51, 105). In contrast to human and mouse, very few such duplications have been found in dog (77), again in agreement with the data indicating low SINE-SINE recombination activities in dog. Nevertheless, a direct relationship between germline-specific transmission and stability of TEs remains to be shown.

POTENTIAL ROLE IN SPECIATION

In 1984, McClintock proposed that species could originate due to sudden, TE-induced chromosomal reorganizations followed by a complex genomic response (87). Recently, this idea was synthetically reviewed in the context of classical paleontological data and current knowledge of TEs (80). McClintock also noted that TEs can be activated suddenly due to abiotic stress, and concluded "that stress, and the genome reaction to it, may underlie many formations of new species" (87). Growing evidence confirms the role of stress in generating and unmasking genetic diversity (reviewed in 3). Activation of TEs under stress may produce beneficial outbursts of mutations at critical evolutionary junctions when populations face a choice between extinction and rapid change. Apart from potential contributions to speciation, activation of TEs under stress may also increase their chance of survival because host populations under stress are likely to shrink, and TEs are more likely to spread by genetic drift in small populations than in large ones. Other early ideas on the role of TEs in speciation (102) were inspired in part by the discovery of hybrid dysgenesis in fruit fly induced by the *P* element (67). The renewed interest in interspecies hybridization as a model for TE activation in speciation is stimulated by more evidence of retroelement upregulation in animal and plant species (34, 50, 74, 93). Recently, the upregulation of TEs in hybrid dysgenesis was linked to disturbance of the RNA silencing system in germ cells (57).

Any model of speciation that includes the participation of TEs must include changes in regulatory systems. The potential significance of repetitive DNA for evolution of eukaryotic regulation was first proposed by Britten & Davidson (13), before the relationship between TEs and repetitive DNA was established. Later, Wilson and colleagues (69, 125) proposed that gene rearrangements, particularly those leading to changes in regulatory regions, may account for the major organismal differences. Since then, more experimental evidence of changes in transcription regulation and their impact on organismal phenotypes at a microevolutionary level became available (8, 126). However, many regulatory changes are likely harmful or lethal, as indicated by the apparent lack of TE insertions in many developmental regulators (107). Yet, as discussed above, certain repeat families are overrepresented in CRMs and conserved regions, whereas others are not (38). This strongly indicates that fixation of conserved TEs in regulatory regions did occur in the past, although the process is not well understood. Some of these conserved repeats were derived from tRNA or 5SRNA elements, which might have contributed useful regulatory signals such as pol promoters. However, there are other tRNA-derived families, such as MIR, that are not overrepresented in conserved regions, in contrast to tRNA-derived MER133 or MARE3 (38, 53). There could be additional factors affecting the fixation, and among them coincidence with speciation is of particular interest. Such fixations need not necessarily come from large outbursts of (retro)transpositions. Instead, any TEs active at the time of speciation could have a better chance to make multiple impacts on regulatory regions, particularly those involved in developmental processes. Recent analysis of selected retropseudogenes revealed that the peaks of their retropositions appear to roughly correspond with major events in the primate phylogenetic history (26). More recently, it was demonstrated that significant genome expansion in three hybrid sunflower species is attributable to proliferation of *Gypsy*-type LTR retrotransposons (117). It remains to be seen if this type of rapid expansion can translate to multiple fixations of repeats in regulatory sites. Successfully modified regulatory sites could later undergo duplications, further increasing the proportions of conserved copies relative to the total number of genomic copies. Major changes during speciation are unlikely to be optimal or stable and may require extensive fine-tuning over time. TEs may play another important role in such processes by affecting DNA methylation patterns and microRNA-mediated regulation.

DISCLOSURE STATEMENT

The authors are not aware of any biases that might be perceived as affecting the objectivity of this review.

ACKNOWLEDGMENTS

This work was supported by National Institutes of Health grant 5 P41 LM006252-09. We thank Jolanta Walichiewicz and Andrew Gentles for help with editing the manuscript.

LITERATURE CITED

1. Altschul SF, Madden TL, Schaffer AA, Zhang J, Zhang Z, et al. 1997. Gapped BLAST and PSI-BLAST: a new generation of protein database search programs. *Nucleic Acids Res.* 25:3389–402
2. Arkhipova IR, Pyatkov KI, Meselson M, Evgen'ev MB. 2003. Retroelements containing introns in diverse invertebrate taxa. *Nat. Genet.* 33:123–24
3. Badyaev AV. 2005. Stress-induced variation in evolution: from behavioural plasticity to genetic assimilation. *Proc. Biol. Sci.* 272:877–86
4. Bandelt HJ, Quintana-Murci L, Salas A, Macaulay V. 2002. The fingerprint of phantom mutations in mitochondrial DNA data. *Am. J. Hum. Genet.* 71:1150–60
5. Bejerano G, Lowe CB, Ahituv N, King B, Siepel A, et al. 2006. A distal enhancer and an ultraconserved exon are derived from a novel retroposon. *Nature* 441:87–90
6. Bejerano G, Pheasant M, Makunin I, Stephen S, Kent WJ, et al. 2004. Ultraconserved elements in the human genome. *Science* 304:1321–25
7. Bernstein E, Allis CD. 2005. RNA meets chromatin. *Genes Dev.* 19:1635–55
8. Biemont C, Vieira C. 2006. Genetics: junk DNA as an evolutionary force. *Nature* 443:521–24
9. Blanchette M, Bataille AR, Chen X, Poitras C, Laganiere J, et al. 2006. Genome-wide computational prediction of transcriptional regulatory modules reveals new insights into human gene expression. *Genome Res.* 16:656–68
10. Blanco L, Salas M. 1996. Relating structure to function in phi29 DNA polymerase. *J. Biol. Chem.* 271:8509–12
11. Borchert GM, Lanier W, Davidson BL. 2006. RNA polymerase III transcribes human microRNAs. *Nat. Struct. Mol. Biol.* 13:1097–101
12. Brandt J, Veith AM, Volff JN. 2005. A family of neofunctionalized Ty3/gypsy retrotransposon genes in mammalian genomes. *Cytogenet. Genome Res.* 110:307–17
13. Britten RJ, Davidson EH. 1969. Gene regulation for higher cells: a theory. *Science* 165:349–57
14. Britten RJ, Kohne DE. 1968. Repeated sequences in DNA. Hundreds of thousands of copies of DNA sequences have been incorporated into the genomes of higher organisms. *Science* 161:529–40
15. Brodersen P, Voinnet O. 2006. The diversity of RNA silencing pathways in plants. *Trends Genet.* 22:268–80
16. Brosius J. 1991. Retroposons–seeds of evolution. *Science* 251:753

17. Buchon N, Vaury C. 2006. RNAi: a defensive RNA-silencing against viruses and transposable elements. *Heredity* 96:195–202
18. Campillos M, Doerks T, Shah PK, Bork P. 2006. Computational characterization of multiple Gag-like human proteins. *Trends Genet.* 22:585–89
19. Cappello J, Handelsman K, Lodish HF. 1985. Sequence of Dictyostelium DIRS-1: an apparent retrotransposon with inverted terminal repeats and an internal circle junction sequence. *Cell* 43:105–15
20. Capy P, Langin T, Higuet D, Maurer P, Bazin C. 1997. Do the integrases of LTR-retrotransposons and class II element transposases have a common ancestor? *Genetica* 100:63–72
21. Carrington JC, Ambros V. 2003. Role of microRNAs in plant and animal development. *Science* 301:336–38
22. Chan SW, Henderson IR, Jacobsen SE. 2005. Gardening the genome: DNA methylation in *Arabidopsis thaliana*. *Nat. Rev. Genet.* 6:351–60
23. Craig NL. 1995. Unity in transposition reactions. *Science* 270:253–54
24. de Vega M, Lazaro JM, Salas M, Blanco L. 1998. Mutational analysis of phi29 DNA polymerase residues acting as ssDNA ligands for 3′–5′ exonucleolysis. *J. Mol. Biol.* 279:807–22
25. Devor EJ. 2006. Primate microRNAs miR-220 and miR-492 lie within processed pseudogenes. *J. Hered.* 97:186–90
26. Devor EJ, Moffat-Wilson KA. 2003. Molecular and temporal characteristics of human retropseudogenes. *Hum. Biol.* 75:661–72
27. Doak TG, Doerder FP, Jahn CL, Herrick G. 1994. A proposed superfamily of transposase genes: transposon-like elements in ciliated protozoa and a common "D35E" motif. *Proc. Natl. Acad. Sci. USA* 91:942–46
28. Duncan L, Bouckaert K, Yeh F, Kirk DL. 2002. Kangaroo, a mobile element from Volvox carteri, is a member of a newly recognized third class of retrotransposons. *Genetics* 162:1617–30
29. Eickbush TH, Malik HS. 2002. Origins and evolution of retrotransposons. In *Mobile DNA II*, ed. NL Craig, R Craigie, M Gellert, AM Lambowitz, pp. 1111–144. Washington, DC: ASM Press
30. Evgen'ev MB, Arkhipova IR. 2005. Penelope-like elements—a new class of retroelements: distribution, function and possible evolutionary significance. *Cytogenet. Genome Res.* 110:510–21
31. Feschotte C. 2004. Merlin, a new superfamily of DNA transposons identified in diverse animal genomes and related to bacterial IS1016 insertion sequences. *Mol. Biol. Evol.* 21:1769–80
32. Fowler KJ, Hudson DF, Salamonsen LA, Edmondson SR, Earle E, et al. 2000. Uterine dysfunction and genetic modifiers in centromere protein B-deficient mice. *Genome Res.* 10:30–41
33. Galagan JE, Calvo SE, Cuomo C, Ma LJ, Wortman JR, et al. 2005. Sequencing of *Aspergillus nidulans* and comparative analysis with *A. fumigatus* and A. oryzae. *Nature* 438:1105–15
34. Garcia Guerreiro MP, Biemont C. 1995. Changes in the chromosomal insertion pattern of the copia element during the process of making chromosomes homozygous in *Drosophila melanogaster*. *Mol. Gen. Genet.* 246:206–11
35. Gardner MJ, Hall N, Fung E, White O, Berriman M, et al. 2002. Genome sequence of the human malaria parasite *Plasmodium falciparum*. *Nature* 419:498–511
36. Genetic Information Research Institute. 2007. *Repbase update*. **http://www.girinst.org/repbase/index.html**

37. Gentles AJ, Kohany O, Jurka J. 2005. Evolutionary diversity and potential recombinogenic role of integration targets of non-LTR retrotransposons. *Mol. Biol. Evol.* 22:1983–91
38. Gentles AJ, Wakefield MJ, Kohany O, Gu W, Batzer MA, et al. 2007. Evolutionary dynamics of transposable elements in the short-tailed opossum Monodelphis domestica. *Genome Res.* In press. doi:10.1101/gr.6070707
39. Goodwin TJ, Butler MI, Poulter RT. 2003. Cryptons: a group of tyrosine-recombinase-encoding DNA transposons from pathogenic fungi. *Microbiology* 149:3099–109
40. Goodwin TJ, Poulter RT. 2001. The DIRS1 group of retrotransposons. *Mol. Biol. Evol.* 18:2067–82
41. Goodwin TJ, Poulter RT. 2004. A new group of tyrosine recombinase-encoding retrotransposons. *Mol. Biol. Evol.* 21:746–59
42. Grindley, ND, Sherratt DJ. 1979. Sequence analysis at IS1 insertion sites: models for transposition. *Cold Spring Harb Symp Quant Biol.* 43:1257–61
43. Hackenberg M, Bernaola-Galvan P, Carpena P, Oliver JL. 2005. The biased distribution of Alus in human isochores might be driven by recombination. *J. Mol. Evol.* 60:365–77
44. Hartl DL, Dykhuizen DE, Miller RD, Green L, de Framond J. 1983. Transposable element IS50 improves growth rate of E. coli cells without transposition. *Cell* 35:503–10
45. Hellman-Blumberg U, Hintz MF, Gatewood JM, Schmid CW. 1993. Developmental differences in methylation of human Alu repeats. *Mol. Cell. Biol.* 8:4523–30
46. Howlett SK, Reik W. 1991. Methylation levels of maternal and paternal genomes during preimplantation development. *Development* 113:119–27
47. Irelan JT, Gutkin GI, Clarke L. 2001. Functional redundancies, distinct localizations and interactions among three fission yeast homologs of centromere protein-B. *Genetics* 157:1191–203
48. Jacobs ME, Sanchez-Blanco A, Katz LA, Klobutcher LA. 2003. Tec3, a new developmentally eliminated DNA element in Euplotes crassus. *Eukaryot. Cell* 2:103–14
49. Jordan IK, Matyunina LV, McDonald JF. 1999. Evidence for the recent horizontal transfer of long terminal repeat retrotransposon. *Proc. Natl. Acad. Sci. USA* 96:12621–25
50. Josefsson C, Dilkes B, Comai L. 2006. Parent-dependent loss of gene silencing during interspecies hybridization. *Curr. Biol.* 16:1322–28
51. Jurka J. 2004. Evolutionary impact of human Alu repetitive elements. *Curr. Opin. Genet. Dev.* 14:603–8
52. Jurka J, Kapitonov VV. 1999. Sectorial mutagenesis by transposable elements. *Genetica* 107:239–48
53. Jurka J, Kapitonov VV, Pavlicek A, Klonowski P, Kohany O, Walichiewicz J. 2005. Repbase Update, a database of eukaryotic repetitive elements. *Cytogenet. Genome Res.* 110:462–67
54. Jurka J, Kohany O, Pavlicek A, Kapitonov VV, Jurka MV. 2004. Duplication, coclustering, and selection of human Alu retrotransposons. *Proc. Natl. Acad. Sci. USA* 101:1268–72
55. Jurka J, Kohany O, Pavlicek A, Kapitonov VV, Jurka MV. 2005. Clustering, duplication and chromosomal distribution of mouse SINE retrotransposons. *Cytogenet. Genome Res.* 110:117–23
56. Jurka J, Krnjajic M, Kapitonov VV, Stenger JE, Kohkanyy O. 2002. Active Alu elements are passed primarily through paternal germlines. *Theor. Popul. Biol.* 61:519–30
57. Kalmykova AI, Klenov MS, Gvozdev VA. 2005. Argonaute protein PIWI controls mobilization of retrotransposons in the *Drosophila* male germline. *Nucleic Acids Res.* 33:2052–59
58. Kamal M, Xie X, Lander ES. 2006. A large family of ancient repeat elements in the human genome is under strong selection. *Proc. Natl. Acad. Sci. USA* 103:2740–45

59. Kapitonov VV, Jurka J. 1999. Molecular paleontology of transposable elements from *Arabidopsis thaliana*. *Genetica* 107:27–37
60. Kapitonov VV, Jurka J. 2001. Rolling-circle transposons in eukaryotes. *Proc. Natl. Acad. Sci. USA* 98:8714–19
61. Kapitonov VV, Jurka J. 2003. Molecular paleontology of transposable elements in the *Drosophila melanogaster* genome. *Proc. Natl. Acad. Sci. USA* 100:6569–74
62. Kapitonov VV, Jurka J. 2004. Harbinger transposons and an ancient *HARBI1* gene derived from a transposase. *DNA Cell Biol.* 23:311–24
63. Kapitonov VV, Jurka J. 2005. Helitron-1_SP, a family of autonomous Helitrons in the sea urchin genome. *Repbase Rep.* 5:393
64. Kapitonov VV, Jurka J. 2005. RAG1 core and V(D)J recombination signal sequences were derived from Transib transposons. *PLOS Biol.* 3:e181
65. Kapitonov VV, Jurka J. 2006. Self-synthesizing DNA transposons in eukaryotes. *Proc. Natl. Acad. Sci. USA* 103:4540–45
66. Kapitonov VV, Pavlicek A, Jurka J. 2004. Anthology of human repetitive DNA. In *Encyclopedia of Molecular Cell Biology and Molecular Medicine*, ed. RA Meyers, pp. 251–305. Weinheim: Wiley-VCH Verlag GmbH & Co. KGaA
67. Kidwell MG, Kidwell JF, Sved JA. 1977. Hybrid dysgenesis in *Drosophila melanogaster*: a syndrome of aberrant traits including mutation, sterility and male recombination. *Genetics* 86:813–33
68. Kidwell MG, Lisch DR. 2001. Perspective: transposable elements, parasitic DNA, and genome evolution. *Evol. Int. J. Org. Evol.* 55:1–24
69. King MC, Wilson AC. 1975. Evolution at two levels in humans and chimpanzees. *Science* 188:107–16
70. Kojima KK, Fujiwara H. 2004. Cross-genome screening of novel sequence-specific non-LTR retrotransposons: various multicopy RNA genes and microsatellites are selected as targets. *Mol. Biol. Evol.* 21:207–17
71. Kordis D, Gubensek F. 1997. Bov-B long interspersed repeated DNA (LINE) sequences are present in Vipera ammodytes phospholipase A2 genes and in genomes of Viperidae snakes. *Eur. J. Biochem.* 246:772–79
72. Kulpa DA, Moran JV. 2006. Cis-preferential LINE-1 reverse transcriptase activity in ribonucleoprotein particles. *Nat. Struct. Mol. Biol.* 13:655–60
73. Kunze R, Weil CF. 2002. The hAT and CACTA superfamilies of plant transposons. In *Mobile DNA II*, ed. NL Craig, R Craigie, M Gellert, AM Lambowitz, pp. 565–610. Washington, DC: ASM Press
74. Labrador M, Farre M, Utzet F, Fontdevila A. 1999. Interspecific hybridization increases transposition rates of Osvaldo. *Mol. Biol. Evol.* 16:931–37
75. Lai J, Li Y, Messing J, Dooner HK. 2005. Gene movement by Helitron transposons contributes to the haplotype variability of maize. *Proc. Natl. Acad. Sci. USA* 102:9068–73
76. Lander ES, Linton LM, Birren B, Nusbaum C, Zody MC, et al. 2001. Initial sequencing and analysis of the human genome. *Nature* 409:860–921
77. Lindblad-Toh K, Wade CM, Mikkelsen TS, Karlsson EK, Jaffe DB, et al. 2005. Genome sequence, comparative analysis and haplotype structure of the domestic dog. *Nature* 438:803–19
78. Liu W, Seto J, Sibille E, Toth M. 2003. The RNA binding domain of Jerky consists of tandemly arranged helix-turn-helix/homeodomain-like motifs and binds specific sets of mRNAs. *Mol. Cell Biol.* 23:4083–93
79. Llorens C, Marin I. 2001. A mammalian gene evolved from the integrase domain of an LTR retrotransposon. *Mol. Biol. Evol.* 18:1597–600

80. Lonnig WE, Saedler H. 2002. Chromosome rearrangements and transposable elements. *Annu. Rev. Genet.* 36:389–410
81. Lorenzi HA, Robledo G, Levin MJ. 2006. The VIPER elements of trypanosomes constitute a novel group of tyrosine recombinase-enconding retrotransposons. *Mol. Biochem. Parasitol.* 145:184–94
82. Lux A, Beil C, Majety M, Barron S, Gallione CJ, et al. 2005. Human retroviral gag- and gag-pol-like proteins interact with the transforming growth factor-beta receptor activin receptor-like kinase 1. *J. Biol. Chem.* 280:8482–93
83. Lyon MF. 1998. X-chromosome inactivation: a repeat hypothesis. *Cytogenet. Cell Genet.* 80:133–37
84. Lyozin GT, Makarova KS, Velikodvorskaja VV, Zelentsova HS, Khechumian RR, et al. 2001. The structure and evolution of Penelope in the virilis species group of *Drosophila*: an ancient lineage of retroelements. *J. Mol. Evol.* 52:445–56
85. Mah C, Sarkar R, Zolotukhin I, Schleissing M, Xiao X, et al. 2003. Dual vectors expressing murine factor VIII result in sustained correction of hemophilia A mice. *Hum. Gene Ther.* 14:143–52
86. Makova KD, Yang S, Chiaromonte F. 2004. Insertions and deletions are male biased too: a whole-genome analysis in rodents. *Genome Res.* 14:567–73
87. McClintock B. 1984. The significance of responses of the genome to challenge. *Science* 226:792–801
88. Mendez J, Blanco L, Esteban JA, Bernad A, Salas M. 1992. Initiation of phi 29 DNA replication occurs at the second 3′ nucleotide of the linear template: a sliding-back mechanism for protein-primed DNA replication. *Proc. Natl. Acad. Sci. USA* 89:9579–83
89. Miller W, Makova KD, Nekrutenko A, Hardison RC. 2004. Comparative genomics. *Annu. Rev. Genom. Hum. Genet.* 5:15–56
90. Morgante M, Brunner S, Pea G, Fengler K, Zuccolo A, Rafalski A. 2005. Gene duplication and exon shuffling by helitron-like transposons generate intraspecies diversity in maize. *Nat. Genet.* 37:997–1002
91. Morrish TA, Garcia-Perez JL, Stamato TD, Taccioli GE, Sekiguchi J, Moran JV. 2007. Endonuclease-independent LINE-1 retrotransposition at mammalian telomeres. *Nature* 446:208–12
92. Nishihara H, Smit AF, Okada N. 2006. Functional noncoding sequences derived from SINEs in the mammalian genome. *Genome Res.* 16:864–74
93. O'Neill RJ, O'Neill MJ, Graves JA. 1998. Undermethylation associated with retroelement activation and chromosome remodelling in an interspecific mammalian hybrid. *Nature* 393:68–72
94. Ono R, Kobayashi S, Wagatsuma H, Aisaka K, Kohda T, et al. 2001. A retrotransposon-derived gene, *PEG10*, is a novel imprinted gene located on human chromosome 7q21. *Genomics* 73:232–37
95. Ono R, Nakamura K, Inoue K, Naruse M, Usami T, et al. 2006. Deletion of Peg10, an imprinted gene acquired from a retrotransposon, causes early embryonic lethality. *Nat. Genet.* 38:101–6
96. Peaston AE, Evsikov AV, Graber JH, de Vries WN, Holbrook AE, et al. 2004. Retrotransposons regulate host genes in mouse oocytes and preimplantation embryos. *Dev. Cell* 7:597–606
97. Plasterk RHA, Van Luenen HGAM. 2002. The Tc1/mariner family of transposable elements. In *Mobile DNA II*, ed. NL Craig, R Craigie, M Gellert, AM Lambowitz, pp. 519–32. Washington DC: ASM Press

98. Poulter RT, Goodwin TJ. 2005. DIRS-1 and the other tyrosine recombinase retrotransposons. *Cytogenet. Genome Res.* 110:575–88
99. Poulter RT, Goodwin TJ, Butler MI. 2003. Vertebrate helentrons and other novel Helitrons. *Gene* 313:201–12
100. Qi Y, He X, Wang XJ, Kohany O, Jurka J, Hannon GJ. 2006. Distinct catalytic and noncatalytic roles of ARGONAUTE4 in RNA-directed DNA methylation. *Nature* 443:1008–12
101. Richards EJ. 2006. Inherited epigenetic variation: revisiting soft inheritance. *Nat. Rev. Genet.* 7:395–401
102. Rose MR, Doolittle WF. 1983. Molecular biological mechanisms of speciation. *Science* 220:157–62
103. Schatz DG, Oettinger MA, Baltimore D. 1989. The V(D)J recombination activating gene, RAG-1. *Cell* 59:1035–48
104. Seitz H, Youngson N, Lin SP, Dalbert S, Paulsen M, et al. 2003. Imprinted microRNA genes transcribed antisense to a reciprocally imprinted retrotransposon-like gene. *Nat. Genet.* 34:261–62
105. Sharp AJ, Cheng Z, Eichler EE. 2006. Structural variation of the human genome. *Annu. Rev. Genom. Hum. Genet.* 7:407–42
106. Siepel A, Bejerano G, Pedersen JS, Hinrichs AS, Hou M, et al. 2005. Evolutionarily conserved elements in vertebrate, insect, worm, and yeast genomes. *Genome Res.* 15:1034–50
107. Simons C, Pheasant M, Makunin IV, Mattick JS. 2006. Transposon-free regions in mammalian genomes. *Genome Res.* 16:164–72
108. Smalheiser NR, Torvik VI. 2005. Mammalian microRNAs derived from genomic repeats. *Trends Genet.* 21:322–26
109. Smalheiser NR, Torvik VI. 2006. Alu elements within human mRNAs are probable microRNA targets. *Trends Genet.* 22:532–36
110. Smit AF, Riggs AD. 1996. Tiggers and DNA transposon fossils in the human genome. *Proc. Natl. Acad. Sci. USA* 93:1443–48
111. Sun FJ, Fleurdepine S, Bousquet-Antonelli C, Caetano-Anolles G, Deragon JM. 2007. Common evolutionary trends for SINE RNA structures. *Trends Genet.* 23:26–33
112. Tomascik-Cheeseman L, Marchetti F, Lowe X, Shamanski FL, Nath J, et al. 2002. CENP-B is not critical for meiotic chromosome segregation in male mice. *Mutat. Res.* 513:197–203
113. Tonegawa S. 1983. Somatic generation of antibody diversity. *Nature* 302:575–81
114. Toth M, Grimsby J, Buzsaki G, Donovan GP. 1995. Epileptic seizures caused by inactivation of a novel gene, *jerky*, related to centromere binding protein-B in transgenic mice. *Nat. Genet.* 11:71–75
115. Truniger V, Blanco L, Salas M. 1999. Role of the "YxGG/A" motif of Phi29 DNA polymerase in protein-primed replication. *J. Mol. Biol.* 286:57–69
116. Tudor M, Lobocka M, Goodell M, Pettitt J, O'Hare K. 1992. The pogo transposable element family of *Drosophila melanogaster*. *Mol. Gen. Genet.* 232:126–34
117. Ungerer MC, Strakosh SC, Zhen Y. 2006. Genome expansion in three hybrid sunflower species is associated with retrotransposon proliferation. *Curr. Biol.* 16:R872–73
118. Vaughn MW, Martienssen R. 2005. It's a small RNA world, after all. *Science* 309:1525–26
119. Volff J, Korting C, Schartl M. 2001. Ty3/Gypsy retrotransposon fossils in mammalian genomes: Did they evolve into new cellular functions? *Mol. Biol. Evol.* 18:266–70
120. Volff JN. 2006. Turning junk into gold: domestication of transposable elements and the creation of new genes in eukaryotes. *Bioessays* 28:913–22

121. Volff JN, Hornung U, Schartl M. 2001. Fish retroposons related to the Penelope element of *Drosophila* virilis define a new group of retrotransposable elements. *Mol. Genet. Genom.* 265:711–20
122. Walbot V, Rudenko GN. 2002. MuDR/Mu transposable elements of maize. In *Mobile DNA II*, ed. NL Craig, R Craigie, M Gellert, AM Lambowitz, pp. 533–64. Washington DC: ASM Press
123. Walter J, Hutter B, Khare T, Paulsen M. 2006. Repetitive elements in imprinted genes. *Cytogenet. Genome Res.* 113:109–15
124. Weiner AM, Deininger PL, Efstratiadis A. 1986. Nonviral retroposons: genes, pseudogenes, and transposable elements generated by the reverse flow of genetic information. *Annu. Rev. Biochem.* 55:631–61
125. Wilson AC, Sarich VM, Maxson LR. 1974. The importance of gene rearrangement in evolution: evidence from studies on rates of chromosomal, protein, and anatomical evolution. *Proc. Natl. Acad. Sci. USA* 71:3028–30
126. Wray GA. 2003. Transcriptional regulation and the evolution of development. *Int. J. Dev. Biol.* 47:675–84
127. Xie X, Kamal M, Lander ES. 2006. A family of conserved noncoding elements derived from an ancient transposable element. *Proc. Natl. Acad. Sci. USA* 103:11659–64
128. Zilberman D, Henikoff S. 2005. Epigenetic inheritance in *Arabidopsis*: selective silence. *Curr. Opin. Genet. Dev.* 15:557–62

Congenital Disorders of Glycosylation: A Rapidly Expanding Disease Family

Jaak Jaeken[1] and Gert Matthijs[2]

[1]Department of Pediatrics, Center for Metabolic Disease, University of Leuven, Leuven, Belgium B-3000; email: Jaak.Jaeken@uz.kuleuven.ac.be

[2]Center for Human Genetics, University of Leuven, Leuven, Belgium B-3000; email: Gert.Matthijs@uz.kuleuven.ac.be

Annu. Rev. Genomics Hum. Genet. 2007. 8:261–78

First published online as a Review in Advance on May 23, 2007.

The *Annual Review of Genomics and Human Genetics* is online at genom.annualreviews.org

This article's doi: 10.1146/annurev.genom.8.080706.092327

1527-8204/07/0922-0261$20.00

Key Words

CDG, CDG-x, hyperglycosylation, lipid hypoglycosylation, N-glycan, O-glycan

Abstract

Congenital disorders of glycosylation (CDG) are a large family of genetic diseases resulting from defects in the synthesis of glycans and in the attachment of glycans to other compounds. These disorders cause a wide range of human diseases, with examples emanating from all medical subspecialties. Since our 2001 review on CDG (36), this field has seen substantial growth: The number of N-glycosylation defects has doubled (from 6 to 12), five new O-glycosylation defects have been added to the two previously known ones, three combined N- and O-glycosylation defects have been identified, the first lipid glycosylation defects have been discovered, and a new domain, that of the hyperglycosylation defects, has been introduced. A number of CDG are due to defects in enzymes with a putative glycosyltransferase function. There is also a growing group of patients with unidentified defects (CDG-x), some with typical clinical presentations and others with presentations not seen before in CDG. This review focuses on the clinical, biochemical, and genetic characteristics of CDG and on advances expected in their future study and clinical management.

CDG: congenital disorders of glycosylation

N-glycosylation: the assembly of a glycan on and in the ER and its attachment to a particular asparagine of target proteins, followed by a remodeling of this glycan mainly in the Golgi

O-glycosylation: the assembly of a glycan and its attachment to a serine or threonine of target proteins, or the attachment of a monosaccharide to one of these amino acids

INTRODUCTION

Congenital disorders of glycosylation (CDG) are genetic diseases owing to deficient or increased glycosylation of glycoconjugates (mainly glycoproteins and glycolipids). The glycans on proteins are either N-linked (to the amide group of asparagine via an N-acetylglucosamine residue) or O-linked (to the hydroxyl group of serine or threonine via an N-acetylgalactosamine or other monosaccharide residue). The synthesis of the N-glycans encompasses a much longer pathway than that of the O-glycans because N-glycosylation, apart from an assembly pathway, also comprises a processing pathway that is lacking in O-glycosylation. N-glycosylation needs three cellular compartments: the cytosol, the endoplasmic reticulum (ER), and the Golgi. In the cytosol, the mannose donor, GDP-mannose, is synthesized from fructose 6-phosphate, an intermediate of the glycolytic pathway. N-glycan assembly is carried out first on the cytoplasmic and then on the lumenal side of the ER. This results in the formation of the lipid-linked oligosaccharide dolichylpyrophosphate-$GlcNAc_2Man_9Glc_3$. The oligosaccharide moiety of this compound is then transferred to selected asparagines of the nascent proteins. Finally, the glycan of the newly formed glycoprotein is processed first in the ER, by trimming off the three glucoses, and further in the Golgi, by trimming off six mannoses and replacing these with two residues each of N-acetylglucosamine, galactose, and sialic acid. O-glycan assembly usually starts in the Golgi. Contrary to N-glycan biosynthesis, it has no processing.

The phenotypic spectrum of glycosylation disorders ranges from severe to mild disease and from multisystem to mono-organ disease (for reviews, see 18, 21, 22, 24, 28, 32, 35–39, 43, 44, 46, 48, 50, 56, 83). The CDG field is expanding rapidly. The first patients were reported in 1980. Twenty eight disorders are currently known: 16 in protein N-glycosylation, 6 in protein O-glycosylation, 4 in both protein N- and O-glycosylation, and 2 in lipid glycosylation (**Table 1**).

Isoelectrofocusing of serum transferrin is the most widely used screening test for N-glycosylation disorders accompanied by sialic acid deficiency; sialic acid is an endstanding, negatively charged monosaccharide. In disorders of N-glycan assembly (CDG-I), a type 1 pattern is obtained, characterized by a decrease of anodal fractions and an increase of disialo- and asialotransferrin. The type 2 pattern also shows an increase of the trisialo- and/or monosialofractions as seen in defects of N-glycan processing (CDG-II) (**Figure 1**).

This review focuses on the 16 CDG (indicated in bold in **Table 1**) reported since our 2001 review on CDG (36) as well as on protein hyperglycosylation disorders.

DISORDERS OF N-GLYCAN ASSEMBLY

Mannosyltransferase VIII Deficiency (CDG-Ig)

Phenotype. Six patients have been described with this mannosyltransferase deficiency. Common clinical features include psychomotor retardation, hypotonia, facial dysmorphy, and decreased factor XI. In four patients decreased serum IgG was reported. Variable features included feeding problems (requiring gastrostomy in one), convulsions, microcephaly, frequent respiratory tract infections, convergent strabism, inverted nipples, micropenis, undescended testes, club foot, wide lateral ventricles, and low serum IgA, IgM, IGF-I, and IGF-binding protein 3. The serum sialotransferrin pattern showed only a small increase of asialotransferrin. The typical biochemical feature of this disorder is an accumulation of dolichyl pyrophosphate-$Man_7GlcNAc_2$ in fibroblasts (10, 14, 16, 27, 67).

Genotype. CDG-Ig is due to a deficiency in dolichyl pyrophosphate-$Man_7GlcNAc_2$

Table 1 Human genetic disorders of glycosylation

Disorders	Defective protein	Defective gene	OMIM
Defects in protein N-glycosylation			
CDG-Ia	Phosphomannomutase II	*PMM2*	212065
CDG-Ib	Phosphomannose isomerase	*MPI*	602579
CDG-Ic	Dol-P-Glc: Man_9-$GlcNAc_2$-P-P-Dol glucosyltransferase (glucosyltransferase I)	*ALG6*	603147
CDG-Id	Dol-P-Man: Man_5-$GlcNAc_2$-P-P-Dol mannosyltransferase (mannosyltransferase VI)	*ALG3*	601110
CDG-Ie	GDP-Man: Dol-P-mannosyltransferase (Dol-P-Man synthase I)	*DPM1*	603503
CDG-If	Lec35 (Man-P-Dol utilization 1)	*MPDU1*	608799
CDG-Ig	**Dol-P-Man: Man_7-$GlcNAc_2$-P-P-Dol mannosyltransferase (mannosyltransferase VIII)**	***ALG12***	**607143**
CDG-Ih	**Dol-P-Glc: Glc_1-Man_9-$GlcNAc_2$-P-P-Dol glucosyltransferase (glucosyltransferase II)**	***ALG8***	**608104**
CDG-Ii	**GDP-Man: Man_1-$GlcNAc_2$-P-P-Dol mannosyltransferase (mannosyltransferase II)**	***ALG2***	**607906**
CDG-Ij	**UDP-GlcNAc: Dol-P-GlcNAc-P transferase**	***DPAGT1***	**608093**
CDG-Ik	**GDP-Man: $GlcNAc_2$-P-P-Dol mannosyltransferase (mannosyltransferase I)**	***ALG1***	**608540**
CDG-Il	**Dol-P-Man: Man_6-and Man_8-$GlcNAc_2$-P-P-Dol mannosyltransferase (mannosyltransferase VII-IX)**	***ALG9***	**608776**
CDG-IIa	N-acetylglucosaminyltransferase II	*MGAT2*	212066
CDG-IIb	Glucosidase I	*GLS1*	606056
CDG-IIc	GDP-fucose transporter	*SLC35C1/FUCT1*	266265
CDG-IId	**β-1,4 galactosyltransferase**	***B4GALT1***	**607091**
Defects in protein O-glycosylation			
O-xylosylglycan defects - multiple cartilaginous exostoses - Ehlers-Danlos syndrome	Glucuronyltransferase/N-acetylglucosaminyltransferase β-1,4-galactosyltransferase 7	*EXT1/EXT2* *B4GALT7*	133700/133701 130070
O-N-acetylgalactosaminylglycan defect - familial tumoral calcinosis	**Polypeptide N-acetylgalactosaminyltransferase 3**	***GALNT3***	**211900**
O-mannosylglycan defects - Walker-Warburg syndrome - muscle-eye-brain disease	**O-mannosyltransferase 1 O-mannose β-1,2-N-acetylglucosaminyltransferase**	***POMT1/POMT2* *POMGNT1***	**236670 253280**
O-fucosylglycan defect - spondylocostal dysostosis type 3	**O-fucose-specific β-1,3-N-acetylglucosaminyltransferase**	***SCD03***	**602576**
Combined protein N- and O-glycosylation defects			
Hereditary inclusion body myopathy	UDP-GlcNAc epimerase/kinase	*GNE*	600737
CDG-II/COG7	**Conserved oligomeric Golgi complex subunit 7**	***COG7***	**608779**
CMP-sialic acid transporter deficiency	**CMP-sialic acid transporter**	***SLC35A1***	**605634**
CDG-II/COG1	**Conserved oligomeric Golgi complex subunit 1**	***COG1***	**606973**
Defects in lipid glycosylation			
Amish infantile epilepsy	**Lactosylceramide α-2,3 sialyltransferase (GM3 synthase)**	***SIAT9***	**609056**
Glycosylphosphatidylinositol deficiency	**Phosphatidylinositolglycan, class M**	***PIGM***	**610273**

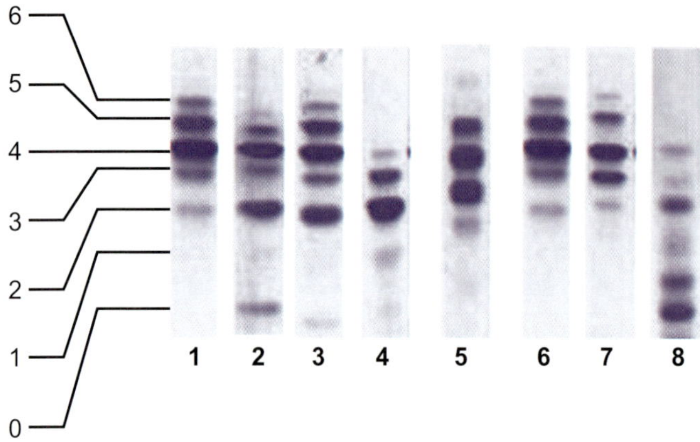

Figure 1

Basic patterns of serum transferrin isoelectrofocusing. Lanes 1 and 6, normal pattern; lanes 2 and 3, type 1 pattern; lanes 4 and 5, type 2 pattern; lane 7, transferrin protein variant; lane 8, the artifactual pattern obtained when using normal EDTA plasma. Numbers on the left indicate sialotransferrin fractions 0 to 6. Reprinted from Reference 36.

CMP: cytidine monophosphate

COG: conserved oligomeric Golgi (complex)

GM3 synthase: sialyltransferase that is required for the synthesis of the ganglioside GM3 (sialic acid-2,3-galactose-β1,4-glucose-ceramide) from lactosylceramide

mannosyltransferase (dolichyl-P-Man) caused by mutations in the *hALG12* gene. To date, 10 different mutations have been identified. Two patients are homozygous for a missense mutation [c.424T>G (p.F142V) or c.241G>A (p.A81T)] and the others are compound heterozygous for either a missense [c.824G>A (p.S275N) or c.473T>C (p.L158P)] and a nonsense mutation [c.1242C>G (p.Y414X)] or one single base pair deletion (c.29delG), or for two missense mutations [c.688T>G (p.Y230D) and c.931C>T (p.R311C); c.200C>T (p.T67M) and c.437G>A (p.R146Q)] (10, 14, 16, 27, 67).

Glucosyltransferase II Deficiency (CDG-Ih)

Phenotype. This CDG has been reported in five children from four families with a clinical spectrum from mild to very severe. The first patient was described in 2003 with protein-losing enteropathy, hypoalbuminemia, ascites, edema, moderate hepatomegaly, a slightly abnormal electroretinogram, and low factor XI, protein C, and antithrombin. After 18 months of treatment with a low-fat diet and essential fatty acid supplements, the protein-losing enteropathy resolved; at three years psychomotor development was normal. In 2004, three other patients (from two families) were reported with dysmorphy, thorax and lung hypoplasia, cardiorespiratory problems, hypotonia, multiple cystic bile ducts, bilateral renal micro cysts, absent extramedullary hematopoiesis, anemia, thrombocytopenia, and early fatal outcome (3 d–3 m). The fifth patient, reported in 2005, showed central nervous system involvement (severe macrocrania at birth), psychomotor retardation and seizures with diffuse leukoencephalopathy), cataracts, osteopenia, and pancytopenia, and died at 16 months from massive bleeding (9, 17, 59).

Genotype. Mutations in the *hALG8* gene lead to a deficiency in dolichyl-P-Glc: Glc$_1$Man$_9$GlcNAc$_2$ glycosyltransferase, with an accumulation of dolichyl pyrophosphate-linked Glc$_1$Man$_9$GlcNAc$_2$ and Man$_9$GlcNAc$_2$. Six different mutations have been described. One patient is homozygous, the others are compound heterozygous. One patient is compound heterozygous for two truncating mutations, 396insA and 413delC (9). The other patients are compound heterozygous for a splice defect (c.96–2A>C in intron 1 or c.672+4 A>G in intron 6) and a missense mutation [c.139A>G (p.T47P) in exon 2 or c.824 G>A (p.G275D) in exon 8]. One patient is homozygous for the p.T47P mutation (17, 59).

Mannosyltransferase II Deficiency (CDG-Ii)

Phenotype. In 2003, Thiel et al. (68) reported for the first time a defect in one of the early steps of dolichol-linked oligosaccharide biosynthesis at the cytosolic side of the endoplasmic reticulum. The patient was a three-year-old girl. She had bilateral coloboma of the iris, unilateral cataract, and an irregular nystagmus. At four months the child started infantile spasms with hypsarrhythmia. An MRI of the brain showed pronounced dysmyelination. She had brisk tendon reflexes. Psychomotor development was severely retarded. The activated partial thromboplastin time and clotting factor IX were strongly reduced. Isoelectrofocusing of serum transferrin showed a type 1 pattern. By using the more hydrophobic chloroform-methanol (3:2) as well as the usual chloroform-methanol-water (10:10:3) for extraction, an accumulation of $Man_1GlcNAc_2$-PP-dolichol and $Man_2GlcNAc_2$-PP-dolichol was observed in skin fibroblasts of the patient. A severely reduced activity of the mannosyltransferase-elongating $Man_1GlcNAc_2$-PP-dolichol was found. This enzyme had not yet been characterized in higher eukaryotes. It acts as an α-1,3-mannosyltransferase. The resulting Manα1,3-$Man_1GlcNAc_2$-PP-dolichol is further elongated by an as-yet-unknown α1,6-mannosyltransferase.

Genotype. This α-1,3-mannosyltransferase is encoded by *hALG2*. Interestingly, two transcripts have been identified, which should be kept in mind during mutation detection. The only patient described is compound heterozygous for c.393G>T (p.K131N) and c.1040delG, causing a frameshift after amino acid 346 and a premature translation stop after amino acid 372 (68).

UDP-GlcNAc: Dolichol Phosphate N-acetylglucosamine 1-phosphate Transferase Deficiency (CDG-Ij)

Phenotype. In 2003, Wu et al. (78) described another defect in the very early steps of the dolichol-linked oligosaccharide biosynthesis. The patient, a girl, showed intractable infantile spasms at four months. Her psychomotor development was severely retarded and she was very hypotonic. Dysmorphic features such as microcephaly, esotropia, arched palate, micrognathia, fifth finger clinodactyly, single palmar creases, and dimples on the upper part of the thighs were present. An MRI of the brain at six months was normal, but positron emission tomography showed multifocal areas of decreased metabolic activity. Isoelectrofocusing of serum transferrin revealed a mild type 1 pattern. In the microsomal fractions of cultured fibroblasts a 10% residual activity of uridine diphosphate (UDP)-GlcNAc: dolichol phosphate N-acetylglucosamine 1-phosphate transferase was found.

Genotype. The transferase is encoded by *DPAGT1*. This patient is a compound heterozygote at the *DPAGT1* locus with a missense mutation c.660A>G (p.Y170C) and a complex splice mutation. It has been shown that the latter allele produces about 12% of normal mature messenger RNA (mRNA), with the remainder showing a complex exon skipping pattern, all leading to truncated (if any), nonfunctional enzyme (78).

Mannosyltransferase I Deficiency (CDG-Ik)

Phenotype. In 2004, four male patients with this severe syndrome were reported. They showed epilepsy (refractory in three, and starting as early as 2 h in one), severe psychomotor retardation, and early death in three (2 weeks to 10 months). There were many associated variable features, including dysmorphy, microcephaly, hypotonia, stupor, liver dysfunction, coagulation abnormalities, cardiomyopathy, fetal hydrops, nephrotic syndrome, hypogonadism, severe infections, and depletion of β cells. In three patients, an MRI of the brain revealed cerebral atrophy in two and normal findings in the other patient. On isoelectrofocusing of serum transferrin there was only a very small increase of tetrasialotransferrin. Lipid-linked oligosaccharide analysis showed an accumulation of $GlcNAc_2$-PP-dolichol in fibroblasts and enzymatic investigation a deficiency of mannosyltransferase I (26, 40, 60).

Genotype. Ik is caused by a deficiency in ALG1. This β-1,4 mannosyltransferase is encoded by *ALG1*. It adds the first mannose residue to lipid-linked oligosaccharide. All described patients share the missense mutation p.S258L (c.733C>T). Two patients are homozygous for this mutation. They are the most severely affected patients, with death at 2 and 11 weeks of age. The other two patients are compound heterozygotes with this allele and either c.1025A>C (p.E342P) or c.450C>G (p.S150R) plus c.1287T>A (p.D429E).

Mannosyltransferase VII/IX Deficiency (CDG-IL)

Phenotype. CDG-IL has been reported in two patients in 2004 and 2005, respectively. Clinical features of both were psychomotor retardation, hypotonia, seizures, microcephaly, and hepatomegaly. The second patient also showed a failure to thrive, strabismus, inverted nipples, pericardial effusion, renal cysts, and focal partial villous atrophy. A defect in mannosyltransferase VII/IX was suggested by the accumulation of $Man_6GlcNAc_2$-PP-dolichol and $Man_8GlcNAc_2$-PP-dolichol and confirmed by mutation analysis (20, 72).

Genotype. Two missense mutations have been described: p.E523K (c.1567G>A) (72) and p.Y286C (c.860 A>G) (20). Both patients are homozygous for the pathogenic mutation.

DISORDER OF N-GLYCAN PROCESSING

β-1,4-glactosyltransferase I Deficiency (CDG-IId)

Phenotype. Only one patient with a deficiency of the isoenzyme 1 of β-1,4-galactosyltransferase has been reported. This boy showed psychomotor retardation, Dandy-Walker malformation, severe spontaneous bleeding, a transient cholestatic syndrome, hydrocephalus due to aqueductus stenosis, and myopathy. Laboratory investigation revealed an increase of serum glutamate-oxaloacetate transaminase (with normal glutamate-pyruvate transaminase) and of creatine kinase as well as a marked reduction of IGF-I- and IGF-binding protein-3 despite normal growth. The isoelectrofocusing type 2 pattern of serum sialotransferrins was characterized by a prominent asialotransferrin and monosialotransferrin and strongly reduced tetrasialotransferrin (29, 57). The existence of this phenotype indicates that the function of isoenzyme 1 does not overlap, or only partially overlaps, with that of the six other β-1,4-galactosyltransferase isoenzymes (32).

Genotype. The only patient with CDG-IId is homozygous for the insertion of a C between positions 1031 and 1032 in *B4GALT1*, leading to a C-terminal truncation of the β-1,4-galactosyltransferase I. The truncated protein is catalytically inactive and is retained in the endoplasmic reticulum (29).

O-N-ACETYLGALACTOSAMINYL-GLYCAN DEFECT

O-N-acetylgalactosaminyltransferase 3 Deficiency (Hyperphosphatemic Familial Tumoral Calcinosis; Hyperostosis-hyperphosphatemia Syndrome)

Phenotype. Familial tumoral calcinosis is characterized by recurrent, painful subcutaneous calcium depositions that can cause secondary infections and incapacitating mutilation. There is a hyperphosphatemic and a normophosphatemic type. The hyperphosphatemic type was identified by Topaz et al. (69) as the first defect in mucin type O-glycosylation, more specifically in one of the 24 isoenzymes of O-N-acetylgalactosaminyltransferase. This disease is allelic with the hyperphosphatemia-hyperostosis syndrome characterized by elevated serum phosphate levels and repeated attacks of acute, painful swellings of the long bones with radiological evidence of periosteal reaction and cortical hyperostosis (25).

Genotype. The gene involved in this disease is *GALNT3* (UDP-N-acetyl-α-D-galactosamine: polypeptide N-acetylgalactosaminyltransferase 3), a glycosyltransferase responsible for initiating mucin-type O-glycosylation. Mutations in *GALNT3* were identified in at least eight families. Members of three families carried a homozygous splice site mutation (c.1524+1G>A) disrupting a donor splice site consensus sequence of intron 7, leading to an inframe deletion of 44 amino acid residues (25, 69). One family showed two heterozygous mutations: a nonsense mutation, c.484C>T (p.R162X), and a splice site mutation, c.1524+5G>A (69). This nonsense mutation (p.R162X) was also found in combination with another splice site mutation, c.515–2A>T (34). Two additional families with different homozygous nonsense mutations were described: c.1387 A>T (p. K463X) (8) and c.1774C>T (p.Q592X) (65). In a 25-year-old patient with tumoral calcinosis, two missense mutations, c.815 C>A (T272K) and c.1076 G>A (T359K), were identified (33).

O-MANNOSYLGLYCAN DEFECTS

Protein O-mannosyltransferase 1 Deficiency (Walker-Warburg Syndrome, Limb-girdle Muscular Dystrophy 2K, Intermediate Phenotype)

Phenotype. Three different phenotypes have been associated with this defect. The first is Walker-Warburg syndrome, a rare genetic disorder with brain and eye dysgenesis associated with congenital muscular dystrophy. This neuronal migration disorder usually runs a fatal course before the age of one year. Psychomotor development is absent. The brain lesions consist of cobblestone lissencephaly, agenesis of the corpus callosum, cerebellar hypoplasia, hydrocephaly, and sometimes encephalocoele. There is an aberrant glycosylation of α-dystroglycan, an external membrane protein expressed in muscle, brain, and other tissues. Most glycans of this heavily glycosylated protein seem to be O-linked via mannose and control the interaction with extracellular matrix proteins. Disrupted glycosylation of α-dystroglycan (and probably other glycoproteins) results in loss of this interaction and, hence, in progressive muscle degeneration and abnormal neuronal migration in the brain. The enzyme catalyzes the first step in the synthesis of the O-mannose-linked core Galβ1-4GlcNAcβ1-2Man-O-Ser/Thr. This defect is only present in a subgroup (ca. 20%) of patients with Walker-Warburg syndrome (4). Another, much milder phenotype caused by this defect is limb-girdle muscular dystrophy with mental retardation and microcephaly, but without eye and brain abnormalities (3). The third phenotype shows a severity in between the two others: a congenital muscular dystrophy with early onset, severe motor disability, microcephaly, and mental retardation not associated with structural brain changes on neuroimaging (13).

Genotype. Most mutations in *POMT1* involved in Walker-Warburg syndrome lead to amino acid changes in the highly conserved protein mannosyltransferase (PMT) and mannosyltransferase IP3R and RyR (MIR) domains. Frameshifts and nonsense mutations have also been found. A survey of the different mutations is out of the scope of this review (see e.g., 4).

The mutation that causes limb-girdle muscular dystrophy 2 (LGMD2) is of the missense type: In five Turkish patients, a homozygous p.A200P (c.598G>C) was found (3). In all five patients this mutant allele was part of a conserved core haplotype, which suggests a ancestral founder mutation. The patient with the intermediate phenotype is a Japanese child, homozygous for deletion of a single amino acid (p.L421del) (3).

Protein O-mannosyltransferase 2 Deficiency (Walker-Warburg Syndrome)

Phenotype. Together with protein O-mannosyltransferase 1 deficiency, fukutin deficiency and fukutin-related protein deficiency, this disorder accounts for about one third of patients with Walker-Warburg syndrome. Both POMT1 and POMT2 are required to achieve protein O-mannosyltransferase activity (70).

Genotype. The *POMT2* gene contains 21 exons and shares 36% sequence identity with *POMT1*. Three mutations were described that severely impacted the protein (70). A homozygous nonsense mutation, p.R638X (c.1912C>T), was found in a Moroccan patient born of consanguineous parents. Another homozygous splice mutation (c.1005+1G>A) in intron 8 was found in a Pakistani patient born of consanguineous parents. A homozygous 1-bp deletion (c1261delC) was identified in two siblings, again in a consanguineous family.

O-mannosyl-β1,2-N-acetylglucosaminyltransferase 1 Deficiency (Muscle-eye-brain Disease)

Phenotype. This is a neuronal migration/congenital muscular dystrophy syndrome similar to Walker-Warburg syndrome, but less severe. The defective protein catalyzes the second step in the synthesis of the O-mannosylglycan core (80).

Genotype. The *POMGnT1* gene contains 22 exons and encodes a transmembrane protein of 660 amino acids. Currently, more than 20 mutations associated with muscle-eye-brain disease have been reported. These mutations are located throughout the gene, with no obvious mutational hot spots. The mutations either affect amino acids in the catalytic domain or predict truncated proteins.

This disease was first described in 1977 in Finland, where it was enriched because of a founder effect. The most common single *POMGnT1* mutation in the Finnish population underlying muscle-eye-brain disease is the c.1539+1G>A splice mutation.

O-FUCOSYLGLYCAN DEFECT

O-fucose-specific β1,3-N-acetylglucosaminyltransferase Deficiency (Spondylocostal Dysostosis Type 3)

Phenotype. The spondylocostal dysostoses are a group of vertebral malsegmentation disorders that arise during embryonic development due to a disruption of somitogenesis. Three types with defects in different components of the Notch signaling pathway are known. The most severe type, SCD3, is the result of defective lunatic fringe, a fucose-specific β1,3-N-acetylglucosaminyltransferase that adds N-acetylglucosamine residues to O-fucose on the EGF-like repeats of Notch receptors. This illustrates that proper regulation of the Notch signaling pathway is necessary for the correct patterning of the axial skeleton (64).

Genotype. In a Lebanese family with this severe autosomal-recessive form of spondylocostal dysostosis (SCD3), a homozygous mutation was found in LUNATIC FRINGE (LFNG). *LFNG* encodes a glycosyltransferase (O-fucosylpeptide β1,3-N-acetylglucosaminyl-transferase). A homozygous mutation in exon 3 (c.564 C>A) results in an alteration from the highly conserved phenylalanine to leucine (p.F188L) (64).

COMBINED PROTEIN N- AND O-GLYCOSYLATION DEFECTS

CDG-II Due to COG7 Deficiency

Phenotype. The discovery of this disease opens a new domain in the CDG field because the affected protein is involved in trafficking the glycosylation machinery. The first reported patients, two siblings, showed dysmorphy, particularly of the face, encephalopathy. and cholestatic liver disease. They died at 5 and 10 weeks, respectively. Serum transferrin isoelectrofocusing showed a type 2 pattern. Studies of fibroblast glycoproteins showed a partial N- and O-glycosylation defect caused by a decreased transport of CMP-sialic acid and UDP-galactose into the Golgi as well as a reduced activity of two glycosyltransferases involved in the galactosylation and sialylation of O-glycans. This disruption of multiple glycosylation pathways resembles the situation in Chinese hamster ovary cell mutants with mutations in genes encoding the COG1 and COG2 subunits of the conserved oligomeric Golgi (COG) complex. This eight-subunit complex is involved in protein transport between the endoplasmic reticulum and the Golgi, and within the Golgi. Wu et al. (63, 66, 79) investigated the localization of the COG subunits by indirect immunofluorescence and found an abnormal distribution of COG5, COG6, and COG8. They eventually identified the basic defect in the COG7 subunit. Recently, a second family with two siblings was reported. Their phenotype was similar to that of affected individuals in the first family, except for less pronounced liver involvement. The siblings also showed pre- and postnatal growth retardation and episodes of hyperthermia (82).

Genotype. Sequencing the *hCOG7* gene has revealed a homozygous intronic mutation (c.169+4A>C) in all patients reported thus far. This mutation impairs splicing at the canonical donor splice site and activates a cryptic splice site, resulting in the deletion of 19 base pairs at the complementary DNA (cDNA) level (79).

CDG-II Due to COG1 Deficiency

Phenotype. Contrary to the COG7-deficient patients, the single patient with COG1 deficiency had a mild clinical presentation. She had feeding problems from birth and was hospitalized twice. Generalized hypotonia, mainly of the proximal limbs, was associated with normal muscular strength. She had antimongoloid eyelids and small hands and feet. There was a slight left ventricular hypertrophy. At the age of 7 months she developed rhizomelic-type growth retardation. At the last observation (21 months), she had only mild psychomotor retardation, a slightly enlarged liver and spleen, but progressive microcephaly. An MRI of the brain revealed only slight cerebral and cerebellar hypotrophy. The isoelectrofocusing pattern of serum sialotransferrin was a type 2 and serum apoC-III showed a hypoglycosylation profile, which pointed to a combined N- and O-glycosylation defect. Mass spectrometric analysis of the N-linked glycans from serum proteins revealed a reduction in sialic acid and galactose residues. Quantitative Western blotting of the COG subunits in the patient's fibroblasts showed a significant decrease of the COG2, -3, -4, and -8 subunits. A mutation was found in the *COG1* gene, which destabilizes several other COG subunits and alters their subcellular localization and, hence, the global effect on the overall integrity of the COG-complex. This results in reduced levels and/or altered Golgi localization of α-mannosidase II and β-1,4-galactosyltransferase 1 (19).

Genotype. The only patient described so far has a homozygous single nucleotide insertion in the *hCOG1* gene (c.2659–2660insC). The insertion shifts the open reading frame from

Sialyl-Le^x antigen: a branched O-linked glycan, in which fucose is bound to GlcNAc; functions as a ligand for selectin, which is important for leukocyte trafficking

Gangliosides: glycolipids composed of ceramide and sialic-acid containing oligosaccharides

codon 888 onward, thereby introducing a stop codon at position 900. This results in the production of a truncated Cog1 protein, lacking the normal C-terminal 80 amino acids (19).

CMP-sialic Acid Transporter Deficiency

Phenotype. This newly recognized syndrome was reported in a four-month-old boy with macrothrombocytopenia, neutropenia, and a complete lack of the sialyl-Lex antigen on polymorphonuclear cells. The initial clinical presentation was a spontaneous massive bleeding of the posterior chamber of the right eye and cutaneous hemorrhages, followed by more severe hemorrhages due to pronounced thrombocytopenia and by recurring infections. Bone marrow transplantation was performed at the age of 34 months, but numerous complications finally led to death at 37 months. The bone marrow was normocellular with megakaryocytic hyperplasia and dysmorphy. The isoelectrofocusing pattern of serum transferrin was only slightly abnormal whereas serum apoC-III was clearly hypoglycosylated. As all the glycosyltransferases involved in the biosynthesis of sialyl-Le^x were normal, an analysis was performed of the CMP-sialic acid transporter gene (see Genotype, below). It is not clear why sialic acid deficiency seems to affect O-glycosylation much more than N-glycosylation (49).

Genotype. This male patient has a complex genotype. He inherited a CACT insertion in exon 6 in the *SLC35A1* gene from his mother. This insertion creates a novel U2 snRNA site that competes with the normal site and induces a deletion in exon 6. However, like many CDG mutations, this is a leaky mutation because the mother of the patient and two normal blood donors (out of 53 screened) were homozygous for this variation, without any clinical manifestation. The second allele had two microdeletions (G277 and C281), combined with a 147T>C substitution, leading to a completely inactive gene (49).

LIPID HYPOGLYCOSYLATION DEFECTS

Lactosylceramide α-2,3 Sialyltransferase Deficiency

Phenotype. This is the first reported genetic defect in lipid glycosylation. It was identified in a large Old Order Amish pedigree with an infantile-onset epilepsy syndrome. The first symptoms were irritability and poor feeding. Seizures (grand mal as well as other types) started within the first year of life. The patients showed profound developmental stagnation with regression. An MRI of the brain revealed diffuse atrophy at an older age. The defective enzyme, also called GM3 synthase, is a member of the sialyltransferase family and catalyzes the initial step in the biosynthesis of most complex gangliosides from lactosylceramide. Plasma glycosphingolipid analysis showed an increase of lactosylceramide and a deficiency of GM3 ganglioside (62).

Genotype. The GM3 synthase is encoded by *SIAT9* (sialyltransferase 9). In the members of the described Amish pedigree, a base substitution was found in exon 8 of the *SIAT9* gene (c.694 C>T), causing a premature termination (R232X) of the GM3 synthase (62).

Glycosylphosphatidylinositol Deficiency

Phenotype. This is the first reported genetic defect in the glycosylation of the GPI anchor. GPI-anchored proteins have heterogeneous functions (as enzymes or adhesion molecules, among others). The disease has been reported in two patients in unrelated families who presented with portal and hepatic vein thrombosis by the age of 2 years. Subsequently, they developed persistent absence seizures. No other vascular abnormalities were found on magnetic resonance angiography. A thrombophilia screen revealed a positive hereditary erythroblastic multinuclearity with positive acidified-serum (HAM)

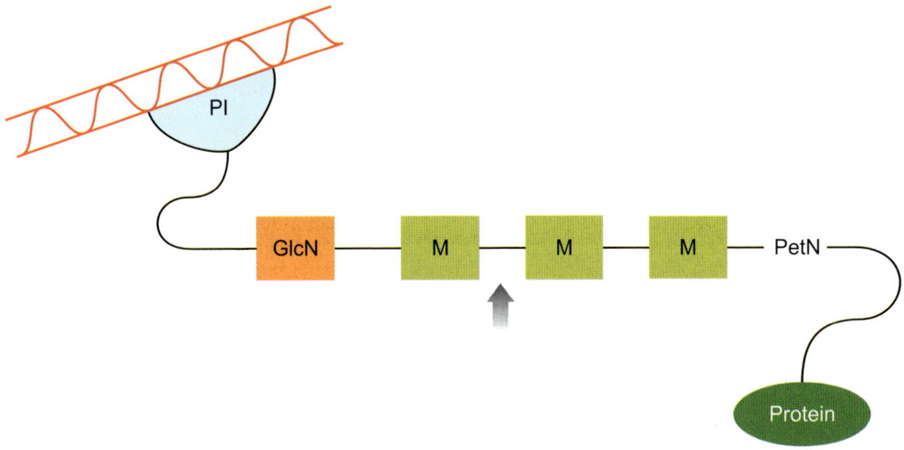

Figure 2

Schematic representation of a glycosylphosphatidylinositol (GPI) anchor. Arrow indicates the phosphatidylglycan, class M (PIGM) defect. PI, phosphatidylinositol; GlcN, glucosamine; M, mannose; PetN, hosphoethanolamine.

test, indicating absence from the erythrocyte surface of the GPI-linked complement inhibitor CD59. Expression of GPI on blood cells was severely reduced. Eventually, a defect was found in the addition of the first mannose of the GPI glycan (**Figure 2**), a step catalyzed by the mannosyltransferase complex comprising two proteins, phosphatidylinositol glycan, class M (PIGM) and phosphatidylinositolglycan, class X (PIGX). Genetic analysis identified *PIGM* as the disease gene (1). This disorder is different from another GPI-anchor glycosylation defect, paroxysmal nocturnal hemoglobinuria, that is due to somatic mutations in hematopoietic stem cells in the X-linked *PIGA* gene that disrupt the incorporation of glucosamine on phosphatidylinositol (1).

Genotype. In both unrelated families, a homozygous point mutation was found in the promotor region of the *PIGM* gene (c.1-270C>G). This mutation disrupts binding of the transcription factor Sp1 to its cognate promoter motif, resulting in a strongly decreased transcription of *PIGM* (virtually zero) and blockage of the mannosylation of GPI, leading to a severe (even if partial) deficiency in GPI (1).

PROTEIN HYPERGLYCOSYLATION

Mutations involving gains of glycosylation are rare, and the pathogenic role of the new carbohydrate chains has never been formally established. Vogt et al. (71) identified three children with genetic susceptibility to mycobacterial disease who were homozygous for a missense mutation in the gene for interferon γ receptor ligand-binding chain (IFNγR2), creating a new N-glycosylation site in the IFNγR2 chain. The authors showed that the resulting additional carbohydrate moiety was necessary and sufficient to abolish the cellular response to interferon γ. They also searched the Human Gene Mutation Database for potential gain-of-N-glycosylation missense mutations; of 10047 mutations in 577 genes encoding proteins trafficked through the secretory pathway, they identified 142 candidate mutations ($\approx$1.4%) in 77 genes ($\approx$13.3%). This indicates that an unexpectedly high proportion of disease gene mutations might lead to the creation of new N-glycosylation sites.

Glycosylphosphatidylinositol (GPI) anchor: composed of a mannose- and glucosamine-containing glycan coupled to phosphatidylinositol; this glycolipid is transferred to C-terminal regions of proteins and anchors them into membranes

Their pathogenic effects may be a direct consequence of adding N-linked carbohydrate.

PUTATIVE CONGENITAL DISORDERS OF GLYCOSYLATION

These are CDG in which the mutated gene has been identified, but the function of the encoded protein is unknown. These proteins have a putative glycosyltransferase signature, but no transferase activity has been demonstrated (**Table 2**).

CDG-x

This term indicates all syndromes with strong evidence for a glycosylation defect, but in which the defective gene has not yet been identified. This is a rapidly growing, clinically very heterogeneous group with a number of patients with various combinations of rather aspecific symptoms (psychomotor retardation, epilepsy, etc.) (2, 11, 12, 52, 53, 58, 76, 81) as well as several clinically recognizable groups. Examples of the latter are: congenital dyserythropoietic anemia type II (41), isolated liver disease (7, 47), cutis laxa syndromes in patients in whom the known cutis laxa defects have been excluded (54, 75, 77), and a syndrome with glomerulopathy and buphtalmia (73).

CONCLUSION AND PERSPECTIVES

The CDG field continues to expand at a rapid pace. Major advances are expected in the following areas:

- the generalized application of screening techniques for CDG (10, 74) and the introduction of new diagnostic techniques;
- the elucidation of putative CDG;
- organ-specific CDG: there is no doubt that, besides liver-specific CDG (7, 47), specific glycanopathies of other organs (brain, heart, kidney, etc.) will be discovered;
- defects in dolichol metabolism (still to be discovered): because dolichol-PP is the anchor for the growing glycan outside and inside the ER, defects in this field will be associated with glycosylation defects (6);
- defects in flippases (31) and oligosaccharyltransferase (43) (still to be discovered);
- defects in transportproteins besides COG (COPI, COPII, etc.);
- understanding the pathogenesis of CDG by investigating animal models (23), by proteome analysis (7, 42) and by other methods (61).

Table 2 Putative human congenital disorders of glycosylation

Disorder	Defective gene	Defective enzyme	Clinical summary	Ref.
Fukuyama muscular dystrophy	*FCMD*	Fukutin, a putative glycosyltransferase	neuronal migration disorder, psychomotor retardation, epilepsy, severe muscle weakness	(51)
Congenital muscular dystrophy type 1C (MDC1C)	*FKRP*	Fukutin-related protein, a putative glycosyltransferase	psychomotor retardation, variable muscle weakness, cardiomyopathy	(5)
Congenital muscular dystrophy type 1D (MDC1D)	*LARGE*	Large, a putative glycosyltransferase	profound mental retardation, muscular dystrophy	(45)
Peters Plus syndrome	*B3GALTL*	β1,3-galactosyltransferase-like protein, a putative glycosyltransferase	anterior eye-chamber defects, psychomotor retardation, growth retardation, other dysmorphy	(55)

The most important issue for the patient is treatment. Progress has been frustratingly slow as only CDG-Ib is efficiently treatable. Current experiments are using membrane-permeant derivatives of mannose 1-phosphate as a possible treatment for the most common N-glycosylation disease, CDG-Ia (15, 30).

SUMMARY POINTS

1. Genetic glycosylation disorders cause a remarkably wide range of phenotypes, with examples from all medical specialities.
2. Many well-known diseases such as Walker-Warburg syndrome and muscle-eye-brain disease are caused by a glycosylation defect.
3. Some clinical pictures such as hyperphosphatemic tumoral calcinosis and spondylocostal dystosis type 3 are totally unexpected on the basis of their glycosylation defect.
4. Defects in protein complexes functioning in transport in and between ER and Golgi (such as COG) promise to become an important new group of glycosylation disorders.
5. Lipid glycosylation defects are an emerging domain in the field of CDG.
6. Particular attention has to be given to another new field, the hyperglycosylation disorders.
7. Screening methods covering all (or most) CDG are urgently needed.
8. It is proposed to classify all hypo- and hyperglycosylation disorders under the umbrella name CDG.

DISCLOSURE STATEMENT

The authors are not aware of any biases that might be perceived as affecting the objectivity of this review.

ACKNOWLEDGMENTS

We thank our coworkers for their assistance in the preparation of this manuscript and for their contributions to the CDG Web site and mutation database (**http://www.euroglycanet.org**): François Foulquier, Lililane Geyskens, Liesbeth Keldermans, Ellen Reynders, Els Schollen, Wendy Vleugels, and Renate Zeevaert. This work was supported by the Sixth Framework program of the European Union (Euroglycanet: contract LSHM-2005-512131) and by the Körber Foundation.

LITERATURE CITED

1. **Almeida AM, Murakami Y, Layton DM, Hillmen P, Sellick GS, et al. 2006. Hypomorphic promotor mutation in *PIGM* causes inherited glycosylphosphatidylinositol deficiency. *Nat. Med.* 12:846–51**
2. Assmann B, Hackler R, Peters V, Schaefer JR, Arndt T, et al. 2001. A new subtype of a congenital disorder of glycosylation (CDG) with mild clinical manifestations. *Neuropediatrics* 32:313–18

1. The first reported human genetic defect in the glycosylation of the GPI anchor.

3. Balci B, Uyanik G, Dincer P, Gross C, Willer T, et al. 2005. An autosomal recessive limb girdle muscular dystrophy (LGMD2) with mild mental retardation is allelic to Walker-Warburg syndrome (WWS) caused by a mutation in the *POMT1* gene. *Neuromuscular Disord.* 15:271–75

4. Beltrán-Valero de Bernabé D, Currier S, Steinbrecher A, Celli J, van Beusekom E, et al. 2002. Mutations in the O-mannosyltransferase gene *POMT1* give rise to the severe neuronal migration disorder Walker-Warburg syndrome. *Am. J. Hum. Genet.* 71:1033–43

4. A minority of patients with the well-known Walker-Warburg syndrome have protein O-mannosyltransferase 1 deficiency, the first reported genetic defect in O-mannosylglycan synthesis.

5. Brockington M, Blake DJ, Prandini P, Brown SC, Torelli S, et al. 2001. Mutations in the fukutin-related protein gene (FKRP) cause a form of congenital muscular dystrophy with secondary laminin alpha2 deficiency and abnormal glycosylation of alpha-dystroglycan. *Am. J. Hum. Genet.* 69:1198–209

6. Burda P, Aebi M. 1999. The dolichol pathway of N-linked glycosylation. *Biochim. Biophys. Acta* 1426:239–57

7. Butler M, Quelhas D, Critchley AJ, Carchon H, Hebestreit HF, et al. 2003. Detailed glycan analysis of serum glycoproteins of patients with congenital disorders of glycosylation indicates the specific defective glycan processing step and provides an insight into pathogenesis. *Glycobiology* 13:601–22

8. Campagnoli MF, Pucci A, Garelli E, Carando A, Defilippi C, et al. 2006. Familial turmoral calcinosis and testicular microlithiasis associated with a new mutation of GALNT3 in a white family. *J. Clin. Pathol.* 59:440–42

9. Chantret I, Dancourt J, Dupré T, Delenda C, Bucher S, et al. 2003. A deficiency in dolichyl-P-glucose: $Glc_1Man_9GlcNAc_2$-PP-dolichyl alpha3-glucosyltransferase defines a new subtype of congenital disorders of glycosylation. *J. Biol. Chem.* 278:9962–71

10. Chantret I, Dupré T, Delenda C, Bucher S, Dancourt J, et al. 2002. Congenital disorders of glycosylation type Ig is defined by a deficiency in dolichyl-P-mannose: $Man_7GlcNAc_2$-PP-dolichyl mannosyltransferase. *J. Biol. Chem.* 277:25815–22

11. Carchon HA, Chevigné R, Falmagne JB, Jaeken J. 2004. Diagnosis of congenital disorders of glycosylation by capillary zone electrophoresis of serum transferrin. *Clin. Chem.* 50:101–11

11. Shows that capillary zone electrophoresis of serum transferrin is a rapid and reliable method for screening sialic acid-deficient N-glycosylation disorders.

12. Cheillan D, Cognat S, Dorche C, Jaeken J, Vianey-Saban C, Guffon N. 2004. CDG IIx with unusual phenotype. *J. Inherit. Metab. Dis.* 27:103–4

13. D'Amico A, Tessa A, Bruno C, Petrini S, Biancheri R, et al. 2006. Expanding the clinical spectrum of *POMT1* phenotype. *Neurology* 66:1564–67

14. Di Rocco M, Hennet T, Grubenmann CE, Pagliardini S, Allegri AE, et al. 2005. Congenital disorder of glycosylation (CDG) Ig: report on a patient and review of the literature. *J. Inherit. Metab. Dis.* 28:1162–64

15. Eklund EA, Merbouh N, Ichikawa M, Nishikawa A, Clima JM, et al. 2005. Hydrophobic Man-1-P derivatives correct abnormal glycosylation in type I congenital disorder of glycosylation fibroblasts. *Glycobiology* 15:1084–93

16. Eklund EA, Newell JW, Sun L, Seo NS, Alper G, et al. 2005. Molecular and clinical description of the first US patients with congenital disorder of glycosylation Ig. *Mol. Genet. Metab.* 84:25–31

17. Eklund EA, Sun L, Westphal V, Northrop JL, Freeze HH, Scaglia F. 2005. Congenital disorder of glycosylation (CDG)-Ih patient with a severe hepato-intestinal phenotype and evolving central nervous system pathology. *J. Pediatr.* 147:847–50

18. Eklund EA, Freeze HH. 2006. The congenital disorders of glycosylation: a multifaceted group of syndromes. *NeuroRx* 3:254–63

19. Foulquier F, Vasile E, Schollen E, Callewaert N, Raemaekers T, et al. 2006. Conserved oligomeric Golgi complex subunit 1 deficiency reveals a previously uncharacterized congenital disorder of glycosylation type II. *Proc. Natl. Acad. Sci. USA* 103:3764–69
20. Frank CG, Grubenmann CE, Eyaid W, Berger EG, Aebi M, Hennet T. 2004. Identification and functional analysis of a defect in the human ALG9 gene: definition of congenital disorder of glycosylation type IL. *Am. J. Hum. Genet.* 75:146–50
21. Freeze HH. 2001. Congenital disorders of glycosylation and the pediatric liver. *Semin. Liver Dis.* 21:501–15
22. Freeze HH. 2001. Update and perspectives on congenital disorders of glycosylation. *Glycobiology* 11:129R–43R
23. Freeze HH. 2002. Human disorders in N-glycosylation and animal models. *Biochim. Biophys. Acta* 1573:388–93
24. Freeze HH, Aebi M. 2005. Altered glycan structures: the molecular basis of congenital disorders of glycosylation. *Curr. Opin. Struct. Biol.* 15:490–98
25. Frishberg Y, Topaz O, Bergman R, Behar D, Fisher D, et al. 2005. Identification of a recurrent mutation in GALNT3 demonstrates that hyperostosis-hyperphosphatemia syndrome and familial tumoral calcinosis are allelic disorders. *J. Mol. Med.* 83:33–38
26. Grubenmann CE, Frank CG, Hulsmeier AJ, Schollen E, Matthijs G, et al. 2004. Deficiency of the first mannosylation step in the N-glycosylation pathway causes congenital disorder of glycosylation type Ik. *Hum. Mol. Genet.* 13:535–42
27. Grubenmann CE, Frank CG, Kjaergaard S, Berger EG, Aebi M, Hennet T. 2002. ALG12 mannosyltransferase defect in congenital disorder of glycosylation type Ig. *Hum. Mol. Genet.* 11:2331–39
28. Grünewald S, Matthijs G, Jaeken J. 2002. Congenital disorders of glycosylation: a review. *Pediatr. Res.* 52:618–24
29. Hansske B, Thiel C, Lubke T, Hasilik M, Honing S, et al. 2002. Deficiency of UDP-galactose: N-acetylglucosamine beta-1, 4-galactosyltransferase I causes the congenital disorder of glycosylation type IId. *J. Clin. Invest.* 109:725–33
30. Hardre R, Khaled A, Willemetz A, Dupre T, Moore S, et al. 2007. Mono, di and trimannopyranosyl phosphates as mannose-1-phosphate prodrugs for potential CDG-Ia therapy. *Bioorg. Med. Chem. Lett.* 17:152–55
31. Helenius J, Ng DTW, Marolda CL, Walter P, Valvano MA, Aebi M. 2002. Translocation of lipid-linked oligosaccharides across the ER membrane requires Rft1 protein. *Nature* 415:447–50
32. Hennet T. 2002. The galactosyltransferase family. *Cell. Mol. Life Sci.* 59:1081–95
33. Ishikawa S, Imel EA, Sorenson AH, Severe R, Knudson P, et al. 2006. Tumoral calcinosis presenting with eyelid calcifications due to novel missense mutations in the glycosyl transferase domain of the GALNT3 gene. *J. Clin. Endocrin. Metab.* 91:4472–75
34. Ishikawa S, Lyles KW, Econs MJ. 2005. A novel GALNT3 mutation in a pseudoautosomal dominant form of tumoral calcinosis: evidence that this disorder is autosomal recessive. *J. Clin. Endocrin. Metab.* 90:2420–23
35. Jaeken J, Carchon H. 2001. Congenital disorders of glycosylation: the rapidly growing tip of the iceberg. *Curr. Opin. Neurol.* 14:811–15
36. Jaeken J, Matthijs G. 2001. Congenital disorders of glycosylation. *Annu. Rev. Genomics. Hum. Genet.* 2:129–51
37. Jaeken J. 2003. Komrower Lecture. Congenital disorders of glycosylation (CDG): it's all in it! *J. Inherit. Metab. Dis.* 26:99–118
38. Jaeken J. 2004. Congenital disorders of glycosylation (CDG): update and new developments. *J. Inherit. Metab. Dis.* 27:423–26

39. Jaeken J, Carchon H. 2004. Congenital disorders of glycosylation: a booming chapter of pediatrics. *Curr. Opin. Pediatr.* 16:434–39
40. Kranz C, Denecke J, Lehle L, Sohlbach K, Jeske S, et al. 2004. Congenital disorder of glycosylation type Ik (CDG-Ik): a defect of mannosyltransferase I. *Am. J. Hum. Genet.* 74:545–51
41. Lanzara C, Ficarella R, Totaro A, Chen X, Roberto R, et al. 2003. Congenital dyserythropoietic anemia type II: exclusion of seven candidate genes. *Blood Cells Mol. Dis.* 30:22–29
42. Lecca MR, Wagner U, Patrignani A, Berger EG, Hennet T. 2005. Genome-wide analysis of the unfolded protein response in fibroblasts from congenital disorders of glycosylation type-I patients. *FASEB J.* 19:240–42
43. Lehle L, Strahl S, Tanner W. 2006. Protein glycosylation, conserved from yeast to man: a model organism helps elucidate congenital human diseases. *Angew. Chem. Int. Ed.* 45:6802–18
44. Leroy JG. 2006. Congenital disorders of glycosylation including diseases associated with O- as well as N-glycosylation defects. *Pediatr. Res.* 60:643–56
45. Longman C, Brockington M, Torelli S, Jimenez-Mallebrera C, Kennedy C, et al. 2003. Mutations in the human LARGE gene cause MDC1D, a novel form of congenital muscular dystrophy with severe mental retardation and abnormal glycosylation of alpha-dystroglycan. *Hum. Mol. Genet.* 12:2853–61
46. Lowe JB, Marth JD. 2003. A genetic approach to mammalian glycan function. *Annu. Rev. Biochem.* 72:643–91
47. Mandato C, Brive L, Miura Y, Davis JA, Di Cosmo N, et al. 2006. Cryptogenic liver disease in four children: a novel congenital disorder of glycosylation. *Pediatr. Res.* 59:293–98
48. Marquardt T, Denecke J. 2003. Congenital disorders of glycosylation: review of their molecular bases, clinical presentations and specific therapies. *Eur. J. Pediatr.* 162:359–79
49. Martinez-Duncker I, Dupré T, Piller V, Piller F, Candelier JJ, et al. 2005. Genetic complementation reveals a novel human congenital disorder of glycosylation of type II, due to inactivation of the Golgi CMP-sialic acid transporter. *Blood* 105:2671–76

49. Contrary to expectations, CMP-sialic acid transporter deficiency is associated with a very restricted, hematological phenotype, at least in this first patient.

50. Matthijs G, Schollen E, Van Schaftingen E. 2004. The prenatal diagnosis of congenital disorders of glycosylation (CDG). *Prenat. Diagn.* 24:114–16
51. Michele DE, Barresi R, Kanagawa M, Saito F, Cohn RD. 2002. Post-translational disruption of dystroglycan-ligand interactions in congenital muscular dystrophies. *Nature* 418:417–22
52. Mills PB, Mills K, Mian N, Winchester BG, Clayton PT. 2003. Mass spectrometric analysis of glycans in elucidating the pathogenesis of CDG type IIx. *J. Inherit. Metab. Dis.* 26:119–34
53. Miura Y, Tay SK, Aw MM, Eklund EA, Freeze HH. 2005. Clinical and biochemical characterization of a patient with congenital disorder of glycosylation (CDG) IIx. *J. Pediatr.* 147:851–53
54. Morava E, Wopereis S, Coucke P, Gillessen-Kaesbach G, Voit T, et al. 2005. Defective protein glycosylation in patients with cutis laxa syndrome. *Eur. J. Hum. Genet.* 13:414–21
55. Oberstein SAJL, Kriek M, White SJ, Kalf ME, Szuhai K, et al. 2006. Peters Plus syndrome is caused by mutations in B3GALTL, a putative glycosyltransferase. *Am. J. Hum. Genet.* 79:562–66
56. Ohtsubo K, Marth JD. 2006. Glycosylation in cellular mechanisms of health and disease. *Cell* 126:855–67
57. Peters V, Penzien JM, Reiter G, Körner C, Hackler R, et al. 2002. Congenital disorder of glycosylation IId (CDG-IId) – a new entity: clinical presentation with Dandy-Walker malformation and myopathy. *Neuropediatrics* 33:27–32

58. Prietsch V, Peters V, Hackler R, Jakobi R, Assmann B, et al. 2002. A new case of CDG-x with stereotyped dystonic hand movements and optic atrophy. *J. Inherit. Metab. Dis.* 25:126–30

59. Schollen E, Frank CG, Keldermans L, Reyntjens R, Grubenmann CE, et al. 2004. Clinical and molecular features of three patients with congenital disorders of glycosylation type Ih (CDG-Ih) (ALG8 deficiency). *J. Med. Genet.* 41:550–56.

60. Schwarz M, Thiel C, Lubbehusen J, Dorland B, de Koning T, et al. 2004. Deficiency of GDP-Man: $GlcNAc_2$-PP-dolichol mannosyltransferase causes congenital disorder of glycosylation type Ik. *Am. J. Hum. Genet.* 74:472–81

61. Shang J, Körner C, Freeze H, Lehrman MA. 2002. Extension of lipid-linked oligosaccharides is a high-priority aspect of the unfolded protein response: endoplasmic reticulum stress in type 1 congenital disorder of glycosylation fibroblasts. *Glycobiology* 12:307–17

62. Simpson MA, Cross H, Proukakis C, Priestman DA, Neville DCA, et al. 2004. Infantile-onset symptomatic epilepsy syndrome caused by a homozygous loss-of-function mutation of GM3 synthase. *Nat. Genet.* 36:1225–29

62. Reports the first human genetic defect in lipid glycosylation, more specifically in the glycan moiety of gangliosides.

63. Spaapen LJ, Bakker JA, Van Der Meer SB, Sijstermans HJ, Steet RA, et al. 2005. Clinical and biochemical presentation of siblings with COG-7 deficiency, a lethal multiple O- and N-glycosylation disorder. *J. Inherit. Metab. Dis.* 28:707–14

64. Sparrow DB, Chapman G, Wouters MA, Whittock NV, Ellard S, et al. 2006. Mutation of the LUNATIC FRINGE gene in humans causes spondylocostal dysostosis with a severe vertebral phenotype. *Am. J. Hum. Genet.* 78:28–37

64. First reported genetic defect in O-fucosylglycan synthesis, which is shown to cause a vertebral malsegmentation disorder, namely spondylocostal dysostosis type 3.

65. Specktor P, Cooper JG, Indelman M, Sprecher E. 2006. Hyperphosphatemic familial tumoral calcinosis caused by a mutation in GALNT3 in a European kindred. *Hum. Genet.* 51:487–90

66. Steet R, Kornfeld S. 2006. COG-7-deficient human fibroblasts exhibit altered recycling of Golgi proteins. *Mol. Biol. Cell* 17:2312–21

67. Thiel C, Schwarz M, Hasilik M, Grebe U, Hanefeld F, et al. 2002. Deficiency of dolichol-P-Man: $Man_7GlcNac_2$-PP dolichyl mannosyltransferase causes congenital disorder of glycosylation type Ig. *Biochem. J.* 367:195–201

68. Thiel C, Schwarz M, Peng J, Grzmil M, Hasilik M, et al. 2003. A new type of congenital disorders of glycosylation (CDG-Ii) provides new insights into the early steps of dolichol-linked oligosaccharide biosynthesis. *J. Biol. Chem.* 278:22498–505

69. Topaz O, Shurman DI, Bergman R, Indelman M, Ratajezak P, et al. 2004. Mutations in *GALNT3*, encoding a protein involved in O-linked glycosylation, cause familial tumoral calcinosis. *Nat. Genet.* 36:579–81

69. First report of a defect in mucin-type O-glycan synthesis.

70. van Reeuwijk J, Janssen M, Van Den Elzen C, Beltram-Valero de Bernabé D, Sabatelli P, et al. 2005. *POMT2* mutations cause α-dystroglycan hypoglycosylation and Walker-Warburg syndrome. *J. Med. Genet.* 42:907–12

71. Vogt G, Chapgier A, Yang K, Chuzhanova N, Feinberg J, et al. 2005. Gains of glycosylation comprise an unexpectedly large group of pathogenic mutations. *Nat. Genet.* 37:692–700

71. Indicates that an unexpectedly high proportion of mutations that cause human genetic disease might lead to the creation of new N-glycosylation sites, and that the addition of N-linked carbohydrates may be pathogenic.

72. Weinstein M, Schollen E, Matthijs G, Neupert C, Hennet T, et al. 2005. CDG-IL: an infant with a novel mutation in the ALG9 gene and additional phenotypic features. *Am. J. Med. Genet. A* 136:194–97

73. Wilson M, Lewis D, Arbuckle S, Jaeken J, McKenzie F, et al. 2003. A new type II congenital disorder of glycosylation associated with progressive glomerulopathy, retinal dystrophy and neurological abnormalities. *J. Inherit. Metab. Dis.* 26 (Suppl. 2):128

74. Introduces a screening method for some O-glycan synthesis defects.

79. First report on a new class of glycosylation disorders, caused by defect in a subunit of a Golgi-associated protein complex.

74. **Wopereis S, Grünewald S, Morava E, Penzien JM, Briones P, et al. 2003. Apolipoprotein C-III isofocusing in the diagnosis of genetic defects in O-glycan biosynthesis. *Clin. Chem.* 49:1839–45**
75. Wopereis S, Morava E, Coucke P, Gilessen-Kaesbach G, Voit T, et al. 2005. Defective protein glycosylation in patients with cutis laxa syndrome. *Eur. J. Hum. Genet.* 13:414–21
76. Wopereis S, Morava E, Grünewald S, Adamowicz M, Huijben KM, et al. 2005. Patients with unsolved congenital disorders of glycosylation type II can be subdivided in six distinct biochemical groups. *Glycobiology* 15:1312–19
77. Wopereis S, Morava E, Grünewald S, Mills PB, Winchester BG, et al. 2005. A combined defect in the biosynthesis of N- and O-glycans in patients with cutis laxa and neurological involvement: the biochemical characteristics. *Biochim. Biophys. Acta* 1741:156–64
78. Wu X, Rush JS, Karaoglu D, Krasnewich D, Lubinsky MS, et al. 2003. Deficiency of UDP-GlcNAc: dolichol phosphate N-acetylglucosamine-1 phosphate transferase (DPAGT1) causes a novel congenital disorder of glycosylation type Ij. *Hum. Mutat.* 22:144–50
79. **Wu X, Steet RA, Bohorov O, Bakker J, Newell J, et al. 2004. Mutation of the COG complex subunit gene COG7 causes a lethal congenital disorder. *Nat. Med.* 10:518–23**
80. Yoshida A, Kobayashi K, Manya H. 2001. Muscular dystrophy and neuronal migration disorder caused by mutations in a glycosyltransferase, POMGnT1. *Rev. Cell* 1:717–24
81. Zentilin Boyer M, de Lonlay P, Seta N, Besnard M, Pelatan C, et al. 2003. Failure to thrive and intestinal diseases in congenital disorders of glycosylation. *Arch. Pediatr.* 10:590–95
82. Zeevaert R, Morava E, Korsch E, Lefeber D, Wopereis S, et al. 2006. A syndrome of growth retardation, dysmorphy and episodes of hyperthermia in children with a *COG7* mutation. *J. Inherit. Metab. Dis.* 29 (Suppl. 1):62
83. Zeevaert R, Schollen E, Carchon H, Matthijs G, Jaeken J. 2006. Congenital disorders of glycosylation (CDG): update and perspectives. *Curr. Pediatr. Rev.* 2:323–30

Annotating Noncoding RNA Genes

Sam Griffiths-Jones

Wellcome Trust Sanger Institute, Wellcome Trust Genome Campus, Hinxton CAMBS CB10 1SA, United Kingdom

Present address: Faculty of Life Sciences, University of Manchester, Manchester M13 9PT, United Kingdom; email: sam.griffiths-jones@manchester.ac.uk

Annu. Rev. Genomics Hum. Genet. 2007. 8:279–98

First published online as a Review in Advance on May 21, 2007.

The *Annual Review of Genomics and Human Genetics* is online at genom.annualreviews.org

This article's doi:
10.1146/annurev.genom.8.080706.092419

1527-8204/07/0922-0279$20.00

Key Words

transcriptome, microRNA, snoRNA, gene prediction, genome annotation

Abstract

Noncoding RNA genes produce a functional RNA product rather than a translated protein. More than 1500 homologs of known "classical" RNA genes can be annotated in the human genome sequence, and automatic homology-based methods predict up to 5000 related sequences. Methods to predict novel RNA genes on a whole-genome scale are immature at present, but their use hints at tens of thousands of such genes in the human genome. Messenger RNA-like transcripts with no protein-coding potential are routinely discovered by high-throughput transcriptome analyses. Meanwhile, various experimental studies have suggested that the vast majority of the human genome is transcribed, although the proportion of the detected RNAs that is functional remains unknown.

INTRODUCTION

The human genome is about 30 times larger than that of the worm *Caenorhabditis elegans* or the fly *Drosophila melanogaster*, and 250 times larger than that of the yeast *Saccharomyces cerevisiae*. However, the human genome contains only a few thousand more protein-coding genes [23,244 according to the EnsEMBL database (**http://www.ensembl.org/**)] than worm (20,060) or fly (14,039), and only around 3.5 times as many as yeast (6680). The protein-coding gene count is thus not well correlated with organism complexity (at least from our human-centric viewpoint). The nonprotein-coding portion of the human genome is often considered "junk." However, although only around 1.2% of bases reside in protein-coding exons, estimates put conserved (and by implication functional) sequence at 5–10% of the genome (60, 99). Further, it now appears that most of the human genome may be transcribed (reviewed in 35), and that the number of genes that do not code for protein is much larger than expected. A subset of these genes encodes functional RNA products; these are called RNA genes or noncoding RNAs (ncRNAs). Some ncRNAs have well-understood and essential functions [for example, transfer RNA (tRNA) and ribosomal RNA (rRNA)]; other large classes have been described more recently [such as small nucleolar RNAs (snoRNAs) and microRNAs (miRNAs)]; and a rapidly increasing number have no known function. Finally, thousands of mRNA-like transcripts with little or no protein-coding potential have been identified in high-throughput studies of the transcriptome (13).

This review discusses the state of RNA annotation, with a focus on the human genome. The most useful computational methods are highlighted, and whole-genome analyses of families of "classical" RNA genes are discussed. Finally, the functional annotation of RNA sequences is considered.

THE HUMAN TRANSCRIPTOME

The cell's transcriptional output includes protein-coding RNA and non-protein-coding RNA, both of which can have many alternative splice variants, transcription start sites, and termination signals. Only around 1.2% of the human genome is contained in protein-coding exons, but 30–40% is transcribed as protein-coding mRNAs, the vast majority of which is intronic sequence. However, studies using genome-tiling arrays identify substantially more transcription than can be explained by current gene annotations (reviewed in 42, 100). The use of a variety of experimental techniques is adding weight to the idea that the majority of the mammalian genome is transcribed, including massively parallel signature sequencing (MPSS), mapping 5′ ends by cap analysis of gene expression (CAGE) and 3′ ends by serial analysis of gene expression (SAGE), and high-throughput cDNA sequencing technologies (reviewed in 31). The Functional Annotation of the Mouse Genome (FANTOM) project has sequenced 102,281 full-length cDNAs (13), of which 32,129 are annotated as protein-coding transcripts (2222 encode previously unknown proteins) and 34,030 as ncRNAs. Although these approaches focus on cytosolic polyadenylated RNA, there is evidence that as much as half of all transcriptional output might never leave the nucleus, and that a significant proportion is not polyadenylated (15).

CLASSICAL AND mRNA-LIKE ncRNAs

ncRNA transcripts fall into at least two main types, which necessarily require very different annotation approaches. "Classical" ncRNA genes encode small (typically tens to hundreds of bases) RNA products, which are expected to be highly structured through the adoption of a base-paired secondary structure. This group includes most well-understood RNA families, including tRNAs, rRNAs (the largest of which is several kilobases in length), miRNAs, and others discussed here. These are distinct from a possibly much-larger class of RNAs, which have all the properties of messenger RNA, but lack protein-coding potential. These RNAs have been termed mRNA-like ncRNAs (mlncRNAs) (83). mlncRNAs include alternative transcripts emanating from protein-coding genes, the products of antisense transcription, and a growing number of annotated transcriptionally-active loci of unknown function [e.g., as discovered by the FANTOM project (13)]. mlncRNAs are long and likely transcribed by RNA polymerase II, and may be spliced, capped, and polyadenylated. mlncRNAs probably do not conserve a base-paired secondary structure and compact three-dimensional form.

Well-studied mlncRNAs include the 17-kb mammalian Xist transcript, which has a vital role in X-dosage compensation. Xist is selectively expressed from one X chromosome in the female, causing its inactivatation (reviewed in 84); Xist was recently suggested to have evolved from a protein-coding pseudogene (20). 7H4 RNA is expressed at the neuromuscular junction (95), and may be a primary miRNA transcript (83). H19 RNA is maternally expressed in eutherian mammals, likely plays an as-yet-unknown role in the imprinting process (78), and expresses a miRNA gene (11). Growth arrest-specific gene 5 (GAS5) and U22 host gene (U22HG) are snoRNA host genes (88). The functions of almost all mlncRNAs are unknown.

Some mRNA transcripts contain predicted base-paired secondary structures in their untranslated regions. Many of these function by diverse mechanisms to regulate transcription and translation. For example, metazoan histone mRNAs do not have a 3′ poly-A tail; instead, their processing depends on a short hairpin structure that is bound by a stem-loop binding protein (SLBP) and a sequence motif that is bound by U7 RNA (19). The selenocysteine-insertion sequence (SECIS element) is a short hairpin that causes an in-frame TGA codon to be recognizezd by a selenocysteinyl tRNA rather than as a stop codon (96). Riboswitch elements are mainly located in 5′ untranslated regions (UTRs) of bacterial genes to regulate expression of the mRNA in response to metabolite binding (reviewed in 92). The presence of the thiamine riboswitch in eukaryotes highlights the possibility of a more widespread mechanism. The Rfam database of RNA families contains more than 120 known *cis*-regulatory UTR structures, including more than 20 internal ribosome-entry sites (IRES), 12 riboswitches, and a handful of frame-shift elements and thermo-regulators (27, 29).

The following discussion concentrates on the annotation of "classical" ncRNAs in mammalian genomes. These approaches are equally appropriate for the study of *cis*-regulatory mRNA structures.

ANNOTATING CLASSICAL ncRNAs IN SEQUENCED GENOMES

Until recently, many genome-annotation resources have ignored the gene complement that does not code for protein. This is not surprising, as only recently have reliable approaches for annotating RNA gene families become available, and an accurate RNA gene finding method is still lacking. De novo computational prediction of structured RNA genes is difficult. For example, ncRNAs lack the signatures that make protein-coding gene prediction possible (e.g., codons, splice signals, and other features such as length and sequence bias). Any ncRNA-gene finding method must therefore use information such as potential to form base-paired secondary structure and sequence and structural homology with known RNA families.

De Novo ncRNA Prediction

The discovery of novel functional RNAs by de novo gene prediction is a primary focus of computational RNA research. To date, no method is sufficiently mature to reliably identify ncRNAs in a genome-wide fashion. Two studies used sequence composition to identify relatively G+C (GC)-rich ncRNA genes in the A+T (AT)-rich genomes of thermophilic bacteria (47, 86). Otherwise, ncRNA prediction has relied largely on the potential of a sequence to adopt a secondary structure. Unfortunately, it has been demonstrated that predicted structures of RNA genes are not significantly more stable than predicted structures of random RNA sequences, at least for reliable discrimination on a genome-wide scale (80). Thus, a predicted stable RNA structure does not indicate functional significance. There are exceptions to this, for example, miRNA precursors do appear to adopt significantly more stable structures than does random RNA sequence (10). However, the accuracy of single-sequence structure prediction is limited in practice by incomplete models and ad hoc scoring schemes, even when the sequence is known to be a structured RNA (reviewed in 63).

More recent and successful approaches exploit the idea that functionally significant RNA structures will be conserved in related species. Computational simulation has shown that a small number of mutations will likely significantly change a secondary structure, thereby giving a conserved structure an implied function (41). In addition, the secondary structure may be maintained without conservation of the primary sequence by compensatory base mutations. These changes can be used as statistical evidence for base pairs at those positions. Predicting consensus secondary structure across a multi-species alignment of sequences is thus far more useful than for a single sequence.

A number of software tools attempt to provide a measure of probability that a given alignment of sequences adopts a conserved RNA fold. QRNA analyzes a pairwise alignment using three models of sequence evolution—protein coding, structured RNA, and other—and reports the highest-scoring model (81). RNAz calculates the probability that a multi-sequence alignment represents a conserved structured RNA by predicting the thermodynamic stability of a consensus secondary structure compared with the stability of a shuffled alignment (98). A derived z-score is combined with a structure conservation index (SCI) that measures the average stability of each sequence compared with the consensus in a support vector machine approach. EvoFold uses a probabilistic model of RNA structure and sequence evolution, called a phylogenetic stochastic context-free grammar (phylo-SCFG), to evaluate how well a substitution pattern in an alignment matches a secondary structure annotation (77).

RNAz and EvoFold have been applied independently to identify putative conserved RNA structures in the human genome (77, 97). Both find tens of thousands of candidate RNA-structural regions. Based on estimated false-positive rates, Pedersen et al. (77) predict approximately 10,000 structured RNA transcripts in the human genome (from an initial set of more than 48,000 structured regions), whereas Washietl et al. (97) estimate that more than 35000 structured RNAs are conserved in mammals. RNAz has also been used to identify candidate ncRNAs in the *C. elegans* (68) and *Ciona* (67) genomes.

The false-positive rates (and therefore the specificity) associated with these methods are subject to large and essentially unknown errors. A recent comparison of EvoFold and RNAz predictions shows that the overlap between the predictions made by the two methods is very small—less than 10% (S. Washietl & J.S. Pedersen, unpublished). The authors also estimate high false-discovery rates of 50–70%; note that the proportion of predictions that can be experimentally validated is so far undetermined. In earlier QRNA analyses of the *Escherichia coli* and *S. cerevisiae* genomes, expression could only be demonstrated for 10–20% of the RNA predictions (65, 82). One might expect the prediction specificity (and thus the validation rates) to be lower in vertebrates due to their larger genomes.

There are two alternative explanations for such poor verification rates: a) the predictions include large numbers of false-positive results; and b) many RNAs are expressed in very tight spatial and/or temporal patterns, making experimental verification difficult. Certainly, some RNA-expression profiles are highly specific. *C. elegans* miRNA lsy-6, which is involved in specifying bilateral asymmetry of chemoreceptor gene expression, is expressed in just a few neuronal cells at a specific developmental time (43). However, such low expression levels are extremely rare among known ncRNAs. In fact, some ncRNAs are among the most highly expressed RNAs in the cell (101). It is also important to note that these predictions make no distinction between bona fide ncRNA genes and structured regions of mRNA transcripts.

Homology Search

When performed on a genome-wide scale, reliable annotation of ncRNAs is currently restricted to searching for homologs of known RNA families. The nucleotide sequence databases and dedicated resources, such as RNAdb (**http://research.imb.uq.edu.au/rnadb/**) (76) and NONCODE (**http://www.bioinfo.org.cn/NONCODE/**) (58), contain thousands of annotated ncRNA sequences. Some ncRNA homologs are readily detected by sequence similarity alone. For example, rRNA sequences are well conserved across phylogenetic kingdoms, such that a simple BLAST search using the sequence of human 18S rRNA will detect all annotated 16S rRNAs in the bacterium *Bacillus subtilis* and the archaeon *Methanococcus jannaschii*. However, RNase P RNA cannot be identified in the *B. subtilis* genome sequence by BLAST using the sequence of the *E. coli* ortholog, even when using parameters that yield maximum sensitivity; for this, more complex models of RNA sequence and structure are required.

Algorithms for detecting homologs of structured RNAs can be divided into two classes: a) those specific to a particular RNA class [e.g., tRNAscan-SE (59) and ARAGORN (52) for tRNAs, miRscan (57) and miRseeker (49) for miRNAs, and snoscan and snoGPS (87) for snoRNAs]; and b) general approaches applicable to all structured RNAs that use patterns or motifs [e.g., PatSearch (30) and RNAMotif (61)] or profile-based methods [e.g., INFERNAL (21) and ERPIN (50)]. The advantages and disadvantages of each approach are clear. A specific tool uses fast, family-specific heuristics to maximize speed and sensitivity, but requires a new approach for each new RNA class. A general tool can be used to detect any RNA, but is likely to be slower and less accurate than a specific tool.

Currently, the most successful general approach for detecting homologs of known ncRNAs involves the use of statistical models, called profile-stochastic context-free grammars (profile-SCFGs) also known as covariance models (CMs). These models are analogous to profile hidden Markov models for protein sequence. An SCFG can be trained using a multi-sequence alignment of related RNA sequences annotated with the consensus secondary structure, and statistically represents the sequence and structure conservation. The model can then be used to analyze a sequence for its similarity to the training set; this can include scanning a whole genome sequence. Model building, sequence analysis, searching, and alignment are implemented in the INFERNAL package (21). Note that these tools are extremely computationally intensive. The Rfam database provides profile-SCFGs for more than 500 families of ncRNA sequences, together with precomputed sequence matches in the nucleotide sequence databases, allowing automatic detection of homologs in whole genome sequences (27, 29).

The Pseudogene Problem

The automatic annotation of structured ncRNA homologs in eukaryotic genomes is complicated by the presence of large numbers of repeats and pseudogenes. These presumed nonfunctional sequences largely swamp out the relatively small number of authentic functional ncRNA. All active SINE repeats in both the human (Alu) and rodent genomes (B1, B2, ID, B4) were derived from RNA genes: Alu/B1 repeats were derived from signal recognition particle (SRP) RNA in human/rodents; B2 evolved from an Ala-tRNA in mouse and rat; ID is related to both the neuronally expressed BC1 RNA and an Ala-tRNA; and B4 represents a fusion of B1 and ID repeats (23, 99). There are over 1 million copies of the Alu repeat in the human genome, accounting for roughly 10% of its bases (51). Other relevant families in the human genome include U6 spliceosomal RNA, 7SK, and Y RNA—all with hundreds of predicted homologs, most of which are likely pseudogenes. Many of the affected classes share characteristics: tRNA, 7SK, Y, and U6 RNAs are all transcribed by RNA polymerase III and have internal promoter sequences, facilitating retention of transcriptional activity in duplicated copies. Some of these are among the most abundant RNA molecules in the cell: there are 10^5 to 10^6 copies of 7SK and U6 RNAs per cell (101). tRNAs, SRP RNAs, and small nuclear RNAs (snRNAs) (particularly U6 RNA) are packaged into retroviral virions (14, 24) and undergo retrotranscription and recombination.

Automated Versus Manual Annotation

The most appropriate automated ncRNA-annotation approach combines the strengths of general and specific ncRNA-homolog detection methods. For example, a reasonable candidate set of human ncRNAs can be generated using tRNAscan-SE (59) to predict tRNAs, BLAST to identify known miRNAs and snoRNAs and find the small and large subunit rRNAs, and the Rfam library of covariance models (29) to annotate homologs of other known ncRNAs. However, the pseudogene problem discussed above necessitates careful use of the resulting data. Several intensive manual efforts have produced the best currently available ncRNA predictions for the human genome, starting from automated homology predictions (23, 51, 89, 99). Note that manual annotation has largely relied on a pragmatic definition of a pseudogene, which is likely to result in conservative annotation of authentic genes. The predicted ncRNA gene sets for various genomes are listed and compared in **Table 1**.

CLASSICAL ncRNA FAMILIES

Table 1 provides the total numbers of classical RNA genes in the human and other genomes. Each of the major families is briefly reviewed below.

Transfer RNAs

Automated annotation of tRNA genes using tRNAscan-SE yields significantly different results among eukaryotic genomes, predicting around 280 tRNAs in chicken and over 175,000 tRNAs and pseudogenes in rat (23, 38). tRNA-prediction programs attempt to distinguish real genes versus pseudogenes, but it is likely that even the most carefully curated rodent tRNA sets contain a large number of pseudogenes and repeat-derived sequences. The pseudogene problem is smaller, but still significant, with other eukaryotic genomes. For example, tRNAscan-SE annotates more than 500 tRNAs in the *C. elegans* genome, and the manually curated annotation of the human genome contains 496 tRNA genes. Both the *Drosophila* and chicken genomes appear to have low numbers of pseudogenes (33, 38) and contain 287 and 280 predicted tRNAs, respectively; both sets contain all tRNA anticodons that are predicted to be required by wobble rules (32) and include a single selenocysteine tRNA. The agreement between these two numbers strongly suggests that the minimal functional tRNA set in animals likely contains around 300 members.

Table 1 Manually annotated whole-genome ncRNA sets in human (51), chicken (38), *C. elegans* (89), and *D. melanogaster* (J. Daub, C. Bergman, D. Ardell & S. Griffiths-Jones, unpublished). A summary of automated analyses of human and mouse ncRNAs is also provided (EnsEMBL)

	H. sapiens (*automated*)	*H. sapiens* (*manual*)	*M. musculus* (*automated*)	*G. gallus* (*manual*)	*C. elegans* (*manual*)	*D. melanogaster* (*manual*)
tRNA	513	496	3278	280	592	297
5S	294	14	147	12	15	99
5.8S	8	0	13	3	2	2
18S	0	0	0	0	3	0
26S	0	0	0	0	1	0
U1	142	14	174	18	12	5
U2	99	14	42	6	19	6
U4	119	2	58	4	5	3
U5	36	4	13	9	13	7
U6	823	49	512	15	23	3
U4atac	0	1	0	1	0	2
U6atac	0	5	0	4	0	1
U11	0	1	0	1	0	1
U12	2	1	6	1	0	1
miRNA	563	474*	549	121	117	78
snoRNA	574	375*	526	83	8	250
SRP	83	3	4	3	5	2
RNase P	2	1	4	1	1	1
RNase MRP	1	1	0	0	0	0
Telomerase	1	1	2	1	1	0
Y	806	32	20	2	1	0
7SK	164	1	22	4	0	0
U7	157	1	41	1	0	0
Vault	5	0	2	0	0	0

*Updated numbers of miRNAs and snoRNAs are derived from the miRBase and snoRNABase databases.

Ribosomal RNAs

The eukaryotic ribosome contains four rRNA species: 5S rRNA, 5.8S rRNA, and small (18S) and large (26S) subunit rRNAs. The atomic structure of prokaryotic ribosome subunits shows that RNA has both major structural and catalytic roles. RNA is responsible for the arrangement of the A- and P-sites and their substrates, and the ribosome itself is a ribozyme, with RNA performing the catalytic function of peptide-bond formation (reviewed in 73).

In higher eukaryotes, 5.8S, 18S, and 26S rRNA genes are usually arranged in tandem arrays that contain many hundreds of copies. The rRNA genes are separated by internally transcribed spacer (ITS) sequences in the arrangement 18S-ITS1-5.8S-ITS2-26S (39). However, complete 18S and 26S rRNA sequences are often missing from assembled genome sequences because such tandem-repeat regions are selected against both during the shotgun sequencing and assembly processes. In human, around 150–200 copies of the 44-kb rDNA repeat reside on the short arms of chromosomes 13, 14, 15, 21, and 22 (51), but the annotated genome sequence has no single intact copy of the repeat. Eight dispersed copies of 5.8S RNA are easily annotated using the Rfam/INFERNAL approach.

5S rRNA is also arranged in tandem repeats, the largest of which is located in the subtelomeric region of human chromosome 1q (51). The Rfam model identifies close to 300 5S rRNA sequences, mostly dispersed throughout the genome and thus likely to be nonfunctional pseudogenes. Only 14 human 5S sequences are classified as authentic rRNAs by manual annotation.

Spliceosomal RNAs

The spliceosome is a large ribonucleoprotein complex containing five RNA species [snRNAs U1, U2, U4, U5, and U6] together with over 200 proteins (94). The functions of the RNAs are well described (101). Interactions between pre-mRNA and snRNA are responsible for splice-site (U1) and branch-site (U2) recognition. Base pairing with a conserved loop of U5 snRNA is partially responsible for aligning the two flanking exons. A di-snRNA complex of U4 and U6 forms prior to activation of the spliceosome, from which U4 is subsequently removed. During the splicing reaction, U6 binds U2, replacing U1. Like rRNAs, snRNAs are heavily modified by pseudouridylation, 2′-O methylation, and base methylation, with such modifications being essential for the proper formation of snRNP complexes and splicing (93). It is suggested that the catalytic steps in the splicing reaction are mediated by spliceosomal RNA (94).

Plants and some animals also have a second, so-called minor or U12-dependent splicesome. The minor spliceosome is involved in the splicing of U12 introns, with characteristic AT/AC splice sites, as opposed to the canonical GT/AG sites. It also utilizes U5 snRNA, along with U4atac, U6atac, U11, and U12 species. The U4/U4atac, U6/U6atac, U1/U11, and U2/U12 pairs are structural and functional homologs (101).

Automated homology-based annotation of spliceosomal RNAs in the human genome identifies well over 800 U6-related sequences. In contrast, similar analyses of the chicken genome sequence suggests a minimal functional set of only 10-20 U6 genes. Thus, the vast majority of human homologs are likely to be pseudogenes. Other spliceosomal RNA classes have similar, albeit less extreme, numbers of pseudogenes in mammals. Rfam identifies U12 snRNA homologs in all analyzed mammalian and fish genomes, and minor splicesomal RNAs have also been annotated in *Drosophila* (74). The absence of minor spliceosomal snRNA homologs in *C. elegans* is consistent with its expected lack of U12 introns.

A recent study confirmed the expression of three U1 human snRNA variants that lack complementarity to the canonical 5′ splice site (48). These sequences can be identified by homology searches with previously known U1 sequences, but are classified as putative pseudogenes in the manually curated set (**Table 1**). These results raise the possibility that a number of other snRNA-related sequences may represent functional RNAs that bind noncanonical recognition sites.

Small Nucleolar RNAs

snoRNAs direct the site-specific modification of ribosomal RNAs and other ncRNAs in Eukaryota and Archaea. Two classes of snoRNAs, called C/D box and H/ACA box, guide 2′-O-ribose methylation and pseudouridylation modifications, respectively (see **Figure 1**). The RNAs act as guides, base pairing with complementary regions of the target RNA, while the catalytic function resides with proteins in the snoRNP complex—specifically, fibrillarin for methylation and dyskerin for pseudouridylation (18).

A third family of guide RNAs is termed the small Cajal body-specific RNAs (scaRNAs). scaRNAs accumulate in Cajal bodies, subnuclear organelles where the final steps of spliceosome assembly take place, where they guide the modification of the RNA-polymerase-II-transcribed spliceosomal RNAs U1, U2, U4, U5, and U12 (17).

scaRNAs are often composed of both H/ACA and C/D box domains, but some have two H/ACA domains or one or two C/D box domains (55).

The snoRNABase database catalogs the presence of 257 C/D box snoRNAs, 94 H/ACA box snoRNAs, and 25 scaRNAs in the human genome. These sequences are often well conserved—about 65 have reported homologs in *S. cerevisiae* (55). Mammalian snoRNAs are usually located in the introns of protein-coding mRNA transcripts. More than 60% of human sno/scaRNAs (230/375) reside within introns associated with transcripts in the EnsEMBL database. A small number of snoRNAs appear to be transcribed independently by RNA polymerase II (93). An increasing number of transcripts that host snoRNAs appear not to encode proteins; some may be functional mlncRNAs, whereas others may be so-called "inside-out" genes, with the intronic snoRNA the sole functional product. snoRNABase contains target sites for 215/375 (57%) of human sno/scaRNAs. A growing number have no obvious complementary sites in known rRNAs/snRNAs, and may guide the modification of other classes of RNAs, including mRNAs. For example, the C/D box snoRNA HBII-52 regulates alternative splicing of the serotonin receptor 2C mRNA (45). Several clusters of snoRNAs reside within imprinted regions of the human genome, but no imprinting function has yet been defined.

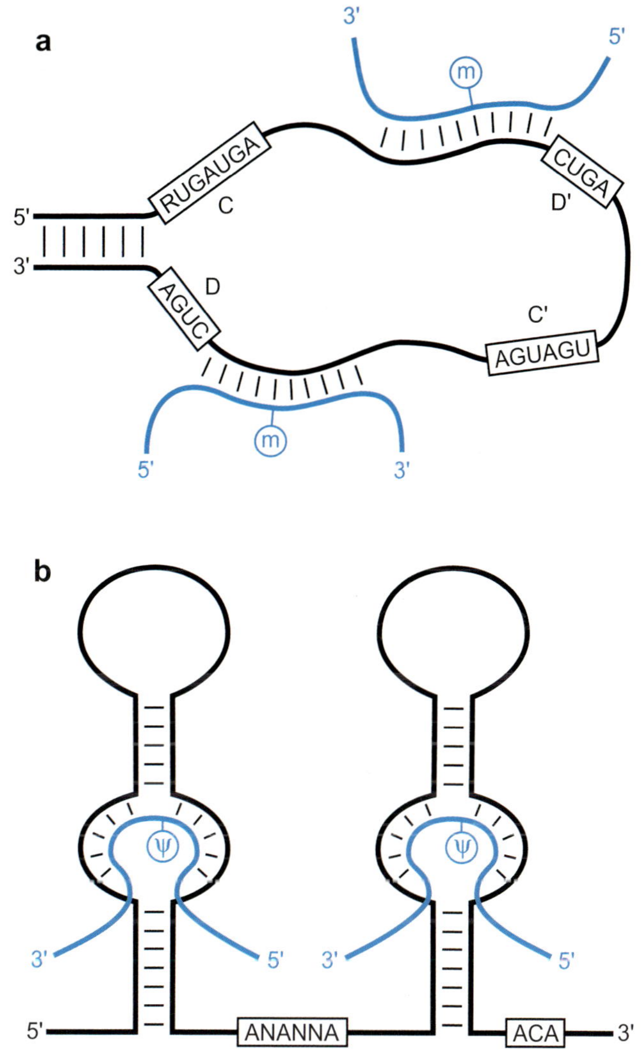

Figure 1

Secondary structures of (*a*) C/D box and (*b*) H/ACA box snoRNAs (*black*). Complementary binding to rRNA (*blue*) is shown, just 5′ of the D (and sometimes D′) boxes in C/D small nucleolar RNAs (snoRNAs), and in the bulge loops of H/ACA snoRNAs. Characteristic sequence motifs are boxed, and the position of ribosomal RNA (rRNA) modifications circled.

microRNAs

The most startling recent development in the ncRNA field has been the widespread importance of miRNAs. miRNAs are short ~22-nt sequences that inhibit the expression of target genes by binding to complementary regions in the 3′ UTRs of transcripts, triggering translational repression or transcript degradation. The history of their discovery and details of their biogenesis pathways are reviewed elsewhere (44, 53). In summary, the mature ~22-nt miRNA is excised from a stem loop called the precursor miRNA (pre-miRNA) by the ribonuclease enzyme DICER. In mammals, the pre-miRNA is around 70 nt long, and is in turn processed from the primary transcript (pri-miRNA) by the DROSHA enzyme. The duplex of mature miRNA sequence and the sequence from the opposite arm of the precursor hairpin (called the miR* sequence) is recruited into the RNA-induced silencing complex (RISC). Identical mature miRNA products can be processed from more than one hairpin precursor, expressed from multiple genomic loci. A scheme for miRNA-gene nomenclature is described elsewhere (1, 26).

The miRBase database contains 474 human precursor miRNAs, which give rise to 443 distinct mature products (26, 28). The genomic context of these sequences is varied and interesting (see **Figure 2**), with 212 (45%) appearing to be expressed on the same strand within introns of EnsEMBL protein-coding genes. An additional 22 have evidence of intronic expression based on expressed sequence tag (EST) data. The available evidence suggests that these human intronic miRNAs are processed from host transcripts, rather than as autonomous transcription units (4, 83). Twenty one overlap exons of annotated transcripts, and 14 map to UTRs, although in many cases, alternative splicing serves to locate these miRNAs in introns. Forty miRNAs overlap transcripts in the opposite orientation, with over two thirds of these having an intron at the same position on the opposite strand.

The nature of intergenic miRNA primary transcripts is poorly understood. Only a very small number have been experimentally characterized (12), with the available evidence suggesting that the pri-miRNA may be long (perhaps 10 kb or more), is transcribed by RNA polymerase II (54), is capped and polyadenylated, and may contain more than one pre-miRNA. Roughly 46% (216/474) of human pre-miRNAs are located within 5000 bases of another miRNA sequence, making up 52 clusters. The largest of these contains 36 closely related pre-miRNAs (the mir-512 family) in a region of human chromosome 19 that spans over 70 kb. A well-studied cluster of six miRNAs on human chromosome 13, with paralogous copies on chromosomes 7 and X (**Figure 2*a***) (90), was implicated as a human oncogene, with established roles in B-cell lymphomas (36) as well as lung (34) and breast cancers (40).

Other Small ncRNA Families

Both manual and automated analyses of the human genome sequence identify tens to hundreds of homologs of a further eight classes of classical ncRNA genes. The related RNase P and RNase MRP sequences are enzymatically active RNAs involved in processing precursor tRNAs and rRNAs, respectively (102). SRP RNA is an essential component of the signal-recognition particle, and the telomerase RNA carries the telomere template, together with an H/ACA box snoRNA domain. 7SK RNA inhibits the kinase activity of the CDK9/cyclin T complex, leading to reduced transcription by RNA polymerase II (71). The vault is a huge ribonucleoprotein complex of unknown function.

Two recent findings add to the complexity of the human RNA repertoire. Pollard et al. (79) reported a computationally predicted, novel RNA gene that is very rapidly evolving in the human lineage (based on human/chimpanzee genome-sequence comparisons) and is expressed during cortical development. This finding suggests an intriguing role for RNA in the expansion of brain function in humans. A class of short RNAs, called piwi-interacting RNAs (piRNAs), was recently discovered to be associated with piwi-domain proteins in the mammalian germline (2, 25). Tens of thousands of candidate piRNAs reside in large clusters in the mouse genome. Orthologous regions are conserved in the human genome, but appear to give rise to distinct piRNA sequences. Their function is currently unknown, but a role in gametogenesis has been proposed (2, 25).

HOW MANY ncRNAs RESIDE IN THE HUMAN GENOME?

Conservative manual approaches identify around 1700 classical ncRNAs in the human genome. Automated homology search methods predict over 4000 ncRNAs, with many of these likely reflecting pseudogenes

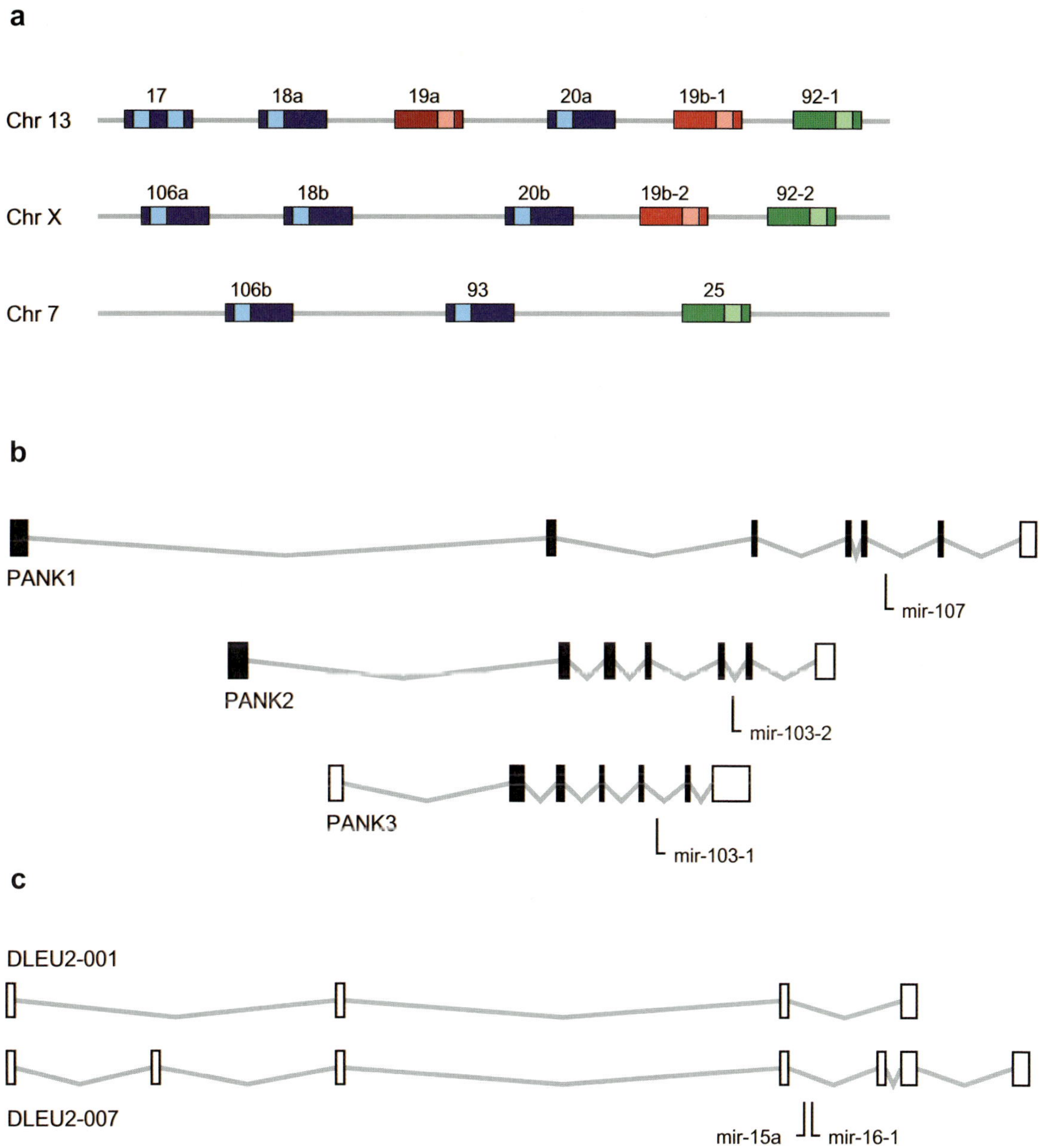

Figure 2

microRNA (miRNA) gene contexts. (*a*) The mir-17 cluster has paralogous copies on human chromosomes 7 and X. Colors depict three families of miRNA precursors, with mature miRNAs indicated by the light bands. (*b*) Paralogous miRNAs mir-107 and mir-103 are expressed from introns of paralogous panthenoate kinase (PANK) genes on human chromosomes 10, 20, and 5 (EnsEMBL accessions ENST00000371774, ENST00000316562, and ENST00000239231). (*c*) The noncoding DLEU2 locus on human chromosome 13 (two representative transcripts shown, VEGA accessions OTTHUMT00000044954 and OTTHUMT00000044960) contains mir-15a and mir-16-1. Filled exons represent open reading frames. (*a*) is adapted from (90) with permission.

and repeat-derived sequences. De novo gene-prediction programs identify upwards of 30,000 genomic sequences whose conservation patterns suggest structured RNA, but the false-positive rates for these predictions are likely to be high. mlncRNAs appear to be prevalent, with over 30,000 non-protein-coding full-length cDNA sequences catalogued, but the proportion that are functional is essentially unknown. Although such numbers are interesting, they are preliminary at best, and any discussion about the total ncRNA count in the human genome is largely speculative.

More informed discussion about specific classes of RNA is possible. The miRBase database contains 474 human miRNA sequences, already far more than the early estimates that put the total number at no more than 255 (56). Necessarily, the current miRNA set is biased toward those that are conserved (and have thus been amenable to prediction by computational approaches) and highly expressed (and thus easily detected experimentally). Recent evidence suggests that significant numbers of miRNAs may be primate specific (5, 7), and have specific temporal and spatial expression patterns. Indeed, second-generation miRNA-detection methods [e.g., amplification cloning protocols (91), microarray assays (5, 8), and deep sequencing] are identifying hundreds of candidate miRNAs. A phylogenetic-profiling method predicts close to 1000 miRNA sequences that are conserved between human and rodent (6). Speculation that the human genome contains tens of thousands of miRNAs is not yet well-supported by the data, but the presence of 1000 human miRNAs is fast becoming a realistic estimate.

FUNCTIONAL PRODUCTS OR NOISY TRANSCRIPTION

Understanding the function of all ncRNAs in the vertebrate genome is a far more pertinent problem. The known functional repertoire of ncRNAs is large. In addition to the functions described above, RNAs have roles in gene silencing, antisense regulation, imprinting, DNA methylation, chromosome maintenance and segregation, and processing of the 3′ ends of transcripts (9, 64, 69, 85). However, some known classical RNAs are still without an assigned function. For example, the vast majority of experimentally annotated miRNAs have unknown targets, and computational target prediction is a difficult problem. An increasing number of snoRNAs are also without identified targets. Y RNAs, components of the Ro ribonucleoprotein complex, were identified in 1981 (37), yet were only found to be involved in chromosomal DNA replication in 2006 (16). The functions of almost all mlncRNAs are unknown.

The question of how much non-protein-coding transcription is functional remains actively debated. Such prolific transcription can be viewed in two ways: that the majority of the human genome is functional through the action of transcribed, mostly noncoding, sequences, or that most noncoding transcription is biological noise. The latter could result from background levels of transcription that may result from initiation by randomly distributed cryptic promoter signals and transcriptional readthrough of termination signals. Only 5–10% of the human genome is conserved at thresholds commonly used to infer evolutionary constraint (51, 60, 99), and only 3–4% of mouse FANTOM cDNAs are conserved in human (13). However, functional RNAs may be conserved at a structural level in the absence of conservation at the primary-sequence level (75). Indeed, the primary sequences of several well-described mlncRNAs are not well conserved among mammals, including Xist and Air (70). It is difficult to rule out the possibility that so-called noncoding transcripts do in fact encode small proteins (22). It is also important to consider that the act of transcription may be a function in itself (for example, keeping a chromosomal region "open" for processes such as replication, other transcription, or epigenetic modification).

A paradigm shift relating to RNA function is not well supported by genetic screens, which have historically provided many more mutations with functional consequences in protein-coding transcripts than in ncRNAs (66). Possible explanations for this include a bias toward the publishing of protein-coding data and the tolerance of RNA function to point mutations. However, deleting megabase-sized regions of the mouse genome, selected as deserts with respect to protein-coding genes but containing more than 1000 noncoding sequences, resulted in viable mice with no obvious phenotypes (72). Although it is tempting to believe that increasingly sensitive transcription assays will eventually suggest that every base in the (euchromatic) genome is transcribed at some time in some cell (100), the relevance of such widespread transcription to the functional repertoire of RNA is not yet clear.

IMPROVING THE FUNCTIONAL ANNOTATION OF RNA SEQUENCES

How then can we improve our ability to predict classical ncRNA function? Better homology-detection methods can directly improve the assignment of function based on relationships among related sequences. Currently, many known ncRNA families are restricted to narrow taxonomic ranges. Despite intensive searches, telomerase RNA is largely unknown outside ciliates and vertebrates, yet is expected to be present in most higher eukaryotes; the *C. elegans* homolog was recently detected by improved computational methods (46). RNase P is ubiquitous and well described, yet the *C. elegans* homolog was only recently identified in the same study, and this gene has yet to be identified in the *Aquifex* genome (62). Many experimentally validated, small bacterial RNAs are known only in *E. coli* and its close relatives. We often do not know whether RNA functions are truly so specific, or whether we are simply unable to recognize remote homologs. For example, the *E. coli* 6S RNA family was, until recently, unknown outside γ-proteobacteria. Computational analyses have now identified homologs in diverse bacterial groups, including previously identified ncRNAs of unknown function in *B. subtilis* and cyanobacteria (3).

RNA gene identification, both homolog detection and de novo predicton, currently relies almost exclusively on RNA sequence and structure. Improved methods would likely utilize a much wider range of available signals. For example, ribosomal-modification maps improve the sensitivity of snoRNA-prediction methods. Prediction of promoter sequences (including internal RNA polymerase III signals) may enable better distinction between real ncRNAs and pseudogenes. Protein-binding signals, such as for Sm proteins in many snRNAs, may also be of use. It is interesting to note that ncRNA genes do not appear to be conserved with respect to gene order and orientation, in marked contrast to protein-coding genes (38). Synteny is therefore not a useful signal for ncRNA detection.

It appears that most novel functional information about ncRNAs will come directly from experimental studies. However, increased understanding about how ncRNA structure relates to function may lead to improved functional predictions for novel ncRNAs. Many classical RNAs demonstrate a pattern of sequence conservation with lower mutation rates in base-paired regions and higher rates in loops, as a consequence of selection pressure to maintain the overall structure. However, structure may be conserved without conserving the primary sequence, and the sites in loops may be highly conserved for binding or catalysis (see **Figure 3**). For example, miRNA stems are highly conserved, both because the mature miRNA needs to be heavily sequence-constrained to retain complementarity to its targets and because the structure must be conserved for proper processing. Orthologous snoRNAs have conserved loops for complementary binding to the target RNA, and many other RNAs form an intermolecular base-paired structure. Alternate structures can be

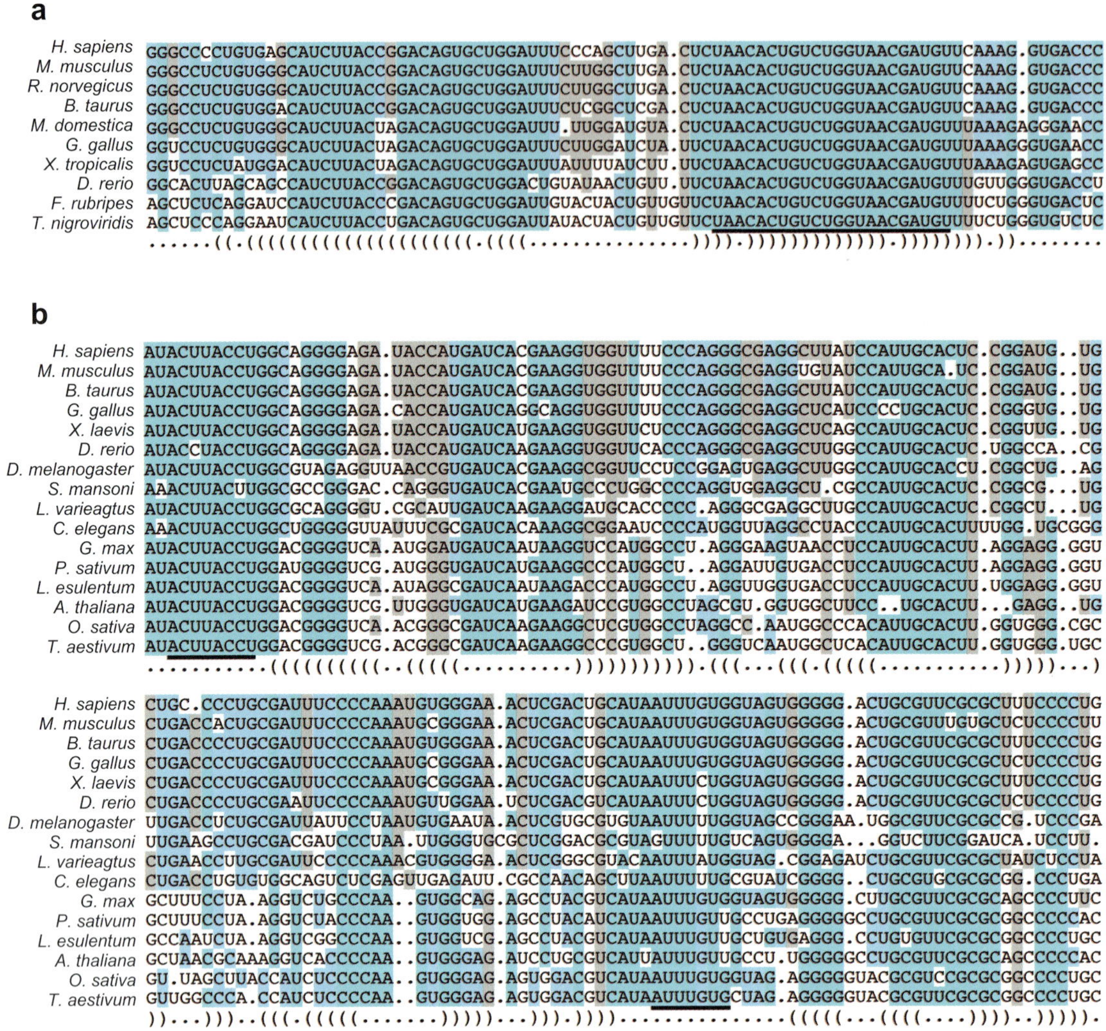

Figure 3

Multi-sequence alignments of (*a*) vertebrate mir-200a precursor-flanking sequences and (*b*) animal and plant U1 spliceosomal RNA sequences. Color represents the degree of sequence conservation, and the consensus secondary structure is shown below in dot-bracket notation. The helical region of the microRNA (miRNA) precursor includes the mature miRNA product (*underlined*) and is highly conserved. The secondary structure of the U1 RNA is conserved without high sequence conservation, whereas the loop regions, including a known 5′ splice site complementary region and Sm protein-binding sequence (*underlined*), are highly conserved.

used as regulatory switches (for example, by riboswitch elements). Sequence- and structure-conservation patterns are therefore likely to be highly family specific (I. Holmes & S. Griffiths-Jones, manuscript in preparation). In summary, although our functional understanding of the non-protein-coding portion of the human transcriptome is rudimentary, there is hope that computational analyses can be effectively combined with experimental studies to improve the annotation of functional RNAs.

SUMMARY POINTS

1. Although the robustness of de novo RNA gene prediction programs is improving rapidly, reliable genome-wide ncRNA annotation is currently restricted to homologs of known structured RNA families.
2. More than 1500 classical ncRNA genes have to date been manually annotated in the human genome, with over 4000 related sequences detected by automated methods.
3. High-throughput cDNA-sequencing efforts have revealed more than 34,000 sequences that are annotated as mlncRNAs. A number of experimental techniques suggest that the majority of the human genome is transcribed, but the functional significance of this transcription remains to be established.

DISCLOSURE STATEMENT

The author is not aware of any biases that might be perceived as affecting the objectivity of this review.

ACKNOWLEDGMENTS

I would like to thank Paul Gardner and Alex Bateman for useful comments, and Sean Eddy and Ian Holmes for discussions. S.G.J. was funded by the Wellcome Trust.

LITERATURE CITED

1. Ambros V, Bartel B, Bartel DP, Burge CB, Carrington JC, et al. 2003. A uniform system for microRNA annotation. *RNA* 9:277–79
2. Aravin A, Gaidatzis D, Pfeffer S, Lagos-Quintana M, Landgraf P, et al. 2006. A novel class of small RNAs bind to MILI protein in mouse testes. *Nature* 442:203–7
3. Barrick JE, Sudarsan N, Weinberg Z, Ruzzo WL, Breaker RR. 2005. 6S RNA is a widespread regulator of eubacterial RNA polymerase that resembles an open promoter. *RNA* 11:774–84
4. Baskerville S, Bartel DP. 2005. Microarray profiling of microRNAs reveals frequent coexpression with neighboring miRNAs and host genes. *RNA* 11:241–47
5. Bentwich I, Avniel A, Karov Y, Aharonov R, Gilad S, et al. 2005. Identification of hundreds of conserved and nonconserved human microRNAs. *Nat. Genet.* 37:766–70
6. Berezikov E, Plasterk RH. 2005. Camels and zebrafish, viruses and cancer: a microRNA update. *Hum. Mol. Genet.* 14(Suppl. 2):R183–90
7. Berezikov E, Thuemmler F, van Laake LW, Kondova I, Bontrop R, et al. 2006. Diversity of microRNAs in human and chimpanzee brain. *Nat. Genet.* 38:1375–77
8. Berezikov E, van Tetering G, Verheul M, van de Belt J, van Laake L, et al. 2006. Many novel mammalian microRNA candidates identified by extensive cloning and RAKE analysis. *Genome Res.* 16:1289–98
9. Bernstein E, Allis CD. 2005. RNA meets chromatin. *Genes Dev.* 19:1635–55
10. Bonnet E, Wuyts J, Rouzé P, Van de Peer Y. 2004. Evidence that microRNA precursors, unlike other noncoding RNAs, have lower folding free energies than random sequences. *Bioinformatics* 20:2911–17

11. Cai X, Cullen BR. 2007. The imprinted H19 noncoding RNA is a primary microRNA precursor. *RNA* 13:313–6
12. Cai X, Hagedorn CH, Cullen BR. 2004. Human microRNAs are processed from capped, polyadenylated transcripts that can also function as mRNAs. *RNA* 10:1957–66
13. Carninci P, Kasukawa T, Katayama S, Gough J, Frith MC, et al. 2005. The transcriptional landscape of the mammalian genome. *Science* 309:1559–63
14. Chen PJ, Cywinski A, Taylor JM. 1985. Reverse transcription of 7SL RNA by an avian retrovirus. *J. Virol.* 54:278–84
15. Cheng J, Kapranov P, Drenkow J, Dike S, Brubaker S, et al. 2005. Transcriptional maps of 10 human chromosomes at 5-nucleotide resolution. *Science* 308:1149–54
16. Christov CP, Gardiner TJ, Szüts D, Krude T. 2006. Functional requirement of noncoding Y RNAs for human chromosomal DNA replication. *Mol. Cell. Biol.* 26:6993–7004
17. Darzacq X, Jády BE, Verheggen C, Kiss AM, Bertrand E, et al. 2002. Cajal body-specific small nuclear RNAs: a novel class of 2′-O-methylation and pseudouridylation guide RNAs. *EMBO J.* 21:2746–56
18. Decatur WA, Fournier MJ. 2003. RNA-guided nucleotide modification of ribosomal and other RNAs. *J. Biol. Chem.* 278:695–98
19. Dominski Z, Marzluff WF. 1999. Formation of the 3′ end of histone mRNA. *Gene* 239:1–14
20. Duret L, Chureau C, Samain S, Weissenbach J, Avner P. 2006. The Xist RNA gene evolved in eutherians by pseudogenization of a protein-coding gene. *Science* 312:1653–55
21. Eddy SR. 2002. A memory-efficient dynamic programming algorithm for optimal alignment of a sequence to an RNA secondary structure. *BMC Bioinformatics* 3:18
22. Frith MC, Forrest AR, Nourbakhsh E, Pang KC, Kai C, et al. 2006. The abundance of short proteins in the mammalian proteome. *PLoS Genet.* 2:e52
23. Gibbs RA, Weinstock GM, Metzker ML, Muzny DM, Sodergren EJ, et al. 2004. Genome sequence of the Brown Norway rat yields insights into mammalian evolution. *Nature* 428:493–521
24. Giles KE, Caputi M, Beemon KL. 2004. Packaging and reverse transcription of snRNAs by retroviruses may generate pseudogenes. *RNA* 10:299–307
25. Girard A, Sachidanandam R, Hannon GJ, Carmell MA. 2006. A germline-specific class of small RNAs binds mammalian Piwi proteins. *Nature* 442:199–202
26. Griffiths-Jones S. 2004. The microRNA registry. *Nucleic Acids Res.* 32:D109–11
27. Griffiths-Jones S, Bateman A, Marshall M, Khanna A, Eddy SR. 2003. Rfam: an RNA family database. *Nucleic Acids Res.* 31:439–41
28. Griffiths-Jones S, Grocock RJ, van Dongen S, Bateman A, Enright AJ. 2006. miRBase: microRNA sequences, targets and gene nomenclature. *Nucleic Acids Res.* 34:D140–44
29. Griffiths-Jones S, Moxon S, Marshall M, Khanna A, Eddy SR, et al. 2005. Rfam: annotating noncoding RNAs in complete genomes. *Nucleic Acids Res.* 33:D121–24
30. Grillo G, Licciulli F, Liuni S, Sbisa E, Pesole G. 2003. PatSearch: a program for the detection of patterns and structural motifs in nucleotide sequences. *Nucleic Acids Res.* 31:3608–12
31. Gustincich S, Sandelin A, Plessy C, Katayama S, Simone R, et al. 2006. The complexity of the mammalian transcriptome. *J. Physiol. (Lond.)* 575:321–32
32. Guthrie C, Abelson J. 1982. Organization and expression of tRNA genes in *Saccharomyces cerevisiae*. In *The Molecular Biology of the Yeast Saccharomyces: Metabolism and Gene Expression*, ed. J. Strathern, J. Broach, pp. 487–528 New York: Cold Spring Harbor
33. Harrison PM, Milburn D, Zhang Z, Bertone P, Gerstein M. 2003. Identification of pseudogenes in the *Drosophila melanogaster* genome. *Nucleic Acids Res.* 31:1033–37

34. Hayashita Y, Osada H, Tatematsu Y, Yamada H, Yanagisawa K, et al. 2005. A polycistronic microRNA cluster, miR-17–92, is overexpressed in human lung cancers and enhances cell proliferation. *Cancer Res.* 65:9628–32
35. Hayashizaki Y, Carninci P. 2006. Genome network and FANTOM3: assessing the complexity of the transcriptome. *PLoS Genet.* 2:e63
36. He L, Thomson JM, Hemann MT, Hernando-Monge E, Mu D, et al. 2005. A microRNA polycistron as a potential human oncogene. *Nature* 435:828–33
37. Hendrick JP, Wolin SL, Rinke J, Lerner MR, Steitz JA. 1981. Ro small cytoplasmic ribonucleoproteins are a subclass of La ribonucleoproteins: further characterization of the Ro and La small ribonucleoproteins from uninfected mammalian cells. *Mol. Cell. Biol.* 1:1138–49
38. Hillier LW, Miller W, Birney E, Warren W, Hardison RC, et al. 2004. Sequence and comparative analysis of the chicken genome provide unique perspectives on vertebrate evolution. *Nature* 432:695–716
39. Hillis DM, Dixon MT. 1991. Ribosomal DNA: molecular evolution and phylogenetic inference. *Q. Rev. Biol.* 66:411–53
40. Hossain A, Kuo MT, Saunders GF. 2006. Mir-17–5p regulates breast cancer cell proliferation by inhibiting translation of AIB1 mRNA. *Mol. Cell. Biol.* 26:8191–201
41. Huynen M, Gutell R, Konings D. 1997. Assessing the reliability of RNA folding using statistical mechanics. *J. Mol. Biol.* 267:1104–12
42. Johnson JM, Edwards S, Shoemaker D, Schadt EE. 2005. Dark matter in the genome: evidence of widespread transcription detected by microarray tiling experiments. *Trends Genet.* 21:93–102
43. Johnston RJ, Hobert O. 2003. A microRNA controlling left/right neuronal asymmetry in *Caenorhabditis elegans*. *Nature* 426:845–49
44. Kim VN. 2005. MicroRNA biogenesis: coordinated cropping and dicing. *Nat. Rev. Mol. Cell Biol.* 6:376–85
45. Kishore S, Stamm S. 2006. The snoRNA HBII-52 regulates alternative splicing of the serotonin receptor 2C. *Science* 311:230–32
46. Klein RJ, Eddy SR. 2003. RSEARCH: finding homologs of single structured RNA sequences. *BMC Bioinformatics* 4:44
47. Klein RJ, Misulovin Z, Eddy SR. 2002. Noncoding RNA genes identified in AT-rich hyperthermophiles. *Proc. Natl. Acad. Sci. USA* 99:7542–47
48. Kyriakopoulou C, Larsson P, Liu L, Schuster J, Söderbom F, et al. 2006. U1-like snRNAs lacking complementarity to canonical 5′ splice sites. *RNA* 12:1603–11
49. Lai EC, Tomancak P, Williams RW, Rubin GM. 2003. Computational identification of *Drosophila* microRNA genes. *Genome Biol.* 4:R42
50. Lambert A, Fontaine JF, Legendre M, Leclerc F, Permal E, et al. 2004. The ERPIN server: an interface to profile-based RNA motif identification. *Nucleic Acids Res.* 32:W160–65
51. Lander ES, Linton LM, Birren B, Nusbaum C, Zody MC, et al. 2001. Initial sequencing and analysis of the human genome. *Nature* 409:860–921
52. Laslett D, Canback B. 2004. ARAGORN, a program to detect tRNA genes and tmRNA genes in nucleotide sequences. *Nucleic Acids Res.* 32:11–16
53. Lee R, Feinbaum R, Ambros V. 2004. A short history of a short RNA. *Cell* 116:S89–92
54. Lee Y, Kim M, Han J, Yeom KH, Lee S, et al. 2004. MicroRNA genes are transcribed by RNA polymerase II. *EMBO J.* 23:4051–60
55. Lestrade L, Weber MJ. 2006. snoRNA-LBME-db, a comprehensive database of human H/ACA and C/D box snoRNAs. *Nucleic Acids Res.* 34:D158–62

56. Lim LP, Glasner ME, Yekta S, Burge CB, Bartel DP. 2003. Vertebrate microRNA genes. *Science* 299:1540
57. Lim LP, Lau NC, Weinstein EG, Abdelhakim A, Yekta S, et al. 2003. The microRNAs of *Caenorhabditis elegans*. *Genes Dev.* 17:991–1008
58. Liu C, Bai B, Skogerbo G, Cai L, Deng W, et al. 2005. NONCODE: an integrated knowledge database of noncoding RNAs. *Nucleic Acids Res.* 33:D112–15
59. Lowe TM, Eddy SR. 1997. tRNAscan-SE: a program for improved detection of transfer RNA genes in genomic sequence. *Nucleic Acids Res.* 25:955–64
60. Lunter G, Ponting CP, Hein J. 2006. Genome-wide identification of human functional DNA using a neutral indel model. *PLoS Comput. Biol.* 2:e5
61. Macke TJ, Ecker DJ, Gutell RR, Gautheret D, Case DA, et al. 2001. RNAMotif, an RNA secondary structure definition and search algorithm. *Nucleic Acids Res.* 29:4724–35
62. Marszalkowski M, Teune JH, Steger G, Hartmann RK, Willkomm DK. 2006. Thermostable RNase P RNAs lacking P18 identified in the *Aquificales*. *RNA* 12:1915–21
63. Mathews DH. 2006. Revolutions in RNA secondary structure prediction. *J. Mol. Biol.* 359:526–32
64. Mattick JS, Makunin IV. 2006. Non-coding RNA. *Hum. Mol. Genet.* 15(Suppl. 1):R17–29
65. McCutcheon JP, Eddy SR. 2003. Computational identification of noncoding RNAs in *Saccharomyces cerevisiae* by comparative genomics. *Nucleic Acids Res.* 31:4119–28
66. Mendes Soares LM, Valcarcel J. 2006. The expanding transcriptome: the genome as the "Book of Sand." *EMBO J.* 25:923–31
67. Missal K, Rose D, Stadler PF. 2005. Non-coding RNAs in *Ciona intestinalis*. *Bioinformatics* 21(Suppl. 2):ii77–78
68. Missal K, Zhu X, Rose D, Deng W, Skogerbø G, et al. 2006. Prediction of structured noncoding RNAs in the genomes of the nematodes *Caenorhabditis elegans* and *Caenorhabditis briggsae*. *J. Exp. Zool. B. Mol. Dev. Evol.* 306:379–92
69. Munroe SH, Zhu J. 2006. Overlapping transcripts, double-stranded RNA and antisense regulation: a genomic perspective. *Cell. Mol. Life Sci.* 63:2102–18
70. Nesterova TB, Slobodyanyuk SY, Elisaphenko EA, Shevchenko AI, Johnston C, et al. 2001. Characterization of the genomic Xist locus in rodents reveals conservation of overall gene structure and tandem repeats but rapid evolution of unique sequence. *Genome Res.* 11:833–49
71. Nguyen VT, Kiss T, Michels AA, Bensaude O. 2001. 7SK small nuclear RNA binds to and inhibits the activity of CDK9/cyclin T complexes. *Nature* 414:322–25
72. Nóbrega MA, Zhu Y, Plajzer-Frick I, Afzal V, Rubin EM. 2004. Megabase deletions of gene deserts result in viable mice. *Nature* 431:988–93
73. Noller HF. 2005. RNA structure: reading the ribosome. *Science* 309:1508–14
74. Otake LR, Scamborova P, Hashimoto C, Steitz JA. 2002. The divergent U12-type spliceosome is required for pre-mRNA splicing and is essential for development in *Drosophila*. *Mol. Cell* 9:439–46
75. Pang KC, Frith MC, Mattick JS. 2006. Rapid evolution of noncoding RNAs: lack of conservation does not mean lack of function. *Trends Genet.* 22:1–5
76. Pang KC, Stephen S, Engstrom PG, Tajul-Arifin K, Chen W, et al. 2005. RNAdb: a comprehensive mammalian noncoding RNA database. *Nucleic Acids Res.* 33:D125–30
77. Pedersen JS, Bejerano G, Siepel A, Rosenbloom K, Lindblad-Toh K, et al. 2006. Identification and classification of conserved RNA secondary structures in the human genome. *PLoS Comput. Biol.* 2:e33
78. Pfeifer K, Leighton PA, Tilghman SM. 1996. The structural H19 gene is required for transgene imprinting. *Proc. Natl. Acad. Sci. USA* 93:13876–83

79. Pollard KS, Salama SR, Lambert N, Lambot MA, Coppens S, et al. 2006. An RNA gene expressed during cortical development evolved rapidly in humans. *Nature* 443:167–72
80. Rivas E, Eddy SR. 2000. Secondary structure alone is generally not statistically significant for the detection of noncoding RNAs. *Bioinformatics* 16:583–605
81. Rivas E, Eddy SR. 2001. Noncoding RNA gene detection using comparative sequence analysis. *BMC Bioinformatics* 2:8
82. Rivas E, Klein RJ, Jones TA, Eddy SR. 2001. Computational identification of noncoding RNAs in *E. coli* by comparative genomics. *Curr. Biol.* 11:1369–73
83. Rodriguez A, Griffiths-Jones S, Ashurst JL, Bradley A. 2004. Identification of mammalian microRNA host genes and transcription units. *Genome Res.* 14:1902–10
84. Rougeulle C, Avner P. 2004. The role of antisense transcription in the regulation of X-inactivation. *Curr. Top. Dev. Biol.* 63:61–89
85. Royo H, Bortolin ML, Seitz H, Cavaille J. 2006. Small noncoding RNAs and genomic imprinting. *Cytogenet. Genome Res.* 113:99–108
86. Schattner P. 2002. Searching for RNA genes using base-composition statistics. *Nucleic Acids Res.* 30:2076–82
87. Schattner P, Brooks AN, Lowe TM. 2005. The tRNAscan-SE, snoscan and snoGPS web servers for the detection of tRNAs and snoRNAs. *Nucleic Acids Res.* 33:W686–89
88. Smith CM, Steitz JA. 1998. Classification of gas5 as a multi-small-nucleolar-RNA (snoRNA) host gene and a member of the 5′-terminal oligopyrimidine gene family reveals common features of snoRNA host genes. *Mol. Cell. Biol.* 18:6897–909
89. Stricklin SL, Griffiths-Jones S, Eddy SR. 2005. *C. elegans* noncoding RNA genes. In *Wormbook*, ed. The *C. elegans* Research Community. **http://www.wormbook.org**
90. Tanzer A, Stadler PF. 2004. Molecular evolution of a microRNA cluster. *J. Mol. Biol.* 339:327–35
91. Takada S, Berezikov E, Yamashita Y, Lagos-Quintana M, Kloosterman WP, et al. 2006. Mouse microRNA profiles determined with a new and sensitive cloning method. *Nucleic Acids Res.* 34:e115
92. Tucker BJ, Breaker RR. 2005. Riboswitches as versatile gene control elements. *Curr. Opin. Struct. Biol.* 15:342–48
93. Tycowski KT, Aab A, Steitz JA. 2004. Guide RNAs with 5′ caps and novel box C/D snoRNA-like domains for modification of snRNAs in metazoa. *Curr. Biol.* 14:1985–95
94. Valadkhan S. 2005. snRNAs as the catalysts of pre-mRNA splicing. *Curr. Opin. Chem. Biol.* 9:603–8
95. Velleca MA, Wallace MC, Merlie JP. 1994. A novel synapse-associated noncoding RNA. *Mol. Cell. Biol.* 14:7095–804
96. Walczak R, Westhof E, Carbon P, Krol A. 1996. A novel RNA structural motif in the selenocysteine insertion element of eukaryotic selenoprotein mRNAs. *RNA* 2:367–79
97. Washietl S, Hofacker IL, Lukasser M, Hüttenhofer A, Stadler PF. 2005. Mapping of conserved RNA secondary structures predicts thousands of functional noncoding RNAs in the human genome. *Nat. Biotechnol.* 23:1383–90
98. Washietl S, Hofacker IL, Stadler PF. 2005. Fast and reliable prediction of noncoding RNAs. *Proc. Natl. Acad. Sci. USA* 102:2454–59
99. Waterston RH, Lindblad-Toh K, Birney E, Rogers J, Abril JF, et al. 2002. Initial sequencing and comparative analysis of the mouse genome. *Nature* 420:520–62
100. Willingham AT, Gingeras TR. 2006. TUF love for "junk" DNA. *Cell* 125:1215–20
101. Yu YT, Scharl EC, Smith CM, Steitz JA. 1999. The growing world of small nuclear riboproteins. In *The RNA World*, ed. RF Gesteland, TR Cech, JF Atkins. New York: Cold Spring Harbor

102. Zhu Y, Stribinskis V, Ramos KS, Li Y. 2006. Sequence analysis of RNase MRP RNA reveals its origination from eukaryotic RNase P RNA. *RNA* 12:699–706

RELATED RESOURCES

C. elegans ncRNA annotations: **http://selab.wustl.edu/people/sls/WBhtml/**
EnsEMBL 41: **http://www.ensembl.org/**
miRBase 9.0: **http://microrna.sanger.ac.uk/**
NONCODE: **http://www.bioinfo.org.cn/NONCODE/**
Rfam 7.0: **http://www.sanger.ac.uk/Software/Rfam/**
RNAdb: **http://research.imb.uq.edu.au/rnadb/**
snoRNABase v3: **http://www-snorna.biotoul.fr/**

Using Genomics to Study How Chromatin Influences Gene Expression

Douglas R. Higgs, Douglas Vernimmen, Jim Hughes, and Richard Gibbons

MRC Molecular Haematology Unit, Weatherall Institute of Molecular Medicine, John Radcliffe Hospital, Headington, Oxford OX3 9DS, United Kingdom; email: doug.higgs@imm.ox.ac.uk, douglas.vernimmen@imm.ox.ac.uk, jim.hughes@imm.ox.ac.uk, richard.gibbons@imm.ox.ac.uk

Annu. Rev. Genomics Hum. Genet. 2007. 8:299–325

First published online as a Review in Advance on May 21, 2007.

The *Annual Review of Genomics and Human Genetics* is online at genom.annualreviews.org

This article's doi: 10.1146/annurev.genom.8.080706.092323

1527-8204/07/0922-0299$20.00

Key Words

transcriptional program, epigenetic program, chromosome structure, microarrays

Abstract

A postgenome challenge is to understand how the code in DNA is converted into the biological processes underlying various cell fates. By establishing the appropriate technical tools, we are moving from an era in which such questions have been asked by studying individual genes to one in which large domains, whole chromosomes, and the entire human genome can be investigated. These developments will allow us to study in parallel the transcriptional program and components of the epigenetic program (nuclear position, timing of replication, chromatin structure and modification, DNA methylation) to determine the hierarchy and order of events required to switch genes on and off during differentiation and development.

Epigenetics: the study of heritable changes in gene function that occur without a change in the DNA sequence

***Cis*-acting:** a *cis*-acting site only affects the activity of sequences that lie on the same DNA molecule

Promoter: sequences at the 5′ end of a gene that direct the initiation of transcription

Enhancer: a *cis*-acting regulatory DNA sequence that greatly enhances the transcription of a gene

INTRODUCTION

Since the discovery of the structure of DNA in 1953, biology has progressed to the point, 50 years later, whereby we now have an essentially complete reference sequence of the human genome (~3 × 10^9 bases) providing the code for life. Because we now realize that the phenotype of any cell and its developmental fate ultimately depends on which proteins are, or have been, expressed in such cells, a postgenome challenge is to understand how the information encoded in DNA directs the specification, development, and fate of hundreds of different cell types. In other words, how is gene expression controlled in time and space?

Over the past 20 years it has gradually emerged that, although the code inherited via the germ line may lie entirely in the DNA sequence, the way in which DNA is modified by methylation and is packaged into chromatin during development plays a critical role by indexing the information. When wrapped into chromatin, some information is available and other parts of the code are "hidden." This indexing changes during development. In multipotent cells it appears that a large proportion of the code is accessible, whereas in fully differentiated cells a great deal of information is hidden. An important aspect of this process is that when a cell committed to a particular lineage divides, its daughters can inherit the same indexing. In this way a cellular memory is passed from one cell to its progeny.

This process is referred to as epigenetic programming, a term that includes those nuclear processes that can alter gene expression in a stable, heritable manner without any change in the DNA code. The complexity arises because, as the epigenetic program emerges, there are associated changes in many interdependent phenomena including nuclear position, the timing of DNA replication, chromatin structure, chromatin modifications, the association of nonhistone proteins on chromatin, and DNA methylation. In turn, these features may be influenced by the expression of noncoding RNAs, the availability of the enzymes that bring about some of these changes, the transcription factors available to bind DNA, and the cofactors with which they associate. At present, we do not understand the hierarchy of these events in the emerging programs of gene expression. The next step is to use genomics to obtain an integrated picture involving all of these phenomena as cells go through development and differentiation, and this requires the development and application of new tools for analyzing the genome.

COMPONENTS OF THE CHALLENGE

The components of the challenge are set out in **Figure 1**. First, great progress has been made in characterizing protein-coding sequences such that we can now identify and, to some extent, classify ~20,000–25,000 genes that orchestrate and/or contribute to cell structure and function (49) (**Figure 1**, box labelled A). A current challenge is the identification of sequences which control whether a gene is switched on or off. These involve cis-acting elements (promoters, enhancers, silencers, splicing regulators, chromatin boundary/insulator elements, and locus control regions) that regulate when and where particular genes, and the proteins they encode, are expressed during development and differentiation. Identifying all of the *cis*-elements controlling a single mammalian gene can be a complex task because these sequences may be distributed across 10–1000 kb, and they may even be located within apparently unrelated genes. The ultimate aim is to identify and define the roles of all such sequences in the genome, and the tools for this are currently being developed by intensive analysis of 1% (~30 Mb) of the genome in the Encyclopedia of DNA Elements (ENCODE) Project (**http://www.genome.gov/10005107**).

An active area of research revolves around understanding how classical, DNA-binding

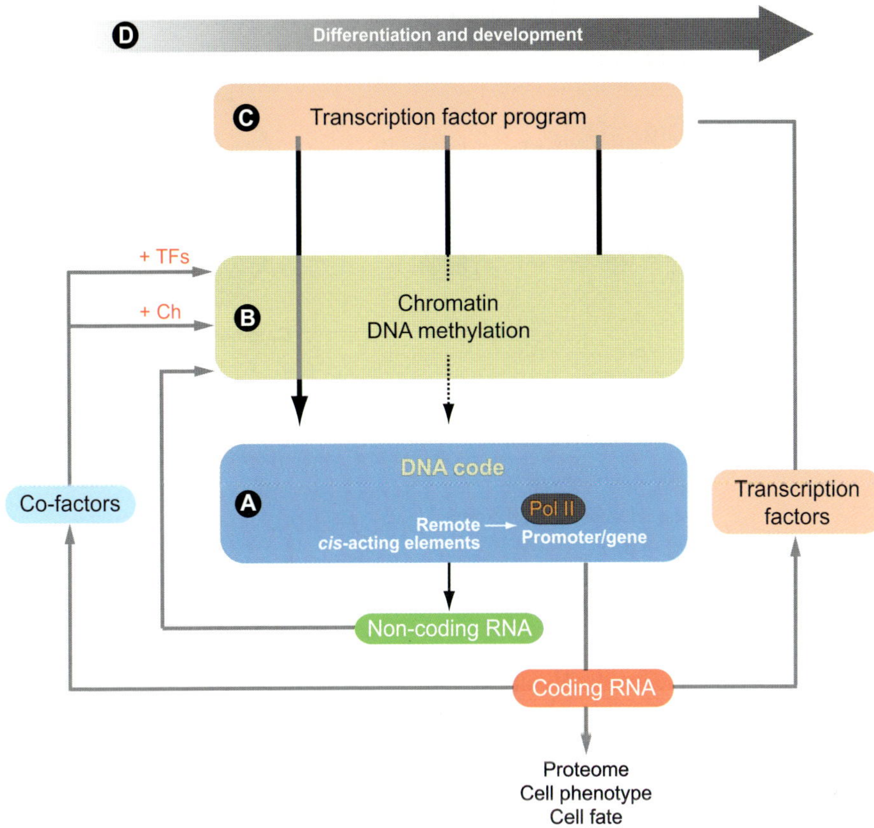

Figure 1

The components of the postgenome challenge. The annotation +TFs refers to cofactors that are recruited via DNA-binding transcription factors; the annotation +Ch denotes factors that directly recognize chromatin modifications. The three arrows leading from the transcription factor program indicate three possible outcomes dictated by the state of chromatin. The DNA may be fully accessible (*solid arrow*), partially accessible (*dashed arrow*) or not accessible (*blocked line*).

transcription factors, acting collectively as the transcription factor program (**Figure 1**, box labelled C), interact with *cis*-acting sequences to control gene expression. At some level, this is already understood. The *cis*-acting elements provide a platform to assemble transcription factors and cofactors that either positively or negatively influence the recruitment and activation of RNA Polymerase II (Pol II) at the promoter (**Figure 1**, box labelled A). Ultimately, to switch a gene "on," Pol II must be recruited and activated (for clearance and elongation) to the gene promoter. When a gene is "off," Pol II is either not recruited or is recruited but paused on the promoter (43, 90, 96).

Regulation of gene expression is further complicated by the way in which DNA (~2 m per cell) is packaged within each nucleus (**Figure 1**, box labelled B, and **Figure 2**), where it is compacted more than 10,000-fold by being wrapped into chromatin. The core unit of chromatin is the nucleosome, which consists of 147 base pairs (bp) of DNA wrapped around a histone octamer, comprised of a centrally located histone $(H3/H4)_2$ tetramer associated with two histone H2A/H2B dimers. Consecutive nucleosomes are assembled into 11-nm fibers (resembling beads on a string), which are then compacted in association with the incorporation of linker histones (H1) into poorly defined, higher-order structures and eventually into individual chromosomes (32). Although for many years chromatin was regarded merely as a structural device for packaging DNA, it is now known to play an important role in regulating gene expression, mainly by influencing the accessibility of transcription factors and cofactors to the *cis*-acting elements

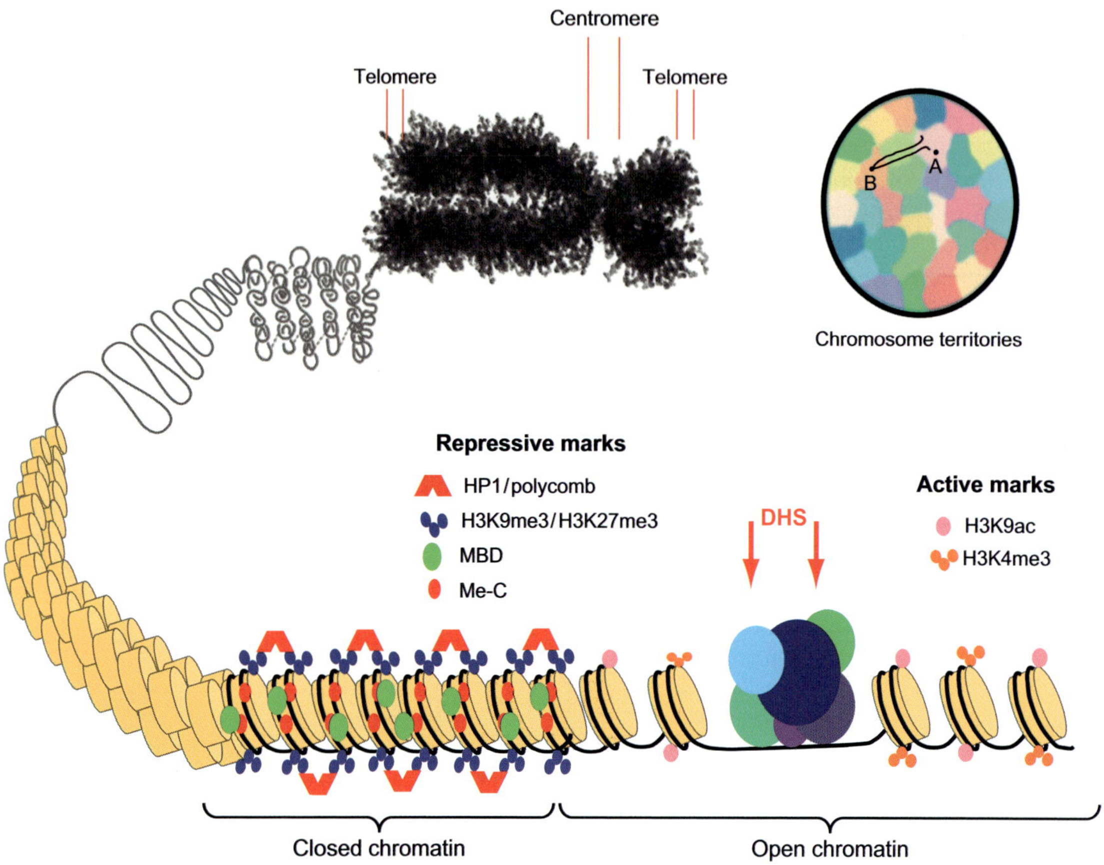

Figure 2

The structure and modifications of chromatin, showing the progression from condensed inactive chromatin to unfolded, active chromatin. Chromosome modifications are annotated and DHS refers to a DNAse1-hypersensitive site. At present, the higher-order structures of chromosomes are unknown. In situ fluorescence hybridization studies show that in interphase nuclei (*top right*), chromosomes occupy distinct territories (*colored areas*). Genes can lie within their own chromosome territory (*A*) or may lie outside on extended loops of chromatin (*B*). Although it was originally thought that the relationship of a gene to its chromosome territory was related to its expression status, it now seems likely that it is more closely related to the structure of the chromatin associated with the gene.

that control gene expression. As a general rule, the more compacted chromatin is, the less active it becomes (**Figure 2**). We now know that the establishment and maintenance of the patterns of gene expression, which determine a cell's phenotype, can be influenced by how DNA is packaged into chromatin. This, in turn, is determined by covalent modification of the nucleosomes, the presence of histone variants, and the binding of nonhistone proteins that associate with chromatin. Although this review concentrates on the role of chromatin in regulating gene expression, it is also becoming clear that chromatin structure and modification may influence all nuclear processes (e.g., replication, recombination, DNA repair, and chromosome segregation).

Until the past few years, most of our understanding of how chromatin influences mammalian gene expression has come from the careful analysis of a handful of well-characterized genes. However, now that the entire genome sequence is available and technologies to interrogate this sequence are being developed, it is becoming possible to rapidly study any gene or locus of interest. Furthermore, these developments make it possible to study many genes in parallel and are changing the investigation of chromatin from a specialized pursuit to a more general and straightforward activity.

Clearly, if we aim to investigate the dynamic changes of emerging epigenetic and transcriptional programs within defined lineages, the purification and definition of specific subsets of primary cells as they go through development and differentiation are required (**Figure 1**, box labelled D). Most preliminary work to date has been carried out either in static cell lines or in cell lines induced to recapitulate aspects of differentiation. Although this is suboptimal, even these early studies are allowing us to begin to address the relationship between the transcriptional program, the epigenetic program, and the resulting cell phenotypes.

In this review we concentrate on how genomics can be applied to study any single locus to investigate how it is expressed from within its natural chromosomal environment. We focus on the human genome and, where appropriate, illustrate principles from work carried out using the α-globin gene locus [one of the ENCODE Project loci (101)] as a model. We also discuss how these approaches are being applied to the whole genome.

THE LONG-RANGE STRUCTURE OF THE GENOME

Each of the 23 pairs of chromosomes that make up the human genome contains a centromere and is capped at each end by telomeres (**Figure 2**). Both structures are predominantly composed of tandem arrays of repetitive DNA. The DNA between the telomeres contains genes (2%), regulatory elements (1–2%), and noncoding DNA, including origins of replication, chromosome structural elements, repetitive elements (50%), and other sequences (45%). Whole-genome analysis has shown that the DNA composition and distribution of these chromosome elements is nonrandom. For example, some regions of the genome are gene-rich and others are gene-poor (49). Early studies of human chromosomes suggested that each different region was associated with recognizably different chromatin structures (9). In addition to telomeres and centromeres, regions thought to represent "open" (euchromatin) and "closed" (constitutive and facultative heterochromatin) structures were recognized (see Sidebar: Different Types of Chromatin). Preliminary studies lend support to the notion that the types of chromatin that form may be determined by the associated DNA (GC-rich as opposed to AT-rich, and repetitive DNA as opposed to unique DNA). Now that chromatin structure can be analyzed and compared in detail with the underlying sequence (see below), the dauntingly complex relationship between DNA sequence, chromatin structure, and function is emerging (58, 65). We do not discuss telomeres and centromeres, although they may play important roles in regulating both normal and abnormal gene expression (11, 15); rather, we examine the relationships between DNA sequence, gene regulation, and chromatin within the body of a chromosome.

DEFINING REGULATORY DOMAINS

It was originally proposed that the genome might be organized into discrete, self-contained chromosomal domains or units (50–1000 kb), each containing a gene (or coregulated gene cluster) surrounded by specialized boundary elements to isolate these sequences from the effects of the surrounding chromosomal environment (114). However, full annotation of the genome, demonstrating overlapping genes and regulatory elements, has shown that this concept is an oversimplification (27). Nevertheless, to understand how a gene is expressed from its natural chromosomal environment, it is still necessary to

DIFFERENT TYPES OF CHROMATIN

Heterochromatin was originally observed by Emil Heitz (46) as darkly staining aggregates at the periphery of the interphase nucleus. It represents a tightly packed form of DNA. Constitutive heterochromatin is stable and conserves its heterochromatic properties during all stages of development and in all tissues. It is located at centromeres, telomeres, and sometimes pericentromerically. It is largely comprised of tandemly repeated sequences, is gene-poor, and is late-replicating. Facultative heterochromatin is reversible and its heterochromatic state depends on the stage of development or cell type. The inactive X chromosome, which is seen as the Barr body in the interphase cell, is the clearest example of facultative heterochromatin. Facultative heterochromatin is relatively poor in genes and these genes are usually not transcribed when packaged in heterochromatin. Euchromatin appears as lightly stained material in the interphase nucleus and this reflects its less compact structure; it only becomes condensed during mitosis. It is gene-rich and is often, but not always, under active transcription. It is usually early-replicating (46).

define the segment of the chromosome that contains all of the *cis*-sequences required for its full gene regulation.

It is possible to address this issue from comparative genomics. During evolution there has been considerable reassortment of many segments of the genome. Presumably, when this occurs, all (or most) appropriate sequences required to fully regulate the associated gene must remain linked together, otherwise the gene will be expressed suboptimally (e.g., 48, 98). Therefore, defining the extent of conserved synteny around any gene of interest should delimit the region containing the gene and its associated regulatory elements. For example, the human α-globin cluster of most species is located close to the telomere. However, in the mouse this locus has translocated into an interstitial locus, within the body of the chromosome. By comparing the segment of DNA that is conserved in 22 species (spanning a 500 million year evolution) a region of ~135 kb of synteny has been defined, thus delimiting the region containing all sequences required for fully regulated expression (**Figure 3**). This approach may not be so informative at some loci where the breaks in synteny lie far from the locus being studied. In this case, additional information may be obtained from local chromosomal rearrangements and the breakpoints of chromosomal translocations (**http://agserver01.azn.nl:8080/ecaruca/ecaruca.jsp**) that either do, or do not, affect expression of the locus in question (also see below).

Synteny: Conservation of synteny is said to exist when the order of genes on a chromosomal segment is preserved between related organisms

Transgene: a piece of DNA that is introduced into cells or organisms to modify the genome

Currently, the ultimate test of such regulatory units is whether or not one obtains fully controlled expression when these relatively large chromosomal segments are introduced in transgenic animals in either a bacterial or yeast artificial chromosome (BAC or YAC) construct. However, even large constructs may not contain all cis-acting elements, which may be dispersed throughout >1000 kb of DNA. Furthermore, expression from such transgenes may be compromised by chromosomal position effects, variation in copy number, and rearrangements, which are often difficult to detect. The recent development of recombinase-mediated genomic replacement to swap an endogenous mouse domain for the corresponding region of human synteny offers a well-controlled experimental approach to determine whether a defined segment of a chromosome contains all sequences required for full regulation of a locus (111).

IDENTIFYING KEY REGULATORY ELEMENTS

Once a regulatory unit has been defined, the next aim is to identify all of the key *cis*-elements within it. Because the promoters of over 50% of human genes are associated with CpG islands (49), which can be defined bioinformatically, these provide a simple and effective guide to many promoters (e.g., **Figure 3**). Many, but not all, regulatory elements, including promoters, can be identified within fully annotated DNA as conserved, noncoding sequences using

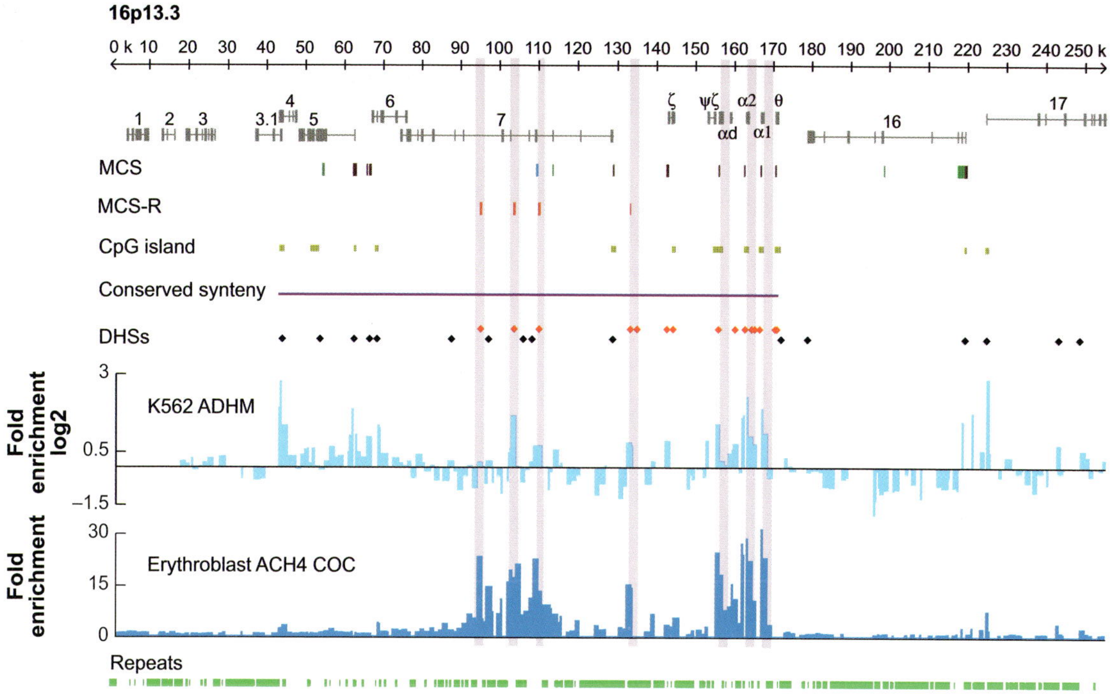

Figure 3

An example of in-depth analysis of a single region of the genome: the human α-globin cluster, which is one region being studied in the ENCODE Project. The genes located in the telomeric region of chromosome 16 are numbered as before (22) and the globin genes are labeled. All multispecies conserved sequences (MCS) and a subset associated with erythroid-specific regulatory elements (MCS-R) are shown. Yellow boxes indicate the positions of CpG islands. A CpG island is a region rich in cystine and guanine dinucleotides that usually remain unmethylated in all tissues. These regions are often found covering the transcriptional start sites of eukaryotic genes. The region of conserved synteny (48) is shown as a purple line. The positions of all previously mapped DNAse1-hypersensitive sites (including erythroid-specific sites, *red diamonds*) are below the purple line. The two charts show examples of techniques used for global analysis of DNAse1-hypersensitive site (DHS) K562 approach to DHS mapping (ADHM) (33) and ChIP on chip analysis [AcH4 COC ChIP/chip] (M. De Gobbi, D. Higgs, manuscript in preparation). The locations of repeats are also shown.

comparative sequence analysis [so-called multi-species conserved sequences (MCS)]. The underlying principle is that regulatory elements contain conserved sequence motifs (6–20 bp) for the various transcription factors that bind them. The sequence context and spacing between binding sites may also be conserved, giving rise to more extended regions of conservation (200–300 bp) (48). Not all regulatory elements are equally well conserved, and some may have been recently recruited in evolution and may even be species specific (13). In addition, not all conserved noncoding sequences are proven transcriptional enhancers.

Current programs to identify regulatory elements based solely on their DNA sequence (without using comparative genomics) have been disappointing. Therefore, an efficient search for regulatory elements should always incorporate a comparison between different species and include the identification of nuclease-hypersensitive sites (**Figure 3**), which are the hallmarks of interactions between transcription factors and DNA in chromatin in vivo (**Figure 2**).

***Trans*-acting:** a *trans*-acting factor, usually a diffusible protein or RNA, can affect the activity of sequences that lie on a different DNA molecule

DNAse1-hypersensitive sites (DHSs) occur at regions of increased chromatin accessibility, corresponding to where some, but not all, trans-acting factors bind chromatin (32, 42). The underlying chromatin is usually modified (e.g., H3 and H4 acetylated, see below) and the local nucleosomal structure disrupted, making the associated DNA (~300–500 bp) more accessible to cleavage by nucleases. The conventional method used to identify DHSs involves digesting intact nuclei with DNAse1, extracting the DNA, and mapping the DHSs by Southern blotting. This is a very sensitive, but very time-consuming, method when characterizing large segments of the genome. Nevertheless, we used this approach to map DHSs across the 275-kb human α-globin ENCODE region (**Figure 3**) and found that all of the known key regulatory elements (enhancers and promoters) in this region, including most of the MCSs and all of the CpG islands, are associated with either constitutive or tissue-specific DHSs (48). This demonstrates the value of using information from MCS, CpG islands, and DHSs together. Such detailed analysis of a single locus (**Figure 3**) provides an important test case for studying the high-throughput assays required to efficiently analyze all other loci.

Over the past two years, four approaches have been developed to map DHSs throughout the genome with variable degrees of sensitivity and specificity. The first used multiple, real-time polymerase chain reaction (PCR) assays across a region of interest to compare the amplification of DNA from nuclei digested with DNAse1 with that from undigested nuclei where an amplimer spans a DHS. Using this assay, the expectation is that there will be reduced product in the digested template compared with the undigested template (28). This method was very sensitive and specific but could not be considered high throughput because the authors performed 855 PCR reactions to study 30.4-kb DNA in just a single cell type. In the second method (33), DNA from DNAse1-digested nuclei was captured via biotin-labeled linkers that were ligated to the cut DNAse1 site. This DNA was compared to nonselected, input DNA by hybridization to a tiled microarray. Although this useful approach successfully identified many DHSs, it did not identify all sites with equal confidence (**Figure 3**). A third method used high-throughput cloning and sequencing of libraries generated from DNAse1-digested nuclei. This approach was sensitive but not very specific, with more than 70% of sequences thought to represent "noise" resulting from random shearing (21), although this could be compensated for by screening for clustered, rather than single, sites. Most recently, Sabo et al. (87) exploited the fact that a characteristic feature of DHSs is the occurrence of multiple cleavage events over a short distance generating small DNA fragments. By separating, and differentially labeling, small (multiple-hit) and large (random-cleavage) DNA fragments from DNAse1-treated nuclei, they identified DHSs after hybridization to a tiled microarray. This resulted in both a high sensitivity (92%) and specificity (>99%) and is a promising approach. For targeted high-resolution analysis, classical DHS analysis remains the gold standard, but the global approaches being developed now provide an excellent guide for devising more focused analyses of specific loci.

ANALYZING CHROMATIN MODIFICATIONS

Changes in both local and long-range chromatin structure can influence gene expression (**Figure 1**, box labelled B, and **Figure 2**). This epigenetic information may be determined either by introducing distinct histone variants (89) or as covalent modifications of the underlying nucleosomes (50, 106), particularly within the accessible, unstructured N-terminal histone tails. The most intensively studied post-translational modifications include phosphorylation (S and T), acetylation (K), and methylation (K and R), although an increasing number of other modifications are being recognized (**Figure 4**). Using mass

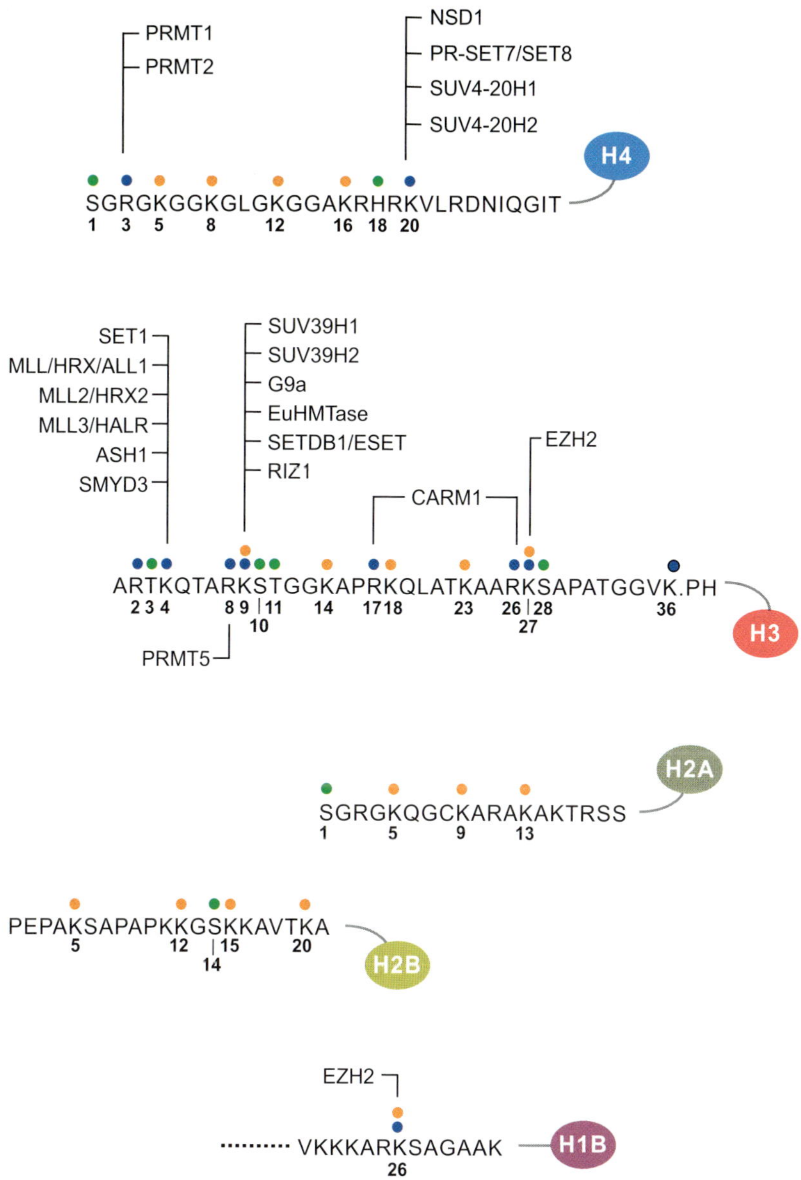

Figure 4

Covalent, post-translational modifications of chromatin. Histones can be modified on the residues indicated by methylation (*blue dots*), acetylation (*orange dots*), or phosphorylation (*green dots*). The enzymes responsible for histone methylation are shown above or below the residues that they are known to modify. Adapted from Reference 65.

spectrometry it is possible to show linked modification on the same N-terminal tail, giving still more complexity to the histone code. In general, histone H3 and H4 acetylation and H3K4 methylation correlate with active chromatin, whereas histone H3K9 methylation and H3K27 methylation frequently correlate with repression (**Figures 2** and **4**). However, with only limited studies on relatively few cell types in mammals, there are still exceptions and important controversies around these observations that demand further investigation.

To date, most global studies of histone modifications have been performed in yeast (6) and Drosophila (93), and relatively few large-scale studies have been carried out in complex mammalian genomes. Until recently, most studies in mammalian genomes used chromatin immunoprecipitation (ChIP) (**Figure 5**) analyzed by selected, quantitative

ChIP on Chip

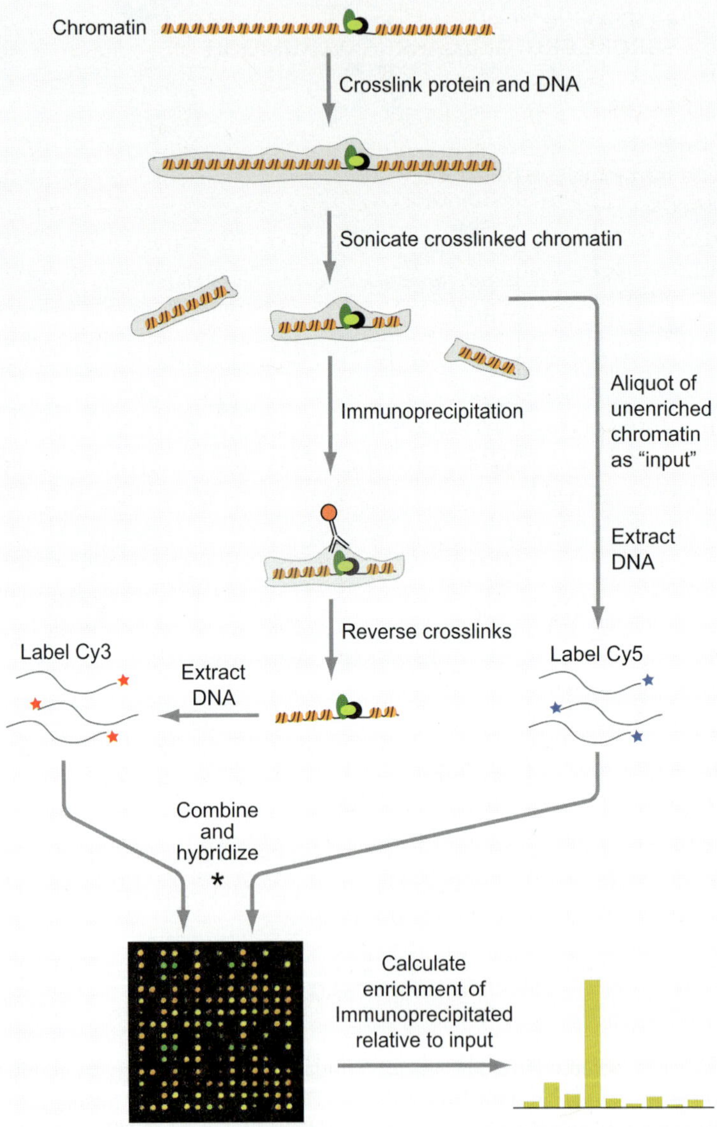

* On single color microarrays the immunoprecipitated fractions can be hybridized to a separate array and the results compared computationally.

PCR analyses across a few loci of interest. These studies added significantly to the characterization of MCSs and DHSs because, for example, within the ENCODE model globin loci (**http://www.genome.gov/10005107**), all peaks of chromatin activation corresponding to the MCSs and DHSs can be ascribed a function (**Figure 3**). When the locus is fully activated the chromatin modifications may spread more extensively throughout the entire domain from peaks at the regulatory elements (**Figure 3**). Defining the patterns of histone modifications at *cis*-elements in cells that either express or silence the genes they regulate can give further insight into the potential roles of the underlying sequences. The more recent development of ChIP analyzed using tiled microarrays (53 and **Figure 5**), serial analysis of gene expression (SAGE) technology (84), or exogenous adenine methylation (e.g., DamID 109) have enabled a few relatively large chromosomal segments to be analyzed and we found that these analyses truly reflect extensive quantitative PCR (qPCR) analysis of the same regions (see legend of **Figure 3**). Clearly, these validated approaches now enable high-throughput analyses of chromatin modifications in more extensive regions of the human genome. For example, large-scale studies across human chromosomes 21 and 22 using HepG2 cells (7) and activated human T cells (83) have generally confirmed the yeast studies and shown that histone H3K4 methylation and H3/H4 acetylation typically affect short stretches of active (<1 kb) chromatin. Many of these correspond either to the 5′ ends of genes (presumably coinciding with promoters), but in mammals many also colocalize with regulatory elements lying some distance from the genes they control.

During development, chromatin may package DNA into a state of permanent quiescence (constitutive heterochromatin) or a state in which a gene is silenced but capable of being activated (facultative heterochromatin) (see Sidebar: Different Types of Chromatin). Chromatin modifications that associate with repressed chromatin may differ from one intrachromosomal region to another, indicating that different mechanisms may silence activity. There are at least nine repressive histone methylation marks (H3K9, H3K27, and H4K20, each of which may be mono-, di-, or trimethylated). In an extensive cluster analysis in mouse, stable modifications of H3K9Me3 and H4K20Me3 were found concentrated at pericentric and centric heterochromatin that contained tandem repeats (67). In addition, H3K27Me3 was concentrated at the inactive X chromosome (which contains a very high density of repeats), providing an example of facultative heterochromatin (86). However, patterns of repressive chromatin at short and long interspersed retrotransposable elements (SINEs and LINEs) and DNA transposons and retroposons varied from one cell type to another, suggesting that they might play some important, as yet undefined, functional role in vivo (67). Note that, to date, no study has comprehensively investigated intrachromosomal modifications at individual repetitive elements within their normal chromosomal context. This is technically difficult in mammalian genomes because hybridization

Transposons: mobile genetic elements that can move from one site in the genome and reinsert at another site

Retroposons: a form of transposon that has an intermediate RNA stage and is inserted into DNA after reverse transcription

Figure 5

Chromatin immunoprecipitation (ChIP) is used for analyzing modifications of chromatin and the association of specific proteins [including transcription factors (*light green, dark green, and black spheres*)] within defined chromosomal regions in vivo. In these experiments, chromatin is cross-linked by adding formaldehyde to living cells. Cross-linked chromatin is then fragmented (usually by sonication or by micrococcal nuclease) to oligo- and sometimes mononucleosomes. These small chromatin fragments are then immunoprecipitated with an antibody specific for a particular chromatin modification or protein. After reversal of the cross-links, deproteinization, and DNA extraction, DNA sequences that associate with a given protein or a particular chromatin modification are analyzed by quantitative PCR (qPCR) (not shown) or microarray analysis.

methods cannot distinguish one set of repeats from another.

How can genes be silenced in the context of otherwise open euchromatin? H3K27Me3 has been implicated in Polycomb-mediated gene silencing (86), which, in Drosophila, occurs via Polycomb response elements, which also contain short repetitive elements. Profiles for patterns of active and repressive histone lysine methylation have been performed on individual mammalian genomic regions (63, 71), parts of chromosomes (41), and with genome-wide, but promoter-based, sequences on microarrays (88, 93). Again, note that most methods for analyzing tiled arrays only interrogate unique DNA and a great deal of potentially very interesting information from repetitive elements may thus be lost.

So far, only a few cell types have been analyzed. However, this approach has revealed a fascinating result: Two groups have found that in pluripotent embryonic stem (ES) cells, highly conserved regulatory elements (8) and lineage-specific genes (3) have both repressive (H3K27Me3) and active (H3K4Me3 and H3K9Ac) chromatin modifications. This has been provocatively interpreted as showing that these genes are poised for activation but are held in a repressed state prior to lineage commitment. When the cells differentiate, either the repressive (silent) mark or the active (expressed) mark persists, but not both.

Another level of complexity comes in the guise of histone variants. There are eight known isoforms of the linker histone H1, four isoforms of H3 (H3.1, H3.2, H3.3, and CENPA) and four H2A variants (H2AX, H2AZ, macroH2A, and H2ABBD). These variants and the mechanisms by which they become incorporated into chromatin have been reviewed in detail elsewhere (89). Of the H3 variants, CENPA is localized exclusively to centromeres, where it plays a critical role in kinetochore assembly, whereas H3.3 is present at transcriptionally active loci. The H2A variants play specialized roles in DNA repair and recombination (H2AX), gene regulation (H2AZ and H2ABBD), and X-chromosome inactivation (macroH2A). To date, only a few studies have used ChIP assays to examine the distribution of these variants, mainly because there are major limitations in the ability of antibodies to distinguish the variant isoforms from the canonical, core structures.

IDENTIFYING TRANSCRIPTION FACTORS BOUND TO CHROMATIN

It seems possible that although the epigenetic program may play an important role in determining the probability that any gene is activated, and provide a way to transmit cellular memory, the primary driving force behind cell fates lies in the transcription factor program and its targets (**Figure 1**). For example, in hematopoietic cells, overexpression of just one or another of the key lineage-affiliated transcription factors (TFs) is sufficient to change cell fate (55). Furthermore, when carefully examined, TF binding often seems to precede maximal chromatin modification and it is thought that chromatin-modifying enzymes are recruited to chromatin by the TFs when bound to their cis-acting elements.

Over the past few years, the global transcriptional program has been established in many cell lines and primary cells. However, the presence of mRNAs corresponding to the TFs is only a part of the story. Still to be determined are the levels of protein, their modifications, and, most importantly, their targets. The same techniques (ChIP and qPCR, ChIP/chip, and methods based on high-throughput sequencing) used to analyze chromatin modifications at single loci, chromosomes, and the entire genome can also be applied to characterize the patterns of transcription factors binding to *cis*-acting sequences within chromatin (**Figure 5**). Within carefully studied individual loci, transcription factors bind specifically at the known, key *cis*-acting elements (usually MCSs) and not elsewhere throughout the locus (2). What determines specific binding to these sites (often

6–20 bp) and no others is an important, unresolved question.

In the past, genes have been studied one by one and a great deal has been learned about how individual mammalian genes are regulated and the order in which trans-acting factors are recruited to these individual loci. More extensive ChIP microarray studies have focused on promoter regions (73) and CpG islands (113), and preliminary studies looking at the ENCODE loci have shown that, in principle, the ChIP/chip approach can now be used for whole-genome studies. However, there is no doubt that it is more difficult to use ChIP analysis to look for TF binding than it is for histone modification. Success is very dependent on the specificity and efficiency of the antibodies used. Using a global approach, the binding of estrogen receptor (16) c-myc, Sp1, and p53 (18) throughout chromosomes 21 and 22 has shown binding sites at distal regulatory sequences (enhancers) and promoters identifying discrete chromatin elements involved in transcriptional regulation. ChIP/chip studies to characterize transcription factor binding to the entire genome have been performed in yeast (reviewed in 4, 44) and are underway in human. Together with microarray studies profiling patterns of gene expression, these studies highlighting the targets of transcription factors will add significantly to the characterization of the entire set of transcriptional programs underlying cell-fate decisions.

Ultimately, expression depends on the entire preinitiation complex (PIC), including Pol II and the general transcription factors (GTFs), being recruited and transcribing from the promoter of the gene in question. Understanding the mechanisms underlying this process is a major issue in current biology and, to date, the focus has been on individual loci. ChIP analyses using antibodies against Pol II in the ENCODE Project nicely identify Pol II (and other components of the PIC) bound to the promoters of active genes. Individual studies show that Pol II is recruited to these promoters at different stages of differentiation (43, 96) depending on the gene studied, implying that the recruitment and elongation of Pol II may be regulated differently at different genes. In addition, analysis of several individual loci has shown that Pol II may also be recruited to the distal regulatory elements that play a role in the activation of gene expression activation from these sites (60, 100). The preliminary studies from the ENCODE Project and a few more extensive surveys (59) show that complete genome analysis for Pol II binding is feasible and will enable us to understand the common features and general principles by which genes are activated in vivo.

THE BINDING OF COFACTORS AND OTHER NONHISTONE PROTEINS

The classical transcription factors (repressors and activators) described in the previous section are DNA-binding proteins that are thought to bind directly to key *cis*-elements. However, they are often components of multiprotein complexes and thus recruit other transcription factors and cofactors to the regions they bind (**Figure 1**, box labelled B, +TFs). These cofactors include a wide variety of proteins that mediate changes in local and long-range chromatin by imposing modifications (e.g., histone acetylation and methylation) that make access to the *cis*-elements more or less likely (**Figures 1** and **5**). These chromatin modifications may be passed on to daughter cells.

Over the past few years it has emerged that many cofactors and other nonhistone proteins share a common subset of protein domains that specifically recognize different combinations of chromatin modifications (**Figure 1**, box labelled B, +Ch). For example, bromodomains (e.g., found in histone acetyl transferases) recognize acetylated chromatin and chromodomains (e.g., found in Polycomb proteins), and plant homeodomains (PHD) (found in a wide variety of chromatin-associated proteins) recognize methylated chromatin. Therefore, these proteins will recognize and "read" these modifications and reassemble particular epigenetic states in the next generation of cells, thereby providing a memory of gene expression from one cell generation to the next (summarized in **Figure 1**, box labelled A, and **Figure 6**). Once this process is set in place, the pattern may be propagated even in the absence of the transcription factor(s) that initiated it.

Chromatin remodeling factors comprise an important group of nonhistone, multiprotein complexes involved in reading these modifications. These are multiprotein complexes that use the energy of ATP hydrolysis (via ATPase subunits) to remodel or remove nucleosomes. Components of all such complexes include one of the domains set out in **Figure 6** and **Table 1**.

Some of these nonhistone proteins have been analyzed at individual loci, but it is often more difficult to cross-link and immunoprecipitate these protein complexes than DNA-binding transcription factors (see above), even using modified cross-linking protocols. To date, although some proteins [e.g., histone acetyl transferases (HATs) and histone deacetylases (HDACs)] have been analyzed in yeast with variable success, there have been no global surveys of their distribution at individual mammalian loci across the mammalian genome.

THE IMPORTANCE OF DNA METHYLATION

Although the DNA code comprises four bases, the modification of cytosine by methylation gives additional scope for encoding new information. A DNA molecule.... ACGT.... is clearly distinguishable from ....A^{m}CGT.... This modification is present in most vertebrates and in somatic cells 70–80% of CpGs are methylated (31). With the exception of imprinted genes (80), the pattern of DNA methylation is newly acquired and established early in embryogenesis by the de novo DNA methyltransferases DNMT3a and DNMT3b, which exhibit their highest activity at this time (74). Maintaining the pattern of methylation mainly depends on DNMT1 (reviewed in 10). In most organisms that exhibit CpG methylation, it is predominantly found within repetitive elements including centromeric and pericentromeric satellites (69) and interspersed repeats such as LINEs and SINEs (118). In such regions of the genome, methylated DNA provides the target for methyl-binding domain proteins (MBDs), which have several duties, including transcription repression (10).

A major pursuit over the past 20 years has been to determine whether DNA methylation plays a primary role in the regulation of gene expression. There is certainly a very good correlation between DNA methylation of promoters and regulatory elements, and gene expression. This is particularly true of promoters associated with CpG islands, which are always active (or capable of being activated) when unmethylated and always silenced when methylated, as seen on the inactive X chromosome (37), at imprinted loci, (79) and at some tissue-specific genes (19). The relationships between DNA methylation and associated chromatin changes are complex and under intense investigation (36). Whether chromatin modifications determine the underlying pattern of DNA methylation and binding of MBDs or vice-versa has not been resolved. However, in general, in regions of active chromatin (histones acetylated) DNA is unmethylated, and in repressed regions (histones methylated) DNA is methylated, but there are many exceptions to this.

Genome-wide studies of DNA methylation, in combination with analysis of gene expression, chromatin modification, and distribution of MBDs and DNMTs in appropriate cells, may help resolve the issues surrounding the role of DNA methylation in vivo. In the meantime, covalent modification of the DNA code by methylation is arguably the most robust way of monitoring changes in emerging epigenetic programs. Consequently, the Human Epigenome Project was established to determine the

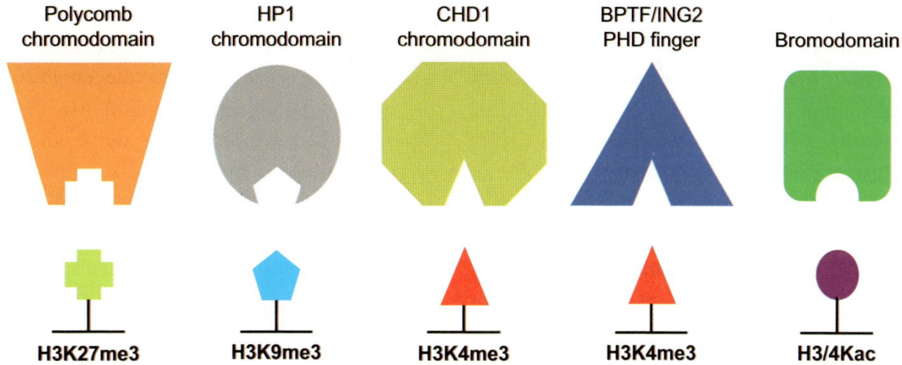

Figure 6
Examples of reading histones in which particular domains occurring in many of the nonhistone-associated proteins are known to bind specific modifications of chromatin.

methylation pattern across the entire genome in appropriately informative cell types (**http://www.epigenome.org/index.php?page=project**).

There are three main methods by which to establish the pattern of methylation across a segment of the genome. The first involves the use of methylation-sensitive restriction enzymes (MSREs). These have long been used in conjunction with Southern blotting to interrogate specific CpGs that lie within the enzymes' recognition sequences. The degree of cutting at any site can be easily determined, enabling accurate quantitation of methylation. However, the resolution of this approach is low and, even using enzymes that cut frequently (e.g., HpaII, CCGG), is measured in hundreds of base pairs of DNA. By combining MSREs with linkers ligated to the cut sites, it has recently become possible to enrich or amplify methylated or unmethylated fractions (19, 35, 94), which may then be identified by hybridization to microarrays bearing the target sequences.

A second approach is to use antimethylcytosine antibodies. In this case, methylated

Table 1 Summary of the currently known groups of proteins that associate with and modify chromatin structure and/or function

Nonhistone chromatin protein	Example	Review	Disease*	Reference
DNA methyltransferases	DNMT1, DNMT3A, DNMT3B	(54)	ICF Syndrome (DNMT3B)	(116)
Methyl-binding proteins	MBD1-4, MeCP2 Kaiso	(54)	Rett Syndrome (MeCP2)	(1)
Histone acetyltransferases	CBP, p300	(85)	Rubenstein-Taybi Syndrome (CBP, p300)	(76) (82)
Histone deacetylases	HDAC1-9	(66)		
Histone methyltransferases	Suv39H1/2, EZH2	(51)	Sotos Syndrome (NSD1)	(56)
Histone kinase	RSK2	(72)	Coffin-Lowry Syndrome (RSK2)	(102)
Histone demethylases	LSD1, JHDM1	(104)		
Chromatin remodeling proteins	BRG, BRM, CHD1-9	(24)	Schimke Immunosseous Dysplasia (SMARCAL1) Cockayne Syndrome (ERCC6) ATR-X Syndrome (ATRX) CHARGE Syndrome (CHD7)	(12) (103) (38) (108)
Histone binding	Hp1α,β,γ	(45)		
Polycomb group proteins	Pc, Suz12	(81)		

*The emerging group of human diseases known to result from mutations in some of the genes encoding such proteins.

sequences are enriched by immunoprecipitation. The enriched fraction can then be identified by hybridization to microarrays bearing the target sequences (112, 119). Resolution (usually >500 bp) is limited by the size of the sonicated DNA used for the immunoprecipitation and separately by the length of the DNA fragments spotted on the arrays. This approach was recently used to generate methylation profiles of all human chromosomes at 80-kb resolution (112). Using primary cells, broad regions of differential methylation were identified throughout the genome, but, at present, the resolution of such analyses does not adequately address the role of DNA methylation.

Finally, it is possible to analyze DNA methylation at the single-nucleotide level using bisulphite modification and sequence analysis. The bisulphite treatment of DNA leads to a chemical conversion of unmethylated cytosine to uracil; methylation of cytosines blocks this conversion. After modification, the DNA can be amplified by PCR and positions converted to uracil appear as thymine in the product (**Figure 7**). The product may then be analyzed to determine the ratio of cytosine to thymine at any position. This may be done by direct sequencing (with specialized software to allow quantitation from

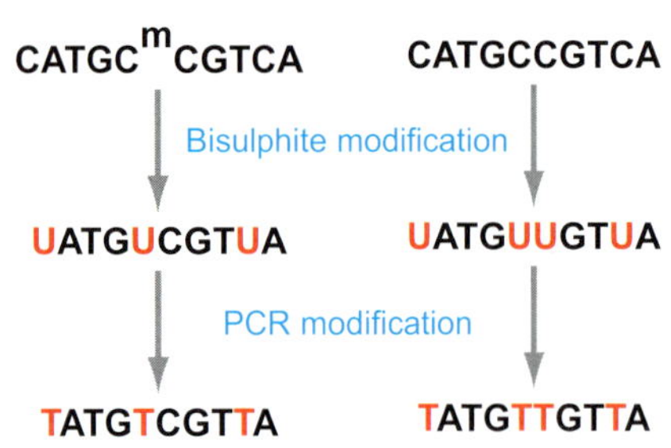

Figure 7

Bisulphite modification. Two identical sequences are shown. One has a methylated cytosine (*left*). Treatment with bisulphite leads to the chemical conversion of unmethylated cytosine to uracil, whereas methylated cytosines remain unconverted. Subsequent amplification of the DNA leads to the positions converted to uracil appearing as thymine.

the sequence trace) (61), by primer extension and analysis of fragments by mass spectrometry (MS) (78), by pyrosequencing (20), or by base-specific cleavage and MS (91). These methods are expensive and labor intensive as they require analysis of individual amplimers across the region. Nevertheless, it is possible to resolve the pattern of methylation to the single-nucleotide level and, unlike hybridization-based methods, they allow the analysis of some repetitive sequences. By using bead-based massive parallel sequencing, genome-wide analysis of methylation may be possible without having to generate specific amplimers covering the genome.

CHARACTERIZING CHANGES IN CHROMOSOME CONFORMATION

The relationship between the transcriptional program, the epigenetic program, and the position of a gene within the interphase nucleus is not understood but is currently under intense investigation (34, 40). The question is whether nuclear position is an independent and functionally important epigenetic variable or whether it merely reflects other aspects of the transcriptional and epigenetic programs. It is clear that chromosomes occupy specific locations (chromosome territories) in interphase nuclei (**Figure 2**) and genes located on that chromosome may be positioned within the territory or may extend from it (110). This seems to depend as much on whether the gene is present in a region of "open" or "closed" chromatin as on whether or not it is expressed (39). Some silenced genes may be sequestered with heterochromatin, but others are not (14). Active genes may congregate within subnuclear structures that contain the proteins required for their expression (see Sidebar: Transcription Factories). Inevitably, by the nature of the packaging of chromatin into a nucleus, any segment of a chromosome will be juxtaposed to other segments of the genome, both in *cis* and in *trans*, and these associations may change with the activity of the underlying genes. It appears that some of these associations are nonrandom (75), but even so, whether these gross changes in relative position play any role in regulating gene expression is not yet known.

By contrast, several lines of evidence suggest that smaller-scale intrachromosomal interactions between cis-acting elements may play a role or reflect very important functional aspects of gene regulation. There are multiple mechanisms by which remote, *cis*-acting elements may communicate with their target genes to activate or silence transcription within their natural chromatin environment. Currently, most evidence favors a chromosomal looping model involving direct physical interaction between *cis*-acting protein/DNA complexes (30, 57, 97, 99, 107). Various methods have been developed to study such interactions.

The first method [RNA-tagging and recovery of associated proteins (TRAP)] was used to study interactions between the active β-globin promoters and their upstream regulatory elements [β-globin locus control region (LCR)] (17). In this assay, oligonucleotides complementary to nascent RNA transcripts were used to target horseradish peroxidase (HRP) to the site of active β-globin transcription. Localized HRP activity deposits biotin on chromatin that is in close proximity to the site of active transcription. The chromatin was subsequently fragmented and the labeled chromatin isolated and quantified by PCR. This assay is probably only suitable for the analysis of small, highly expressed genes such as β-globin and therefore has not been pursued for the analysis of any other genes.

A second method (**Figure 8**) (26) was devised to measure the frequency with which two defined chromosomal regions interact [chromosome conformation capture (3C)], and it has been used to analyze the chromatin structure of an increasing number of loci. The 3C method is described in **Figure 8**, section A. As for the ChIP assay, 3C uses formaldehyde to cross-link protein/DNA interactions (including long-distance chromosomal interactions). The cross-linked chromatin is then digested by a restriction enzyme followed by ligation. If an interaction between a remote regulatory sequence and a promoter occurs, new fragments containing these two elements are generated and a variety of carefully designed qPCR reactions can be used to detect these newly combined elements.

3C was initially used to study the spatial organization of chromosome III in yeast (26), but has since been applied to analyze several mammalian loci including ENCODE loci such as the β-globin locus (25, 107),

TRANSCRIPTION FACTORIES

In the interchromatin compartment of the nucleus are subnuclear structures that contain high concentrations of RNA polymerase II. These sites, referred to as transcription factories, are where transcription occurs. It has been suggested that genes move into these transcription factories and are "reeled through" the factory as they are transcribed, rather than recruitment of RNA polymerase II to the gene itself. The concept of RNA polymerase II foci emerged from studies of nucleoli that contain one or more clusters of RNA polymerase I at their core. Each is dedicated to the synthesis of rRNA. In the same way, a cluster of RNA polymerase II molecules might form a factory where all the proteins required for class II transcription units work together. An analogous cluster of RNA polymerase III molecules would work only on class III units. Thus, the three types of nuclear RNA polymerases (i.e., I, II, and III) are each concentrated in their own dedicated factories, and the different types of transcription units (i.e., classes I, II, and III) are only transcribed once they associate with an appropriate factory. Genome analysis has shown that coregulated genes are often clustered on eukaryotic chromosomes, and it has been proposed that groups of genes may be drawn to particular factories because they use a common set of transcription factors (e.g., β-globin and α-hemoglobin stabilizing gene) (75). This concept of a subnuclear factory can also be applied to sites of replication, transcription, and repair of damaged DNA in eukaryotic cells (5). The current goal is to understand whether these factories and their associations are of functional significance or whether they merely reflect the association of factors involved in a common purpose.

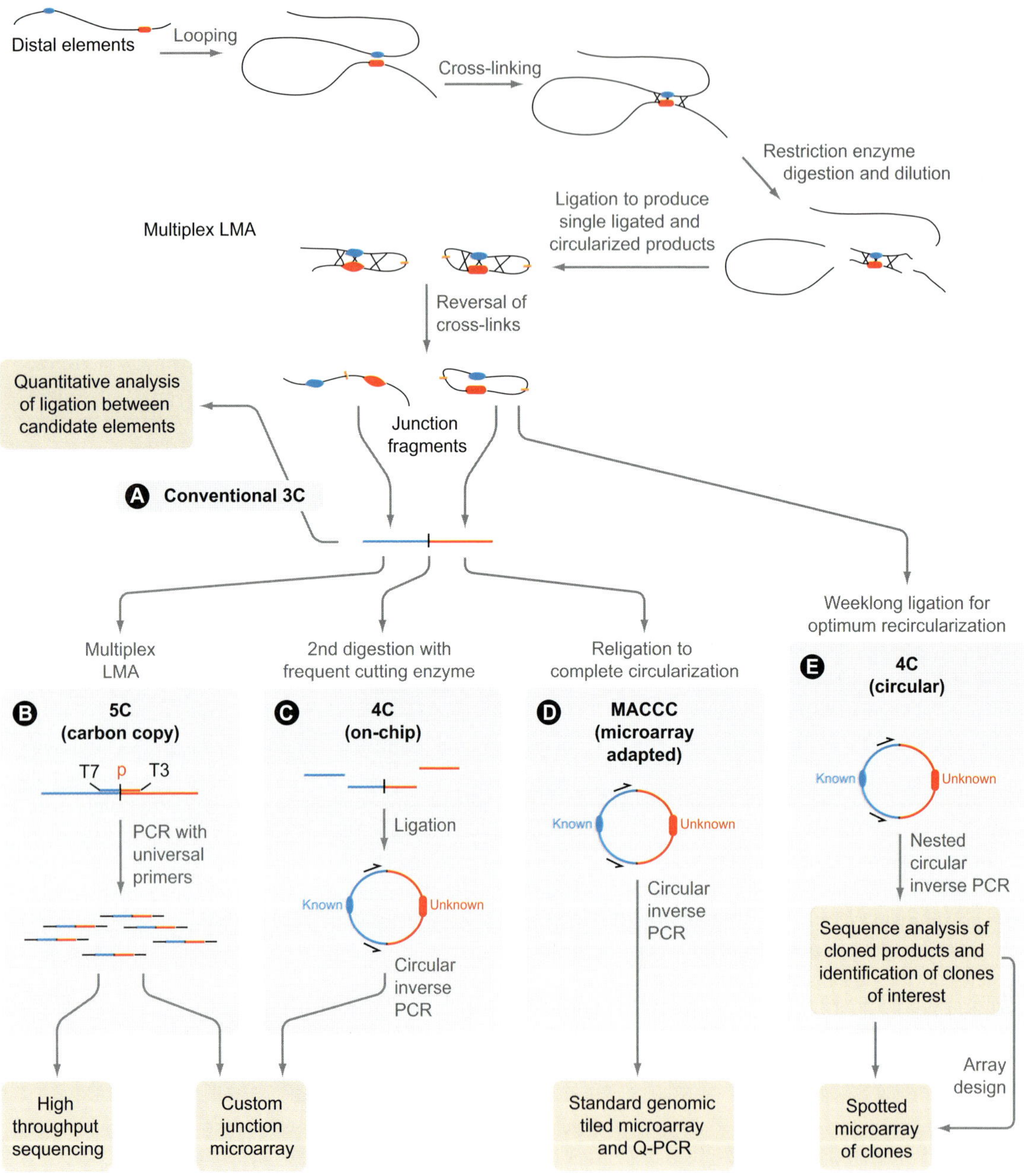

Figure 8

Methods for studying chromosome conformation. The initial stages of these assays are identical to the chromatin immunoprecipitation (ChIP) protocol, with cross-linking of protein/protein and DNA interactions. Subsequently, cross-linked structures are digested with a restriction enzyme. The choice of enzymes is limited to those that are active under the appropriate buffer conditions. Cross-linked, cut products are next ligated under dilute conditions to favor intramolecular ligations. Once ligated, the cross-links are reversed and DNA is extracted. These molecules can then be analyzed by the different methods (*a–e*), as shown in the figure and described in the text.

the α-globin locus (121) (D. Vernimmen, D. Higgs, in preparation), the T-helper type 2 cytokine locus (97), the immunoglobulin locus (64), and the Igf 2 locus (70). These studies have demonstrated direct interactions between elements located in *cis* and have also revealed *trans* interactions (26, 62, 97, 117).

RNA-TRAP and 3C were designed to test the hypothesis that intrachromosomal interactions occurred between known *cis*-acting elements. However, more recently, the 3C method was modified to identify all sequences that interact with any given DNA element. Dostie et al. (29) developed a method referred to as 5C (chromosome conformation capture carbon copy) technology, which uses a conventional 3C template (**Figure 8**, section B) but then applies multiplex ligation-dependent probe amplification (MLPA) analysis to detect ligated oligonucleotide pairs that anneal to the newly apposed fragments resulting from the 3C protocol. These primers, designed to abut the restriction sites cut in the 3C protocol, also contain universal sequences (T7 and T3) that can be used for subsequent PCR amplification to create a 5C library of interacting sequences. The resulting products can be analyzed either by microarrays specifically designed to detect the newly ligated amplified oligonucleotides or by high-throughput sequencing.

Two recently described methods exploit the fact that ligated 3C products can be converted into circular molecules thus bringing together the free ends. All interacting fragments captured in the 3C protocol can be identified by inverse PCR from the test fragment. Whereas the method described by Zhao et al. (120) still requires specifically designed arrays to identify hybrid molecules (**Figure 8**, section C), the method of Simonis et al. (95) involves cloning the inverse PCR products, then sequencing them and using the sequences to create a customized microarray (**Figure 8**, section E). We (J. Hughes, D. Higgs, in preparation) recently used a similar approach, microarray-adapted 3C (MACCC), to study the globin gene clusters to show that these products can be effectively analyzed using tiled microarrays (**Figure 8**, section D).

In general, these methods are proving to be extremely informative when analyzing the interactions of *cis*-acting elements. They are also identifying much longer-range interactions in *cis* and *trans* (26, 75). It remains to be seen which of these interactions reflect the inevitable, albeit nonrandom, juxtaposition of the region analyzed next to another region of the genome, and which reflect interactions of genuine functional relevance.

CHROMATIN STRUCTURE AND THE TIMING OF REPLICATION

A final consideration in the relationship between the epigenetic program, the transcriptional program, and the regulation of gene expression concerns when a region of the genome replicates in the S-phase of the cell cycle. As for other phenomena discussed here, it is not clear where replication lies in the hierarchy of the pathway regulating gene expression. In general, the accessible, acetylated chromatin associated with genes that are either poised for expression or expressed replicates early in the S-phase of the cell cycle, whereas silenced genes and constitutive heterochromatin (centromeres and telomeres) replicate late in the cycle. However, there are many exceptions to this rule and the precise relationship between chromatin structure, gene expression, and DNA replication is largely unresolved. Recently, genome-wide studies of replication were undertaken in yeast (77) and Drosophila (92) using DNA microarrays. In Drosophila, there was a correlation between the expression of a gene and the likelihood of it being replicated early in the S-phase (92). Over the past two years, experiments using human, cosmid-based arrays (115) and high-density Affymetrix arrays (52) also showed that regions of high gene density tend to replicate early in the S-phase.

At each cell division, as DNA is replicated and distributed to daughter cells, the associated epigenetic templating must also be faithfully copied to ensure that this aspect of cellular memory is passed on. This requires a close coordination between the DNA replication machinery, histone- and DNA-modifying enzymes, and other nonhistone proteins. It is of interest that not only DNA polymerases, but also chromatin assembly factors, histone-modifying enzymes, and DNA methyltransferases, all colocalize to replication sites. This suggests that there is some coordination between DNA replication and the duplication of the epigenetic modifications (reviewed in 68). Whether or not this represents an important point regulating subsequent gene expression remains unknown.

PUTTING IT ALL TOGETHER

In this review we summarized how new high-throughput technology is evolving from methods established to analyze single-gene loci and is being applied to analyze global transcriptional and epigenetic programs. This will allow us to produce a fully integrated, genome-wide picture of how the transcriptional program is read out via the key *cis*-elements and influenced by the epigenetic program as cells go through different cell fates. These approaches will not only tell us how cells are programmed, but may ultimately provide the means to reprogram cells (47). An important aspect of these approaches is to ensure that carefully purified and well-characterized populations of cells are analyzed so that the data truly reflect the behavior of cells in vivo. Because early, uncommitted progenitors are rare, this will require either expanding such cells or reducing the numbers of cells required for analysis.

In addition to understanding the basic biology of development and differentiation, a major aim of the postgenome venture is to understand how the underlying processes go awry in human genetic diseases. There is an ever-burgeoning repository of molecular defects (**http://www.hgmd.cf.ac.uk/ac/index.php**) known to directly affect the structure and expression of critical genes. Furthermore, there are now many examples of disease-causing mutations in local and remote cis-acting elements, including some mutations that are thought to influence epigenetic templating (e.g., 105) or long-range chromosomal interactions (e.g., 23). However, perhaps of greatest relevance to this review are the increasing number of human genetic diseases that result from mutations in the cofactors and proteins that play a major role in the epigenetic program. These include mutations in chromatin remodeling factors, DNA methyl transferases and methyl-binding proteins, proteins involved in the Polycomb complex, and the enzyme pathways underlying chromatin modification (**Table 1**).

The postgenome challenge to decode the 3-gigabase human DNA sequence is a much larger and more open-ended program than the Human Genome Project. However, the tools to do the job are being developed, and there is every reason to be confident that when complete, it will give us an unprecedented understanding of the nature of life.

SUMMARY POINTS

1. A major goal of the postgenome challenge is to understand how mammalian genes are switched on and off during development and differentiation.
2. Understanding mammalian gene regulation requires an understanding and integration of the transcriptional program with the epigenetic program.
3. Many of the techniques originally developed to characterize these programs at individual loci are now being applied to chromosomal domains, whole chromosomes, and the entire genome by using microarray technology.

4. Analysis of the transcriptional and epigenetic programs in cell lines and primary cells from normal individuals and those harboring mutations in key players will allow us to gain insights into the regulation of these programs and understand how they are perturbed in human genetic disease.

LITERATURE CITED

1. Amir RE, Van den Veyver IB, Wan M, Tran CQ, Francke U, Zoghbi HY. 1999. Rett syndrome is caused by mutations in X-linked MECP2, encoding methyl-CpG-binding protein 2. *Nat. Genet.* 23(2):185–88
2. Anguita E, Hughes J, Heyworth C, Blobel GA, Wood WG, Higgs DR. 2004. Globin gene activation during hemopoiesis is driven by protein complexes nucleated by GATA-1 and GATA-2. *EMBO J.* 23(14):2841–52
3. Azuara V, Perry P, Sauer S, Spivakov M, Jorgensen HF, et al. 2006. Chromatin signatures of pluripotent cell lines. *Nat. Cell Biol.* 8(5):532–38
4. Banerjee N, Zhang MQ. 2002. Functional genomics as applied to mapping transcription regulatory networks. *Curr. Opin. Microbiol.* 5(3):313–17
5. Bartlett J, Blagojevic J, Carter D, Eskiw C, Fromaget M, et al. 2006. Specialized transcription factories. *Biochem. Soc. Symp.* (73):67–75
6. Bernstein BE, Humphrey EL, Erlich RL, Schneider R, Bouman P, et al. 2002. Methylation of histone H3 Lys 4 in coding regions of active genes. *Proc. Natl. Acad. Sci. USA* 99(13):8695–700
7. Bernstein BE, Kamal M, Lindblad-Toh K, Bekiranov S, Bailey DK, et al. 2005. Genomic maps and comparative analysis of histone modifications in human and mouse. *Cell* 120(2):169–81
8. Bernstein BE, Mikkelsen TS, Xie X, Kamal M, Huebert DJ, et al. 2006. A bivalent chromatin structure marks key developmental genes in embryonic stem cells. *Cell* 125(2):315–26
9. Bickmore WA, Sumner AT. 1989. Mammalian chromosome banding—an expression of genome organization. *Trends Genet.* 5(5):144–48
10. Bird A. 2002. DNA methylation patterns and epigenetic memory. *Genes Dev.* 16(1):6–21
11. Blasco MA. 2005. Telomeres and human disease: ageing, cancer and beyond. *Nat. Rev. Genet.* 6(8):611–22
12. Boerkoel CF, Takashima H, John J, Yan J, Stankiewicz P, et al. 2002. Mutant chromatin remodeling protein SMARCAL1 causes Schimke immuno-osseous dysplasia. *Nat. Genet.* 30(2):215–20
13. Boffelli D, Nobrega MA, Rubin EM. 2004. Comparative genomics at the vertebrate extremes. *Nat. Rev. Genet.* 5(6):456–65
14. Brown KE, Amoils S, Horn JM, Buckle VJ, Higgs DR, et al. 2001. Expression of alpha- and beta-globin genes occurs within different nuclear domains in hemopoietic cells. *Nat. Cell Biol.* 3(6):602–6
15. Carroll CW, Straight AF. 2006. Centromere formation: from epigenetics to self-assembly. *Trends Cell Biol.* 16(2):70–78
16. Carroll JS, Liu XS, Brodsky AS, Li W, Meyer CA, et al. 2005. Chromosome-wide mapping of estrogen receptor binding reveals long-range regulation requiring the forkhead protein FoxA1. *Cell* 122(1):33–43

17. Carter D, Chakalova L, Osborne CS, Dai YF, Fraser P. 2002. Long-range chromatin regulatory interactions in vivo. *Nat. Genet.* 32(4):623–26
18. Cawley S, Bekiranov S, Ng HH, Kapranov P, Sekinger EA, et al. 2004. Unbiased mapping of transcription factor binding sites along human chromosomes 21 and 22 points to widespread regulation of noncoding RNAs. *Cell* 116(4):499–509
19. Ching TT, Maunakea AK, Jun P, Hong C, Zardo G, et al. 2005. Epigenome analyses using BAC microarrays identify evolutionary conservation of tissue-specific methylation of SHANK3. *Nat. Genet.* 37(6):645–51
20. Colella S, Shen L, Baggerly KA, Issa JP, Krahe R. 2003. Sensitive and quantitative universal pyrosequencing methylation analysis of CpG sites. *Biotechniques* 35(1):146–50
21. Crawford GE, Holt IE, Whittle J, Webb BD, Tai D, et al. 2006. Genome-wide mapping of DNase hypersensitive sites using massively parallel signature sequencing (MPSS). *Genome Res.* 16(1):123–31
22. Daniels RJ, Peden JF, Lloyd C, Horsley SW, Clark K, et al. 2001. Sequence, structure and pathology of the fully annotated terminal 2 Mb of the short arm of human chromosome 16. *Hum. Mol. Genet.* 10(4):339–52
23. De Gobbi M, Viprakasit V, Hughes JR, Fisher C, Buckle VJ, et al. 2006. A regulatory SNP causes a human genetic disease by creating a new transcriptional promoter. *Science* 312(5777):1215–17
24. de la Serna IL, Ohkawa Y, Imbalzano AN. 2006. Chromatin remodelling in mammalian differentiation: lessons from ATP-dependent remodellers. *Nat. Rev. Genet.* 7(6):461–73
25. de Laat W, Grosveld F. 2003. Spatial organization of gene expression: the active chromatin hub. *Chromosome Res.* 11(5):447–59
26. Dekker J, Rippe K, Dekker M, Kleckner N. 2002. Capturing chromosome conformation. *Science* 295(5558):1306–11
27. Dillon N, Sabbattini P. 2000. Functional gene expression domains: defining the functional unit of eukaryotic gene regulation. *Bioessays* 22(7):657–65
28. Dorschner MO, Hawrylycz M, Humbert R, Wallace JC, Shafer A, et al. 2004. High-throughput localization of functional elements by quantitative chromatin profiling. *Nat. Methods* 1(3):219–25
29. Dostie J, Richmond TA, Arnaout RA, Selzer RR, Lee WL, et al. 2006. Chromosome Conformation Capture Carbon Copy (5C): a massively parallel solution for mapping interactions between genomic elements. *Genome Res.* 16(10):1299–309
30. Drissen R, Palstra RJ, Gillemans N, Splinter E, Grosveld F, et al. 2004. The active spatial organization of the beta-globin locus requires the transcription factor EKLF. *Genes Dev.* 18(20):2485–90
31. Ehrlich M, Gama-Sosa MA, Huang LH, Midgett RM, Kuo KC, et al. 1982. Amount and distribution of 5-methylcytosine in human DNA from different types of tissues of cells. *Nucleic Acids Res.* 10(8):2709–21
32. Felsenfeld G, Groudine M. 2003. Controlling the double helix. *Nature* 421(6921):448–53
33. Follows GA, Dhami P, Gottgens B, Bruce AW, Campbell PJ, et al. 2006. Identifying gene regulatory elements by genomic microarray mapping of DNaseI hypersensitive sites. *Genome Res.* 16(10):1310–19
34. Francastel C, Schubeler D, Martin DI, Groudine M. 2000. Nuclear compartmentalization and gene activity. *Nat. Rev. Mol. Cell Biol.* 1(2):137–43
35. Frigola J, Ribas M, Risques RA, Peinado MA. 2002. Methylome profiling of cancer cells by amplification of intermethylated sites (AIMS). *Nucleic Acids Res.* 30(7):e28
36. Fuks F. 2005. DNA methylation and histone modifications: teaming up to silence genes. *Curr. Opin. Genet. Dev.* 15(5):490–95

18. The first demonstration of the use of tiled arrays of human genomic sequence coupled with chromatin immunoprecipitation to map transcription factor binding sites.

26. The analysis and identification of *cis*-acting elements and promoters was transformed by this ingenious technique, by which interacting elements were captured by cross-linking and then ligating these elements.

37. Gartler SM, Dyer KA, Goldman MA. 1992. Mammalian X chromosome inactivation. *Mol. Genet. Med.* 2:121–60
38. Gibbons RJ, Picketts DJ, Villard L, Higgs DR. 1995. Mutations in a putative global transcriptional regulator cause X-linked mental retardation with alpha-thalassemia (ATR-X syndrome). *Cell* 80(6):837–45
39. Gilbert N, Boyle S, Fiegler H, Woodfine K, Carter NP, Bickmore WA. 2004. Chromatin architecture of the human genome: gene-rich domains are enriched in open chromatin fibers. *Cell* 118(5):555–66
40. Gilbert N, Gilchrist S, Bickmore WA. 2005. Chromatin organization in the mammalian nucleus. *Int. Rev. Cytol.* 242:283–336
41. Greil F, van der Kraan I, Delrow J, Smothers JF, de Wit E, et al. 2003. Distinct HP1 and Su(var)3–9 complexes bind to sets of developmentally coexpressed genes depending on chromosomal location. *Genes Dev.* 17(22):2825–38
42. Gross DS, Garrard WT. 1988. Nuclease hypersensitive sites in chromatin. *Annu. Rev. Biochem.* 57:159–97
43. **Hatzis P, Talianidis I. 2002. Dynamics of enhancer-promoter communication during differentiation-induced gene activation. *Mol. Cell* 10(6):1467–77**
44. Hawkins RD, Ren B. 2006. Genome-wide location analysis: insights on transcriptional regulation. *Hum. Mol. Genet.* 15(Spec. No. 1):R1–7
45. Hediger F, Gasser SM. 2006. Heterochromatin protein 1: don't judge the book by its cover? *Curr. Opin. Genet. Dev.* 16(2):143–50
46. Heitz E. 1928. Das heterochromatin der Moose. *Jahrb. Wiss. Bot.* 69:762–818
47. Hochedlinger K, Jaenisch R. 2006. Nuclear reprogramming and pluripotency. *Nature* 441(7097):1061–67
48. Hughes JR, Cheng JF, Ventress N, Prabhakar S, Clark K, et al. 2005. Annotation of *cis*-regulatory elements by identification, subclassification, and functional assessment of multispecies conserved sequences. *Proc. Natl. Acad. Sci. USA* 102(28):9830–35
49. **International Human Genome Sequencing Consortium. 2004. Finishing the euchromatic sequence of the human genome. *Nature* 431(7011):931–45**
50. Jenuwein T, Allis CD. 2001. Translating the histone code. *Science* 293(5532):1074–80
51. Jenuwein T. 2006. The epigenetic magic of histone lysine methylation. *FEBS J.* 273(14):3121–35
52. Jeon Y, Bekiranov S, Karnani N, Kapranov P, Ghosh S, et al. 2005. Temporal profile of replication of human chromosomes. *Proc. Natl. Acad. Sci. USA* 102(18):6419–24
53. Kim TH, Barrera LO, Zheng M, Qu C, Singer MA, et al. 2005. A high-resolution map of active promoters in the human genome. *Nature* 436(7052):876–80
54. Klose RJ, Bird AP. 2006. Genomic DNA methylation: the mark and its mediators. *Trends Biochem. Sci.* 31(2):89–97
55. Kulessa H, Frampton J, Graf T. 1995. GATA-1 reprograms avian myelomonocytic cell lines into eosinophils, thromboblasts, and erythroblasts. *Genes Dev.* 9(10):1250–62
56. Kurotaki N, Imaizumi K, Harada N, Masuno M, Kondoh T, et al. 2002. Haploinsufficiency of NSD1 causes Sotos syndrome. *Nat. Genet.* 30(4):365–66
57. Kurukuti S, Tiwari VK, Tavoosidana G, Pugacheva E, Murrell A, et al. 2006. CTCF binding at the H19 imprinting control region mediates maternally inherited higher-order chromatin conformation to restrict enhancer access to Igf2. *Proc. Natl. Acad. Sci. USA* 103(28):10684–89
58. Lachner M, O'Sullivan RJ, Jenuwein T. 2003. An epigenetic road map for histone lysine methylation. *J. Cell Sci.* 116(Pt. 11):2117–24

43. Describes the tracking of an entire DNA-protein complex formed on an upstream enhancer along the intervening DNA until it encountered the proximal promoter.

49. This landmark paper represents the near completion of the human genome sequencing project covering over 99% of euchromatin and represents the first vertebrate genome sequence.

59. Lee TI, Jenner RG, Boyer LA, Guenther MG, Levine SS, et al. 2006. Control of developmental regulators by Polycomb in human embryonic stem cells. *Cell* 125(2):301–13
60. Levings PP, Zhou Z, Vieira KF, Crusselle-Davis VJ, Bungert J. 2006. Recruitment of transcription complexes to the beta-globin locus control region and transcription of hypersensitive site 3 prior to erythroid differentiation of murine embryonic stem cells. *FEBS J.* 273(4):746–55
61. Lewin J, Schmitt AO, Adorjan P, Hildmann T, Piepenbrock C. 2004. Quantitative DNA methylation analysis based on four-dye trace data from direct sequencing of PCR amplificates. *Bioinformatics* 20(17):3005–12
62. Ling JQ, Li T, Hu JF, Vu TH, Chen HL, et al. 2006. CTCF mediates interchromosomal colocalization between Igf2/H19 and Wsb1/Nf1. *Science* 312(5771):269–72
63. Litt MD, Simpson M, Gaszner M, Allis CD, Felsenfeld G. 2001. Correlation between histone lysine methylation and developmental changes at the chicken beta-globin locus. *Science* 293(5539):2453–55
64. Liu Z, Garrard WT. 2005. Long-range interactions between three transcriptional enhancers, active Vkappa gene promoters, and a 3′ boundary sequence spanning 46 kilobases. *Mol. Cell Biol.* 25(8):3220–31
65. Margueron R, Trojer P, Reinberg D. 2005. The key to development: interpreting the histone code? *Curr. Opin. Genet. Dev.* 15(2):163–76
66. Marks PA, Miller T, Richon VM. 2003. Histone deacetylases. *Curr. Opin. Pharmacol.* 3(4):344–51
67. Martens JH, O'Sullivan RJ, Braunschweig U, Opravil S, Radolf M, et al. 2005. The profile of repeat-associated histone lysine methylation states in the mouse epigenome. *EMBO J.* 24(4):800–12
68. McNairn AJ, Gilbert DM. 2003. Epigenomic replication: linking epigenetics to DNA replication. *Bioessays* 25(7):647–56
69. Miller OJ, Schnedl W, Allen J, Erlanger BF. 1974. 5-Methylcytosine localised in mammalian constitutive heterochromatin. *Nature* 251(5476):636–37
70. Murrell A, Heeson S, Reik W. 2004. Interaction between differentially methylated regions partitions the imprinted genes Igf2 and H19 into parent-specific chromatin loops. *Nat. Genet.* 36(8):889–93
71. Noma K, Allis CD, Grewal SI. 2001. Transitions in distinct histone H3 methylation patterns at the heterochromatin domain boundaries. *Science* 293(5532):1150–55
72. Nowak SJ, Corces VG. 2004. Phosphorylation of histone H3: a balancing act between chromosome condensation and transcriptional activation. *Trends Genet.* 20(4):214–20
73. Odom DT, Zizlsperger N, Gordon DB, Bell GW, Rinaldi NJ, et al. 2004. Control of pancreas and liver gene expression by HNF transcription factors. *Science* 303(5662):1378–81
74. Okano M, Bell DW, Haber DA, Li E. 1999. DNA methyltransferases Dnmt3a and Dnmt3b are essential for de novo methylation and mammalian development. *Cell* 99(3):247–57
75. Osborne CS, Chakalova L, Brown KE, Carter D, Horton A, et al. 2004. Active genes dynamically colocalize to shared sites of ongoing transcription. *Nat. Genet.* 36(10):1065–71
76. Petrij F, Giles RH, Dauwerse HG, Saris JJ, Hennekam RC, et al. 1995. Rubinstein-Taybi syndrome caused by mutations in the transcriptional coactivator CBP. *Nature* 376(6538):348–51
77. Raghuraman MK, Winzeler EA, Collingwood D, Hunt S, Wodicka L, et al. 2001. Replication dynamics of the yeast genome. *Science* 294(5540):115–21

78. Rakyan VK, Hildmann T, Novik KL, Lewin J, Tost J, et al. 2004. DNA methylation profiling of the human major histocompatibility complex: a pilot study for the human epigenome project. *PLoS Biol.* 2(12):e405
79. Reik W, Dean W, Walter J. 2001. Epigenetic reprogramming in mammalian development. *Science* 293(5532):1089–93
80. Reik W, Walter J. 2001. Genomic imprinting: parental influence on the genome. *Nat. Rev. Genet.* 2(1):21–32
81. Ringrose L, Paro R. 2004. Epigenetic regulation of cellular memory by the Polycomb and Trithorax group proteins. *Annu. Rev. Genet.* 38:413–43
82. Roelfsema JH, White SJ, Ariyurek Y, Bartholdi D, Niedrist D, et al. 2005. Genetic heterogeneity in Rubinstein-Taybi syndrome: mutations in both the CBP and EP300 genes cause disease. *Am. J. Hum. Genet.* 76(4):572–80
83. Roh TY, Cuddapah S, Zhao K. 2006. Active chromatin domains are defined by acetylation islands revealed by genome-wide mapping. *Genes Dev.* 19:542–52
84. Roh TY, Ngau WC, Cui K, Landsman D, Zhao K. 2004. High-resolution genome-wide mapping of histone modifications. *Nat. Biotechnol.* 22(8):1013–16
85. Roth SY, Denu JM, Allis CD. 2001. Histone acetyltransferases. *Annu. Rev. Biochem.* 70:81–120
86. Rougeulle C, Chaumeil J, Sarma K, Allis CD, Reinberg D, et al. 2004. Differential histone H3 Lys-9 and Lys-27 methylation profiles on the X chromosome. *Mol. Cell Biol.* 24(12):5475–84
87. Sabo PJ, Kuehn MS, Thurman R, Johnson BE, Johnson EM, et al. 2006. Genome-scale mapping of DNase I sensitivity in vivo using tiling DNA microarrays. *Nat. Methods* 3(7):511–18
88. Santos-Rosa H, Schneider R, Bannister AJ, Sherriff J, Bernstein BE, et al. 2002. Active genes are tri-methylated at K4 of histone H3. *Nature* 419(6905):407–11
89. Sarma K, Reinberg D. 2005. Histone variants meet their match. *Nat. Rev. Mol. Cell Biol.* 6(2):139–49
90. Sawado T, Halow J, Bender MA, Groudine M. 2003. The beta-globin locus control region (LCR) functions primarily by enhancing the transition from transcription initiation to elongation. *Genes Dev.* 17(8):1009–18
91. Schatz P, Dietrich D, Schuster M. 2004. Rapid analysis of CpG methylation patterns using RNase T1 cleavage and MALDI-TOF. *Nucleic Acids Res.* 32(21):e167
92. Schubeler D, Scalzo D, Kooperberg C, van Steensel B, Delrow J, Groudine M. 2002. Genome-wide DNA replication profile for *Drosophila melanogaster*: a link between transcription and replication timing. *Nat. Genet.* 32(3):438–42
93. Schubeler D, MacAlpine DM, Scalzo D, Wirbelauer C, Kooperberg C, et al. 2004. The histone modification pattern of active genes revealed through genome-wide chromatin analysis of a higher eukaryote. *Genes Dev.* 18(11):1263–71
94. Schumacher A, Kapranov P, Kaminsky Z, Flanagan J, Assadzadeh A, et al. 2006. Microarray-based DNA methylation profiling: technology and applications. *Nucleic Acids Res.* 34(2):528–42
95. Simonis M, Klous P, Splinter E, Moshkin Y, Willemsen R, et al. 2006. Nuclear organization of active and inactive chromatin domains uncovered by chromosome conformation capture-on-chip (4C). *Nat. Genet.* 38(11):1348–54
96. Soutoglou E, Talianidis I. 2002. Coordination of PIC assembly and chromatin remodeling during differentiation-induced gene activation. *Science* 295(5561):1901–4
97. Spilianakis CG, Lalioti MD, Town T, Lee GR, Flavell RA. 2005. Interchromosomal associations between alternatively expressed loci. *Nature* 435(7042):637–45

90. Demonstrates that without the β-globin LCR, RNA Pol II is still recruited to the proximal promoter.

98. Spitz F, Herkenne C, Morris MA, Duboule D. 2005. Inversion-induced disruption of the Hoxd cluster leads to the partition of regulatory landscapes. *Nat. Genet.* 37(8):889–93
99. Splinter E, Heath H, Kooren J, Palstra RJ, Klous P, et al. 2006. CTCF mediates long-range chromatin looping and local histone modification in the beta-globin locus. *Genes Dev.* 20(17):2349–54
100. Szutorisz H, Canzonetta C, Georgiou A, Chow CM, Tora L, Dillon N. 2005. Formation of an active tissue-specific chromatin domain initiated by epigenetic marking at the embryonic stem cell stage. *Mol. Cell Biol.* 25(5):1804–20
101. The ENCODE Project Consortium. 2004. The ENCODE (ENCyclopedia Of DNA Elements) Project. *Science* 306(5696):636–40

101. The ENCODE Project is an ambitious NIH-funded enterprise to identify all the functional elements in the human genome sequence.

102. Trivier E, De Cesare D, Jacquot S, Pannetier S, Zackai E, et al. 1996. Mutations in the kinase Rsk-2 associated with Coffin-Lowry syndrome. *Nature* 384(6609):567–70
103. Troelstra C, van Gool A, de Wit J, Vermeulen W, Bootsma D, Hoeijmakers JH. 1992. ERCC6, a member of a subfamily of putative helicases, is involved in Cockayne's syndrome and preferential repair of active genes. *Cell* 71(6):939–53
104. Trojer P, Reinberg D. 2006. Histone lysine demethylases and their impact on epigenetics. *Cell* 125(2):213–17
105. Tufarelli C, Stanley JA, Garrick D, Sharpe JA, Ayyub H, et al. 2003. Transcription of antisense RNA leading to gene silencing and methylation as a novel cause of human genetic disease. *Nat. Genet.* 34(2):157–65
106. Turner BM. 2005. Reading signals on the nucleosome with a new nomenclature for modified histones. *Nat. Struct. Mol. Biol.* 12(2):110–12
107. Vakoc CR, Letting DL, Gheldof N, Sawado T, Bender MA, et al. 2005. Proximity among distant regulatory elements at the beta-globin locus requires GATA-1 and FOG-1. *Mol. Cell* 17(3):453–62
108. Vissers LE, van Ravenswaaij CM, Admiraal R, Hurst JA, de Vries BB, et al. 2004. Mutations in a new member of the chromodomain gene family cause CHARGE syndrome. *Nat. Genet.* 36(9):955–57
109. Vogel MJ, Guelen L, de Wit E, Peric-Hupkes D, Loden M, et al. 2006. Human heterochromatin proteins form large domains containing KRAB-ZNF genes. *Genome Res.* 16(12):1493–504
110. Volpi EV, Chevret E, Jones T, Vatcheva R, Williamson J, et al. 2000. Large-scale chromatin organization of the major histocompatibility complex and other regions of human chromosome 6 and its response to interferon in interphase nuclei. *J. Cell Sci.* 113(Pt. 9):1565–76
111. Wallace HAC, Marques-Kranc F, Richardson M, De Luna-Crespo F, Sharpe JA, et al. 2007. Engineering precise syntenic replacements in the mouse for functional analysis of human regulatory domains. *Cell* 128:197–209

111. The first experiment to define a segment of chromosome that contains all the sequences required for full regulation of a locus by replacing a mouse locus with the corresponding human locus.

112. Weber M, Davies JJ, Wittig D, Oakeley EJ, Haase M, et al. 2005. Chromosome-wide and promoter-specific analyses identify sites of differential DNA methylation in normal and transformed human cells. *Nat. Genet.* 37(8):853–62
113. Weinmann AS, Yan PS, Oberley MJ, Huang TH, Farnham PJ. 2002. Isolating human transcription factor targets by coupling chromatin immunoprecipitation and CpG island microarray analysis. *Genes Dev.* 16(2):235–44
114. West AG, Gaszner M, Felsenfeld G. 2002. Insulators: many functions, many mechanisms. *Genes Dev.* 16(3):271–88
115. Woodfine K, Fiegler H, Beare DM, Collins JE, McCann OT, et al. 2004. Replication timing of the human genome. *Hum. Mol. Genet.* 13(2):191–202

116. Xu GL, Bestor TH, Bourc'his D, Hsieh CL, Tommerup N, et al. 1999. Chromosome instability and immunodeficiency syndrome caused by mutations in a DNA methyltransferase gene. *Nature* 402(6758):187–91
117. Xu N, Tsai CL, Lee JT. 2006. Transient homologous chromosome pairing marks the onset of X inactivation. *Science* 311(5764):1149–52
118. Yoder JA, Walsh CP, Bestor TH. 1997. Cytosine methylation and the ecology of intragenomic parasites. *Trends Genet.* 13(8):335–40
119. Zhang X, Yazaki J, Sundaresan A, Cokus S, Chan SW, et al. 2006. Genome-wide high-resolution mapping and functional snalysis of DNA methylation in *Arabidopsis*. *Cell* 126:1189–1201
120. Zhao Z, Tavoosidana G, Sjolinder M, Gondor A, Mariano P, et al. 2006. Circular chromosome conformation capture (4C) uncovers extensive networks of epigenetically regulated intra- and interchromosomal interactions. *Nat. Genet.* 38(11):1341–47
121. Zhou GL, Xin L, Song W, Di LJ, Liu G, et al. 2006. Active chromatin hub of the mouse alpha-globin locus forms in a transcription factory of clustered housekeeping genes. *Mol. Cell Biol.* 26(13):5096–105

Multistage Sampling for Genetic Studies*

Robert C. Elston,[1] Danyu Lin,[2] and Gang Zheng[3,**]

[1]Department of Epidemiology and Biostatistics, Case Western Reserve University, Cleveland, Ohio 44106; email: rce@darwin.case.edu

[2]Department of Biostatistics, University of North Carolina at Chapel Hill, Chapel Hill, North Carolina 27599; email: lin@bios.unc.edu

[3]Office of Biostatistics Research, National Heart, Lung and Blood Institute, Bethesda, Maryland 20892; email: zhengg@nhlbi.nih.gov

Annu. Rev. Genomics Hum. Genet. 2007. 8:327–42

First published online as a Review in Advance on May 22, 2007.

The *Annual Review of Genomics and Human Genetics* is online at genom.annualreviews.org

This article's doi:
10.1146/annurev.genom.8.080706.092357

1527-8204/07/0922-0327$20.00

*The U.S. Government has the right to retain a nonexclusive, royalty-free license in and to any copyright covering this paper.

**The authors are ordered alphabetically.

Key Words

association, DNA pooling, linkage analysis, genome-wide, segregation analysis, self-replication

Abstract

In the past, to study Mendelian diseases, segregating families have been carefully ascertained for segregation analysis, followed by collecting extended multiplex families for linkage analysis. This would then be followed by association studies, using independent case-control samples and/or additional family data. Recently, for complex diseases, the initial sampling has been for a genome-wide linkage analysis, often using independent sib-pairs or nuclear families, to identify candidate regions for follow-up with association studies, again using case-control samples and/or additional family data. We now have the ability to conduct genome-wide association studies using 100,000–500,000 diallelic genetic markers. For such studies we focus especially on efficient two-stage association sampling designs, which can retain nearly optimal statistical power at about half the genotyping cost. Similarly, beginning an association study by genotyping pooled samples may also be a viable option if the cost of accurately pooling DNA samples outweighs genotyping costs. Finally, we note that the sampling of family data for linkage analysis is not a practice that should be automatically discontinued.

INTRODUCTION

In the past 25 years, new methods for genetic epidemiology have been mainly focused on answering whether or not a trait shows Mendelian segregation, cosegregation in families with a genetic marker, and finally association in the population with a genetic marker (10). Segregation analysis has been used to determine an appropriate genetic model for model-based linkage analysis. Segregation analysis has been an efficient initial approach for simple Mendelian diseases, but rarely successful for a complex disease because, when the mode of inheritance requires numerous parameters to be estimated, the estimates have typically been imprecise. Linkage analysis can be done assuming a known genetic model for the trait being studied or in a model-free manner, which may initally be more appropriate for a complex trait. Often, the demonstration of linkage has provided the ultimate proof of a genetic component in the etiology of a disease, whether it is one that segregates in a Mendelian fashion or one that shows a complex pattern of inheritance. As a large number of single nucleotide polymorphisms (SNPs) are now available at relatively low cost, it has been proposed that testing for association between a disease and a large panel of SNPs on unrelated individuals is a more efficient design for gene discovery when gene effects are small. However, association can arise as a result of chance, of population stratification/heterogeneity, of very close linkage, or of pleiotropy. The essential problem for association analysis is to distinguish the first two of these causes from the second two.

The more recent designs and methods of analysis for linkage and association studies have been driven by the advances of molecular technology, which provide more genetic markers at ever-reduced cost. In 1980, Botstein et al. (2) proposed the use of restriction fragment length polymorphisms to find cosegregation of a trait of interest to any point on the genome. Later, microsatellites (63, 64, 68) became the genetic markers used for a genome-wide linkage study, usually numbering about 400, roughly equally spaced across the genome. Today, SNPs are the genetic markers of choice for either linkage or association studies. They are mostly diallelic and abundant, current platforms covering with good potential power about 80% of the Caucasian genome. With the HapMap project (1), millions of SNPs have been identified throughout the genome. In fact, genome-wide association studies with 500,000 SNPs are already being published.

However, there are challenges in performing large-scale association studies to investigate the genetics of complex diseases (3). First, the cost of genotyping is still high. For a genome-wide association study with 500,000 SNPs and a couple of thousand individuals, the genotyping cost alone lies between a quarter and half a million dollars. Second, a large-scale genetic study has a potentially high false positive rate and it is often difficult to replicate results in later independent studies. Multistage designs incorporating both linkage and association components, or cost-efficient association designs with multiple stages of genotyping, are essential to maintain statistical power at an affordable cost with limited false positive rates. In this article, we review the classical paradigm of multistage sampling designs, using both families and unrelated individuals, that have been successful for Mendelian diseases; a two-stage linkage-association design for more complex diseases; and several cost-efficient multistage designs for large association studies. We conclude by noting that there are advantages to collecting family data and conducting a linkage analysis prior to any genome-wide association study.

CLASSICAL PARADIGM

The classical paradigm for genetic analysis in humans has been a multistage procedure. First, after showing that a trait is familial, segregating families have been collected, segregation analysis performed to verify Mendelian segregation and define the best segregating model for a model-based linkage analysis, and then, after such a linkage analysis, association studies undertaken. This procedure involves multistage sampling of families and individuals. How to ascertain segregating families depends on the disease. For an autosomal dominant disease, families may be ascertained through the parents' phenotypes. For a rare autosomal recessive disease, heterozygote × heterozygote matings are the informative mating types that are most commonly available; however, because the phenotypes of individuals carrying one risk allele and no risk allele are indistinguishable, these mating types can only be ascertained through the presence of an affected offspring (45).

Following a segregation analysis, a model-based linkage study is conducted with marker loci using a sample of multiplex family data. During meiosis, owing to the crossing over of a pair of chromosomes, a parent may transmit a recombinant gamete or a nonrecombinant gamete to an offspring. A genetic marker is linked to a disease locus if the fraction of gametes recombining between them is less than one half. The recombination fraction measures the genetic distance between two loci and a model-based linkage analysis uses family data to test whether or not the recombination fraction is one half. Usually, three or more generations are used to estimate the fraction of gametes that are recombinant. The base 10 logarithm of a likelihood ratio (LOD), often misinterpreted as the logarithm of the odds for linkage (6), is commonly used to test the null hypothesis of no linkage and we reject the null hypothesis when the recombination fraction is significantly less than one half. For genome-wide linkage analysis, in a large sample and the simplest of situations, a maximum LOD score of 3 corresponds to a pointwise P-value of $\sim 10^{-4}$ (6) and indicates a good linkage signal. In other more general situations, a LOD of 3 or more may have much less statistical significance, which has led to the recommendation that LODs be abandoned in favor of P-values (7, 29). Multipoint linkage analysis simultaneously uses the information from all the markers on a chromosome to find positions where trait loci may lie. Here again LODs are usually reported, but their significance is not always clear (69).

When linkage signals are detected, further studies for allelic association due to linkage disequilibrium (LD) can focus on these chromosomal regions. An association study to test LD is complementary to linkage analysis. LD between two loci causes the alleles at these two loci to be associated in the population. Testing for association between the alleles at two loci is most easily done on a random population sample, but both family data and case-control samples can be used.

The study of complex diseases, e.g., diseases caused by segregation at multiple loci with incomplete penetrance, has typically started with linkage analysis followed by association analysis. Linkage analysis can be performed using a model-free method on a sample of family units as small as independent sib-pairs. For example, linkage can be tested using concordantly affected sib-pairs (ASPs), where families with an ASP are ascertained and genotyped to determine the number of alleles that each sib-pair has identical by descent (IBD) at any genomic location; more generally, other affected pairs of relatives may also be studied (30). A sib-pair can share 0, 1, or 2 alleles IBD at a locus, most other relative pairs share 0 or 1 allele IBD. For some mating types, IBD cannot be uniquely determined, in which case the expected IBD sharing is estimated using all the available marker data.

For a quantitative trait locus (QTL), the Haseman-Elston (HE) regression model can be used. The original HE method uses independent sib-pairs to test for linkage of a QTL. The squared measured trait difference between sibs is regressed on the proportion of alleles the sib-pair shares IBD, and testing for linkage is equivalent to testing that this regression model has a negative slope (17). The HE model has been extended to many situations, e.g., larger sib-ships (49) and

Multiplex family: a family containing multiple individuals affected with a disease

LOD: the logarithm (usually to base 10) of a ratio of two likelihoods, that of linkage to that of free recombination

Linkage disequilibrium (LD): population association between the alleles at linked loci

Identical by descent (IBD): alleles at a locus inherited from the same chromosome of an ancestor

Table 1 Parental transmission data for a diallelic SNP

	Transmitted		
Not transmitted	*A*	*B*	Total
A	a	b	a + b
B	c	d	c + d
Total	a + c	b + d	n

Table 2 Genotype data for a diallelic SNP in a case-control design

	Genotypes			
	AA	*AB*	*BB*	Total
Case	R_0	R_1	R_2	R
Control	S_0	S_1	S_2	S
Total	N_0	N_1	N_2	N

extended pedigrees (60). Provided unaffected sibs are included in the sample, binary affected/unaffected traits can also be analyzed as quantitative data by this same method (9).

If a candidate gene is already suspected, linkage to it can be tested from a sample of parents and an affected offspring (trios) using the transmission/disequilibrium test (TDT) (54). The original TDT compares the frequency that a particular allele is transmitted to an affected offspring from a heterozygous parent to the frequency that it is not so transmitted. **Table 1** presents the parental transmission data for a diallelic marker with alleles *A* and *B*. In *n* trios, *b* heterozygous parents transmitted allele *B* and *c* heterozygous parents transmitted allele *A*. The test statistic, TDT = $(b - c)^2/(b + c)$, has a large sample chi-square distribution with one degree of freedom either when the marker and disease locus are not linked or when they are not in LD. The TDT was proposed as a test of linkage, but it can also be used to test for LD, or both LD and linkage, depending on assumptions and the sampling scheme (47, 53). Whereas it can always be used validly to test for linkage, it is only a valid test for LD when independent trios are sampled or the test properly allows for the nonindependence of sibs. In linkage analysis, several sampling designs may be applied together. For example, if one has both ASPs and parent-offspring trios, then linkage can be tested by testing IBD patterns in the ASPs and by using the TDT for the trios.

Hardy-Weinberg equilibrium (HWE): HWE proportions that remain stable over generations (no selection)

Hardy-Weinberg equilibrium (HWE) proportions: genotypic frequencies that result from random mating

After candidate genes have been detected by linkage analysis, we can conduct an association study focusing on these candidate genes with densely spaced SNPs that is either family based or population based. The simplest sampling scheme for family data comprises trios (**Table 1**) for TDT analysis. Schaid & Sommer (42) considered more general score tests for candidate gene association using trios, and the TDT is a special case of these score tests. Note that the TDT and similar tests are family based, i.e., require family data, and are also designed to protect against detecting spurious associations that are due to population stratification, which leads to departure from Hardy-Weinberg equilibrium (HWE) proportions. However, more power is available from the same family-based data if it is not necessary to protect against spurious association (8, 15). In a population-based association study, independent cases and controls are typically sampled. For a SNP with alleles *A* and *B*, the data for a case-control sample are presented in **Table 2** by genotypes and in **Table 3** by alleles. The total sample size is *N*, with *R* cases and *S* controls. Association between the disease and the SNP can be tested by comparing the genotype frequencies (genotype-based analysis) or the allele (*A*, *B*) frequencies (allele-based analysis) (p, q) between cases and controls (13, 37, 51). Let *T* represent all the samples. In **Table 2**, the frequencies of genotypes (*AA*, *AB*, *BB*) in cases and controls are estimated by $(p_0, p_1, p_2) = (R_0/R, R_1/R, R_2/R)$ and $(q_0, q_1, q_2) = (S_0/S,$

Table 3 Allele-based case-control samples for a diallelic SNP in a case-control design

	Alleles		
	A	*B*	Total
Case	$2R_0 + R_1$	$2R_2 + R_1$	$2R$
Control	$2S_0 + S_1$	$2S_2 + S_1$	$2S$
Total	$2N_0 + N_1$	$2N_2 + N_1$	$2N$

S_1/S, S_2/S), respectively. In **Table 3**, the frequencies for allele A in cases and controls are estimated by $p = (2R_0 + R_1) / (2R)$ and $q = (2S_0 + S_1)/(2S)$, respectively. The trend test statistic and the allele-based test statistic are then respectively given by

$$Z_T(T) = \frac{(RSN)^{1/2}\{(p_2 + p_1/2) - (q_2 + q_1/2)\}}{\{N(N_2 + N_1/2) - (N_2 + N_1/2)^2\}^{1/2}}$$

and

$$Z_A(T) = \frac{(8RSN)^{1/2}(p - q)}{\{(2N_0 + N_1)(2N_2 + N_1)\}^{1/2}}.$$

For the test using $Z_A(T)$, HWE proportions must hold in the population, and then the two tests have similar power. Both test statistics in large samples follow a standard normal distribution if there is no association.

A two-stage linkage-association sampling design has several advantages compared with an association study alone. Linkage analysis can identify candidate gene regions at a discovery stage, so a subsequent association study can focus on candidate genes at a confirmatory stage. If linkage to a genomic region is known to exist, it is less likely that an association to a variant in that region is spurious. Linkage analysis results based on the HE regression model may also be used to help select samples to design an optimal subsequent association study. Wang & Elston (61) define a quantitative linkage score (QLS) that captures the linkage signals; the QLS can guide the selection of subsamples from a linkage study to increase the power of a subsequent association study.

Thus, the classical paradigm has always started with sampling data for segregation analysis, followed by sampling for linkage analysis and then association analysis, the last using case-control data and/or further family data. This multistage method will find multiple loci and multiple variants within each (multiallelic) locus if the disease being studied is due to many rare variants. However, while this approach might be the only method to find rare mutations at a reasonable cost, sampling family data is expensive.

TWO-STAGE DESIGNS FOR ASSOCIATION

Because family data are required to conduct linkage analysis, it might be more efficient to perform a case-control association study without a preceding linkage study when sampling family data is expensive, especially for a large-scale association study (11, 23, 35, 36). Moreover, for late-onset diseases or diseases related to newly developed technology and medical devices, family data are usually not available. For example, in-stent restenosis is a complication related to stent therapy for atherosclerotic coronary artery disease (14). To study any genetic effect on in-stent restenosis, a case-control sample would be far simpler to collect than family data. Case-control samples are also useful for genome-wide association studies with 100,000–500,000 SNPs, because it is much less costly to recruit thousands of unrelated, rather than related, cases and controls.

A cost-effective two-stage design was first developed for case-control association studies in a candidate gene setting. The genotype-based case and control samples for the two-stage design are given in **Table 4**, where the total sample size is $N_{(1)} + N_{(2)}$. Satagopan et al. (39) proposed to genotype all the SNPs under

Table 4 Genotype data for a diallelic SNP, selected for advancement to the second stage, in the first and second stages of a two-stage case-control design

First-stage data T_1				
	Genotypes			
	AA	*AB*	*BB*	Total
Case	R_{01}	R_{11}	R_{21}	$R_{(1)}$
Control	S_{01}	S_{11}	S_{21}	$S_{(1)}$
Total	N_{01}	N_{11}	N_{21}	$N_{(1)}$
Second-stage data T_2				
	Genotypes			
	AA	*AB*	*BB*	Total
Case	R_{02}	R_{12}	R_{22}	$R_{(2)}$
Control	S_{02}	S_{12}	S_{22}	$S_{(2)}$
Total	N_{02}	N_{12}	N_{22}	$N_{(2)}$

study on a fraction, $f_s = R_{(1)}/(R_{(1)} + R_{(2)}) = S_{(1)}/(S_{(1)} + S_{(2)})$, of the cases and controls in the first stage, denoted T_1 (**Table 4**), and test each SNP for association using the test statistic $Z_A(T_1)$. The most significant small fraction of SNPs is then genotyped on the rest of the cases and controls at the second stage, denoted T_2 (**Table 4**). Here again the testing for association uses $Z_A(T)$, but this is now calculated from all the samples. For a powerful two-stage design with a fixed total cost, the general guideline given was to genotype about 25% of the cases and controls ($f_s = 0.25$) in the first stage and then to genotype the 10% ($f_m = 0.10$) most-promising SNPs in the second stage on the rest (75%) of the samples. Given equal cost for the single-stage and two-stage designs, this two-stage approach allows one to genotype 325% more cases and controls than in a single-stage association study ($N_{(1)} + N_{(2)} = 4.25N$), resulting in a substantial increase in power to detect susceptibility SNPs for a disease. In practice, however, the total number of cases and controls may not be arbitrary, but rather prespecified as the sample size of the study. Thus, two-stage designs may be compared with the single-stage design with $N_{(1)} + N_{(2)} = N$.

Satagopan & Elston (38) then modified this two-stage design, assuming that markers are in linkage equilibrium, to find the optimal design parameters that minimize the cost of the study while retaining a prespecified Type I error and power. This approach was further extended (40) to allow for LD among the SNPs, with the result that genotyping 50% of the samples in the first stage and evaluating the most significant 10% of the SNPs in the second stage yields power close to that of the single-stage design. Following the same designs (38–40), Thomas et al. (56) considered estimating relative risks: the risk of having a disease with genotype *AA* relative to that with *BB*, and the risk of having a disease with genotype *AB* relative to that with *BB*—in other words genotype *BB* is used as the reference genotype. When the SNP and the disease locus are not associated (which could result from linkage equilibrium), the RRs are equal to 1. In the second stage they incorporated the data T_1 of the unselected SNPs to enhance power. They demonstrated that a two-stage design with appropriate tag SNP selection can be very cost-efficient compared with the single-stage design.

These two-stage sampling designs and analyses were developed for candidate gene studies or association studies with hundreds, or thousands, of markers. The same designs and analysis strategies can be applied to genome-wide association studies with 100,000–500,000 SNPs. To test a total of $m = 500{,}000$ SNPs, the significance level for each SNP is set to $\alpha = 0.05/m = 1 \times 10^{-7}$. Following (38), Wang et al. (59) studied two-stage designs using allele-based tests and odds ratios for such genome-wide scans. They considered more focused fine mapping in the second stage using the HapMap data and a model with different unit costs of genotyping in the two stages. Let f_s be the fraction of samples genotyped in the first stage and f_m be the fraction of SNPs selected in the first stage for carryover to the second stage in a two-stage design. Let c_1 and c_2 be the per-sample costs of genotyping a SNP at the first and second stages, respectively. The per-sample genotyping cost of the two-stage design is then $c_1 f_s + c_2 (1 - f_s) f_m$, whereas that for a single-stage design is just c_1. If $c_1 > c_2 f_m$, then $c_1 f_s + c_2 (1 - f_s) f_m < c_1$, indicating that a two-stage design reduces the genotyping cost.

A two-stage design is optimal when the design parameters are determined under some optimality criterion, e.g., for minimizing the total cost when the Type I error and power are fixed, or for maximizing the power when the cost and Type I error are fixed. Wang et al. (59) derived optimal configurations for the design parameters, along with sample size and power calculations. Based on 1000 dollars per subject (currently only 250–500 dollars) for 500,000 SNPs in the first stage ($c_1 = 0.2$ cents), and $c_2 = 3.5$ cents in the second stage, they showed that, to minimize the cost, the optimal two-stage design for a range of risk allele frequencies (0.1–0.9) and realistic odds ratios (1.35–2.0) is to genotype about 30% of all the samples in the first stage at significance levels $\alpha_1 = 0.00362 - 0.00388$, achieving about 90.7% power in the first stage. Then the significance level in the second stage, α_2, should be about 1.6×10^{-7} or 1.7×10^{-7}, compared with the single-stage design with a more stringent level $\alpha = 1 \times 10^{-7}$. To minimize the total sample size under the same allele frequencies and odds ratios, while constraining the cost to be within 110% of the minimum cost, about 40% of the samples are genotyped in the first stage with $\alpha_1 = 0.00510 - 0.00533$, giving 95% power for this stage, and the second-stage level reduces to $\alpha_2 = 1.1 \times 10^{-7}$ or 1.2×10^{-7}, closer to α. In practice, which optimality criterion should be used to design an optimal two-stage study depends on many factors, including disease prevalence and budget.

For the second stage, they proposed a notable extension: to genotype an average of five more densely spaced SNPs, determined from HapMap data, around each SNP significant in the first stage (59). Assuming the same unit costs of genotyping in the two stages, genotyping five more SNPs in the second stage for each selected first-stage SNP may be prohibitive if a large fraction of the SNPs is selected and a small fraction of samples is used in the first stage. In fact, an optimal two-stage design to minimize the total cost, when genotyping five extra SNPs at the second stage for each significant first-stage SNP, entails $\alpha_1 = 0.0005$—much smaller than $\alpha_1 = 0.00362 - 0.00388$ for the same design, but without genotyping the extra SNPs. In addition, about 50% of the samples would be allocated to the second stage, compared with 70% if these extra SNPs were not genotyped. Because 500,000 SNPs may not capture all the genetic variation across the genome, there is a distinct advantage to using more densely spaced SNPs in the second stage to increase coverage of the genome. With this optimal design, the measure of genetic variation covered (see 55) is increased by more than 66% when the extra SNPs are genotyped in the second stage. The disadvantage, however, is that the total cost, although minimized, is almost double that of the optimal two-stage design without genotyping these extra SNPs. When genotyping costs go down even further—currently, only a quarter of what was assumed in Reference 57—genotyping more densely spaced SNPs in the second stage could be promising. However, although one could then genotype extra case-control samples in the second stage for the same cost, and the increased sample size would lead to increased power, it should be noted that, to the extent that genetic variation is not covered by the first-stage SNPs, no amount of extra SNPs at the second stage will help if they are chosen only on the basis of SNPs showing association at the first stage.

Skol et al. (50) studied a simple procedure to combine the test statistics in the two stages under various configurations of the design parameters. When test statistics $Z_A(T_1)$ and $Z_A(T_2)$ are applied to the two stages, respectively, a weighted test statistic for a joint analysis can be written as $Z_\omega(T) = \omega_1 Z_A(T_1) + \omega_2 Z_A(T_2)$, where the weights ω_1 and ω_2 are proportional to the sample sizes used in the two stages. Since the number of markers tested in the second stage is much smaller than the total number of markers, one might be tempted to view stage two as a replication study and assess statistical significance based on the stage-two data alone. However, this strategy almost always results in decreased power to detect genetic association, as compared with analyzing the data from both stages together, despite the fact that less-stringent-significance criteria are used.

It is not obvious how to evaluate statistical significance in two-stage association studies. A common practice is to use the Bonferroni correction based on the total number of markers tested in the first stage (56). This strategy can be highly conservative, especially when the markers are in strong LD, because it assumes that none of the markers eliminated after the first-stage testing would reach statistical significance if they were genotyped and tested in the second stage, and that the test statistics are independent across all markers. Lin (24) proposed a Monte Carlo procedure that properly accounts for the two-stage sampling and the correlated nature of the SNP data. Through simulation studies, he demonstrated that this procedure provides accurate control of Type I error and can be substantially more powerful than using the Bonferroni correction.

A comparison of various optimal two-stage sampling designs and analyses is summarized in **Table 5**, showing that most studies use no more than 50% of the samples in the first stage and select less than 25% of the SNPs for genotyping in the second stage. In particular, for genome-wide association studies, as the number of SNPs increases, the proportion of SNPs selected for the second stage decreases. One limitation of any two-stage design is that a disease susceptibility SNP may not be selected in the first stage if only a small fraction of SNPs is carried forward to the second stage. For example, Thomas et al. (57) used the level $\alpha_1 = 0.005$ at the first stage for a total of 500,000 SNPs. On average, about 2500 SNPs would then be selected for genotyping and analysis in the second stage. Although the probability that at least one SNP with true association is in the selected 2500 SNPs is unknown, it appears that it may be low (see table 2 in 70). For example, for a genome-wide association study with 200,000 null SNPs and one SNP with true association, the probability that the most significant 20,447 (9931) SNPs would contain the true SNP is 95% when the power for detecting this true SNP is 90% (95%). This transforms to selecting about 5–10% of the total SNPs in the first stage, within the range discussed by Skol et al. (50).

With advances in technology, the genotyping cost is dramatically reduced. When the genotyping cost is of no concern, especially for candidate gene analysis, a single-stage design using all samples in the analysis is more powerful than a two-stage design.

Table 5 Comparisons of two-stage designs and analyses

Comparison items	Satagopan et al. (2002)	Satagopan & Elston (2003)	Satagopan et al. (2004)	Thomas et al. (2004)	Wang et al. (2006)	Skol et al. (2006)
Number of SNPs	Candidate genes	Candidate genes	Candidate genes	Candidate genes	500,000	300,000
Optimality for the design parameters (α_1, α_2, and f)	Maximizing power for given total cost, unrestricted sample size	Minimizing cost for given power	Maximizing power for given total cost, fixed sample size	Fixed design parameters	Various optimality criteria	Fixed design parameters
Statistical procedure	Allele-based analysis	Trend test	Trend test	Estimation of relative risks	Allele-based analysis and odds ratios	Allele-based analysis
Proportion of samples in the first stage	25%	Calculated to achieve 95% power	50%	No	30–40%	10–50%
Proportion of SNPs in the second stage	10%	15–25%	10%	No	0.4–0.5%	1–10%

A single-stage design also allows one to examine multiple phenotypes and to perform multimarker testing. On the other hand, a large-scale case-control association study is costly. Among 500,000 SNPs, only a few of the SNPs are, or are in LD with, true susceptibility SNPs. Genotyping all SNPs using all samples is a waste of resources. With an appropriate choice of design parameters, a two-stage design would reduce genotyping costs up to 50%, yet have power close to the corresponding single-stage design. Therefore, a two-stage sampling design is cost efficient and particularly useful for genome-wide association studies.

In the two-stage designs shown in **Table 5**, the same test statistic is applied to both stages, e.g., the trend test statistic $Z_T(T)$. However, different procedures and test statistics can be applied to the two stages. For example, at the first stage we could pool a fraction of all the DNA samples separately for cases and controls. If the statistic used simply compares the frequencies of an allele in cases and controls, pooling DNA is an efficient way to reduce the genotyping cost in the first stage. Comparing the frequencies of alleles A and B in pooled cases and pooled controls for each SNP is equivalent to the first-stage test of Skol et al. (50). In the second stage, the selected SNPs are separately genotyped for each of the remaining samples. If the genotyping cost in the second stage is much higher than that in the first stage—in particular if more densely spaced SNPs are to be genotyped in the second stage—pooling DNA could be done in the second stage, rather than in the first stage. Pooling DNA in only one stage has the advantage that the individual genotype data can be adjusted for covariates. Pooling DNA has been applied to SNP and haplotype analyses (e.g., 4, 26, 48, 67, 71, 74; for a review, see 46). For quality control, several sets of pools within cases and controls are usually used as replications and, if the pools are not too large, it is possible to obtain good estimates of haplotype frequencies from them (34). However, optimal two-stage designs with pooling DNA require further study and the number of pools is a further design parameter to be determined. As another example, following the same two-stage design as in **Table 4**, we can apply a multimarker analysis to the first stage and select clusters of SNPs for the second stage (25). In this case, the two-stage design is based on clusters of SNPs rather than on single SNPs. This approach is similar to the two-stage design for haplotype analysis proposed by Thomas et al. (57). The optimal configurations of design parameters for two-stage sampling with multimarker analysis are unknown and need further research.

Research on multistage sampling has mainly focused on a binary outcome. For quantitative traits, we can also apply the multistage sampling designs used for binary outcomes by defining two thresholds to determine cases and controls (e.g., for blood pressure, cases could have pressure greater than 160/90 and normal controls pressure less than 120/80). For a QTL, the power to reject the null hypothesis is substantially increased when subjects with more extreme trait values are sampled. Therefore, sampling only the more extreme trait values in each stage may be useful to reduce the cost of genotyping when the cost of measuring trait values is much cheaper than the cost of genotyping.

One disadvantage of case-control association studies is that population stratification or allelic correlation may lead to spurious association (25). Population stratification occurs when subjects with different genetic backgrounds are enrolled. For example, individuals with different ethnic backgrounds may be defined as subpopulations. The subpopulations can cause spurious associations if they have different disease prevalences as well as different allele frequencies for a SNP. An extreme situation of population stratification occurs when cases and controls are sampled from two different ethnic groups. Allelic correlation, arising from cryptic relatedness, may occur within a population if there is not random mating, with the result that HWE does not hold in the population. This can

Hardy-Weinberg disequilibrium (HWD): departure from Hardy-Weinberg equilibrium proportions, often resulting from selection

distort the test statistics, leading to either an underestimate or overestimate of their variances. Adjustments for either population stratification or cryptic relatedness have been studied in the literature for single-stage design, e.g., using structure analysis and a latent-class modeling approach (32, 33, 41, 43, 62), or a genomic control approach (5, 16, 44, 72). The effect of population stratification and allelic correlation on two-stage designs has not been studied and may be more complicated than for single-stage designs. One possibility is to apply genomic control to correct for variance distortion using what are believed to be null SNPs—SNPs that are not in LD with loci underlying the trait under study. Because only a portion of the SNPs are genotyped in the second stage, the candidate null SNPs should be from among those not selected from the first stage for advancement to the second stage. However, genotypes of these null SNPs are only obtained from a fraction of the case-control samples.

MULTISTAGE DESIGNS FOR ASSOCIATION STUDIES

Most research on optimal cost-efficient designs for association studies has focused on two-stage designs. Multistage designs for association can also be considered. Prentice et al. (31) discussed a three-stage design for an association study with 250,000 SNPs. They suggested dividing the case-control samples into three independent portions of equal size for each stage. To allow for an overall average of 2.5 false positives in this three-stage study, the significance level 0.022 was used for each stage. This design has a two-thirds reduction in the genotyping cost, with only 121 SNPs typed at the third stage, each tested at a significance level of 0.022. This design is simple but not optimal, and its power performance is not clear. A similar three-stage screening procedure for genome-wide studies that divides the samples into three fractions, with decreasing sizes from the first stage to the third stage, has also been discussed (19). The significance levels used at the first two stages were liberal, but a more stringent level was proposed for the third stage. The optimality of this design is unknown. Optimal three-stage designs based on different optimality criteria can be obtained, analogous to the optimal two-stage designs. However, because more design parameters are involved, the optimal solutions will be more complicated and difficult to define.

Statistical power can be further enhanced if a "self-replication" analysis stage is employed. This idea of self-replication was introduced by Van Steen et al. (58) for a single-stage family-based association study and was applied to a genome-wide family-based association study with 100,000 SNPs (18). The idea is to identify and use two statistically independent procedures to test each SNP for association. After the first procedure is performed to test for association, the second can be regarded as an independent replication of the results obtained. The main feature of this procedure is that all samples are used in both analysis stages. In a single-stage case-control design, the power of using self-replication is often greater than that of a single-stage analysis without the self-replication (73).

When the two alleles, A and B, of a SNP are independent, HWE holds at this SNP. HWE is often tested by the Hardy-Weinberg disequilibrium (HWD) coefficient (65). Defining $p_1 = R_1/R$, $p_2 = R_2/R$, $q_1 = S_1/S$ and $q_2 = S_2/S$ (**Table 2**), this coefficient is $\Delta_1 = p_2 - (p_2 + p_1/2)^2$ in cases and $\Delta_0 = q_2 - (q_2 + q_1/2)^2$ in controls. Departure from HWE among cases has been used to test for association (12), but has no power under a multiplicative model (28, 66). After standardization, the resulting statistic can be used because it has a large sample chi-squared distribution with 1 degree of freedom under the null hypothesis. HWD has also been used, in both cases and controls, to indicate genotyping errors as a form of quality control for large association studies). Song & Elston (52) proposed using the difference in HWD coefficients between cases and controls, $\Delta_1 - \Delta_0$, to test for association; we refer to this as the Hardy-Weinberg disequilibrium trend test (HWDTT), denoted $Z_H(T)$. When HWE holds in the population, in large samples $Z_H(T)$ and $Z_T(T)$ are independent under the null hypothesis and so can be used for self-replication. Using all samples, each SNP is first tested by $Z_H(T)$ at the significance level α_1, and those of the m SNPs that are significant at this level are then tested using $Z_T(T)$ at the level α_2, where $\alpha_1\alpha_2 = \alpha/m$ (73).

Using the concept of self-replication, we can introduce into a multisample design the test $Z_H(T)$ as a new analysis stage. For example, in the two-stage design in **Table 4**, we can add this as an extra stage once both sets of data T_1 and T_2 become available: The test statistic $Z_H(T_1 + T_2)$ is calculated on the total sample of $N_{(1)} + N_{(2)}$ individuals and then only those SNPs that are significant (at a predetermined level) are tested using $Z_T(T)$. Thus fewer SNPs are tested, mitigating the multiple testing problem. This self-replication can also be applied to the second stage of any of the two-stage designs in **Table 5** where the trend test is used. Thus, instead of using the trend test in the second stage, we apply the self-replication procedure. In this case, although genotyping costs will not be reduced by introducing a self-replication stage, the power of the two-stage design would be improved (73).

DISCUSSION

In this review, we outlined three types of multistage sampling schemes for genetic studies. The first follows the classic paradigm that has been successfully used for Mendelian diseases: First, familial segregation patterns are examined in families; this is followed by linkage analysis using samples of affected sib-pairs, complete sib-ships, nuclear families, case-parents trios, or large pedigrees; and this is followed by an association study using family and/or individual data.

The second scheme is for more complex diseases: Linkage analysis using family data is first conducted, followed by an association study for candidate genes. This linkage-association design has several advantages. First, results of the prior linkage analysis provide insight into association studies, because markers that are significantly linked to a disease locus can be regarded as providing candidate genes for the association study. Second, testing association between a marker and a disease locus can incorporate results of the linkage analysis to enhance the power of the association study (27, 61). Third, if many rare variants at a single locus can each cause the disease, a linkage study will have more power than an association study to detect that locus. Fourth, a linkage study involves the collection of family data that enables one to detect phenomena such as anticipation and parent-of-origin effects (imprinting).

The third scheme is a multistage design for genome-wide association studies using only case-control samples. This multistage sampling can reduce the genotyping cost while retaining nearly optimal power compared with a single-stage association design. In a two-stage design, a fraction (usually between 25% and 50%) of case-control samples is first genotyped on all 100,000–500,000 SNPs in the first stage, in which association between each SNP and the disease is tested. Only a small fraction of the SNPs (often less than 10%) is selected for genotyping on the rest of the samples. Then the two stages are combined to test the selected SNPs. With high-throughput genotyping technology, this last scheme is most promising for the future, but the advantages of conducting a linkage study prior to a genome-wide association study should not be ignored.

In this review of multistage sampling designs, we focused on designs in which each sample is ascertained conditional on its phenotypes. In pharmarcogenetic or gene-intervention interaction studies of candidate genes, samples are randomized to two treatments (interventions) for a period of time. The genetic characteristics of individuals who develop disease during this period are then compared between the two treatment groups and tested. The genotyping cost may not then be an issue because, in such a prospective case-control study, follow-up of individuals usually costs much more than genotyping. For example, to study whether or not regular use of an inhaled b2-adrenergic agonist worsens airflow and clinical outcomes in asthma patients, a small-scale retrospective case-control study was carried out in a first stage (22). This suggested that adverse events occurred in individuals with homozygosity for arginine (Arg/Arg), rather than glycine (Gly/Gly), at aminoacid residue 16 of the receptor. No significant events were identified with heterozygosity (Arg/Gly). Therefore, in a second stage, only Gly/Gly and Arg/Arg individuals were identified and enrolled, and then regularly scheduled treatment with albuterol or placebo was given to each matched pair (Gly/Gly-Arg/Arg, matching for forced expiratory volume in 1 s, i.e., FEV1) in a masked, crossover design (21). This two-stage genotype-stratified design dramatically reduced the total cost of the prospective trial compared with the same trial that would have enrolled random samples without selective genotyping.

In conclusion, most genetic studies need to have independent replication, in particular for a large study, because there is a danger of a high false positive rate from testing many SNPs. Once the results of a genotype-phenotype association have been replicated, additional sampling may be needed for fine mapping, resequencing, functional studies, and, finally, developing personalized medicine.

SUMMARY POINTS

1. The classical paradigm for the genetic analysis of a familial disease in humans has been to collect family samples for segregation analysis and linkage analysis, followed by collecting further samples for association studies. This multistage sampling procedure has been extremely successful for Mendelian diseases.
2. For complex diseases, the first stage has tended to be a genome-wide linkage study to suggest candidate regions for further fine-mapping and association studies.
3. A well-designed optimal two-stage association study, possibly pooling DNA at one stage, is cost-efficient for genome-wide association studies. It has power close to that of a single-stage design, but with up to a 50% reduction in the genotyping cost.
4. Three-stage association designs are also promising, but need further study of their optimality.
5. Statistical methods need to be developed to control for confounding factors and to test for gene-gene and gene-environment interactions in multistage association studies.
6. Replication and resequencing are important stages of a sampling design for genetic studies.
7. Collecting a sample of families for a linkage study before conducting a genome-wide association study may still be useful, either to reduce the cost when there is allelic heterogeneity or to detect generational effects.

DISCLOSURE STATEMENT

The authors are not aware of any biases that might be perceived as affecting the objectivity of this review.

ACKNOWLEDGMENTS

This work was supported in part by grants from the U.S. Public Health Service, resource grant RR03655 from the National Center for Research Resources, research grants GM-28356 from the National Institute of General Medical Sciences and CA82659 from the National Cancer Institute, and Cancer Center Support Grant P30CAD43703 from the National Cancer Institute. We wish to thank Vernon Chinchilli for providing references on the asthma studies.

LITERATURE CITED

1. Altshuler D, Brooks LD, Chakravarti A, Collins FS, Daly MJ, Donnelly P. 2005. A haplotype map of the human genome. *Nature* 437:1299–320
2. Botstein D, White RL, Skolnick M, Davis RW. 1980. Construction of a genetic linkage map in man using restriction fragment length polymorphisms. *Am. J. Hum. Genet.* 32:314–31
3. Clark AG, Boerwinkle E, Hixson J, Sing CF. 2005. Determinants of the success of whole-genome association testing. *Genome Res.* 15:1463–67
4. Craig DW, Huentelman MJ, Hu-Lince D, Zismann VL, Kruer MC, et al. 2005. Determinants of the success of whole-genome identification of disease causing loci using an array-based genotyping approach on pooled DNA. *BMC Genomics* 6:138
5. Devlin B, Roeder K. 1999. Genomic control for association studies. *Biometrics* 55:997–1004
6. Elston RC. 1994. P values, power, and pitfalls in the linkage analysis of psychiatric disorders. In *Genetic Approaches to Mental Disorders*, ed. ES Gershon, CR Cloninger, pp. 3–21. Am. Psych Press, Washington, D.C.
7. Elston RC. 1998. Methods of linkage analysis—and the assumptions underlying them. *Am. J. Hum. Genet.* 63(4):931–34
8. Elston RC, George VT, Severtson F. 1992. The Elston-Stewart algorithm for continuous genotypes and environmental factors. *Hum. Hered.* 41:16–27
9. Elston RC, Song D, Iyengar SK. 2005. Mathematical assumptions versus biological reality: myths in affected sib-pair linkage analysis. *Am. J. Hum. Genet.* 76:152–56
10. Elston RC, Spence MA. 2006. Advances in statistical human genetics over the last 25 years. *Statist. Med.* 25:3049–80
11. Evangelou E, Trikalinos TA, Salanti G, Ioannodos JPA. 2006. Family-based versus unrelated case-control designs for genetic associations. *PLoS Genet.* 2:1147–55
12. Feder JN, Gnirke A, Thomas W, Tsuchihashi Z, Ruddy DA, et al. 1996. A novel MHC class I-like gene is mutated in patients with hereditary hemochromatosis. *Nat. Genet.* 13(4):399–408
13. Freidlin B, Zheng G, Li Z, Gastwirth JL. 2002. Trend tests for case-control studies of genetic markers: power, sample size and robustness. *Hum. Hered.* 53:146–52
14. Ganesh SK, Skelding KA, Mehta L, O'Neill K, Joo J, et al. 2004. Rationale and study design of the CardioGene Study: genomics of in-stent restenosis. *Pharmacogenomics* 5:949–1004
15. George VT, Elston RC. 1987. Testing the association between polymorphic markers and quantitative traits in pedigrees. *Genet. Epidemiol.* 4:193–202
16. Gorroochurn P, Heiman GA, Hodge SE, Greenberg DA. 2006. Centralization the noncentral chi-square: a new method to correct for population stratification in genetic case-control association studies. *Genet. Epidemiol.* 30:277–89
17. Haseman JK, Elston RC. 1972. The investigation of linkage between a quantitative trait and a marker locus. *Behav. Genet.* 2:3–19

18. Herbert A, Gerry NP, McQueen MB, Heid IM, Pfeufer A, et al. 2006. A common genetic variant is associated with adult and childhood obesity. *Science* 312:279–83
19. Hirschhorn JN, Daly MJ. 2005. Genome-wide association studies for common diseases and complex traits. *Nat. Rev. Genet.* 6:95–108
20. Hoh J, Wille A, Ott J. 2001. Trimming, weighting, and grouping SNPs in human case-control association studies. *Genome Res.* 11(12):2115–19
21. Israel E, Chinchilli VM, Ford JG, Boushey HA, Cherniack RM, et al. 2004. Use of regularly scheduled albuterol treatment in asthma: genotype-stratified, randomised, placebo-controlled cross-over trial. *Lancet* 364:1505–12
22. Israel E, Drazen JM, Liggett SB, Boushey HA, Cherniack RM, et al. 2000. The effect of polymorphisms of the beta (2)-adrenergic receptor on the response to regular use of albuterol in asthma. *Am. J. Respir. Crit. Care Med.* 162:75–80
23. Klein RJ, Zeiss C, Chew EY, Tsai JY, Sackler RS, et al. 2005. Complement factor H polymorphism in age-related macular degeneration. *Science* 308:385–89
24. Lin DY. 2006. Evaluating statistical significance in two-stage genomewide association studies. *Am. J. Hum. Genet.* 78:505–09
25. Marchini J, Cardon LR, Phillips MS, Donnelly P. 2004. The effects of human population structure on large genetic association studies. *Nat. Genet.* 36:512–17
26. Meaburn E, Butcher LM, Liu L, Fernandes C, Hansen V, et al. 2005. Genotyping DNA pools on microarrays: tackling the QTL problem of large samples and large numbers of SNPs. *BMC Genomics* 6:52
27. Nagelkerke NJD, Hoebee B, Teunis P, Kimman TG. 2004. Combining the transmission disequilibrium test and case-control methodology using generalized logistic regression. *Eur. J. Hum. Genet.* 12:964–97
28. Nielson DM, Ehm MG, Weir BS. 1998. Detecting marker-disease association by testing for Hardy-Weinberg disequilibrium at a marker locus. *Am. J. Hum. Genet.* 63:1531–40
29. Nyholt DR. 2000. All LODs are not created equal. *Am. J. Hum. Genet.* 67(2):282–88
30. Olson JM. 1999. A general conditional-logistic model for affected-relative-pair linkage studies. *Am. J. Hum. Genet.* 65:1760–69
31. Prentice RL, Pettinger M, Anderson GL. 2005. Statistical issues arising in the Women's Health Initiative. *Biometrics* 61:899–941
32. Pritchard JK, Rosenberg NA. 1999. Use of unlinked genetic markers to detect population structure in genetic association studies. *Am. J. Hum. Genet.* 65:220–28
33. Pritchard JK, Stephens M, Rosenberg NA, Donnelly P. 2001. Association mapping in structured populations. *Am. J. Hum. Genet.* 67:170–81
34. Quade SRE, Elston RC, Goddard KAB. 2005. Estimating haplotype frequencies in pooled DNA samples when there is genotyping error. *BMC Genet.* 6:25
35. Risch N. 2000. Searching for genetic determinants in the new millennium. *Nature* 405:847–56
36. Risch N, Merikangas K. 1996. The future of genetic studies of complex human diseases. *Science* 273:1516–17
37. Sasieni PD. 1997. From genotypes to genes: doubling the sample size. *Biometrics* 53:1253–61
38. Satagopan JM, Elston RC. 2003. Optimal two-stage genotyping in population-based association studies. *Genet. Epidemiol.* 25:149–57
39. Satagopan JM, Verbel DA, Venkatramanm ES, Offit KE, Begg CB. 2002. Two-stage design for gene-disease association studies. *Biometrics* 58:163–70
40. Satagopan JM, Veukatraman ES, Begg CB. 2004. Two-stage designs for gene-disease association studies with sample size constraints. *Biometrics* 60:589–97

41. Sattern GA, Flanders WD, Yang Q. 2001. Accounting for unmeasured population substructure in case-control studies of genetic association using a novel latent-class model. *Am. J. Hum. Genet.* 68:466–77
42. Schaid DJ, Sommer SS. 1993. Genotype relative risks: methods for design and analysis of candidate-gene association studies. *Am. J. Hum. Genet.* 53:1114–26
43. Schork NJ, Fallin D, Thiel B, Xu X, Broeckel U, et al. 2001. The future of genetic case-control studies. *Adv. Genet.* 42:191–212
44. Setakis E, Stirnadel H, Balding DJ. 2006. Logistic regression protects against population structure in genetic association studies. *Genome Res.* 16:290–96
45. Sham PC. 1998. *Statistics in Human Genetics*. London: Arnold
46. Sham PC, Bader JS, Craig I, O'Donovan M, Owen M. 2002. DNA pooling: a tool for large-scale association studies. *Nat. Rev. Genet.* 3:862–71
47. Sham PC, Curtis D. 1995. An extended transmission/disequilibrium test (TDT) for multiallele marker loci. *Ann. Hum. Genet.* 59:855–56
48. Shaw SH, Carrasquillo MM, Kashuk C, Puffenberger EG, Chakravarti A. 1998. Allele frequency distributions in pooled DNA samples: applications to mapping complex disease genes. *Genome Res.* 8:111–23
49. Shete S, Jacobs KB, Elston RC. 2003. Adding further power to the Haseman and Elston method for detecting linkage in larger sibships: weighting sums and differences. *Hum. Hered.* 55:79–85
50. Skol AD, Scott LJ, Abecasis GR, Boehnke M. 2006. Joint analysis is more efficient than replication-based analysis for two-stage genome-wide association studies. *Nat. Genet.* 38:209–13
51. Slager SL, Schaid DJ. 2001. Case-control studies of genetic markers: power and sample size approximations for Armitage's test for trend. *Hum. Hered.* 52:149–53
52. Song KJ, Elston RC. 2006. A powerful method of combining measures of association and Hardy-Weinberg disequilibrium for fine-mapping in case-control studies. *Statist. Med.* 25:105–26
53. Spielman RS, McGinnis RE, Ewens WJ. 1994. The transmission/disequilibrium test detects cosegregation and linkage. *Am. J. Hum. Genet.* 54:559–60
54. Spielman RS, McGinnis RE, Ewens WJ. 1993. Transmission test for linkage disequilibrium—the insulin gene region and insulin-dependent diabetes-mellitus (IDDM). *Am. J. Hum. Genet.* 52:506–16
55. Stram DO. 2004. Tag SNP selection for association studies. *Genet. Epidemiol.* 27:365–74
56. Thomas DC, Haile RW, Duggan D. 2005. Recent developments in genomewide association scans: a workshop summary and review. *Am. J. Hum. Genet.* 77:337–45
57. Thomas DC, Xie R, Gebregziabher M. 2004. Two-stage sampling designs for gene association studies. *Genet. Epidemiol.* 27:401–14
58. Van Steen K, McQueen MB, Herbert A, Raby B, Lyon H, et al. 2005. Genomic screening and replication using the same data set in family-based association testing. *Nat. Genet.* 37:683–91
59. Wang H, Thomas DC, Pe'er I, Stram DO. 2006. Optimal two-stage genotyping designs for genome-wide association scans. *Genet. Epidemiol.* 30:356–68
60. Wang T, Elston RC. 2005. Two-level Haseman-Elston regression for general pedigree data analysis. *Genet. Epidemiol.* 29:12–22
61. Wang T, Elston RC. 2006. A quantitative linkage score for an association study following a linkage analysis. *BMC Genet.* 7:5
62. Wang Y, Localio R, Rebbeck TR. 2005. Bias correction with a single null marker for population stratification in candidate gene association studies. *Hum. Hered.* 59:165–75

63. Weber JL, May PE. 1989. Abundant class of human DNA polymorphisms which can be typed using the polymerase chain reaction. *Am. J. Hum. Genet.* 44:388–96
64. Weber JL, Broman KW. 2001. Genotyping for human whole-genome scans: past, present, and future. *Adv. Genet.* 42:77–96
65. Weir BS. 1996. *Genetic Data Analysis II: Methods for Discrete Population Genetic Data.* Sunderland, Massachusetts: Sinauer Assoc.
66. Wittke-Thompson JK, Pluzhnikov A, Cox NJ. 2005. Rational inferences about departures from Hardy-Weinberg equilibrium. *Am. J. Hum. Genet.* 76:967–86
67. Yang YN, Zhang JS, Hoh J, Matsuda F, Xu P, et al. 2003. Efficiency of single-nucleotide polymorphism haplotype estimation from pooled DNA. *Proc. Nat. Acad. Sci. USA* 100:7225–30
68. Yuan B, Vaske D, Weber JL, Beck L, Sheffield VC. 1997. Improved set of short-tandem-repeat polymorphisms for screening the human genome. *Am. J. Hum. Genet.* 60:459–60
69. Xing C, Elston RC. 2006. Distribution and magnitude of Type I error of the model-based multipoint lod score. Implications for multipoint mod score. *Genet. Epidemiol.* 30:447–58
70. Zaykin DV, Zhivotovsky LA. 2005. Ranks of genuine associations in whole-genome scans. *Genetics* 171:813–23
71. Zeng D, Lin DY. 2005. Estimating haplotype-disease associations with pooled genotype data. *Genet. Epidemiol.* 28:70–82
72. Zheng G, Freidlin B, Gastwirth JL. 2006. Robust genomic control for association studies. *Am. J. Hum. Genet.* 78:350–56
73. Zheng G, Song KJ, Elston RJ. 2007. Combining trend tests for genome-wide association studies: a double trend test. Submitted for publication
74. Zuo Y, Zou G, Zhao H. 2006. Two-stage designs in case-control association analysis. *Genetics* 173:1747–60

The Uneasy Ethical and Legal Underpinnings of Large-Scale Genomic Biobanks

Henry T. Greely

Stanford University, Stanford, California 94305; email: hgreely@stanford.edu

Annu. Rev. Genomics Hum. Genet. 2007. 8:343–64

First published online as a Review in Advance on June 5, 2007.

The *Annual Review of Genomics and Human Genetics* is online at genom.annualreviews.org

This article's doi:
10.1146/annurev.genom.7.080505.115721

1527-8204/07/0922-0343$20.00

Key Words

anonymity, confidentiality, informed consent, clinical information

Abstract

Large-scale genomic databases are becoming increasingly common. These databases, and the underlying biobanks, pose several substantial legal and ethical problems. Neither the usual methods for protecting subject confidentiality, nor even anonymity, are likely to protect subjects' identities in richly detailed databases. Indeed, in these settings, anonymity is itself ethically suspect. New methods of consent will need to be created to replace the blanket consent common to such endeavors, with a consent procedure that gives subjects some real control over what they might consider inappropriate use of their information and biological material. Through their use, these biobanks are also likely to yield information that will be of some clinical significance to the subjects, information that they should have access to. Failure to adjust to these new challenges is not only legally and ethically inappropriate, but puts at risk the political support on which biomedical research depends.

"When *I* use a word," Humpty Dumpty said in rather a scornful tone, "it means just what I choose it to mean – neither more nor less."

"The question is," said Alice , 'whether you CAN make words mean so many different things."

"The question is," said Humpty Dumpty, "which is to be master – that's all." (4)

Around the world, large-scale research resources are being created, resources made up of human genetic and phenotypic information—health records, physical or mental traits, and so on. Sometimes these resources include the physical DNA samples; other times they are just databases containing genotypes that were analyzed somewhere else. Some of the resources focus on a particular medical or scientific problem; others are to be used broadly. Some are local, some national; some are private to a particular research project or firm, governments run others. These genotype/phenotype resources, which I will call "genomic biobanks" in this article, all share one characteristic: Without the physical DNA and the phenotypic information contributed by thousands of ordinary (and sometimes extraordinary) people, they would be worthless.

The men, women, and children who contribute to these biobanks are almost never paid and rarely, if ever, can they have realistic hopes that their contribution will lead to an improvement in their own lives, a treatment for their disease. Biobanks result from their goodwill, and, in many places, can only continue to be funded by their direct or indirect political support.

This goodwill, and its concomitant political support, are now being put at risk by the ways some of these biobanks plan to operate, ways that threaten to abuse the trust with which these donations were given. Those decisions are justified by twisting the meanings of words like "confidential," "anonymous," "human subjects," "informed consent," and "justice" in ways that jeopardize, for the sake of administrative convenience and short-term research gains, the interests, the wishes, and, I believe, the rights of those who contributed to these resources. And by failing to respect donors, the biobanks put at risk the long-term interests of biomedical science, which can only prosper with the trust and support of the population.

Specifically, this article demonstrates that patient identity is not, and cannot be, effectively protected in large-scale genomic biobanks. It argues that, as one result, the practice of anonymizing data is not only nearly useless, but is itself unethical. It contends that a new interpretation of the Common Rule and plans for a Genome-Wide Association Studies database by the United States National Institutes of Health (NIH) illegally and unethically strip people of the legally protected status of human subjects. It urges that blanket consent, although extremely convenient for genomic biobanks, is both ethically and legally suspect. And it insists that biomedical research, including genomic biobanks, has a moral obligation, now almost always evaded, to inform research subjects of clinically significant information.

To make these points, I first discuss the new and growing world of large-scale genomic biobanks. I then argue that many of these efforts are not adequately protecting the people who contributed to them in four areas: ensuring the confidentiality of their data, providing those people with the protections of "human subjects," obtaining their informed consent for specific uses of the biobanks' information, and managing clinically significant information about the donors. Some biobanks are making good decisions on some of these issues; many are making bad decisions on most of them. The article ends by calling for large-scale genomic biobanks to be run with much greater attention to the people whose contributions of tissue and information made them possible.

THE RISE OF LARGE-SCALE GENOMIC BIOBANKS AND THEIR EQUIVALENTS

Since the dawn of human genetics, researchers have collected information about subjects' characteristics to try to understand the mechanisms of inheritance and its role in various traits and diseases. Some classical genetics researchers collected and analyzed biological samples, looking at variations in proteins, for example, but the modern understanding of DNA and the ability to analyze it opened up great new possibilities for human genetics research. The first researcher to collect and link both DNA and phenotypic information from human subjects created the first genotype-phenotype resource or genomic biobank. In recent years technical advances have made it possible to create genomic biobanks that include more donors and contain more information and material about each, creating or exacerbating the ethical tensions around genomic biobank efforts. This section describes the first-generation genomic biobanks and the forces encouraging the creation of a new generation. The section ends by describing the efforts thus far at bringing this next generation into existence.

The First Generation of Genomic Biobanks

In the past, researchers collected information about particular health issues along with relevant biological samples, including DNA. A research project might look at breast cancer, autism, or the effectiveness of a particular drug for treating congestive heart failure. Samples would typically be collected from affected and unaffected family members, often in families where the disease was common. Researchers would often establish long-term connections with "their" families and the families would take great interest in the research—research that donors hope might provide useful knowledge or interventions for themselves or their loved ones.

The "biobank" containing any such project's information would have limited uses because of the limited data collected. The phenotypic information would be limited to matters thought to be relevant to studying breast cancer, for example. In the past, even the genotypic information would have likely been limited to a small number of loci, mainly those thought to be linked to the disease being studied. If the project kept additional DNA, that DNA could be genotyped for additional loci, but that would often have been an expensive and time-consuming project. If the project included genome-wide analyses of DNA in a study of, say, breast cancer, those analyses could be searched for genetic variations thought to be correlated with, for example, hypertension, but without any phenotypic data, or, perhaps more insidiously, without good phenotypic data about hypertension, the genotypic data would be useless for hypertension research. The DNA (or the genetic data abstracted from it) might serve some useful purposes beyond breast cancer, such as providing some evidence about the prevalence of particular alleles in the relevant population, but, in general, its value would be sharply limited by the quantity and quality of phenotypic data collected.

These traditional collections were limited in several other ways. The samples and their associated data, phenotypic and genetic, were primarily available to the research group that collected and created them. They might be given or lent to other researchers occasionally, but this was not normally a major use. Researchers would be reluctant to transfer DNA samples because the DNA from subjects was generally a finite resource; 50 milligrams given to another researcher was 50 milligrams that could only be replaced by a new collection from the subject. Similarly, transferring data was somewhat more difficult before high-speed data transfer methods.

Still, some research projects were designed, at least in part, to share samples and data. For example, the U.S. Department of Veterans Affairs sponsored a trial of bucindolol, a beta blocker, for the treatment of congestive heart failure. This trial, known as the Beta-blocker Evaluation of Survival Trial (BEST), collected data to examine the response of different people to the drug, looking to see what characteristics of the subjects might have affected the drug's effectiveness. The BEST trial included the collection of DNA from many of the subjects as one substudy (34). Small amounts of this stored DNA have been made available through a controlled process to several researchers who have genotyped the DNA for variations that they suspect may have affected the subject's reactions to bucindolol. The substudy expressly restricted the access to the DNA to researchers working on heart failure, but the limited nature of the phenotypic data has had the same effect.

A different kind of biobank created cell lines from donor samples and distributed DNA from transformed cell lines and sometimes the cell lines themselves. The cell lines can provide an indefinite amount of DNA from the donor, thus relieving concern about giving away a limited resource. The Coriell Institute has long maintained cell lines and distributed DNA and cells from them to researchers (10); a similar cell line repository is maintained in Paris by the Human Genome Diversity Project (HGDP) and the Foundation Jean Dausset-Centre d'Etude du Polymorphisme Humain (CEPH). (3). These resources distribute DNA but not substantial amounts of phenotypic data.

A New Generation of Genomic Biobanks

By the late 1990s it became clear that although many common diseases had some genetic component, only in a few of them did single genes seem to contribute strongly to disease risk, and even there any strong genetic component was limited to a small percentage of those with the disease. Examples include breast and ovarian cancer and BRCA 1 and BRCA 2 (OMIM +113705, +600185); colon cancer and APC, MSH2, and MLH1 (OMIM +175100, *609309, *120436) and Alzheimer disease and Presenilin 1, Presenilin 2, and APP (OMIM, *104311, *600759, *104760). In each case, mutated versions of the genes are associated with a high risk that the individual carrying them would develop the disease, but somatic mutations in the genes were found in only a very small percentage of people with the disease. Traditional studies of families at high risk for the disease have had little success teasing out what were believed to be the effects of numerous alleles of different genes in increasing disease risk. Instead, studies of the association of different genes (or genomic regions) with disease are expected to require data from a large population. The result has been an effort to create an increasing number of genomic biobanks with more donors, more information, and more users. This effort has met with mixed success.

The creation of a new generation of genomic biobanks is pushed in part by the perceived need for data for association studies, but it is also being pulled in part by technical progress made in genotyping. Instead of genotyping only loci thought to be relevant to, say, breast cancer, today a researcher can, for about the same or less time and money, do a fairly detailed genome-wide scan. Modern mass-produced arrays can now quickly provide information on many single nucleotide polymorphisms (SNPs). In the past few years, the number of SNPs available has grown from 10,000 to 64,000 to 500,000; a one million SNP array may routinely be available by the time this volume is published. The cost of genotyping these SNPs has not increased with the quantity genotyped; at least one firm will genotype human DNA samples for 500,000 SNPs for about $600 per sample.

Unfortunately, "phenotyping" has not experienced any similar technical improvements. Collecting, standardizing, and entering information about an individual subject's characteristics remain time consuming and expensive. Before long, the widespread adoption of electronic medical records may provide researchers with easy and routine access to the health information contained in such records, although those records may not be sufficiently accurate and precise for research purposes. In any event, although electronic health records are beginning to make inroads into some medical systems, such as the Veterans Administration, they are not yet used widely enough, or are sufficiently uniform, to make much difference.

Although collecting phenotypic data and putting it into databases remain expensive, the costs of computerized data storage, analysis, and transfer have all fallen. Statistical methods for assessing the significance of correlations found through data mining have improved. And the rise of high-speed internet connections have also made transferring data anywhere in the world fast, cheap, and easy. As a result, some groups, notably the U.K. Biobank, are now willing to pay for collecting and storing this kind of data, and other groups, such as the NIH, are willing to think about paying for it. A database with broad phenotypic data and extensive genotypes expands enormously the scope of uses of the genomic biobank, making cheap genome wide association studies possible on any trait or condition for which data were collected.

The continuing reductions in the costs of both genotyping and computer memory have made it possible to envision much larger genomic biobanks, with much broader uses, than would have been feasible before. It is now reasonable to think about collecting samples and information from hundreds of thousands of people, analyzing their DNA for millions of features, and making the resulting primary data available for analysis to anyone. The nature of these biobanks, however, has some important implications for the subjects. Given the still-high cost of collecting samples and phenotypic data and the low costs of genotyping samples, it makes sense to construct genomic biobanks that can be used to study a wide range of problems. Thus, the research subjects will no longer be recruited solely, or even in part, because of a particular disease they have or that is found in their families. And the phenotypic data collected will be broad, allowing researchers to search for correlations between the genotype and as many traits or diseases as possible. Subjects, therefore, will not be participating in research aimed solely, or possibly even at all, at diseases they are interested in. Subjects will also be counted in the tens and hundreds of thousands, and not in the hundreds or few thousands. Both of these facts make it less likely that the subjects will have any significant continuing relationship with researchers.

Efforts to Create the New Type of Genomic Biobanks

The first population-wide genomic biobank to be widely discussed was proposed by the Icelandic firm deCODE Genetics in the late 1990s (17). deCODE proposed to create a broad research tool with data from the entire population of Iceland. Its initial plan had three components, each of which was intended to cover all 270,000 Icelanders: a genealogical database, a collection of DNA samples, and a system of electronic health records. At deCODE's urging and with the government's support, in December 1998 the Icelandic parliament passed the Health Sector Database bill authorizing the creation of such a clinical health record. The idea was that a biotechnology company (presumably deCODE) would pay the high price to create this database, which it would use for research, selling the results to other firms to recoup the costs (plus a profit) from its investment, and which the Icelandic health system would use for clinical purposes.

The deCODE plan created much discussion and controversy in Iceland and around the world. In Iceland it led to the formation of a research subjects' rights group, Mannvernd, and litigation against the database (36). The Icelandic government granted deCODE the license to create the Health Sector Database in early 2000. About six months after the license was granted, deCODE had a successful initial public offering, selling over $170 million in stock, but work on the database never started. In the notes to its consolidated financial statements in its 2005 annual report, deCODE reports that "we do not expect to operate the IHD [Icelandic Health Sector Database] under the terms of the Agreement" (12). Instead, deCODE has proceeded in a more conventional way, collecting information and samples from particular Icelandic families to study specific diseases.

Although in the long run the deCODE biobank plan failed, its early efforts prompted others to follow (53). Plans to create large-scale genomic biobanks were announced for northern Sweden (47), Estonia (15), Tonga (2), and Newfoundland (23), as well as a failed effort to create a private firm that would add a genomic component to the long-running Framingham Heart Study (32, 33). The Estonian project has produced, as a pilot project, a genomic biobank of samples and data from 10,000 Estonians, but it has thus far been unable to find financing for an expansion of the project. (14) The other projects have failed, although one project in Quebec may yet succeed (5), as may a currently stalled project in Taiwan (7). At least one such project, however, seems likely to succeed: the 500,000-person U.K. Biobank (54).

This project, funded by four government agencies in the United Kingdom and the philanthropic Wellcome Trust,

> aims to build a major resource that can support a diverse range of medical research intended to improve the prevention, diagnosis, and treatment of illness and the promotion of health throughout society.
>
> Lifestyle and environmental information, medical history, physical measurements, and biological samples are to be collected from about 500,000 people aged 40–69 at presentation and then, with informed consent, their health will be followed for many years through medical and other health-related records. The biological samples will be stored so that they can be used for a wide range of biochemical and genetic analyses in the future (54).

The U.K. Biobank undertook a 3800-person pilot project in Manchester from April to June 2006. It declared the pilot a success and announced plans to begin its general collection of samples and information from British residents in late 2006. The information will include the results of a physical examination, answers to a life-style survey, and clinical information from the donors' medical records. With the financial backing of the wealthy Wellcome Trust, its future success—at least in terms of collecting and analyzing samples and information—seems assured.

It is unclear how many other large-scale genomic biobanks will succeed. The Estonian plan continues to look for funding and Taiwan is planning a 200,000-person genomic biobank. (7) The U.S. NIH floated the idea of a very large U.S. genomic biobank (50). The proposal has been the subject of a cautiously favorable study by the Department of Health and Human Services Secretary's Advisory Committee on Genetics, Health, and Society (50). Pressures on the NIH budget make it unlikely that such a genomic biobank will be created anytime soon.

But the NIH is planning other systems that will have some of the characteristics of genomic biobanks. In 2006 NIH released a "Request for Information" about its plans to encourage (or require) accessible databases of the results of "Genome-Wide Association Studies" (GWASs) (40). NIH wants researchers who have "active research applications" that constitute a "study of human genetic variation across the entire human genome" to submit the data from those studies to NIH-operated databases. Those databases will then provide access to the data, of genotype and phenotype, to any qualified researchers.

This is something less than a genomic biobank, in that the NIH at least will not have physical specimens of DNA and will not be able to undertake additional genetic studies of the subjects. (The submitting researchers may, of course, still have DNA samples that would allow further analysis.) But the "whole genome analysis," which is most likely at this point to be an array-based analysis of many thousands of SNPs, will provide a great deal of genomic data for users who can then analyze it for any purposes they wish. These GWAS databases will still be limited by the fact that most of them will have been collected to study a particular disease or class of diseases, so the phenotypic data will be limited. On the other hand, the GWAS databases will be very inexpensive for NIH. The hard, and costly, work of collecting and analyzing DNA samples and phenotypic information will be done by someone else. The NIH role will be to receive the completed data and make it generally available. Data from many different studies could be made available so that some phenotypic data on a trait or disease of interest may be available on many donors from one or more studies. It would not be as useful as the U.K. Biobank, but should be vastly cheaper.

ETHICAL AND LEGAL CONCERNS

Ethical and legal concerns about the operation of genomic biobanks are not new. One might date the start of the controversy to a December 1995 article in the *Journal of the American Medical Association* questioning the use of previously collected stored tissue samples for unconsented research (8). Criticisms and suggestions for reform have continued (6, 16, 19, 20, 38, 49, 55), but the legal and ethical frameworks have not been reformed. The Common Rule, the system for regulating most human subjects research in the United States, has done a decent job of protecting research subjects from physical risks, but has largely ignored the fact that research subjects may have other interests they care about. The growth of genomic biobanks aggravates these gaps in the Common Rule as it enables biobanks to provide more samples with more associated information to a greater number of researchers for purposes of less interest to the subjects.

Thus far, the main response by the backers of genomic biobanks to these tensions has been either to ignore them or to try to structure their projects to avoid them. This approach is not only wrong, but risky, putting in jeopardy both the legal basis for these expensive endeavors and the political support for research that is necessary both for expensive genomic biobanks and for biomedical research generally. This section looks at four areas where genomic biobanks need to pay more attention to the legitimate interests of research subjects: protecting the subjects' (*a*) identity, (*b*) legal status as human subjects, (*c*) control over research uses, and (*d*) right to clinically important information.

"Confidentiality," "Anonymity," and the Protection of Identity

Protecting the identity of research subjects is a crucial issue in the ethical and legal situation of genomic biobanks, both for ethical reasons and because of the legal requirements of consent. Genomic biobanks use one of two methods to protect the personal information of research subjects: confidentiality through coded identifiers or anonymity through complete removal of all personal identifiers. But neither this "confidentiality" nor this "anonymity" means what most research subjects expect, because neither can be guaranteed to be effective. The problems with each are magnified in genomic biobanks.

Before examining the weaknesses in these systems for protecting identity, it is worthwhile to consider why research subjects may want their information, genomic and phenotypic, kept confidential. Consent documents often seem to assume that subjects are primarily worried about discrimination as a result of disclosure of their genomic (or phenotypic) information. Such discrimination is a worry for many, although possibly an exaggerated fear (18, 21, 25). But people may fear other consequences of disclosure of confidential health or genomic information. An individual may be worried about the consequences among family, friends, or others of disclosure of a past sexually transmitted disease, treatment for illegal drug use, psychiatric treatment, or an elective abortion. Comparing genomic information from two people thought to be related could lead to some unpleasant surprises, breaking families or blighting careers. Or, out of a strong desire for privacy, people may just not want personal information known to anyone beyond those they authorize for reasons having nothing to do with possible consequence of disclosure—the fact of disclosure alone may be profoundly distasteful.

It can be argued that the American public puts too much stress on the privacy of health-related information (22). Most Americans are protected, legally and socially, from health insurance or employment discrimination based on such information (18, 21). And it is unclear how many people have something deeply embarrassing in their health records, let alone their genomes. Personally, I would much rather have my health records publicly disclosed than my credit card or internet search records. My health records reveal little that is not obvious on seeing me (my doctor wants me to lose weight) and what else they reveal has little importance (a kidney stone in 2001, a possible allergy to sulfa drugs). Both the instrumental consequences of lost health information privacy and the social importance of that privacy will vary from individual to individual and from culture to culture. It may be that Americans should say, as Scott McNealy of Sun Microsystems famously did several years ago: "You have zero privacy anyway. Get over it" (51). But Americans have not gotten over it; protection of the privacy of their genotypic and phenotypic information is likely to be important to most potential biobank subjects.

Coded methods of protecting information rely on separating personal identifiers from available genomic and phenotypic data, but keeping a coded link between a subject's file and his identity. Guidance issued by the U.S. Office of Human Research Protections defines coded as

1. identifying information (such as name or social security number) that would enable the investigator to readily ascertain the identity of the individual to whom the private information or specimens pertain has been replaced with a number, letter, symbol, or combination thereof (i.e., the code); and
2. a key to decipher the code exists, enabling linkage of the identifying information to the private information or specimens (43).

Typically, a file on research subject Henry T. Greely would be assigned a meaningless code, such as "ak13dfsn54." The key linking the code to identifying information about Hank Greely would be kept in a secure location and used only as necessary. Researchers using the data in the biobank will not normally have access to the key: For them, the information will be potentially linkable but not linked. The advantage of using coded identifiers is that they make it possible, for good reason, to re-establish the link. The most important reason for that would be to add new information to a file thus preserving the ability to do prospective studies. If there were no way to connect subjects' data files with their identities, no new information about the subject could ever be added.

The problem with coded identifiers is that they cannot be made definitely secure.

Those who are authorized to use the links might use them for improper purposes, either as a result of a decision by a biobank (or the government that controls or regulates it) or as a rogue action by a biobank official. Third parties might steal the links by hacking into a computer file where they are contained, finding a notebook in which they are recorded, or watching an authorized person enter, say, a password that gives access to the links. One source of breaches of such confidential information, which has saturated the U.S. news recently, is the theft or loss of laptop computers that contain the sensitive information. Finally, one could determine a particular subject's file by submitting new information on that person and then searching the coded database for someone with that information. The information could be a unique value—a weight of 60.23539 kilograms, for example—or a unique or unusual characteristic (green-colored hair). (Coding as a way of protecting confidentiality also has all the problems of anonymization, discussed below.)

An alternative to coded identifiers is "anonymization"—stripping the information of all identifiers, trying to make it impossible for anyone, including the biobank managers, to establish individual identities. ["Anonymization" is different from "deidentification," a concept under federal health records privacy regulations required by the Health Insurance Portability and Accountability Act (HIPAA), which focuses on removing some very specific kinds of information. Depending on the context, HIPAA may (or may not) add to confidentiality requirements for research data, but I will not discuss them in this paper (39).] This comes at the price of being unable to add any new information to the file, but for some biobanks, that might be unimportant. Anonymization, though, has another problem. With a rich data set, it will not work—some, and potentially all, of the donors could be re-identified.

Consider a detailed 500,000-person biobank containing both genomic and phenotypic data. Assume I was known to be a subject in the biobank and someone wanted to determine which file had my information. Someone with access to the information in the biobank and to other genomic information about me should be able to establish which single file belonged to me (or my nonexistent, identical twin brother). They might get that genomic data from another database, such as a forensic database, a poorly protected research database, or, to an increasing degree as genetic information is in broader clinical use, poorly protected medical records. Someone really interested could get a DNA sample from me—from a licked stamp, a drinking glass, or some tissue—and have it genotyped for a few hundred dollars, but few will have to go to the genomic data; the phenotypic and demographic data will often be sufficient.

Consider just two pieces of information: date of birth and place of birth. (Files with such information would not be considered "de-identified" under HIPAA but might still be considered "anonymous" by researchers.) I was born on June 25, 1952, in Columbus, Ohio. Someone who wanted to find my file could look for someone in the 500,000-person biobank with my date of birth; if the biobank had roughly equal coverage from 40 years of births, one would expect about 35 people to share my birthdate. Unless, perhaps, I had been born in New York City, one would expect none of them to share my city of birth. But even if there are two or more people in the biobank with the same date and place of birth, it may take only a little investigation to distinguish them. My cousin Mike and I were born on the same day in the same city; he is 5 feet, 8 inches tall, I am 6 feet, 2 inches tall. Similar observable physical characteristics, not just height, but sex, race, weight, hair color, and so on, would almost certainly be sufficient to identify my file.

A similar process could work, with somewhat more difficulty, in reverse. One could start with a file of an unknown person and try to establish his identity from the data, starting with data from other accessible genomic databases that did not protect donor identity well and then looking at other important information. Eliminating name, mailing address, and social security number does not eliminate identifiers; it just eliminates the easiest identifiers, making the search somewhat more difficult and expensive. But note that to the extent that the other information is available online, the search becomes easier and cheaper. Genealogists are keen to put more birth, death, and family records on the Internet, potentially making it much easier to find out who was born in Columbus, Ohio on June 25, 1952.

These weaknesses in anonymity are not just speculation. Professor Latanya Sweeney, a computer scientist at Carnegie Mellon University, did the experiment. William Weld, then governor of Massachusetts, released certain medical records of state employees to researchers. He said privacy would be protected because all identifiers had been eliminated from the records. A few days after his announcement, Sweeney had Governor Weld's records delivered to his office. She used his date of birth and the zip code of his home to locate one likely file. She consulted the voter list for his town and found that he was the only person in his zip code with his date of birth. (This demonstration predated the HIPAA regulations and the information released would not be considered "de-identified" under HIPAA today.)

One could, of course, be more careful about putting possibly identifying data into the file. Date of birth might be eliminated; place of birth might be limited to large regions (states, provinces, or nations). Unfortunately, the omitted or "fuzzed" data might turn out to be important for the research. The more the data is removed or obscured, the more scientific value is lost; the more data is kept, the less real the anonymity (35). Sweeney has proposed some methods that may make most records "pretty" anonymous while keeping them "pretty" useful, but there is no perfect solution (52).

The increase in genomic data, as well as the increase of computerization of other records about individuals, will only make identifying "anonymous" biobank files easier and easier. Most genomic biobanks expect to give broad access to their information to qualified researchers; the broader the access, the greater the chance of the information falling into the hands of someone with the desire and ability to break the biobank's anonymity for some research subject. (Broader access would also lead to less effective methods of tracking and potentially punishing those who made such an attempt.) At least two consequences follow from these weaknesses in protecting identity. Neither one means that human subjects research cannot be ethically done or that efforts to protect personal information have no value, but they both speak to how biobank confidentiality should be described and performed.

First, consent forms need to stress that confidentiality cannot be guaranteed, even with anonymous samples. Consent forms need to be honest and say things like "we will try to keep your identity secret, but we cannot guarantee it. Someone who really wanted to determine the identity of a subject in this database might well be able to do so." Researchers, and biobank promoters, will resist such language because it is likely to increase resistance from prospective subjects and from institutional review boards (IRBs), but anything less is not honest. It may be that research will have to proceed with fewer subjects, with people who are willing to be "information altruists," in the words of Kohane and Altman (31). It should not proceed with people who have been misinformed about the risks that their identities could be determined from the information they give researchers.

Second, anonymizing samples, far from being a simple way to protect subjects, is itself both undesirable and unethical. It should be abandoned as a way to protect donors' identities. If the phenotypic data are sufficiently broad that the person's identity can be deduced with relatively minor effort, anonymity offers little additional protection. On the other hand, anonymity makes it difficult for researchers to recontact donors and provides an excuse for them to avoid trying, which, as argued below, is ethically required in some circumstances. (Researchers could, of course, presumably make the same effort to reconnect data and identities, although they might not be successful.) Anonymity also effectively takes away the right of a research subject to withdraw from research, and anonymity makes it easier for researchers, biobank managers, IRBs, and others to overlook the problems of protecting information. After all, if the information is anonymous, the stakes seem lower. Given that true anonymity precludes adding new data to a subject's file, its costs seem to outweigh its very minor benefits.

"Human Subjects"

"Human subjects research" in the United States is governed by the so-called Common Rule, a regulation adopted by many (but not all) federal agencies and binding on those agencies, on researchers using funds from those agencies, and on researchers working for institutions that have given those agencies "assurances" that all human subjects research at their institutions will follow the regulation (58). A slightly different version of the Common Rule covers research conducted under the authority of the Food and Drug Administration, under, for example, an Investigational New Drug exemption (57). Violation of the Common Rule can make an institution ineligible for federal research funding, a very precarious situation for a research university. (On the other hand, active enforcement of the Common Rule by the federal government is currently of doubtful efficacy.) Regulatory details will vary in other national and professional rules for human subjects research, but each will have to have some way to define what research they cover.

In the United States, most research classified by the Common Rule as human subjects research must be reviewed and approved by an IRB. The IRB is required, among other things, to determine that the potential general benefits of the research outweigh the risks to the individual subjects. It must also assure in most cases that the researchers proceed only with the informed consent of the research subjects. If what researchers are doing does not qualify as "human subjects research," neither these requirements nor any other significant federal rules apply.

Genomic biobanks in the United States may seek to circumvent federal regulation by defining their activities as not involving "human subjects research." Although the subjects of their collections are undeniably human, the biobank's activities may be argued to escape the definitions of the Common Rule. But when research is interpreted as not involving "human subjects," human subjects protections disappear—no weighing of risks and benefits, no obligation of informed consent, no review by an IRB at all.

The GWAS initiative at NIH, as proposed in 2006, seems to be attempting such a circumvention. Biobanks in other nations may try similar methods to avoid their own national (or, in the case of the European Union, supranational) regulations. Because of the limitations of confidentiality and anonymity discussed above, these efforts should not be legal (at least in the United States). In any event, they are, I believe, clearly unethical.

Legal issues. The Common Rule defines "human subject" as "a living individual about whom an investigator (whether professional or student) conducting research obtains (1) Data through intervention or interaction with the individual, or (2) Identifiable private information." Private information is "information which has been provided for specific purposes by an individual and which the individual can reasonably expect will not be made public (for example, a medical record)." And the regulation effectively defines "identifiable" by explaining, "Private information must be individually identifiable (i.e., the identity of the subject is or may readily be ascertained by the investigator or associated with the information) in order for obtaining the information to constitute research involving human subjects." (The Common Rule also contains an exemption from its requirements for some human subjects research, including research on some previously collected samples; the requirements for an exemption are substantially equivalent to the absence of "identifiable private information" and will not be discussed here.)

The Common Rule thus means that human subjects research involves any research that includes collecting information directly from a person as that information will come either from an "intervention," such as obtaining a DNA sample, or an "interaction," such as asking the person questions. Thus, the creation of a genomic biobank by collecting information from subjects will be human subjects research.

The situation is more complicated when someone else collected the information or samples, as with genomic biobanks created with information or samples from other researchers or with researchers who are not part of a genomic biobank but who are using information from it. If the researchers have no contact (no interventions or interactions) with the subjects, the research only falls within the Common Rule if it involves identifiable private information. There seems no question whether the kinds of information in a genomic biobank—DNA analyses, health information, and so on—would constitute "private information." It seems directly akin to the "medical records" singled out in the regulation. The issue then becomes whether that information is "identifiable": "the identity of the subject is or may readily be ascertained by the investigator or associated with the information." If the information is not "identifiable," then the research does not involve human subjects and no IRB approval or other federal requirements (including informed consent) apply.

The meaning of "identifiable" thus becomes quite important. We are not left with only the language of the regulation for guidance. The U.S. Department of Health and Human Services includes an Office for Human Research Protections (OHRP). This office, established in June 2000 to replace the earlier Office for Protection from Research Risks, has broad responsibility for enforcing the human subjects protections in research involving the Department of Health and Human Services. It has no authority over research involving other federal agencies, but its conclusions may be given substantial weight by such agencies. In August 2004 OHRP issued a document entitled "Guidance on Research Involving Coded Private Information or Biological Specimens" (43). This guidance seemed directly applicable to genomic biobanks. In it, OHRP stated that it "does not consider research involving only coded private information or specimens to involve human subjects as defined under 45 CFR 46.102(f) if the following conditions are both met:

> 1. the private information or specimens were not collected specifically for the currently proposed research project through an interaction or intervention with living individuals; and
> 2. the investigator(s) cannot readily ascertain the identity of the individual(s) to whom the coded private information or specimens pertain because, for example:
>
> . . .
>
> (a) the investigators and the holder of the key enter into an agreement prohibiting the release of the key to the investigators under any circumstances, until the individuals are deceased (note that the HHS regulations do not require the IRB to review and approve this agreement); [or]
> (b) there are IRB-approved written policies and operating procedures for a repository or data management center that prohibit the release of the key to the investigators under

> any circumstances, until the individuals are deceased."

Thus, under this Guidance, a researcher receiving only coded information from a genomic biobank would not be doing "human subjects research" and would not need IRB approval (or, as far as federal law requires, informed consent) for his research as long as either he had agreed with the "holder of the key" that the holder would not release the key to him until the subject is dead or the biobank had an IRB-approved policy prohibiting such release.

OHRP's intent to cover such biobanks seems clear:

> This guidance applies to existing private information and specimens, as well as to private information and specimens to be collected in the future for purposes other than the currently proposed research. The following are examples of private information or specimens that will be collected in the future for purposes other than the currently proposed research: (1) medical records; and (2) ongoing collection of specimens for a tissue repository.

Thus, the Guidance seems to mean that neither genomic biobanks assembled from information collected by other researchers nor research undertaken by investigators getting information from such biobanks would be "human subjects research."

Read that way, the OHRP Guidance allows information where the donors' identities can be readily ascertained to be treated as if it does not involve human subjects research. As demonstrated below, this violates the Common Rule, a duly adopted regulation that takes precedence over mere agency guidance. (Alternatively, if one reads the Common Rule's definition broadly, as I argue is appropriate, and reads that definition into the Guidance, the Guidance becomes effectively meaningless.)

The Common Rule defines identifiable private information as the information where "the identity of the subject is or may readily be ascertained by the investigator or associated with the information." As the discussion in the section on confidentiality and anonymity points out, coded information—and even anonymized information—may still allow the subject's identity to be "readily ascertained by the investigator." If the genomic biobank allows submission of new information that will be added to a subject's file, the investigators can "readily identify" a subject by arranging for distinctive information to be added to his file.

In any event, the information in the file will itself, in at least some cases, enable an investigator to readily identify a subject. In such cases, the identity would also be "associated with the information," as it would be deducible from the information itself. Thus, access to the key is irrelevant in those cases whether or not the subjects can be identified; the data itself makes them identifiable. It may be that some of the subjects in the biobank cannot be readily identified through the data in the biobank, but for any of the subjects who could be so identified, the Common Rule would apply. Given the difficulty of determining which collections of information about subjects could or could not easily lead to identification, a biobank should assume that all the research is "human research subjects" protected by the Common Rule.

The broad sweep of this Guidance has not gone unnoticed. NIH, in its Request for Information about its GWAS initiative, states that it expects that

> genotype and phenotype datasets submitted and stored in the GWAS data repository . . . will be provided for research purposes through an NIH Data Access Committee (DAC). Investigators seeking data from the GWAS data repository will be asked to submit a Data Use Certification . . . [which] should include a brief description of the proposed research use of the requested GWAS dataset(s) (40).

The researchers who initially submitted the data to the GWAS must certify that "the identities of research participants will not be disclosed to the GWAS data repository or to secondary users of the coded data without appropriate institutional approvals." The investigators requesting access to the data must make several promises, including a promise not to "attempt to identify individual participants from whom data within a dataset were obtained." At no point does the document say that those investigators must subject their research to IRB review. This is especially disturbing as the proposed GWAS policy does not even meet the already minimal requirements of the OHRP Guidance: It does not require any promise of absolute nondisclosure of the key, but only "assurance" of no disclosure without "appropriate institutional approvals."

Ethical issues. Even if the failure to apply the protections of the Common Rule or similar systems of human subjects protection to genomic biobanks were legal, it would not be ethical. First, as we have seen, coded or anonymized data cannot be guaranteed to be nonidentifiable and, thus, those affected are subject to possible harms from disclosure of personal information of the exact kind the human subjects regulation tries to prevent. That fact alone requires regulation of this research.

But equally importantly, the exemption of truly anonymous information from human subjects protections is flawed. That exemption seems to stem from the idea that subjects cannot be harmed by information not linked to them. A person cannot be subjected to employment or insurance discrimination or be embarrassed by publication of some items of his medical history if he is not identified. This approach, however, takes too narrow a view of less tangible interests of research subjects that deserve protection (49). It is fundamental (as discussed in the next section) that a person has an interest in consenting or not consenting to be part of research. This should apply even if his "participation" is limited to the use of his previously collected data or biological samples.

Similarly, a person may be willing to have her samples or information used in some research, but not in some other research that she finds objectionable. If the activity is not classified as human subjects research, the Common Rule gives her no such protection. (Other laws, however, may provide some protection, but because of the power of the Common Rule, state and other federal law in this area is undeveloped.) One person might be understandably outraged to learn that her DNA and personal data were used, without her knowledge or consent, in research on race, genetics, and violence. Another might be offended to learn that his DNA was used in research on the evolution of humans, a theory that he rejects for religious reasons. (We see in the area of stem cell regulation that both guidance documents and state regulations require explicit donor consent for some kinds of research, such as somatic cell nuclear transfer.) By focusing only on identifiable private information, the Common Rule and similar regulations neglect these interests. As discussed in the next section, working out a practicable scheme for protecting such interests may be complicated, but it cannot ethically be ignored.

"Informed Consent" and Control Over Research Uses

One of the first efforts at a genomic biobank, deCODE's Icelandic Health Sector Database substituted "presumed consent" for "informed consent." At deCODE's request, the Icelandic statute authorizing the database provided that all Icelanders would be in the database unless they filed a government-provided form, within a statutorily defined period, objecting to their inclusion. As even objecting Icelanders died, their data would be included in the database and children or other incompetent persons would be included unless their parents or guardians objected on their behalf. This presumed consent generated much of the Icelandic and international objections to the scheme and triggered some of the litigation that slowed the project.

Informed consent is an essential requirement for human subjects research around the world. Since the Icelandic project, other proposed genomic biobanks have carefully stated that they will require informed consent, but genomic biobanks, by their very nature, find it very hard to satisfy this requirement. It is may be effectively impossible for them to get true informed consent; they can only get a redefined, watered-down version of informed consent. The fault is not one of bad policies, but lies in the nature of genomic biobanks.

The very first directive of the Nuremberg Code, created during trials of Nazi "scientists," states

> The voluntary consent of the human subject is absolutely essential. This means that the person involved should have legal capacity to give consent... *and should have sufficient knowledge and comprehension of the elements of the subject matter involved as to enable him to make an understanding and enlightened decision* (42). [emphasis added]

The World Medical Association's Helsinki Declaration states, "In any research on human beings, each potential subject must be adequately informed of the aims, methods, anticipated benefits and potential hazards of the study and the discomfort it may entail" (56). And the U.S. Common Rule, which governs most human subjects research in the United States, states that "no investigator may involve a human being as a subject in research covered by this policy unless the investigator has obtained the legally effective informed consent of the subject...." The Common Rule defines the "basic elements of informed consent" to include, among other things, a "description of any reasonably foreseeable risks or discomforts to the subject; [and a] description of any benefits to the subject or to others which may reasonably be expected from the research...." (58).

Traditional genetic research projects collected DNA samples and information from people to try to understand one disease or a small number of closely related diseases. Someone participating in research to identify genes linked to familial colon cancer could have explained to them "any reasonably foreseeable risks or discomforts" or "any benefits to the subject or to others which may reasonably be expected from the research...." The U.K. Biobank and its proposed relatives are expected to be used for many years for many different kinds of research. Potential subjects cannot be informed of the specific risks and benefits of the research because the biobanks do not know what those risks or benefits may be; they do not even know what the research topics will be. The Nuremberg Code, Declaration of Helsinki, and Common Rule each envisioned a specific, discrete research project, not a tool for use on unforeseen and unforeseeable future projects.

This problem with so-called blanket consent has long been noted, but has not been satisfactorily resolved (8, 16, 38). One could require actual individual informed consent before making any use of a genomic biobank, but that would make the biobanks nearly useless. On one hand, the genomic biobanks could be created, at enormous cost, to study just one particular issue as to which true informed consent could be given and limited to that issue. This seems economically impracticable, and frankly wasteful, in almost every case. On the other hand, the biobanks could be required to get new informed consent from each subject for each new research project that would use that subject's information. That also seems grossly impracticable, requiring the biobank to remain in regular communication with all its subjects and to interrupt their lives scores, or even hundreds, of times to ask for permission to use the previously collected data and samples for new research projects, most of which would be obscure and confusing to the subjects. This would not only be prohibitively expensive for any large biobank, but probably profoundly annoying to the subjects. One might even argue that government action forcing time-by-time consent could violate a person's right, following from her autonomy, to agree in advance to any and all uses of her information and materials.

Therefore, IRBs and other regulators have been willing to allow blanket or very broad consent in appropriate cases, even though that kind of consent seems to violate Nuremberg, Helsinki, and the Common Rule. Those participating are told that their samples and data may be used by researchers all over the world for a wide range of projects. They get information on general categories of foreseeable problems (discrimination, leaked information) and benefits (better understanding of disease and hoped-for treatments), but they get no information about the specific research that will be done with their samples and information and no real control of how their materials will be used.

There should be a better system, one that gives subjects some more information about and control over their participation in at least some unusually risky or controversial research. It is clear that, at least in some cases, research subjects do care, very much, about the uses to which their information is put.

Consider the case of the Havasupai. The Havasupai are a federally recognized Native American tribe whose reservation is in part of the Grand Canyon. Like many Native American tribes in the United States, the Havasupai suffer from a high rate of type 2 diabetes. In the 1990s, researchers from Arizona State University asked the tribe and many of its members to participate in research aimed at finding genetic associations with type 2 diabetes. Both the tribe and many individuals consented. More than a decade later, both the tribe and a group of individual subjects sued Arizona State University and various researchers, charging that, without their permission and against their expressed wishes, the researchers had used Havasupai samples and data to study both the genetics of schizophrenia and the history and migrations of the Havasupai people (11, 48). This litigation, now in mediation, does not involve blanket consent and it includes other claims of theft and deceit. However, it does indicate that, at least in some circumstances, research subjects can be very unhappy about uses of their samples and data they did not know about and did not like.

In the Havasupai case, at least according to the plaintiffs' allegations, actual fraud was involved. In more benign cases involving blanket consent and biobanks, at least two different strategies, alone or in combination, can help keep research subjects from feeling unhappy and betrayed by unexpected uses of their materials. The first strategy is continuing general communication with research subjects. Something like a newsletter, either physical or electronic, could be sent to research subjects, informing them of the results of past research and the plans for new research. Research that had special risks or that was seen as likely to raise special concerns among some subjects could be highlighted. Subjects could be told to contact the genomic biobank for further information about those specific projects to determine if they wanted to "opt out" of a particular effort. Alternately, for a computer-savvy group of subjects, a regularly updated Web page might serve a similar function.

The second approach is to review specific proposals to see if they are likely to be controversial among subjects. For example, an IRB could be asked to review proposals for research with information from a genomic biobank. If it concluded that there was a reasonable likelihood that a significant number of the research subjects might object to participating in that research, it could require that those investigators reconsent all subjects whose data they wanted to use. A committee of research subjects, acting independently or in conjunction with the IRB, could also be useful for identifying particularly sensitive research topics.

These approaches, separately or in combination, would mean that almost all research projects would go forward with data from almost all research subjects. Additional effort would be required only for those topics that were viewed as likely to be significantly objectionable, or those individuals who saw the newsletter and wanted more information about a particular project. But the advantage of such an approach is that it avoids having research subjects discover too late that their DNA and medical information was used without their knowledge or consent in research they dislike, or despise. Those situations, and the loss of trust that would accompany them, must be guarded against. These methods for some intermediate level of consent to specific projects by research subjects would be useful in preventing them.

"Justice," "Protection," and the Return of Clinically Significant Information

The systems for protecting human subjects require that those people willing to be guinea pigs for the advancement of science and medicine be treated justly and that their interests be protected. Two provisions of the Declaration of Helsinki dealing with nonclinical research seem particularly apropos:

> In the purely scientific application of medical research carried out on a human being, it is the duty of the physician to remain the protector of the life and health of that person on whom biomedical research is being carried out.
>
> In research on man, the interest of science and society should never take precedence over considerations related to the wellbeing of the subject (56).

Yet much biomedical research does not protect the medical well-being of the subject, not through subjecting him to additional risks, but by not warning him of risks discovered in the course of the research.

Consider a situation where an analysis of a subject's DNA showed that the subject carried an allele associated with a very high risk of a serious disease, but a disease for which useful preventive measures were possible. Thus, for example, genotyping might reveal that a subject has a pathogenic mutation in the MLH1 or the MSH2 genes, linked to a very high risk of hereditary nonpolyposis colon cancer. People with such mutations should undergo frequent colonoscopies and should consider prophylactic colectomies. Discovery of such a mutation in the subject was not the purpose of the biobank, but was an incidental, but important, finding. Disclosure of this risk by the biobank to the subject might save the subject's life, but there is, as yet, no regulation or even ethical consensus requiring that researchers return medically significant individual information to research subjects. In fact, currently, researchers will often include in the informed consent for a project the statement that they will not return any individual medical information to the subject.

This seems, at least in extreme situations, immoral, possibly illegal, and certainly unwise. In one sense, this is a special application of a duty to rescue. When one person can, without undue risk or difficulty, save another person, he may have a moral duty to do so. For a third party, such as bystander, that moral duty may or may not be a legal duty; typically, it is not legally required in the United States but is in many countries.

Researchers are not, however, bystanders. They are accepting benefits of information and tissue from research subjects in a situation where those subjects trust them. They have legal duties to protect subjects from risky research. In such a setting, a court might hold that the researcher was not a third party but instead a fiduciary for the patient, with duties to protect the patient's interests that go beyond those of bystanders. This should be clear when the researcher is also the patient's treating physician. A physician owes his patients a duty to advise them of health risks he perceives, whether or not they were the purpose for a particular consultation. A treating physician does not lose that duty by donning the extra hat of researcher (37). There is some American law that holds that researchers are in a fiduciary relationship to patients even when they are not serving as the patient's physician (24), although this point remains uncertain.

Regardless of the moral and legal obligations, it would be extremely unwise for researchers not to disclose such risks. Consider what happens after the first lawsuit by the bereaved family of a research subject whose life would have been saved had researchers revealed a risk they discovered. Whether or not the plaintiffs win, those researchers and their institution will be branded as heartless, interested in subjects only as laboratory animals, and all biomedical research will feel the fallout.

Happily, this issue is beginning to receive the attention it deserves. Several recent articles about genetic research endorse the idea of some kind of obligation to return medically significant results (44, 45). A similar issue is being discussed actively in neuroimaging research, as a surprisingly high percentage (between 18–37%) of healthy research subjects show abnormalities in their brains during brain scanning (27). But consensus will not come easy, in part because the details are very hard (46).

A policy or a regulation that required the return of significant information would need to resolve many difficult issues. How significant must the information be, and significant in what ways? In the case of genomic information, this might involve the seriousness of the disease, the confidence of the association between the genomic variations and the disease (1 study, 2 studies, or 10 studies), the penetrance of the alleles, and the availability of useful interventions. Some subjects might like to get information about their genomes not because of their own medical situation, but to learn something about their relatives or their current or future children. Many research laboratories do not meet the legal requirements for clinical laboratories under the Clinical Laboratory Improvement Amendments Act (CLIA) (9). What should they do with potentially important research findings about an individual? How hard does a researcher have to try to locate the subject, who may well have moved since the research was begun? Does the researcher have to keep checking the literature to see if genomic variations identified in the past in subjects were recently established as linked to important diseases? And how long does the obligation to inform research subjects last: for a few years, for the life of the research project, for the life of the individual? How should the unsettling news—"we're from a research project you participated in and we have some important information about your genetic health risks"—be conveyed and are the researchers responsible for any costs of genetic counseling or medical treatment? And, not least importantly, how can the return of information be offered in a way that preserves the donor's right "not to know"?

All these difficult questions become even harder in the context of genomic biobanks. These biobanks might have to deal with hundreds of thousands of people. The biobanks' intensive genotyping, and possibly ultimately sequencing, will be likely to turn up information about many different diseases or disease risks. And the genomic biobanks are expected to be around for decades, collecting more information, and discovering more health risks, every year. The questions will not get any easier; the time for genomic biobanks to begin confronting them is now.

CONCLUSION

This article has not dealt with all the tough ethical issues that will plague genomic biobanks. In some contexts questions will arise about community consultation or group consent (28–30, 41). Issues of commercialization and the sharing of financial benefits with research subjects, individually or in groups, continue to be controversial (1, 26). And, of course, how well, and how efficiently, any of these ethical prescriptions can be implemented remains a very open issue (13).

Those issues, and the issues discussed in the body of this article, are not new or unique to genomic databases, but they are all magnified in genomic databases. The size, the cost, the breadth, the desired broad researcher access, and the likely high public profile of genomic databases will make these issues especially important to them. Dealing with these issues will be both intellectually and politically difficult, time-consuming, inconvenient, and possibly expensive. But it is not a solution to say that "anonymity" means only "not terribly easy to identify," that "human subjects research" does not include research with data and samples from living humans, or that "informed consent" is satisfied by largely ignorant blanket permission. Humpty Dumpty could make words mean whatever he chose, but the nursery rhyme told his fate: "Humpty Dumpty sat on a wall. Humpty Dumpty had a great fall. And all the king's horses and all the king's men, couldn't put Humpty together again." Genomic biobanks, in particular, and biomedical research more generally, have too much promise to let them, too, have "a great fall."

DISCLOSURE STATEMENT

The author has been for many years an unpaid member of the Ethics Advisory Committee of Affymetrix, which makes tools that genomic biobanks might use. The author is on the Scientific Advisory board of genomic biobanks projects for the Veterans Administration and for Kaiser Permanente, as well as the ethics advisory board for a VA project. The author has participated in workshops and conferences about such biobanks and has occasionally been paid NIH honoraria to do so.

ACKNOWLEDGMENTS

The author would like to thank his research assistant, Sean Rodriguez; his colleagues in Stanford's Center for Integration of Research on Genetics and Ethics; and reviewer Kathy Hudson for their help.

LITERATURE CITED

1. Bear JC. 2004. What's my DNA worth, anyway? A response to the commercialization of individuals' DNA information. *Perspect. Biol. Med.* 47:273–89
2. Burton B. 2002. Proposed genetic database on Tongans opposed. *Br. Med. J.* 324:443
3. Cann HM, de Toma C, Cazes L, Legrande MF, Morel V, et al. 2002. A human genome diversity cell line panel. *Science* 296:261–62
4. Carroll L, (Dodgson CL). 1871. *Through the Looking Glass and What Alice Found There.* New York: Macmillan
5. CARTaGENE. 2005. *A Genetic Map of Quebec.* **http://www.cartagene.qc.ca/index2.cfm?lang=1**
6. Charo RA. 2006. Body of research ownership and use of human tissue. *N. Engl. J. Med.* 355:1517–19
7. Chuang S. 2006. Dilemmas of research. *Taiwan Rev.* 56(7). **http://taiwanreview.nat.gov.tw/ct.asp?xItem=22901&CtNode=128**

8. Clayton EW, Steinberg KK, Khoury MJ, Thomson E, Andrews L, et al. 1995. Informed consent for genetic research on stored tissue samples: consensus statement. *JAMA* 274:1786–92
9. *Clinical Laboratory Improvements Amendments Act of 1988*, codified at 42 U.S. Code §263a
10. Coriell Inst. *Coriell cell repositories.* **http://ccr.coriell.org/**
11. Dalton R. 2004. When two tribes go to war. *Nature* 430:500–2
12. deCODE Genetics. 2005. *Annual Report Form 10-K, Fiscal Year 2004*, p. 73
13. Emanuel EJ, Wood A, Fleischman A, Bowen A, Getz KA, et al. 2004. Oversight of human participants research: identifying problems to evaluate reform proposals. *Ann. Int. Med.* 141:282–91
14. *Estonian Genome Proj. Found.* What is the genome project? **http://www.geenivaramu.ee/index.php?lang=eng&sub=58&PHPSESSID=01c729c136e4132459d1fe53b67c4fd8**
15. Frank L. 2000. Estonia prepares for national DNA database. *Science* 290:31
16. Greely HT. 1999. Breaking the stalemate: a prospective regulatory framework for unforeseen research uses of human tissue samples and health information. *Wake For. Law Rev.* 34:737–66
17. Greely HT. 2000. Iceland's plan for genomics research: facts and implications. *Jurimetrics* 40:153–91
18. Greely HT. 2001. "Genotype discrimination": the complex case for some legislative protection. *Penn. Law Rev.* 149:1483–1505
19. Greely HT. 2001. Human genomics research: new challenges for research ethics. *Perspect. Biol. Med.* 44(2):221–29
20. Greely HT. 2001. Informed consent and other ethical issues in population genetics. *Annu. Rev. Genet.* 35:785–800
21. Greely HT. 2005. Banning genetic discrimination. *N. Engl. J. Med.* 353:865–67
22. Greely HT. 2006. Electronic transfer of health records in clinical care and research: an irresistible force meets a moveable object. In *Biological Material and Electronic Data: Much Ado About Nothing?* ed. BM Knoppers, C Hervé, pp. 35–50. Bordeaux: Les Études Hos.
23. Greenwood J. 2000. The business of genes: Newfoundland hopes to reap the benefits after its genetic heritage has helped decode the human genome. *Natl. Post*, June 24, D.1
24. *Grimes v. Kennedy Krieger Inst., Inc.* 2001. 782 A.2d 807 (Md.)
25. Hall MA, Rich SS. 2000. Laws restricting health insurers' use of genetic information: impact on genetic discrimination. *Am. J. Hum. Genet.* 66:293–307
26. HUGO Ethics Comm. 2000. *Statement on benefit sharing.* **http://www.hugo-international.org/Statement_on_Benefit_Sharing.htm**
27. Illes J, Desmond JE, Huang LF, Raffin TA, Atlas SW. 2002. Ethical and practical considerations in managing incidental findings in functional magnetic resonance imaging. *Brain Cogn.* 50:358–65
28. Int. HapMap Consort. 2004. Integrating ethics and science in the International HapMap Project. *Nat. Rev. Genet.* 5:467–75
29. Juengst ET. 1998. Groups as gatekeepers to genomic research: conceptually confusing, morally hazardous, and practically useless. *Kennedy Inst. Ethics J.* 8:183–200
30. Juengst ET. 1998. Group identity and human diversity: keeping biology straight from culture. *Am. J. Hum. Genet.* 63:673–77
31. Kohane IS, Altman RB. 2005. Health-information altruists—a potentially critical resource. *N. Engl. J. Med.* 353:2074–77
32. Kolata G. 2000. Boston U. founds company to sift leading heart data. *New York Times*, June 17:A10

33. Kolata G. 2000. Company won't get access to data of famed heart study. *New York Times*, Dec. 29:A19
34. Lavori PW, Krause-Steinrauf H, Brophy M, Buxbaum J, Cockroft J, et al. 2002. Principles, organization, and operation of a DNA bank for clinical trials: a Department of Veterans Affairs cooperative study. *Contr. Clin. Trials* 22:222–39
35. Lin Z, Owen AB, Altman RB. 2004. Genomic research and human subject privacy. *Science* 305:183
36. Mannvernd. 2000. *Assoc. Icelanders Ethics Sci. Med.* **http://www.mannvernd.is/English/index.html**
37. *Moore v. Regents Univ. Calif.* 1990. 51 Cal. 3d 120; 793 P.2d 479; 271 Cal. Rptr. 146
38. Natl. Bioethics Advis. Comm. 1999. *The Use of Human Biological Materials in Research.* Bethesda, MD: Natl. Bioethics Advis. Comm.
39. Natl. Inst. Health. 2004. *Research Repositories, Databases, and the HIPAA Privacy Rule.* **http://privacyruleandresearch.nih.gov/research_repositories.asp**
40. Natl. Inst. Health. 2006. *Request for Information (RFI): Proposed policy for sharing of data obtained in NIH supported or conducted Genome-Wide Association Studies (GWAS)*
41. North Am. Reg. Comm., Hum. Genome Divers. Proj. 1997. Proposed model ethical protocol for collecting DNA samples. *Houston Law Rev.* 33:1431–73
42. Nuremberg Code. 1949. *Trials of War Criminals before the Nuremberg Military Tribunals under Control Council Law No. 10, Vol. 2*, pp. 181–82. Washington, DC: USGPO
43. Off. Hum. Res. Protection. 2004. *Guidance on research involving coded private information or biological specimens.* Washington, DC: US Dep. HHS
44. Pullman D, Hodgkinson K. 2006. Genetic knowledge and moral responsibility: ambiguity at the interface of genetic research and clinical practice. *Clin. Genet.* 69:199–203
45. Ravitsky V, Wilfond BS. 2006. Disclosing individual genetic results to research participants. *Am. J. Bioethics* 6(6):8–17
46. Renegar G, Webster CJ, Stuerzebecher S, Harty L, Ide S, et al. 2006. Returning genetic research results to individuals: points to consider. *Bioethics* 20:24–36
47. Rose H. 2003. An ethical dilemma: the rise and fall of UmanGenomics—the model biotech company? *Nature* 425:123–24
48. Rubin R. 2004. Indian givers. *Phoenix New Times*, May 27. **http://www.phoenixnewtimes.com/Issues/2004–05–27/news/feature.html**
49. Saver RS. 2006. Medical research and intangible harm. *Univ. Cinn. Law Rev.* 74:941–1012
50. Secretary's Advis. Comm. Genet. Health Soc. 2007. *Policy issues associated with undertaking a large U.S. population cohort project on genes, environment, and disease.* Final rep. Bethesda, MD: HHS
51. Sprenger P. 1999. Sun on privacy: get over it. *Wired News*, Jan. 26. **http://www.wired.com/news/politics/0,1283,17538,00.html**
52. Sweeney L. 2002. k-Anonymity: a model for protecting privacy. *Int. J. Uncert. Fuzziness Knowl.-Based Syst.* 10:557–70
53. Taubes G. 2001. Your genetic destiny for sale. *Technol. Rev.* 104(2):41–46
54. UK Biobank. 2005. *UK Biobank: Improving the Health of Future Generations.* **http://www.ukbiobank.ac.uk/**
55. Winickoff DE, Winickoff RN. 2003. The charitable trust as a model for genomic biobanks. *N. Engl. J. Med.* 349:1181–84

56. World Med. Assoc. 1964. *Declaration of Helsinki: Recommendations Guiding Physicians in Biomedical Research Involving Human Subjects.* Helsinki., revised Tokyo, Jpn, 1974; Venice, Italy, 1983; Hong Kong, China, 1989; Somerset West, S. Afr. 1996; Edinburgh, UK 2000
57. 21 Code Fed. Reg. Part 50 (FDA version of the Common Rule)
58. 45 Code Fed. Reg. Part 46 (DHHS version of the Common Rule)

Cumulative Indexes

Contributing Authors, Volumes 1–8

A

Abel L, 8:163–92
Agarwal AK, 7:175–99
Ahmad F, 6:185–216
Akey DT, 6:287–312
Albers S, 1:139–77
Albertson DG, 6:331–54
Allard MW, 4:119–41
Altshuler D, 4:257–91
Anderlik MR, 2:401–33
Anderson MG, 6:15–44
Andrews NC, 1:75–98
Anholt RRH, 7:339–67
Annilo T, 6:123–42
Antonarakis SE, 8:17–35
Arkin AP, 3:341–69
Arnold BE, 3:415–32
Ashley T, 5:317–49
Ashurst JL, 4:69–88

B

Badano JL, 7:125–48
Bale AE, 3:47–65
Ballabio A, 6:355–79
Balling R, 2:463–92
Bannert N, 7:149–73
Barbujani G, 5:119–50
Bashirova AA, 7:277–300
Beales PL, 7:125–48
Beaudet AL, 5:479–510
Beck S, 1:117–37
Birney E, 3:293–310
Bisgrove BW, 4:1–32
Blanchette M, 8:193–213
Bonanno JB, 3:242–62
Bordet S, 7:201–21
Borevitz JO, 5:443–77
Bowcock AM, 6:93–121
Bradley A, 7:247–76
Bressler J, 5:479–510
Bross P, 7:103–24
Brown CJ, 6:69–92
Budowle B, 4:119–41
Burke B, 7:369–405
Burke W, 5:1–14
Burley SK, 3:242–62

C

Caldecott KW, 8:37–55
Caplen NJ, 8:81–108
Carrington M, 3:263–92; 7:277–300
Casanova JL, 8:163–92
Cavalli-Sforza LL, 8:1–15
Celniker SE, 4:89–117
Chabas D, 4:459–83
Chace DH, 3:17–45
Chakraborty R, 4:119–41
Chakravarti A, 1:387–407
Chang T-C, 8:215–39
Cheng Z, 7:407–42
Chikhi L, 3:129–52
Childs B, 1:1–19; 6:313–30
Chow JC, 6:69–92
Christensen JH, 7:103–24
Christian SL, 5:379–405
Christianson A, 5:219–65
Clamp M, 3:293–310
Cole S, 5:295–315
Collins JE, 4:69–88
Conrad B, 8:17–35
Cook EH Jr, 5:379–405
Cooper GM, 6:143–64
Corella D, 5:71–118
Crawford DC, 6:287–312
Crow JF, 1:21–40
Cummings CJ, 1:281–328
Cutler DJ, 1:387–407
Cutting GR, 6:237–60

D

Dean M, 3:263–92; 6:123–42
de Hoog CL, 5:267–93
DeSalle R, 1:41–73
Diez-Roux G, 6:355–79
Drayna D, 6:217–35

E

Ecker JR, 5:443–77
Eichler EE, 7:407–42
Ellis NA, 1:409–59
Elston RC, 8:327–42
Enard W, 5:351–78
Evans SK, 7:29–59
Evans WE, 2:9–39

F

Fay JC, 4:213–35
Finch CE, 2:435–62
Fishman MC, 3:311–40
Florez JC, 4:257–91
Fortin A, 8:163–92
Frey N, 1:179–223
Friedman TB, 4:341–402

G

Galitski T, 2:343–72; 5:177–87
Garg A, 7:175–99
Gelb BD, 6:45–68
Giallourakis C, 6:381–406
Gilliland DG, 3:179–98
Gilman A, 3:341–69
Goldstein DB, 3:129–52; 5:119–50
Gould DB, 6:15–44
Gould SJ, 4:165–211
Grabowski GA, 4:403–36
Grady WM, 3:101–28
Green MR, 7:29–59
Gregersen N, 7:103–24
Georges M, 8:131–62
Gibbons R, 8:299–325
Greely HT, 8:343–64
Griffith AJ, 4:341–402
Griffiths-Jones S, 8:279–98
Gros P, 8:163–92
Guttmacher AE, 2:213–33

H

Hansel DE, 4:237–56
Hardison RC, 5:15–56
Harpending HC, 1:361–85
Hassold T, 5:317–49
Henikoff S, 7:61–80
Henson C, 6:381–406
Herman GE, 2:299–341
Higgs DR, 8:299–325
Hill AVS, 2:373–400
Hirschhorn J, 4:257–91
Hood L, 2:343–72
Hopkin RJ, 4:403–36
Hruban RH, 4:237–56
Hubbard T, 3:293–310
Hudson TJ, 7:443–57
Hughes J, 8:299–325
Hunley K, 5:295–315

I

Ideker T, 2:343–72
Inoue K, 3:199–242
Isasi RM, 7:201–21

J

Jaeken J, 2:129–51; 8:261–78
Jiang Y-H, 5:479–510
John SWM, 6:15–44
Johnson JA, 2:9–39
Johnson WE, 6:407–29
Jones SJM, 7:315–38
Jurka J, 8:241–59
Jurka MV, 8:241–59

K

Kalas TA, 3:17–45
Kapitonov VV, 8:241–59
Katsanis N, 7:125–48
Kelley RI, 2:299–341
Kelly LM, 3:179–98
Kere J, 2:103–28
Kern SE, 4:237–56
Kim TH, 7:81–102
Kittles RA, 4:33–67
Knepper JL, 2:153–75
Knoppers BM, 7:201–21
Knudson AG, 6:1–14
Kohany O, 8:241–59
Koonin EV, 1:99–116
Kreitman M, 1:539–59
Kurth R, 7:149–73
Kwok P-Y, 2:235–58

L

Leppert MF, 4:437–57
Levy L, 1:139–77
Libby RT, 6:15–44
Lin D, 8:327–42
Long JC, 5:295–315
Lowrey PL, 5:407–41
Lupski JR, 3:199–242
Lusis AJ, 5:189–218
Lynn A, 5:317–49
Lyon MF, 3:1–16

M

Mackay TFC, 7:339–67
Mahley RW, 1:507–37
Makova KD, 5:15–56
Mann M, 5:267–93
Mar R, 5:189–218
Markianos K, 5:151–75
Markowitz SD, 3:101–28
Martin MP, 7:277–300
Martin SE, 8:81–108
Maston GA, 7:29–59
Matthijs G, 2:129–51; 8:261–78
McCabe ERB, 5:57–69
McCabe LL, 5:57–69
McKinnon PJ, 8:37–55
McKusick VA, 7:1–27
McVicar DW, 7:277–300
Mendell JT, 8:215–39
Mervis CB, 1:461–84
Mignot E, 4:459–83
Miller W, 5:15–56
Mir KU, 1:329–60
Mitsuma N, 7:125–48
Modell B, 5:219–65
Mombaerts P, 2:493–510
Monaco AP, 8:57–79
Mootha VK, 6:381–406
Morelli SH, 4:1–32
Morris CA, 1:461–84
Moses RE, 2:41–68
Moxon ER, 2:259–69
Mulle JG, 8:109–29
Mulligan CJ, 5:295–315

N

Naylor EW, 3:17–45
Nekrutenko A, 5:15–56
Ng PC, 7:61–80
Nicholls RD, 2:153–75
Nickerson DA, 6:287–312
Nicol RL, 1:179–223

O

O'Brien SJ, 3:263–92; 6:407–29
Ogielska-Zei E, 3:415–32

Olson EN, 1:179–223
Ordovas JM, 5:71–118
Ostrander EA, 5:151–75

P

Pääbo S, 5:351–78
Pajukanta P, 5:189–218
Pakendorf B, 6:165–83
Paracchini S, 8:57–79
Penagarikano O, 8:109–29
Pennock RT, 4:143–63
Petit C, 2:271–97
Pfeifer A, 2:177–211
Pinkel D, 6:331–54
Podlisny MB, 3:67–99

R

Rall SC Jr, 1:507–37
Rannala B, 1:225–49
Reich M, 6:381–406
Reilly PR, 1:485–506
Reinke V, 3:153–78
Ren B, 7:81–102
Renier C, 4:459–83
Rogers A, 1:361–85
Rothstein MA, 2:401–33
Rubin GM, 4:89–117
Ruddle F, 2:1–8
Ruvkun G, 2:435–62

S

Scerri T, 8:57–79
Schueler MG, 7:301–13
Scriver CR, 2:69–101
Searls DB, 1:251–79
Seidman CE, 6:185–216
Seidman JG, 6:185–216
Selkoe DJ, 3:67–99
Serre D, 7:443–57
Sharp AJ, 7:407–42
Shin JT, 3:311–40
Sidow A, 6:143–64
Singh NA, 4:437–57
Sklar P, 3:371–413
Slatkin M, 1:225–49
Smukste I, 6:261–86
Southern EM, 1:329–60
Stan R, 1:409–59
Stanford JL, 5:151–75
Stewart CL, 7:369–405
Stockwell BR, 6:261–86
Stone EA, 6:143–64
Stoneking M, 6:165–83
Sullivan BA, 7:301–13

T

Taheri S, 1:159–83
Takahashi JS, 5:407–41
Tang CM, 2:259–69
Tartaglia M, 6:45–68
Thornton JW, 1:41–73
Tishkoff SA, 4:293–340
Trowsdale J, 1:117–37

V

Valle D, 1:1–19; 4:165–211; 6:313–30
van Brabant AJ, 1:409–59
van der Weyden L, 7:247–76
Vang S, 7:103–24
Veenstra-VanderWeele J, 5:379–405
Verma IM, 2:177–211
Vernimmen D, 8:299–325
Verrelli BC, 4:293–340

W

Wang L, 7:223–45
Warren ST, 8:109–29
Weinshilboum RM, 7:223–45
Weiss KM, 4:33–67
Weller S, 4:165–211
White KP, 3:153–78
Wiener C, 6:313–30
Wilson MR, 4:119–41
Wu C-I, 4:213–35

X

Xie X, 6:381–406

Y

Yen Z, 6:69–92
Yost HJ, 4:1–32

Z

Zheng G, 8:327–42
Ziesche SM, 6:69–92
Zoghbi HY, 1:281–328
Zwick ME, 1:387–407

Chapter Titles, Volumes 1–8

Volume 1 (2000)

Genetics, Biology, and Disease	B Childs, D Valle	1:1–19
Two Centuries of Genetics: A View from Halftime	JF Crow	1:21–40
Gene Family Evolution and Homology: Genomics Meets Phylogenetics	JW Thornton, R DeSalle	1:41–73
Iron Metabolism: Iron Deficiency and Iron Overload	NC Andrews	1:75–98
How Many Genes Can Make a Cell: The Minimal-Gene-Set Concept	EV Koonin	1:99–116
The Human Major Histocompatibility Complex: Lessons from the DNA Sequence	S Beck, J Trowsdale	1:117–37
Genetic Screening of Newborns	HL Levy, S Albers	1:139–77
From the Sarcomere to the Nucleus: Role of Genetics and Signaling in Structural Heart Disease	RL Nicol, N Frey, EN Olson	1:179–223
Estimating Allele Age	M Slatkin, B Rannala	1:225–49
Bioinformatics Tools for Whole Genomes	DB Searls	1:251–79
Trinucleotide Repeats: Mechanisms and Pathophysiology	CJ Cummings, HY Zoghbi	1:281–328
Sequence Variation in Genes and Genomic DNA: Methods for Large-Scale Analysis	KU Mir, EM Southern	1:329–60
Genetic Perspectives on Human Origins and Differentiation	H Harpending, A Rogers	1:361–85
Patterns of Genetic Variation in Mendelian and Complex Traits	ME Zwick, DJ Cutler, A Chakravarti	1:387–407

DNA Helicases, Genomic Instability, and Human Genetic Disease	AJ van Brabant, R Stan, NA Ellis	1:409–59
Williams Syndrome and Related Disorders	CA Morris, CB Mervis	1:461–84
Public Concern About Genetics	PR Reilly	1:485–506
Apolipoprotein E: Far More Than a Lipid Transport Protein	RW Mahley, SC Rall Jr.	1:507–37
Methods to Detect Selection in Populations with Applications to the Human	M Kreitman	1:539–59

Volume 2 (2001)

Hundred-Year Search for the Human Genome	F Ruddle	2:1–8
Pharmacogenomics: The Inherited Basis for Interindividual Differences in Drug Response	WE Evans, JA Johnson	2:9–39
DNA Damage Processing Defects and Disease	RE Moses	2:41–68
Human Genetics: Lessons from Quebec Populations	CR Scriver	2:69–101
Human Population Genetics: Lessons from Finland	J Kere	2:103–28
Congenital Disorders of Glycosylation	J Jaeken, G Matthijs	2:129–51
Genome Organization, Function, and Imprinting in Prader-Willi and Angelman Syndromes	RD Nicholls, JL Knepper	2:153–75
Gene Therapy: Promises and Problems	A Pfeifer, IM Verma	2:177–211
Human Genetics on the Web	AE Guttmacher	2:213–33
Methods for Genotyping Single Nucleotide Polymorphisms	P-Y Kwok	2:235–58
The Impact of Microbial Genomics on Antimicrobial Drug Development	CM Tang, ER Moxon	2:259–69
Usher Syndrome: From Genetics to Pathogenesis	C Petit	2:271–97
Inborn Errors of Sterol Biosynthesis	RI Kelley, GE Herman	2:299–341
A New Approach to Decoding Life: Systems Biology	T Ideker, T Galitski, L Hood	2:343–72
The Genomics and Genetics of Human Infectious Disease Susceptibility	AVS Hill	2:373–400
Privacy and Confidentiality of Genetic Information: What Rules for the New Science?	MR Anderlik, MA Rothstein	2:401–33
The Genetics of Aging	CE Finch, G Ruvkun	2:435–62
ENU Mutagenesis: Analyzing Gene Function in Mice	R Balling	2:463–92
The Human Repertoire of Odorant Receptor Genes and Pseudogenes	P Mombaerts	2:493–510

Volume 3 (2002)

A Personal History of the Mouse Genome	MF Lyon	3:1–16
The Application of Tandem Mass Spectrometry to Neonatal Screening for Inherited Disorders of Intermediary Metabolism	DH Chace, TA Kalas, EW Naylor	3:17–45
Hedgehog Signaling and Human Disease	AE Bale	3:47–65
Deciphering the Genetic Basis of Alzheimer's Disease	DJ Selkoe, MB Podlisny	3:67–99
Genetic and Epigenetic Alterations in Colon Cancer	WM Grady, SD Markowitz	3:101–28
Human Migrations and Population Structure: What We Know and Why It Matters	DB Goldstein, L Chikhi	3:129–52
Developmental Genomic Approaches in Model Organisms	V Reinke, KP White	3:153–78
Genetics of Myeloid Leukemias	LM Kelly, DG Gilliland	3:179–98
Molecular Mechanisms for Genomic Disorders	K Inoue, JR Lupski	3:199–242
Structuring the Universe of Proteins	SK Burley, JB Bonanno	3:243–62
Balanced Polymorphism Selected by Genetic Versus Infectious Human Disease	M Dean, M Carrington, SJ O'Brien	3:263–92
Databases and Tools for Browsing Genomes	E Birney, M Clamp, T Hubbard	3:293–310
From Zebrafish to Human: Modular Medical Models	JT Shin, MC Fishman	3:311–40
Genetic "Code": Representations and Dynamical Models of Genetic Components and Networks	A Gilman, AP Arkin	3:341–69
Linkage Analysis in Psychiatric Disorders: The Emerging Picture	P Sklar	3:371–413
Patenting Genes and Genetic Research Tools: Good or Bad for Innovation?	BE Arnold, E Ogielska-Zei	3:415–32

Volume 4 (2003)

Genetics of Human Laterality Disorders: Insights from Vertebrate Model Systems	BW Bisgrove, SH Morelli, HJ Yost	4:1–32
Race, Ancestry, and Genes: Implications for Defining Disease Risk	RA Kittles, KM Weiss	4:33–67
Gene Annotation: Prediction and Testing	JL Ashurst, JE Collins	4:69–88
The Drosophila Melanogaster Genome	SE Celniker, GM Rubin	4:89–117

Forensics and Mitochondrial DNA: Applications, Debates, and Foundations	B Budowle, MW Allard, MR Wilson, R Chakraborty	4:119–41
Creationism and Intelligent Design	RT Pennock	4:143–63
Peroxisome Biogenesis Disorders	S Weller, SJ Gould, D Valle	4:165–211
Sequence Divergence, Functional Constraint, and Selection in Protein Evolution	JC Fay, C-I Wu	4:213–35
Molecular Pathogenesis of Pancreatic Cancer	DE Hansel, SE Kern, RH Hruban	4:237–56
The Inherited Basis of Diabetes Mellitus: Implications for the Genetic Analysis of Complex Traits	JC Florez, J Hirschhorn, D Altshuler	4:257–91
Patterns of Human Genetic Diversity: Implications for Human Evolutionary History and Disease	SA Tishkoff, BC Verrelli	4:293–340
Human Nonsyndromic Sensorineural Deafness	TB Friedman, AJ Griffith	4:341–402
Enzyme Therapy for Lysosomal Storage Disease: Principles, Practice, and Prospects	GA Grabowski, RJ Hopkin	4:403–36
Nonsyndromic Seizure Disorders: Epilepsy and the Use of the Internet to Advance Research	MF Leppert, NA Singh	4:437–57
The Genetics of Narcolepsy	D Chabas, S Taheri, C Renier, E Mignot	4:459–83

Volume 5 (2004)

Genetic Testing in Primary Care	W Burke	5:1–14
Comparative Genomics	W Miller, KD Makova, A Nekrutenko, RC Hardison	5:15–56
Genetic Screening: Carriers and Affected Individuals	LL McCabe, ERB McCabe	5:57–69
Nutritional Genomics	JM Ordovas, D Corella	5:71–118
Africans and Asians Abroad: Genetic Diversity in Europe	G Barbujani, DB Goldstein	5:119–50
Finding Prostate Cancer Susceptibility Genes	EA Ostrander, K Markianos, JL Stanford	5:151–75

Molecular Networks in Model Systems	T Galitski	5:177–87
Genetics of Atherosclerosis	AJ Lusis, R Mar, P Pajukanta	5:189–218
Medical Genetics in Developing Countries	A Christianson, B Modell	5:219–65
Proteomics	CL de Hoog, M Mann	5:267–93
Population Genetics, History, and Health Patterns in Native Americans	CJ Mulligan, K Hunley, S Cole, JC Long	5:295–315
Variation in Human Meiotic Recombination	A Lynn, T Ashley, T Hassold	5:317–49
Comparative Primate Genomics	W Enard, S Pääbo	5:351–78
Autism as a Paradigmatic Complex Genetic Disorder	J Veenstra-VanderWeele, SL Christian, EH Cook Jr.	5:379–405
Mammalian Circadian Biology: Elucidating Genome-Wide Levels of Temporal Organization	PL Lowrey, JS Takahashi	5:407–41
Plant Genomics: The Third Wave	JO Borevitz, JR Ecker	5:443–77
Epigenetics and Human Disease	Y-H Jiang, J Bressler, AL Beaudet	5:479–510

Volume 6 (2005)

A Personal Sixty-Year Tour of Genetics and Medicine	AG Knudson	6:1–14
Complex Genetics of Glaucoma Susceptibility	RT Libby, DB Gould, MG Anderson, SWM John	6:15–44
Noonan Syndrome and Related Disorders: Genetics and Pathogenesis	M Tartaglia, BD Gelb	6:45–68
The Genetics of Psoriasis and Autoimmunity	AM Bowcock	6:93–121
Evolution of the ATP-Binding Cassette (ABC) Transporter Superfamily in Vertebrates	M Dean, T Annilo	6:123–42
Trade-Offs in Detecting Evolutionarily Constrained Sequence by Comparative Genomics	EA Stone, GM Cooper, A Sidow	6:143–64
Mitochondrial DNA and Human Evolution	B Pakendorf, M Stoneking	6:165–83
The Genetic Basis for Cardiac Remodeling	F Ahmad, JG Seidman, CE Seidman	6:185–216
Human Taste Genetics	D Drayna	6:217–35
Modifier Genetics: Cystic Fibrosis	GR Cutting	6:237–60
Advances in Chemical Genetics	I Smukste, BR Stockwell	6:261–86

The Patterns of Natural Variation in Human Genes	DC Crawford, DT Akey, DA Nickerson	6:287–312
A Science of the Individual: Implications for a Medical School Curriculum	B Childs, C Wiener, D Valle	6:313–30
Comparative Genomic Hybridization	D Pinkel, DG Albertson	6:331–54
Sulfatases and Human Disease	G Diez-Roux, A Ballabio	6:355–79
Disease Gene Discovery through Integrative Genomics	C Giallourakis, C Henson, M Reich, X Xie, VK Mootha	6:381–406
Big Cat Genomics	SJ O'Brien, WE Johnson	6:407–29

Volume 7 (2006)

A 60-Year Tale of Spots, Maps, and Genes	VA McKusick	7:1–27
Transcriptional Regulatory Elements in the Human Genome	GA Maston, SK Evans, MR Green	7:29–59
Predicting the Effects of Amino Acid Substitutions on Protein Function	PC Ng, S Henikoff	7:61–80
Genome-Wide Analysis of Protein-DNA Interactions	TH Kim, B Ren	7:81–102
Protein Misfolding and Human Disease	N Gregersen, P Bross, S Vang, JH Christensen	7:103–24
The Ciliopathies: An Emerging Class of Human Genetic Disorders	JL Badano, N Mitsuma, PL Beales, N Katsanis	7:125–48
The Evolutionary Dynamics of Human Endogenous Retroviral Families	N Bannert, R Kurth	7:149–73
Genetic Disorders of Adipose Tissue Development, Differentiation, and Death	AK Agarwal, A Garg	7:175–99
Preimplantation Genetic Diagnosis: An Overview of Socio-Ethical and Legal Considerations	S Bordet, RM Isasi, BM Knoppers	7:201–21
Pharmacogenetics and Pharmacogenomics: Development, Science, and Translation	RM Weinshilboum, L Wang	7:223–45
Mouse: Chromosome Engineering for Modeling Human Disease	L van der Weyden, A Bradley	7:247–76

The Killer Immunoglobulin-like Receptor Gene Cluster: Tuning the Genome for Defense	AA Bashirova, MP Martin, DW McVicar, M Carrington	7:277–300
Structural and Functional Dynamics of Human Centromeric Chromatin	MG Schueler, BA Sullivan	7:301–13
Prediction of Genomic Functional Elements	SJM Jones	7:315–38
Of Flies and Man: Drosophila as a Model for Human Complex Traits	TFC Mackay, RRH Anholt	7:339–67
The Laminopathies: The Functional Architecture of the Nucleus and Its Contribution to Disease	B Burke, CL Stewart	7:369–405
Structural Variation of the Human Genome	AJ Sharp, Z Cheng, EE Eichler	7:407–42
Resources for Genetic Variation Studies	D Serre, TJ Hudson	7:443–57

Volume 8 (2007)

Human Evolution and Its Relevance for Genetic Epidemiology	LL Cavalli-Sforza	8:1–15
Gene Duplication: A Drive for Phenotypic Diversity and Cause of Human Disease	B Conrad, SE Antonarakis	8:17–35
DNA Strand Break Repair and Human Genetic Disease	PJ McKinnon, KW Caldecott	8:37–55
The Genetic Lexicon of Dyslexia	S Paracchini, T Scerri, AP Monaco	8:57–79
Applications of RNA Interference in Mammalian Systems	SE Martin, NJ Caplen	8:81–108
The Pathophysiology of Fragile X Syndrome	O Penagarikano, JG Mulle, ST Warren	8:109–29
Mapping, Fine Mapping, and Molecular Dissection of Quantitative Trait Loci in Domestic Animals	M Georges	8:131–62
Host Genetics of Mycobacterial Diseases in Mice and Men: Forward Genetic Studies of BCG-osis and Tuberculosis	A Fortin, L Abel, JL Casanova, P Gros	8:163–92
Computation and Analysis of Genomic Multi-Sequence Alignments	M Blanchette	8:193–213
Roles of microRNAs in Vertebrate Physiology and Human Disease	T-C Chang, JT Mendell	8:215–39

Repetitive Sequences in Complex Genomes: Structure and Evolution	J Jurka, VV Kapitonov, O Kohany, MV Jurka	8:241–59
Congenital Disorders of Glycosylation: A Rapidly Expanding Gene Family	J Jaeken, G Matthijs	8:261–78
Annotating Noncoding RNA Genes	S Griffiths-Jones	8:279–98
Using Genomics to Study How Chromatin Influences Gene Expression	DR Higgs, D Vernimmen, J Hughes, R Gibbons	8:299–325
Multistage Sampling for Genetic Studies	RC Elston, D Lin, G Zheng	8:327–42
The Uneasy Ethical and Legal Underpinnings of Large-Scale Genomic Biobanks	HT Greely	8:343–64